Microbial Zoonoses

Series Editor

Subhash Chandra Parija, Sri Balaji Vidyapeeth, A Deemed-to-be University
Pondicherry, India

The series on Microbial Zoonoses provides a holistic overview of various zoonotic pathogens including parasitic, fungal, bacterial, and viral. The individual volume provides conceptual knowledge and recent advancements on the various aspects of these pathogens, infectious disease caused by these organisms, control strategies, and their public health importance.

This book series discusses the pathogen life-cycle, disease spectrum and their impact on human health, modern diagnostic methods, treatment, host-pathogen interactions, and the importance of public health control measures including prophylactic measures in human and animal hosts.

Special emphasis on this series is on the immunological response to these pathogens, molecular epidemiology, and genomics, novel diagnostic techniques, and antimicrobial resistance in zoonotic pathogens.

Each book consists of two parts. Part A incorporates section which deals with chapters on the nomenclature of microorganisms and classification of zoonoses, life-cycle of the organism, pathogenesis, pathology, diagnosis, epidemiology and their clinical manifestations.

Part B of the book includes chapters that summarize the host-pathogen interactions, the immunological response of the host, new vaccines and novel treatment regimens against the pathogens. It also discusses strategies for control and prevention of microbial zoonoses.

Given its scope, the individual books in this book series is of interest to professionals, scientists, and postgraduate students in medicine, veterinary medicine, public health, and diagnostic laboratories.

Subhash Chandra Parija •
Shivaprakash M. Rudramurthy
Editors

Textbook of Fungal Zoonoses and Sapronoses

Springer

Editors
Subhash Chandra Parija
Sri Balaji Vidyapeeth University
Pondicherry, India

Shivaprakash M. Rudramurthy
Department of Medical Microbiology
Postgraduate Institute of Medical
Education and Research (PGIMER)
Chandigarh, India

ISSN 2662-5288 ISSN 2662-5296 (electronic)
Microbial Zoonoses
ISBN 978-981-97-3213-5 ISBN 978-981-97-3214-2 (eBook)
https://doi.org/10.1007/978-981-97-3214-2

This Springer imprint is published by the registered company Springer Nature Singapore Pte Ltd.
The registered company address is: 152 Beach Road, #21-01/04 Gateway East, Singapore 189721, Singapore

The Supreme Power Who Has Willed and given us the Strength to take up this work
To my wife Ms Jyotirmayee Parija for self-less support,
my mother Late Smt Nishamani Parija,
my father Late Shri Managovinda Parija,
my sister Late Smt Nirmala Parija, and to
my sister-in-law Late Smt Satyabhama Parija for
their blessings.
Also to my professional colleagues and
mentors for their guidance.
Subhash Chandra Parija

To my wife Dr Dhanalakshmi Rangaiah for her constant support,
my mother Smt. Sarasamma, and my father
my father Late Rudramurthy Bommaiah for their blessings, and finally to
my professional mentors Prof Arunaloke Chakrabarti, Prof Subhash Chandra Parija, and Prof V Rajagopal.
Shivaprakash Mandya Rudramurthy

Foreword

The *Textbook of Fungal Zoonoses and Sapronoses* edited by Drs. Parija and Rudramurthy is a carefully constructed work that through 24 chapters travels among the critical issues of medical mycology today and will become a major reference standard for the medical mycology field. Humans and millions of species of fungi have lived successfully together for millenniums. For instance, it is proposed that during the loss of the dinosaurs, a targeted celestial event created plumes of fungi in which mammals with their "regulated high body temperature" survived the fungal insult but the cold-blooded dinosaurs could not. In the beginning, the fungal–human interactions kept infections at bay. However, certain hosts can be weakened by a variety of causes and their inability to deny certain fungal growth has become an underlying plague of modern medicine. Invasive fungal infections and immunocompromised hosts are no longer rare events and these complicating fungal infections at the host–fungal interface are all too common. In fact, over the last two decades sudden invasive fungal life-threatening outbreaks such as chytridiomycosis attacking our amphibian populations and "white nose" syndrome killing bats (*Pseudogymnoascus destructans*) have appeared. Furthermore, recent clinical experiences have also shown us major human fungal outbreaks such as cryptococcosis in AIDS patients, major outbreaks of contaminated injectable corticosteroids with fungal species infecting tissue spaces, widespread outbreaks of *Candida auris* worldwide with major environmental features and drug resistance, and finally, COVID infection is combined with an invasive fungal infection such as aspergillosis or mucormycosis. These are the dramatic "news-generating" infections. From any recent epidemiological study, the incidence of invasive fungal infections continues to consistently rise in our health care systems affecting both rich and poor and helped by global warming and our medical needs to manipulate the host immune system to treat serious underlying diseases. With the "fungal plague" well-developed in modern medicine in most health care systems, this book dials down with astute clinical advice utilizing outstanding references and adding case studies to quickly get a direct feel for deep-seated mycoses. The work in this book is comprehensive and insightful and can be used by both the advanced and early medical learners. It can help with teaching and understanding of deep-seated mycoses and is so relevant to our evolving clinical practices today. In sum, the message today is: fungal infections, fungal infections, fungal infections everywhere, not a yeast/hyphae to disregard!!!

Division of Infectious Diseases,
Department of Molecular Genetics and Microbiology
Duke University School of Medicine,
Durham, NC, USA
John R. Perfect

Preface

Before anything, the origin of life on Earth comes to one's mind. Earth is around 4.5 billion years old. However, scientists think that about 4.3 billion years ago, suitable conditions were developed to support life on Earth. Dating back to the Cenozoic era, human interaction with animals for the food initiated in the form of hunting and is known to be around 1.5 million years ago. Analysing the plaque on ancient human teeth, scientists revealed about the mushroom consumption by hunter-gatherers in Europe as early as 18,000 years ago, focusing on the fact of an older relationship between humans and fungi in comparison to agriculture.

Earth has seen different ages, even the ice age. 15,000 years back, in the late Pleistocene era, humans started the domestication of dogs, even before they started cultivation. Human dependency on animals of diverse origins has contributed to the process of transfer of microorganisms from animals to humans, leading to infections in the latter. It is no surprise that the history of zoonotic human diseases dates to antiquity.

We animals are joined by the second of the three great multicellular kingdoms, the fungi. At first, it was astonishing that fungi, which appear to be so plant-like, are more closely related to animals than plants, but molecular comparison leaves little doubt. That may not be surprising. From the sun, plants import energy into the biosphere. It is too much to say, but animals and fungi are parasites to plants in different ways.

In healthy and immunocompetent humans and nonhuman vertebrates, fungi are relatively uncommon disease cause, even though there is constant exposure to infectious propagules. However, over the last two decades, there has been an increase in the number of recalcitrant fungal diseases in animals originating from pathogenic and opportunistic fungi. Zoonoses are infections that are naturally transmitted between vertebrate animals and humans. From a global point of view, for many centuries, these zoonotic infections have been recognized and are a reason behind the majority of emerging and re-emerging infectious diseases.

Fungal zoonoses were once considered a rare entity, but this is proven wrong as the incidence is significantly rising. Fungal infections were primarily seen as a plant-infecting agent, challenging agricultural yields and food security until 1990. But lately, there has been an increase in the occurrence of fungal infections among animals. Fungi can survive in the environment for a prolonged duration, but few choose a vertebrate host as part of their life cycle.

In immunocompromised hosts, the systemic infections due to these fungi are fatal compared to immunocompetent hosts.

On the world platform, fungal zoonoses have become an important public health problem as they have the potential to cause outbreaks. Reports prove the increase in fungal infections in the past two decades although they are uncommon among immunocompetent humans and nonhuman vertebrate animals. This increase traces back to the fact that there has also been increased physical contact between humans and animals, usually resulting from rising industrial development and encroachment of human settlements in wildlife habitats.

A textbook, a blend of the latest information, provides a roller-coaster ride of the fungal zoonoses world, providing a great equilibrium of a standard textbook and a review article. *Textbook of Fungal Zoonoses and Sapronoses* is written with the vision that it will be useful for anybody interested in zoonotic fungal diseases and cutting across disciplines. The book aims to furnish the readers with an inevitable understanding of zoonotic fungal diseases for undergraduate and postgraduate students in addition to public health experts, faculties, administrators, and scientists across the disciplines of Medical sciences, Public health sciences, Veterinary sciences, and many other allied health sciences.

We extend a heartfelt thanks you with all sincerity to all our contributors who have worked hard and pitched in this project. Throughout this journey, our publisher, Springer Nature, has shown great support in executing the whole project with immense perfection and dedication. This book will provide a platform of knowledge for all readers across the globe who are interested in knowing about the world of fungal zoonoses.

Pondicherry, India Subhash Chandra Parija
Chandigarh, India Shivaprakash Mandya Rudramurthy

Acknowledgements

We wish to place on record our sincere thanks to all those who helped us in bringing out this book. The authors and co-authors of various chapters have done commendable job in submitting the manuscripts in time despite their preoccupation with multiple tasks.

Subhash Chandra Parija wishes to thank specifically Shri Kailash Chandra Parija, niece Er Kukumina Ray, son-in-law Er Subhasis Ray, nephew Er Rajkumar Parija, daughter-in-law Ms Smrithi Parija, and daughter Dr Madhuri Parija, son-in-law Dr Ajay Halder, Ms Er Mayuri Parija and son-in-law Er Shailesh Nandan, and grandchildren Shri Harihar, Sri Ram, and Ms Shyama Nandan for their support and encouragement.

Shivaprakash Mandya Rudramurthy likes to thank daughter Er Neha S Manru and son Harshavardhan S Manru. He expresses his sincere thanks to colleagues and students, Ms Ritika Harchand, Dr Shreya Singh, Ms Sheetal Thakur, Dr Dipti Sharma, Dr Harsimran Kaur, and Prof Anup Ghosh. He also likes to express his gratitude and support to his brothers Er Prabhakar M Rudramurthy, Keshav Manru, Paramesh Manru, Venkatesh Er Mandya Rudramurthy, and the whole Manru's family.

Pondicherry, India — Subhash Chandra Parija
Chandigarh, India — Shivaprakash Mandya Rudramurthy

About the Book

The Genesis

Fungal diseases have remained on the back foot for a long time, and despite their alarming impact on human health, they have been neglected. In different ways, human fungal diseases differ from several other infections. Fungi, being eukaryotic, share a number of similarities with their host cells, making it difficult to treat disease caused by this. But with the increase in immunocompromised patients, the scenario is changing, and the focus has shifted towards fungal infections. Recently, the World Health Organization, for the first time, released the list of priority fungal pathogens to help public health agencies prioritize research and prevention activities.

Global Action Fund for Fungal Infections (GAFFI) has also called for attention to the devastating effects of focal fungal diseases, often seen in people with intact immune systems. They have estimated that fungal keratitis alone is responsible for around a million eyes going blind. Approximately one billion people have skin mycoses. Overall, GAFFI has estimated that globally, over 300 million people are suffering from serious fungal infections yearly, including people of all age groups, and over 1.5 million of these die from those infections. But still, public health authorities and decision-makers ignore the disease that carries high mortality. Even the financial support compared to the other infectious diseases has remained incredibly low for fungal disease research. For instance, US$1,315 is invested in research and development for each human individual dying from malaria, which is quite a lot in comparison to the investment made for cryptococcal meningitis. Investment per death corresponds to US$276 for diarrheal diseases, US$334 for tuberculosis, and for cryptococcal meningitis, only US$31. Significant gaps in knowledge have existed concerning fungal diseases, and according to Morel, the reason behind the persistence of neglected diseases is the failures in science, market, and public health.

Mycology, an eternal part of Medical Microbiology, remained neglected for a long time. However, this subject is gaining importance with the increased incidence of fungal infections. Many countries have created room for zoonotic fungal infections during the past few years, especially concerning the role of animals in the burden of diseases.

This book, *Textbook of Fungal Zoonoses and Sapronoses*, aims to give out the important and mandatory knowledge of the diagnosis, treatment, prevention, and control of these diseases to undergraduate and postgraduate students, public health experts, faculties, administrators of medicine, veterinary and allied sciences, and other interested individuals who are involved in the management of these zoonotic fungal infections. In our journey towards the opening of the gateway of knowledge, we were wholeheartedly supported by about 61 eminent faculties who willingly contributed the best of their knowledge to these chapters.

The Content Organization

The contents of the book have been organized in a student-friendly way for better understanding. The topics of the book have been divided into five parts.

Part I, General Mycology, provides 'Background Information on Zoonotic Fungal Infections'. This part is an emulsion of chapters on taxonomy, pathogenesis, diagnosis, and antifungal agents. This twenty-first century is 'The era of genomics, proteomics, and other –omics', making it a necessity to provide students with the developments made in this field in relation to fungal zoonoses. Therefore, there is the inclusion of a few more chapters highlighting the progress in the study of genomics, proteomics, and immunology of fungi/fungal infections. In addition, considering the importance of public health for many of these diseases, three different chapters on 'epidemiology', 'antifungal agents and resistance', and 'prevention and control' have been incorporated.

Part II includes 'Superficial Zoonotic Fungal Infections', Part III consists of 'Subcutaneous Zoonotic Fungal Infections', whereas Part IV and V focus on 'Systemic Fungal Zoonotic infections' and 'Opportunistic Fungal Zoonoses'. Each of the chapters in parts II and III, IV and V follows a uniform pattern, including information on history, taxonomy, proteomics, genomics, immunology, fungus morphology, animal models, pathogenesis, epidemiological features, clinical manifestations in humans and animals, diagnostics, therapy, and prevention and control measures.

The Value Addition

Along with the classical features of a standard book, this textbook is provided with two unique additions: a typical case study and some unanswered research questions. The effort is made to help the students brainstorm for a better understanding of the chapters. Case studies give the opportunity to think critically and apply the gained knowledge in clinical practices involving therapy, diagnosis, and more applications. The unanswered questions will help the young minds to evaluate themselves and learn more about these unexplored areas.

In summary, the *Textbook of Fungal Zoonoses and Sapronoses* will help different readers, whether they are beginners or experts, quench their thirst for knowledge. The book contains 24 chapters written by 61 contributors of repute, both young and experienced, from varying fields of medical, veterinary, public health and allied health sciences, and different parts of the globe, with immense knowledge and experience in their respective areas, providing this book a stronger ground.

We hope and wish the readers a knowledgeable experience with intellectual satisfaction!

Editors

Subhash Chandra Parija

Shivaprakash Mandya Rudramurthy

Contents

Editors and Contributors

About the Editors

Subhash Chandra Parija, MBBS, MD, PhD, DSc, FRCPath was the Vice-Chancellor of Sri Balaji Vidyapeeth, a deemed-to-be university, Pondicherry, India, and is the Former Director of the Jawaharlal Institute of Postgraduate Medical Education and Research (JIPMER), Pondicherry, India. With nearly 45 years of teaching and research experience in the field of medical microbiology, Prof. Parija is a Food and Agriculture Organization (FAO) expert and has been consulted to draft guidelines on preventing foodborne parasites. He was on the MD examination boards at Colombo University, Sri Lanka; Sultan Qaboos University, Oman; and the University of Malaya, Malaysia. He received a D.Sc. from Madras University for his contributions to the field of medical parasitology. He holds one granted patent, one published patent, 23 copyrights, and one technology transfer. Prof. Parija has authored or edited 19 books, including the most popular *Textbook of Medical Microbiology & Immunology*, *Effective Medical Communication: The A, B, C, D, E of It*, and *Textbook of Parasitic Zoonoses*, all published by Springer. He has published more than 430 papers in reputable national and international journals.

Shivaprakash Mandya Rudramurthy, MD, PhD, FECMM, MNASc, MNAMS is a Professor at the Department of Medical Microbiology, Postgraduate Institute of Medical Education and Research, Chandigarh, India. Prof. Shivaprakash has dedicated the last 21 years to the field of medical mycology, actively involved in research on fungi and fungal infections and has also trained personnel in this speciality, providing identification service for fungi, supplied cultures for teaching, research, and diagnostic activities, and conducted a national level external quality assurance program for mycology. He has written 16 chapters and has more than 280 publications. He is currently a core member of the Global Antimicrobial Resistance and Use Surveillance System (GLASS) on antifungals.

Contributors

Abhishek Division of Bacteriology and Mycology, ICAR-Indian Veterinary Research Institute, Bareilly, Uttar Pradesh, India

V. Athira Division of Bacteriology and Mycology, ICAR-Indian Veterinary Research Institute, Bareilly, Uttar Pradesh, India

Sayantan Banerjee Department of Microbiology, All India Institute of Medical Sciences (AIIMS), Kalyani, West Bengal, India

Malini R. Capoor Department of Microbiology, VMMC and Safdarjung Hospital, New Delhi, India

Pankaj Chandley Department of Biosciences and Bioengineering, Indian Institute of Technology, Roorkee, Uttar Pradesh, India

Shiv Sekhar Chatterjee Department of Microbiology, All India Institute of Medical Sciences (AIIMS), Kalyani, West Bengal, India

Ariya Chindamporn Mycology Unit, Department of Microbiology, Faculty of Medicine, Chulalongkorn University, Bangkok, Thailand

Mary Kiran Danni Department of Microbiology, Sri Ramachandra Institute of Higher Education and Research, Chennai, Tamil Nadu, India

Ankita Das ICMR Advanced Molecular and Diagnostics Research Centre for Fungi, Department of Microbiology, All India Institute of Medical Sciences, Bhubaneswar, Odisha, India

Marconi Rodrigues de Farias Veterinary Dermatology and Postgraduate Program in Animal Science at the Pontifical Catholic University of Paraná, Paraná, Brazil

Camila Stefanie Fonseca de Oliveira Department of Preventive Veterinary Medicine, Veterinary School, Universidade Federal de Minas Gerais, Belo Horizonte, Minas Gerais, Brazil

Júlia Marcondes Figueiredo Institute of Biomedical Sciences, São Paulo University, São Paulo, Brazil

Manisha Ghosh Laboratory of Fungal Pathogenesis, Centre for DNA Fingerprinting and Diagnostics, Hyderabad, India

Beatriz L. Gómez Studies in Translational Microbiology and Emerging Diseases (MICROS) Research Group, School of Medicine and Health Sciences, Universidad del Rosario, Bogotá, Colombia

Ángel González Basic and Applied Microbiology Research Group (MICROBA), School of Microbiology, Universidad de Antioquia, Medellín, Colombia

Rajendra Guidsa Department of Medical Microbiology, Postgraduate Institute of Medical Education, and Research (PGIMER), Chandigarh, India

Parakriti Gupta Department of Microbiology, Government Medical College and Hospital, Chandigarh, India

Pooja Gurram Division of Infectious Diseases, University of Florida, FL, USA

Vinaykumar Hallur ICMR Advanced Molecular and Diagnostics Research Centre for Fungi, Department of Microbiology, All India Institute of Medical Sciences, Bhubaneswar, Odisha, India

Ritika Harchand Department of Medical Microbiology, Post Graduate Institute of Medical Education and Research (PGIMER), Chandigarh, India

Prasanna Honnavar Department of Microbiology and Immunology, American University of Antigua College of Medicine, St. Johns, Antigua and Barbuda

Shrikrishna Isloor Department of Veterinary Microbiology, Veterinary College, Karnataka Veterinary, Animal and Fisheries Sciences University, Bengaluru, Karnataka, India

Pratibha Kale Department of Clinical Microbiology, Institute of Liver and Biliary Sciences, New Delhi, India

Rimjhim Kanaujia Department of Medical Microbiology, PGIMER, Chandigarh, India

Harsimran Kaur Department of Medical Microbiology, PGIMER, Chandigarh, India

Rupinder Kaur Laboratory of Fungal Pathogenesis, Centre for DNA Fingerprinting and Diagnostics, Hyderabad, India

Anupma Jyoti Kindo Department of Microbiology, Sri Ramachandra Institute of Higher Education and Research, Chennai, Tamil Nadu, India

Brenda Kischkel Department of Internal Medicine and Radboud Institute for Molecular Life Sciences, Radboud University Medical Center, Nijmegen, The Netherlands

Bablu Kumar Division of Biological Standardization, ICAR-Indian Veterinary Research Institute, Bareilly, India

Mendoza Leonel Microbiology and Molecular Genetics, Biomedical Laboratory Diagnostics, Michigan State University, East Lansing, MI, USA

Leila Lopes-Bezerra BIDiagnostics® at the Technology-Based Incubator of São Paulo at São Paulo University (USP), São Paulo, Brazil

Hugo Madrid Departamento de Tecnología Médica, Facultad de Ciencias de la Salud, Universidad de Tarapacá, Sede Iquique, Iquique, Chile

Kornvalee Meesilpavikkai Mycology Unit, Department of Microbiology, Faculty of Medicine, Chulalongkorn University, Bangkok, Thailand

Sanchita Mitra Jhaveri Microbiology Centre and Brien Holden Eye Research Centre, LV Prasad Eye Institute, Hyderabad, India

Fabiana Monti Postgraduate Program in Animal Science at the Pontifical Catholic University of Paraná, Paraná, Brazil

Sonu S. Nair Division of Bacteriology and Mycology, ICAR-Indian Veterinary Research Institute, Bareilly, India

Priyadarshini A. Padaki Department of Microbiology, St. John's Medical College and Hospital, Bengaluru, Karnataka, India

Nishith Kumar Pal Department of Microbiology, Jagannath Gupta Institute of Medical Sciences and Hospital, Kolkatta, West Bengal, India

Subhash Chandra Parija Sri Balaji Vidyapeeth University, Pondicherry, India

Maria Griselda Perona Molecular Biology, Chemistry Institute, São Paulo University, São Paulo, Brazil

Hariprasath Prakash Department of Microbiology, PES Institute of Medical Sciences and Research, Kuppam, Andhra Pradesh, India

Ravikant Ranjan Chanakya University, Bangalore, Karnataka, India

Vilela Raquel Microbiology and Molecular Genetics, Biomedical Laboratory Diagnostics, Michigan State University, East Lansing, MI, USA

Faculty of Pharmacy, Federal University of Minas Gerais, Belo Horizonte, MG, Brazil

D. Rathnamma Department of Veterinary Microbiology, Veterinary College, Karnataka Veterinary, Animal and Fisheries Sciences University, Bengaluru, Karnataka, India

Soma Rohatgi Department of Biosciences and Bioengineering, Indian Institute of Technology, Roorkee, Uttar Pradesh, India

Satwika Rompally Department of Medical Microbiology, Postgraduate Institute of Medical Education and Research, Chandigarh, India

Shivaprakash Mandya Rudramurthy Department of Medical Microbiology, Postgraduate Institute of Medical Education and Research (PGIMER), Chandigarh, India

Palash Samanta BMT and Hematological Malignancy Infectious Diseases Unit, Medicine, Transplant Infectious Diseases, University of Pittsburgh Medical Center, Pittsburgh, PA, USA

Asmita Sarowgi Laboratory of Fungal Pathogenesis, Centre for DNA Fingerprinting and Diagnostics, Hyderabad, India

Graduate Studies, Manipal Academy of Higher Education, Manipal, Karnataka, India

Jayanthi Savio Department of Microbiology, St. John's Medical College and Hospital, Bengaluru, Karnataka, India

Shamanth A. Shankarnarayan Department of Physics, University of Alberta, Edmonton, AB, Canada

R. Sharada Department of Veterinary Microbiology, Veterinary College, Karnataka Veterinary, Animal and Fisheries Sciences University, Bengaluru, Karnataka, India

Sheetal Sharma Department of Microbiology, VMMC and Safdarjung Hospital, New Delhi, India

Dipika Shaw Department of Microbiology, Jawaharlal Nehru Medical College, Datta Meghe Institute of Higher Education and Research, Wardha, Maharashtra, India

Gagandeep Singh Department of Microbiology, AIIMS, New Delhi, India

Pankaj Singh Indian Council of Medical Research-National Institute of Virology, Pune, Maharashtra, India

Shreya Singh Department of Microbiology, Dr B R Ambedkar State Institute of Medical Sciences (AIMS), Mohali, Punjab, India

Prasad Thomas Division of Bacteriology and Mycology, ICAR-Indian Veterinary Research Institute, Bareilly, India

Ángela M. Tobón Internal Medicine, Universidad CES and Colombian Tropical Medicine Institute, Medellín, Colombia

Kalaiselvi Vinayagamoorthy Centre for Public Health (U.I.E.A.S.T.), Panjab University, Chandigarh, India

Navaporn Worasilchai Department of Transfusion Medicine and Clinical Microbiology, Faculty of Allied Health Sciences, Chulalongkorn University, Bangkok, Thailand

Part I

General Mycology

Taxonomy and Classification of Fungi

1

Shivaprakash M. Rudramurthy
and Harsimran Kaur

Learning Objectives
By the end of the chapter, the readers should be able to:

1. Describe the concept of species in fungi and current nomenclature.
2. Understand the current classification of Kingdom Fungi.
3. Learn various databases available for fungal identification.
4. Study challenges in fungal taxonomy.

Introduction

Living things are divided into seven kingdoms of life, namely bacteria, archaea, protozoa, chromista, plantae, fungi, and animalia. Fungi form one of the abundant substantial forms of life. Fungi are heterotrophic eukaryotes with glucan and chitin as the predominant constituents of their cell wall. They exist either as unicellular yeasts or multinucleated mycelia, have an aminoadipic pathway for the biosynthesis of lysine, and possess flattened mitochondrial cristae. Few fungi like *Cryptococcus* have a polysaccharide capsule that is anti-phagocytic and contributes to its virulence. The fungi are ubiquitous and possess a rich species diversity existing in various morphological forms and ecological systems with wide application in biotechnology, medicine, and industry. Currently, of the estimated 2.2–3.8 million fungal species, around 150,000 species are described; amongst them, 700 are pathogenic to humans. Few authors estimate the existing fungal species ranging from 700,000 to 12 million.

An overwhelming rise in human fungal infections over the past two or three decades is of serious concern. The increase is attributed either to the growing number of susceptible hosts in the form of critical illness and immunocompromised conditions or advancement in diagnostic techniques. All these have contributed to the increased identification of previously unnoticed fungi causing infections or systematic taxonomic data expanding the spectrum of pathogenic fungi. This necessitates the understanding of fungal taxonomy and classification, which is a dynamic entity subject to continuous changes in nomenclature. It differs from bacterial taxonomy in terms of variation based on the morphology of both sexual and asexual structures rather than biochemical or nutritional features. The earlier phenotypic classification has increasingly been replaced by phylogenetic methods, which have revolutionised the subject.

There have been major changes in fungal taxonomy when compared with its earliest classification in 'Regnum Vegetabile' by Linnaeus

S. M. Rudramurthy (✉) · H. Kaur
Department of Medical Microbiology, Postgraduate Institute of Medical Education and Research (PGIMER), Chandigarh, India
e-mail: mrshivprakash@gmail.com

S. C. Parija, S. M. Rudramurthy (eds.), *Textbook of Fungal Zoonoses and Sapronoses*, Microbial Zoonoses, https://doi.org/10.1007/978-981-97-3214-2_1

in 1767 and 'Eumycota' by Whittaker in 1969. Eumycota was classified into four phyla, namely Chytridiomycota, Zygomycota, Ascomycota, and Basidiomycota. Currently, the true fungi are differentiated into nine major lineages: Opisthosporidia, Chytridiomycota, Neocallimastigomycota, Blastocladiomycota, Zoopagomycota, Mucoromycota, Glomeromycota, Ascomycota, and Basidiomycota. The improvement in molecular phylogenetics has led to a better understanding of fungal taxonomy. Apart from Eumycota, members of Kingdom Protocista (*Rhinosporidium seeberi*) and Chromista (*Pythium insidiosum*), which evolved similarly to Eumycota and caused infection in humans are now placed in separate kingdoms. The earlier protozoan genus *Microsporidium* was found to have a close affinity to the fungi and was included in Eumycota. Similarly, *Pneumocystis jirovecii* (earlier *Pneumocystis carinii*) was transferred to Eumycota based on phylogenetic analyses of nuclear rRNA sequences. Such dynamic nature of fungal taxonomy necessitates a comprehensive updated knowledge of the subject. This chapter provides here an updated description of fungal taxonomy relevant to medically important fungi.

Formulators of Fungal Taxonomy

All fungi and fungus-related organisms have been named according to the International Code of Nomenclature for algae, fungi, and plants (ICNafp) since 1867 and updated every 6 years. However, since 2018, after collateral discussions, the decisions pertaining nomenclature of all fungi and fungi-like organisms are subject to modification only during the 4-yearly International Mycological Congress (IMC). For the modification in the nomenclature of fungi, IMC has appointed the Nomenclature Committee for Fungi (NCF), which consists of around 20 members who vote on the formal matters of nomenclature (http://www.ima-mycology.org/nomenclature/nomenclature-committee-fungi). After the formal modifications in nomenclature, the guidelines are disseminated and promoted by the International Commission on the Taxonomy of Fungi (ICTF). ICTF comprises around 23 members with start-and-finish task-related working groups and sub-commissions, which also help in drafting proposals for IMC and NCF (https://www.fungaltaxonomy.org).

Phylogeny and Evolution

Lag in evolution in Kingdom Fungi has been mainly attributed to the lack of fossil record, cardinal morphological characteristics, and vast diversity. Previous studies were based on comparative phenotypic characters, ultrastructure, chemotaxonomical features, and the scant fossil record. Recent advances in molecular approaches have provided insight into the controversial aspects and have delineated the Kingdom Fungi as a member of terminal radiation of the major eukaryotic groups. The previous hypothesis that fungi originated from algae has been disregarded, and presently, the nucleotide substitutions in the genome have been noted to be directly proportional to elapsed time and changes in bases. The evolutionary time elapsed has been calculated and noted to be ~550 million years ago for Ascomycota, Zygomycota, and Basidiomycota, with the Ascomycota–Basidiomycota split happening ~400 million years ago.

The Concept of Species in Fungi

'Species' is considered the most basal level in taxonomical classification, which is described in fungi in terms of varying concepts (Fig. 1.1) by mycologists. However, the classification still suffers from huge vagueness attributed to diverse morphological and physiological characteristics. Owing to the ubiquitous presence of fungi, these have been categorised into two main morphological structures: microfungi and macrofungi. Macrofungi encompass ascomata, zygospores, sporangium, and pycnidia, while microscopic fungi include members of Corticiaceae, bracket fungi, *Pezzia* cups, etc. Moreover, the taxonomical attributes are further convoluted by the presence

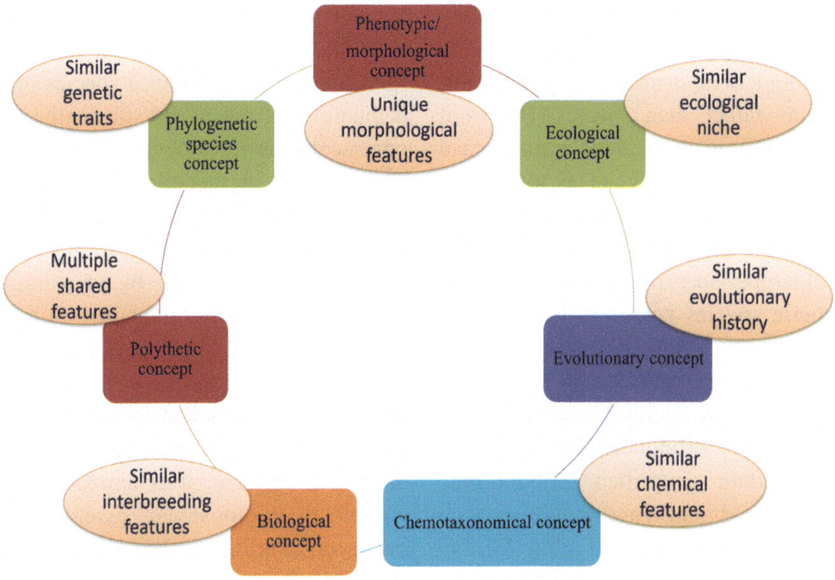

Fig. 1.1 The concept of species in fungi

of two stages in the life cycle of fungi, asexual/anamorph, and sexual/teleomorph stages. These intricate complexities can predispose to the erroneous classification of a rare fungus as a novel species. To address these ambiguities, molecular modalities were introduced to resolve the controversial aspects. Besides the classification of novel species, sequencing has also revolutionised the identification of non-sporulating and anamorphic deuteromycete fungi. ICTF and the International Code of Botanical Nomenclature (ICBN) have delineated species based on morphology, physiology, phenotypic characteristics, chemotaxonomy, ecological alcove, and molecular markers.

Morphological/phenotypic concept: This is the classical approach that catalogues fungi based on morphological features, aided by phase contrast and Nomarski differential interference contrast microscopy. This is simpler in fungi with merely asexual forms, where colony characterisation, conidiogenesis, types of conidia and conidiophore, and their arrangement are exploited for identification. However, in fungi with teleomorph stages, this is intrigued by the presence of varied sporulating bodies like ascus in Ascomycota and fruiting bodies in Basidiomycota. After precisely recording spore dimensions, the structure of mycelium, and the reaction to Melzer's reagent, these characteristics can be analysed.

Polythetic concept: This is a modification of the morphological concept, which combines multiple phenotypic characteristics of the fungus.

Biological concept: This is based on gene exchange within the species via sexual and parasexual reproduction. A species is considered potentially interbreeding if isolated by intrinsic barriers. The concept has its own setbacks as fertility depends on the 'frame of reference', and it is not easy to assess the outcome. Moreover, the concept can be applied to sexually reproducing fungi easily; however, for fungi that reproduce asexually or by other than meiosis, like somatic hybridisation, the theory collapses.

Chemotaxonomy: This is of less significance in the taxonomy of fungi, owing to the plethora of metabolites generated and the inconsistency in production. The same is of supplementary assistance in fungi like *Alternaria*, *Fusarium*, *Penicillium*, etc., simultaneously being insignificant in Zygomycota and Chytridiomycota.

Ecological concept: This concept is mainly used for the classification of plant-pathogenic fungi. The ecological description helps to demarcate the strains rather than species. However, merely relying on an ecological niche might lead to the misidentification of species as many species adapt to the environment and change their preferred niche to a new one if provided with the desired set of temperature, pH, and nutrients.

Phylogenetic species concept: The basis of this concept lies in the molecular modalities and phylogenetic techniques and is also considered the turning point to 'modern systematics'. Modalities like deoxyribonucleic acid (DNA) sequencing, random amplified polymorphic DNA, restriction fragment length polymorphism (RFLP), and Multilocus sequence typing (MLST) have played a dynamic role in transfiguring the taxonomic face of fungal classification.

The foremost revolutionary step in the turnover of fungal taxonomy was the analysis using PCR-amplified ribosomal RNA genes. The study by Hibbett et al. paved the path for this revolutionary approach, in which the authors proposed a new classification based on DNA sequencing. Though many genes viz. internal transcribed spacer (ITS), intergenic spacer (IGS1) region, mtSSU and mtLSU, 18S rRNA small subunit (18S-SSU), 28S ribosomal RNA large subunit (28S-LSU), γ-actin (*act*), b-tubulin (*β-Tub*), ATP synthase (*atp6*), RNA polymerases II (*rpb1* and *rpb2*), elongation factor 1α (*EF*-1α), and calmodulin (*cal*) have been targeted, ITS remains the barcode region and marker for fungal screening. ITS region comprises ITS 1-5.8S-ITS 2, of which the 5.8S region is one of the most conserved regions in all groups, thereby rendering itself of no use in classification. The same is sufficient in the identification of a few genera but insufficient for others like *Scedosporium*, *Cryptococcus*, *Chaetomium*, and *Pseudallescheria boydii*, for which MLST is adopted. Loci with a higher frequency of evolution are targeted for the establishment of the latest evolutionary events, while loci with slow evolution help in reconstructing deep phylogenetic relationships. MLST has also helped in the separation of some cryptic species of *Fusarium oxysporum*, *Histoplasma*, etc. The concept has been further modified by Mayden and Taylor, who considered the same as the 'genealogical concordance species concept' and is the most advanced and substantial basis of taxonomical classification presently. Another substantial leap in this direction was the project 'barcode of life' that aims to delineate the practical method for the identification of ~ten million living species on earth by encouraging varied advanced molecular modalities for preserving biodiversity. 'Darwin Tree of Life' has an ambitious target of barcoding ~66,000+ full genomes and has been launched in the pilot phase from February 2020 to August 2022. During the defined timeframe, it is aimed to generate the complete chromosomal assembly of ~4000 eukaryotic families and to achieve DNA barcoding of 10,000 species from the Natural History Museum, London. All the reviewed data will be published in an open findable accessible interoperable reusable (FAIR) database.

Nomenclature of the Fungi

Though multiple concepts have been recognised for the classification and nomenclature of fungi, for the formal acceptance by taxonomists, it requires description according to international standards using Latin binomials. The biological nomenclature has different codes for bacteria, viruses, plants, and animals, which makes the system more complex. Efforts have been streamlined towards the formulation of a systematic approach to have uniformity. Currently, the nomenclature of fungi is regulated by the International Code of Botanical Nomenclature (ICBN), adopted in the International Botanical Congress. All the changes are made available in the journal *Taxon* and then discussed in the Committee for approval. However, the major controversial issue is the change in nomenclature due to non-scientific reasons, creating confusion among clinicians on how to designate the infection. The Nomenclature Sub-Committee of the International Society for Human and Animal Mycology (ISHAM) came to the rescue for the same and published recent recommendations, which established that the name of an infection/disease should describe a disease. Nomenclature based on the conventional structure has been discouraged since it is vulnerable to taxonomic changes.

One fungus, one name: In the pre-molecular era, fungi were named based on their sexual forms and used to have two names (one each for teleomorph and anamorph). The work of mycologists was complex as it got difficult to select one suitable name for pleomorphic strains like *Penicillium* with *Talaromyces* and *Eupenicillium* teleomorphs. The term 'Holomorph' was used for fungi with teleomorphic sporulation along with anamorphs. Anamorphs that occur alone have been classified into 'Deuteromycotina'. With the advent of molecular modalities, this dual nomenclature was abandoned. It was at the 18th International Botanical Congress held in Melbourne in July 2011 when substantial changes in the fungal nomenclature were made, which is referred to as 'Melbourne Code 2012'. The separate names for anamorphs were discarded, and all approved species names (irrespective of the state they are typified for) were considered acceptable. The earliest legitimate name is most likely the most accepted one. The list of approved names was drafted and submitted to the Nomenclature committee for final approval. Presently, the name of Code has been substituted with International Code of Nomenclature for Algae, Fungi and Plants (ICNafp). The language for validation can be English or Latin. The publication ought to be in portable document format (PDF) and should be in an International Standard Serial Number (ISSN) or an International Standard Book Number (ISBN). However, the same was criticised by a few researchers who consider it mandatory to maintain the deuteromycete taxon names. Consequently, it was agreed upon to conserve the term 'deuteromycetes' (with a small d), but without recognising it as a separate rank, in the Holomorph Conference.

Despite the collaborative efforts to have a unified nomenclature of fungi, the same met other roadblocks when the terms 'anamorph' and 'teleomorph' were replaced with 'mitosporic' and 'meiosporic' fungus. This change in nomenclature is the least accepted one, as many of the clinically pathogenic fungi are ascomycetes, which occasionally depict their teleomorphic forms in culture and instead produce two different anamorphs (synanamorphs). Moreover, the terms 'anamorph' and 'teleomorph' are not merely based on the cellular process of division but also consider cytological processes that precede the events.

Besides the changes in nomenclature, molecular modalities also hold the utmost significance in the identification of unculturable pathogenic as well as environmental fungi. With the increase in the use of sequencing techniques, the number of unidentified sequences is also rising in the available databases. Hence, there is a need to have code for a nomenclature system with high-throughput taxonomy that can simultaneously keep the environmental and pathogenic fungi delineated.

Current Mycological Typing Methods

The identification of fungi is not only clinically significant but is relevant to recognise them in the environment and understanding emerging pathogenic fungi. The modalities available for identification include phenotypic, chemotaxonomic, physiological profiling, and biological and genotypic methods. The phenotypic methods include identification based on morphological features recognised in microscopy and culture. These can be either asexual or sexual structures. Asexual structures include conidia, sporangiospores, conidiophores, sporangiophores, sporangia, apophysis, columella, rhizoids, phialides, and metulae. The sexual structures include ascospores, zygospores, and basidiospores contained in fruiting bodies called ascocarps, basidiocarps, etc. Conidia are commonly noted in *Penicillium* spp., *Aspergillus* spp., and dermatophytes, whereas sporangiospores are seen in Mucorales. The genotypic methods include a screening of barcode region (ITS) or specific species-delineation region targeting actin, beta-tubulin, atp6, rpb 1 and 2, or elongation factor. The most sophisticated techniques available include DNA–DNA hybridisation, sequencing, next-generation sequencing (NGS), and protein profiling.

Molecular Taxonomy

Besides assisting in the classification of fungi, molecular modalities have also been exploited to revisit the species complexes based on a genome-wide approach. The term 'species complex' signifies a group of organisms that are closely related to each other taxonomically and are difficult to differentiate. Two approaches are quite commonly employed for the same, viz., average nucleotide identity and percentage of conserved proteins. The term was first introduced to describe the genus '*Fusarium*' as a substitute for subgeneric 'sections' as the sections described earlier were not in homology to phylogeny. Thereby, the species complexes of *Fusarium* were introduced that encompass all the known species and are monophyletic. Besides *Fusarium*, 'species complexes' is also used to depict some miscellaneous strains of indistinguishable taxonomic eminence like '*Aspergillus viridinutans* species complex', in *Cladosporium* as '*Cladosporium herbarum*' or '*Cladosporium cladosporioides*' complex. Moreover, owing to the lack of distinguishing features between clinical species of *Aspergillus* section *nigri*, these have been included under '*Aspergillus niger* complex'.

Another imperative example is of one of the potentially lethal yeast species, *Cryptococcus*, which was initially separated into two varieties viz. var. *neoformans* and var. *gattii*. However, due to unsettling taxonomic status, these were grouped under '*C. neoformans* complex' based on ITS barcoding. Subsequently, *C. gattii* was staged as a separate species, and now seven separate species are classified under this complex. This species complex is unique from others as the members encompassed in it vary from each other clinically, ecologically, and susceptibility to antifungals but are closely related to each other. Another but less commonly used species complex is *Scedosporium*, which includes *Scedosporium apiospermum* and *Scedosporium boydii*, owing to the complex taxonomic status of sexual and asexual states.

This inconsistency in the classification of 'species complexes' can be overcome by using genes other than ITS as secondary barcode genetic markers like tef1, rbp1, and rbp2. A more advanced and precise approach to having concrete and comprehensive data is a 'whole-genome sequence-based classification and identification system'. This system will be of help not only in classification at the species level but also at the 'below-species' level. Presently, ~500 fungal genome sequences of ~1000 species are available in the NCBI database, and a cut-off of >97% homology of ITS barcoding is used for species of the same operational taxonomic unit (OTU), which is like >97% homology of 16S rRNA region in prokaryotes. With the inclusion of primary and secondary barcoding regions and to delimit the errors in the sequences submitted to databases, a dedicated repository should be there in place for such species complexes. Though NCBI is the best-chosen repository, the errors, misidentifications, and misinformation in genome sequences reported in NCBI demand strict scrutiny and curation of data before submissions. Moreover, whenever a new species is being described, the genomic sequence ought to be compared stringently with all the available sequences in the database in a 'community-agreed standard format'. As more and more molecular modalities are being used for deciphering newer species, the whole-genome sequence data is also increasing. There is a need to extend these databases to higher taxonomic levels with specific homology cut-offs so that these can be precisely grouped under genus order, class, and species.

Current Classification of Clinically Relevant Fungi

The clinically relevant pathogenic fungi may be categorised into three subgroups: (a) obligate parasites like dermatophytes infecting skin and appendages, (b) dimorphic saprophytes existing as different morphotypes in the environment and humans, capable of causing disease in healthy hosts, and (c) opportunistic saprophytes infecting immunocompromised hosts. Currently, all these fungi form a part of four phyla under the kingdom Eumycota, namely, Zygomycota, Zoopagomycota, Ascomycota, and Basidiomycota (Figs. 1.2, 1.3, and 1.4). Apart from these, *R. seeberi* from phylum Mesometazoa (Kingdom Protocista) and *P. insidiosum* from phylum Oomycota (Kingdom Chromista) are known to cause human infections and are referred to as 'Pseudofungi'.

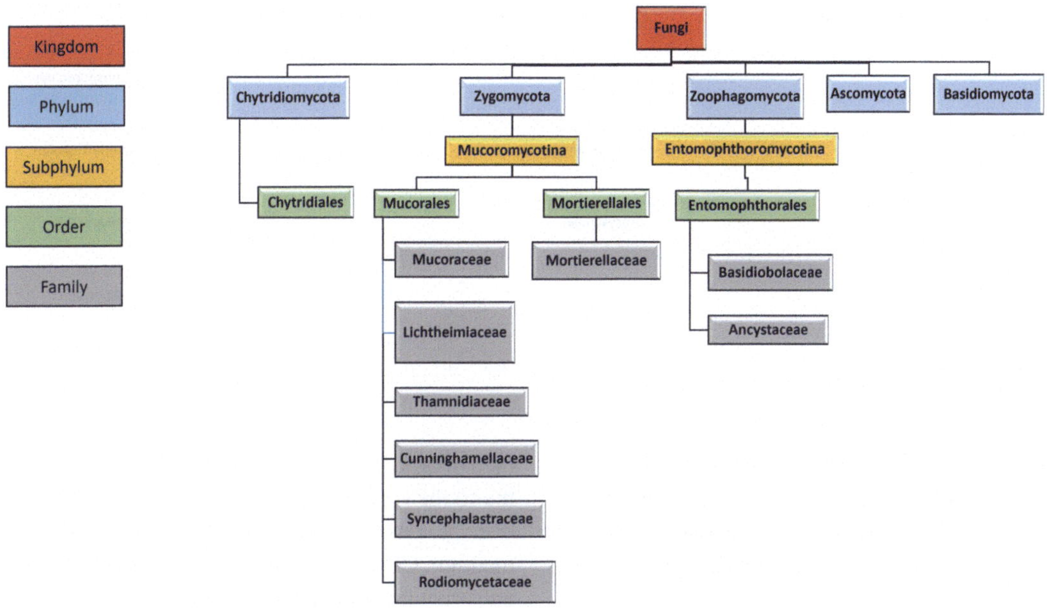

Fig. 1.2 Current classification of clinically relevant fungi (Chytridiomycota, Zygomycota, and Zoopagomycota)

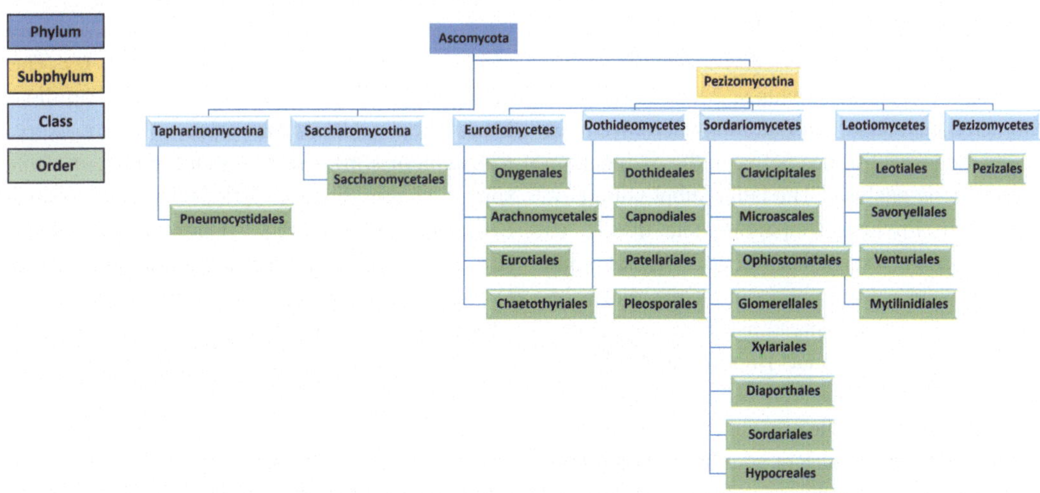

Fig. 1.3 Current classification of clinically relevant fungi (Ascomycota)

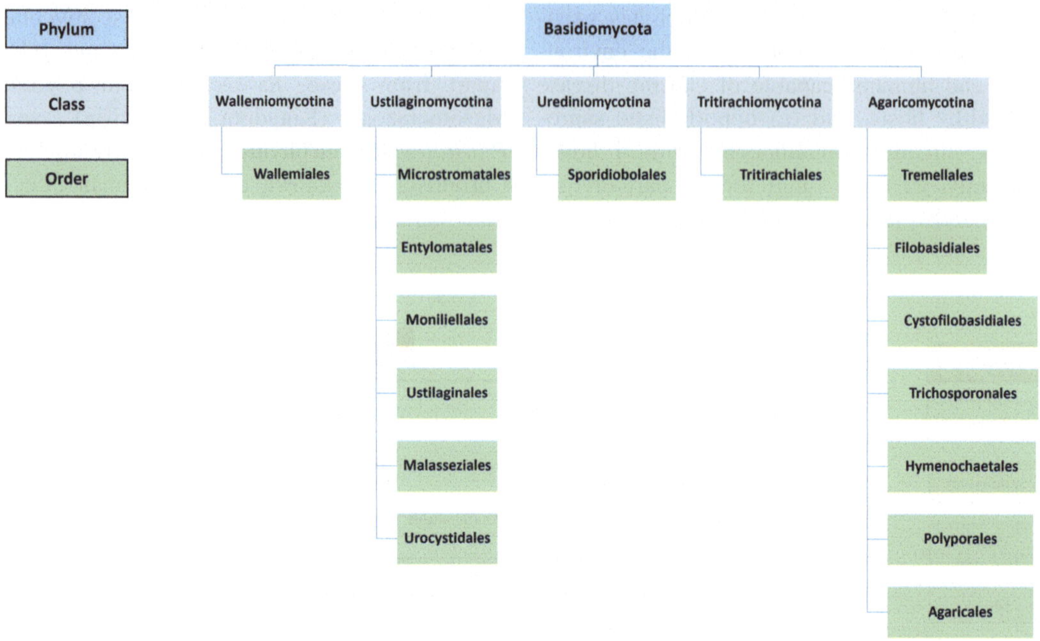

Fig. 1.4 Current classification of clinically relevant fungi (Basidiomycota)

Prototheca, a member of unicellular algae, acts like yeast and can infect a human. Salient features of the most clinically important taxonomic groups of fungi are detailed below.

Zygomycota

This group of lower fungi exhibits coenocytic broad aseptate hyphae with the production of zygospores. The previous class Zygomycota has been replaced by subphylums, Mucoromycotina, Mortierellomycotina, and Entomophthoromycotina, due to the large distance between the members in phylogenetic analysis. The latter subphylum now belongs to Phylum Zoopagomycota due to its phylogenetic diversity. The Orders Mucorales and Mortierellales release spores passively, and colonies are cottony and spreading. They are demonstrated as broad aseptate ribbon-like hyphae in clinical samples. Their cell wall is composed of chitosan and chitin. The asexual structures include sporangiophores ending in sporangia (containing numerous spores) or sporangiolas (≥1 spores), or merosporangia (spores in rows) varying according to the species. They possess stolons to spread over a surface and rhizoids to penetrate the medium. Chlamydospores or oidia are other asexual structures produced by this group of fungi.

The sexual structures include spiny brownish zygospores formed on suspensor cells. Few members can tolerate high temperatures like *Mucor indicus*, *Rhizopus arrhizus*, *Saksenaea vasiformis* (42 °C), *Rhizopus microsporus* (45 °C), and *Rhizomucor pusillus* (50 °C). The order Mucorales further consists of Families, Mucoraceae (*Rhizopus* spp., *Mucor* spp., *Rhizomucor* spp), Lichtheimiaceae (*Lichtheimia* spp), Thamnidiaceae (*Thamnostyllum* spp., *Cokeromyces* spp), Cunninghamellaceae (*Cunninghamella* spp), Syncephalastraceae (*Syncephalastrum* spp), and Rodiomycetaeceae (*Apophysomyces* spp., *Saksenaea* spp). All these members are ubiquitous and causative agents of angioinvasive, highly fatal disease named mucormycosis. The disease presents as rhino-orbito-cerebral (in diabetes mellitus), pulmonary (in immunosuppressed conditions), gastrointestinal, cutaneous (in trauma), renal (in immunocompetent hosts), or disseminated infection. Except for *Mucor irregularis*, all agents cause acute infection. The molecular identification of Mucorales is based on the ITS region of rDNA and specific primers targeting regions of 18S rDNA, 28S rDNA, cytochrome B, mitochondrial gene rnl, or *CotH* genes. The Order Mortierellales comprises *Mortierella polycephala* and *Mortierella wolfii*, mostly pathogenic in animals with rare fatal human infection in stem cell transplant patients.

Zoopagomycota

Zoopagomycota comprises a single subphylum Entomophthoromycotina, Order Entomophthorales, which differs from Orders Mucorales and Mortierellales of Phylum Zygomycota by the ability to release spores forcibly. They may be pauciseptate, produce primary and secondary conidia, and are generally insect pathogens. The zygospores often possess beaks aiding in identification. The colonies are usually smooth and waxy, resembling yeast. The Order comprises Families Basidiobolaceae (*Basidiobolus* spp) and Ancylistaceae (*Conidiobolus* spp). The molecular identification of Entomophthorales is based on the ITS region of rDNA, 18S and partial 28S of rRNA. Entomophthoromycosis is localised mycoses of the face or gastrointestinal tract, which causes firm, progressive swelling of the subcutaneous tissues. *Conidiobolus* usually affects the face or nasal mucosa, while *Basidiobolus* is known to affect the gastrointestinal tract in immunocompetent individuals.

Ascomycota

It comprises the largest group of fungi spread over seven classes, namely, Taphrinomycotina, Saccharomycotina, Eurotiomycetes, Dothideomycetes, Sordariomycetes, Leotiomycetes, and Pezizomycetes. The last five classes belong to Subphylum Pezizomycotina (Fig. 1.3).

- Taphrinomycotina consists of Order Pneumocystidales harbouring a single genus, *Pneumocystis* causing respiratory infections in immunocompromised patients. It consists of a cyst and trophic phase in its life cycle. The targeted genes used for molecular identification are *mtLSU* rRNA, ITS, hsp70, nucRNA, and superoxide dismutase.
- Saccharomycotina forms a group of ascomycetous yeasts with a single order, Saccharomycetales, comprising ascus-forming fungi without any fruiting structures. Mostly the members possess anamorphs structures with budding cells (*Candida*) or with arthroconidia (*Geotrichum*, *Magnusiomyces*). The order includes eight families, Saccharomycetaceae (*Saccharomyces*, *Candida glabrata*, *Candida blankii*, *Candida nivariensis*), Debaryomycetaceae (*C. albicans*, *Candida dubliniensis*, *Candida. africana*, *Merozyma guilliermondii*, *Candida orthopsilosis*, *Candida parapsilosis*, *Candida tropicalis*, *Candida viswanathii*, *Lodderomyces*), Metschnikowiaceae (*Candida auris*, *Clavispora lusitaniae*, *Candida haemulonis*, *Kodamaea*), Pichiaceae (*Pichia kudriavzevii*, *Pichia lambica*, *Pichia inconspicua*, *Pichia norvegensis*), Dipodascaceae (*Geotrichum*, *Magnusiomyces*), Ascoideaceae (*C. lipolytica*), Trichomonasceae (*Trichomonascus ciferii*), and Wickerhamomycetaceae (*Wickerhamomyces anomalus*, *Cyberlindnera jadinii*, *C. fabianii*). They can cause infection in any body site, including disseminated infections originating in the community or hospital. The genes used for identification include the ITS region and the 26S rDNA region.
- Pezizomycotina comprises the filamentous ascomycetes, which possess septate thallus with septa having micropores or Woronin bodies. They grow in culture as yeast, coelomycetes, or hyphomycetes. The asexual cycle harbours a variety of conidia, while the sexual structures include ascocarps containing asci with ascospores. The fruiting structures (ascomata) include gymnothecium (network of hyphae with central asci), cleistothecium

(closed), perithecium (pear-shaped with apical opening), apothecium (flat cells), or naked asci without any fruiting body.

- Class Euromycetes
  - Order Onygenales consists of loosely woven hyphal fruiting structures and thallic conidiation. They include Families Arthrodermataceae (keratinophilic dermatophytes), Onygenaceae (endemic fungi like *Blastomyces*, *Coccidioides*, *Emergomyces*, *Emmonsia*, *Histoplasma*, *Paracoccidiodes*), Spiramastigaceae, Nanniziopsiaceae (*Chrysosporium*), and Gymnoascaceae (*Malbranchea*). The dermatophytes are responsible for dermatophytosis, while endemic fungi cause pulmonary, cutaneous, or disseminated infection.
  - Order Arachnomycetales consists of Family Arachnomycetaceae
  - Order Eurotiales comprises of Families Eremomycetaceae (*Arthrographis*), Monascaceae (*Monascus*), Pseudourotiaceae (*Plasmodium ovale*), Thermoascaceae (*Paecilomyces crustaceous*), Aspergillaceae (*Aspergillus*), and Trichocomaceae (*Talaromyces marneffei*, Rasamsonia). Of all these, Aspergillaceae is most implicated in human infections and causes allergic, cutaneous, invasive, and disseminated infections. *T. marneffei* is also considered one of the endemic mycoses in Southeast Asia and causes disease in HIV-positive and transplant patients.
  - Order Chaetothyriales comprises of Family Herpetrichiellaceae (*Cladophialophora*, *Exophiala*, *Phialophora*, *Rhinocladiella*, *Fonsecaea*) and Trichomeriaceae (*Arthrocladium*). These dematiaceous fungi are being increasingly reported in cutaneous and CNS infections in both immunocompetent and immunosuppressed patients.
- Class Dothideomycetes
  - Order Dothideales comprises of Families Dothideaceae (*Aureobasidium*, *Hortaea werneckii*), and Botryosphericeae (*Lasiodiplodia*, *Neoscytilidium*). These fungi are mainly implicated in infections of nails and hair.
  - Order Capnodiales consists of Families Cladosporiaceae (*Cladosporium cladosporum*, *Cladosporium spherospermum*), Mycosphaerellaceae and Piedraiaceae (*Piedraia hortae*). The fungi are implicated in subcutaneous, nail, hair, and CNS infections.
  - Order Patellariales (*Gloniopsis percutanea*)
  - Order Pleosporales comprises Families Trematosphaeriaceae, Tetraplosphaericeae, Macrodiplodiopsidaceae, Neohendersoniaceae (*Medicopsis romeroi*), Cucurbitariaceae, Montagnuclaceae, Leptosphaeriacae, Phaeosphaericeaea, Pleosporaceae (*Bipolaris*, *Alternaria*, *Curvularia*, *Exserohilum*, *Ulocladium*, *Pappilospora equi*), Lingomycetaecae (*Hongleongmyces pedis*), Roussoellaceae (*Rousella percutanea*), and Didymellaceae (*Epicoccum sorghinum*, *Phoma*). These dematiaceous fungi are known to cause cutaneous infections and keratitis.
- Class Sordariomycetes
  - Order Clavicipitales comprises Families Cordycipitaceae (*Beauveria*), Clavicipitaceae, and Ophiocardycipitaceae (*Purpureocillium lilacinum*).
  - Order Microascales consists of Family Microascaceae (*Scopulariopsis*, *Lomentospora prolificans*, *Scedosporium apiospermum*, *Shigrlla boydii*). These are implicated in onychomycosis, central nervous system (CNS), and pulmonary infections.
  - Order Ophiostomotales is comprised of Family Ophiostomataceae (*Sporothrix*), which causes cutaneous, subcutaneous, and pulmonary infections.
  - Order Glomerellales includes Family Plectosphaerellaceae (*Colletotrichum*), which is implicated in keratitis.
  - Order Xylariales includes family Xylariceae.
  - Order Diaporthales comprises Families Togniniaceae (*Pheoacremonium*), Diporthaceae (*Phomopsis*), and

Calosphariaceae. These agents cause cutaneous infections.
  - Order Sordariales consists of Families Lasiosphaeriaceae, Apiosporaceae (*Nigrospora sphaerica*), Cephalotheceae, Chaetomiaceae (*Chaetomium*, *Madurella*), and Coniochaeticeae (*Acrophialophora fusispora*, *Cladorrhinum*). The agents cause cutaneous, CNS infections and keratitis.
  - Order Hypocreales consists of Families Hypocreaceae (*Acremonium*, *Cylindrocarpon*, *Fusarium*, *Trichoderma*) and Nectriaceae. These can cause cutaneous, ocular, and invasive infections.
- Class Leotiomycetes
  - Order Leotiales consists of Family Leotiaceae.
  - Order Savovyellales consists of Family Savovyellaceae.
  - Order Venturiales consists of Family Sympoventuriaceae (*Ochroconis humicola*, *O. gallopava*, *O. tshawytschae*), implicated in either CNS infections or sinusitis.
  - Order Mytilinidiales consists of Family Mytilinidiaceae.
- Class Pezizimycetes
  - Order Pezizales has Family Ascodesmiaceae.

Basidiomycota

This group (Fig. 1.4) is characterised by a dikaryotic septate thallus, along with the presence of clamp connections and pores in septa called dolipores. Reproduction occurs by plasmogamy, delayed karyogamy, and meiosis, forming basidiospores released with force. The members are either yeasts or filamentous forms. The phylum Basidiomycota comprises classes Wallemiomycotina, Pucciniomycotina, Ustilaginomycotina, and Agarimycotina.

- Wallemiomycotina (Order Wallemiales, Family Wallemiaceae) is a group of osmophilic fungi with a single genus *Wallemia*.
- Pucciniomycotina/Tritirachiomycotina (Order Tritirachiales, Family Tritirachiaceae Genus *Tritirachium* and Order Sporidiobalales, Family Sporidiobalaceace) comprises plant pathogenic fungi including the red yeasts, *Sporobolomyces* and *Rhodotorula* spp. capable of causing infections in humans.
- Ustilaginomycotina comprises six orders, mostly plant pathogens, namely, Entylomatales (Family Entylomataceae), Malasseziales (Family Malasseziaceae), Microstomatales (Family Quambalariaceae), Ustilaginales (Family Ustilaginaceae), Urocystidales (Family Fereydoumiaceae), and Moniliellalaes (Family Moniliellaceae). *Malassezia* spp. in Family Malasseziaceae of order Malasseziales is a commensal and pathogen in humans. The fungus is usually commensal to the scalp but can cause hyperpigmentation or hypopigmentation on the trunk, pityriasis versicolor, folliculitis, and seborrheic dermatitis and is also considered to predispose to pancreatic carcinoma.
- Agarimycotina includes most mushrooms and toadstools. It comprises six orders, namely, Tremellales (Families Tremellaceae, Filobasidiaceae), Cystofilobasidiales (Family Cystofilobasidiaceae), Trichosporonales (Family Trichosporonaceae), Hymenochaetales (Family Hymenochaetaceae), Polyporales (Families Meruliaceae, Phanerochaetaceae, Polyporaceae, Schizophyllaceae), and Agaricales (Families Psathyrellaceae, Pluteaceae). Among the agents pathogenic to humans, *Cryptococcus*, *Trichosporon*, *Inonotus*, *Schizophyllum*, *Phellinus*, and *Hormographiella* belong to Agarimycotina. Of these, *Cryptococcus* is known to cause meningitis, and pulmonary and disseminated disease, while *Inonotus*, *Phellinus*, and *Hormographiella* are more commonly implicated in keratitis. *Schizophyllum* commune is another important member implicated in the cutaneous central nervous system as well as pulmonary infections.
- The selected targets for molecular analysis comprise ITS1, ITS2, LSU, SSU, RPB1, RPB2, and MCM7.

Name Changes and Newly Described Species

Table 1.1 provides a list of changes in the names of some fungi, and Table 1.2 provides a list of a few newer species described as pathogens in humans.

Microbial Databases Used in Fungal Taxonomy

Since molecular techniques have eased the availability of sequences on a larger scale, it becomes imperative to have all these sequences deposited in a unified database. These databases serve as

Table 1.1 List of newer names of clinically relevant fungi

Previous species name	Revised species name
Absidia elegans	*Apophysomyces elegans*
Absidia elegans *Absidia corymbifera*	*Apophysomyces mexicanus*
	Apophysomyces trapeziformis
	Apophysomyces variabilis
	Lichtheimia corymbifera
Acremonium kiliense	*Sarocladium kiliense*
Aureobasidium pullulans	*Aureobasidium melanogenum*
Aureobasidium pullulans *Botryomyces caespitosus*	*Aureobasidium pullulans*
	Alternaria caespitose
Candida catenulata	*Diutina catenulate*
Candida bracarensis	*Nakaseomyces bracarensis*
Candida ciferrii	*Trichomonascus ciferrii*
Candida famata	*Debaryomyces hansenii*
Candida fermentati	*Meyerozyma caribbica*
Candida glabrata	*Nakaseomyces glabrata*
Candida glabrata *Candida guilliermondii*	*Candida nivariensis*
	Candida palmioleophila
	Candida pseudoaaseri
	Meyerozyma guillermondii
Candida guilliermondii var. membranifaciens	*Kodamaea ohmeri*
Candida haemulonis	*Candida pseudoheamulonis*
Candida inconspicua	*Pichia cactophila*
Candida infanticola	*Wickerhamiella infanticola*
Candida kefyr	*Kluyveromyces marxianus*
Candida krusei	*Pichia kudriavzevii*
Candida lambica	*Pichia fermentans*
Candida lipolytica	*Yarrowia lipolytica*
Candida lusitaniae	*Clavispora lusitaniae*
Candida nivariensis	*Nakaseomyces nivariensisaa*
Candida norvegensis	*Pichia norvegensis*
Candida parapsilosis	*Candida parapsilosis*
Candida parapsilosis *Candida pararugosa*	*Candida metapsilosis*
	Candida orthopsilosis
	Wickerhamiella pararugosa
Candida pelliculosa	*Wickerhamomyces anomalus*
Candida pintolopesii	*Kazachstania telluris*
Candida pulcherrima	*Metschnikowia pulcherrima*
Candida rugosa	*Diutina rugosa*
Candida utilis	*Cyberlindnera jadinii*
Chaetomium atrobrunneum	*Amesia atrobrunnea*

(continued)

Table 1.1 (continued)

Previous species name	Revised species name
Cryptococcus albidus	*Naganishia albida*
Cryptococcus curvatus	*Cutaneotrichosporon curvatum*
Cryptococcus diffluens	*Naganishia diffluens*
Cryptococcus laurentii	*Papiliofrema laurentii*
Diutina (Candida) mesorugosa	*Diutina rugosa*
Emmonsia parva	*Blastomyces parvus*
Emmonsia pasteuriana	*Emergomyces pasteurianus*
Fonsecaea compacta	*Fonsecaea pedrosoi*
Fusarium keratoplasticum	*Neocosmospora keratoplastica*
Fusarium lichenicola	*Neocosmospora lichenicola*
Fusarium metavorans	*Neocosmospora metavorans*
Fusarium oxysporum	*Fusarium oxysporum*
Fusarium oxysporum	*Fusarium petroliphilum*
Fusarium petrophila	*Neocosmospora petroliphila*
Fusarium proliferatum	*Fusarium acutatum*
Fusarium proliferatum	*Fusarium andiyazi*
Fusarium solani	*Fusarium anthophilum*
	Fusarium mundagurra
	Fusarium musae
	Fusarium proliferatum
	Fusarium keratoplasticum
Fusarium solani	*Fusarium metavorans*
Fusarium solani species complex 6 (FSSC6)	*Fusarium pseudensiforme*
	Fusarium solani
	Fusarium metavorans
Fusarium solani species complex 7 (FSSC7)	*Neocosmospora gamsii*
Fusarium solani species complex 9 (FSSC9)	*Neocosmospora tonkinensis*
Fusarium solani species complex 9 (FSSC9)	*Cylindrocarpon tonkinense*
Fusarium solani species complex *20 (FSSC20)*	*Fusarium tonkinense*
	Neocosmospora suttoniana
Fusarium solani species complex 43 (FSSC43)	*Neocosmospora catenata*
Fusarium solani species complex 43 (FSSC43)	*Hypocreales*
Geotrichum capitatum	*Magnusiomyces capitatus*
Geotrichum capitatum	*Magnusiomyces clavatus*
Hormographiella aspergillata	*Coprinopsis cinereus*
Madurella grisea	*Trematosphaeria grisea*
Madurella mycetomatis	*Madurella fahalii*
Madurella mycetomatis	*Madurella mycetomatis*
Microsporum gypseum	*Madurella pseudomycetomatis*
	Madurella tropicana
	Nannizzia gypsea
Ochroconis gallopava	*Verruconis gallopava*
Paecilomyces lilacinus	*Purpureocillium lilacinum*
Penicillium marneffei	*Talaromyces marneffei*
Pneumocystis carinii	*Pneumocystis jirovecii*
Pseudallescheria boydii	*Scedosporium apiospermum*
Pseudallescheria boydii	*Scedosporium boydii*
Pseudozyma antarctica	*Moesziomyces antarcticus*
Pseudozyma aphidis	*Moesziomyces aphidis*

(continued)

Table 1.1 (continued)

Previous species name	Revised species name
Pseudozyma parantartica	*Moesziomyces parantarcticus*
Pyrenochaeta romeroi	*Medicopsis romeroi*
Rhizomucor variabilis	*Mucor irregularis*
Rhodotorula minuta	*Cystobasidium minutum*
Rhodotorula slooffiae	*Cystobasidium slooffiae*
Saksenaea vasiformis	*Saksenaea erythrospora*
Saksenaea vasiformis	*Saksenaea oblongispora*
Scopulariopsis acremonium	*Saksenaea vasiformis*
	Acaulium acremonium
Scopulariopsis brumptii	*Microascus paisii*
Scytalidium hyalinum	*Neoscytalidium dimidiatum*
Scedosporium prolificans	*Lomentospora prolificans*
Stephanoascus ciferrii	*Trichomonascus ciferrii*
Trichophyton ajelloi	*Arthroderma uncinatum*
Trichosporon cutaneum	*Cutaneotrichosporon cutaneum*
Trichosporon mucoides	*Cutaneotrichosporon mucoides*

Table 1.2 List of some new fungal taxa isolated from human clinical material

Species	Order	Source(s)
Alternaria anthropophila	Pleosporales	Tissue
Alternaria atrobrunnea	Pleosporales	Exudate
Alternaria guarroi	Pleosporales	Biopsy
Arthroderma chilionensis	Onygenales	Skin scrapings
Aspergillus dobrogensis	Eurotiales	Toenail
Aspergillus microperforatus	Eurotiales	Toenail, lymph node
Aspergillus suttoniae	Eurotiales	Sputum
Blastomyces emzantsi	Onygenales	Various clinical sites
Cunninghamella arunalokei	Mucorales	Skin biopsy
Curvularia coimbatorensis	Pleosporales	Corneal scrapings
Curvularia tamilnaduensis	Pleosporales	Corneal scrapings
Diaporthe oculi	Diaporthales	Cornea scrapings
Diaporthe pseudooculi	Diaporthales	Cornea scrapings
Exophiala arunalokei	Chaetothyriales	Skin biopsy
Fusarium riograndense	Hypocreales	Nasal cavity
Gambiomyces profunda	Pleosporales	Various tissues
Gloniopsis percutanea	Hysteriales	Various tissues
Gloniopsis pneumoniae	Hysteriales	Lung tissue
Knoxdaviesia dimorphospora	Ophiostomatales	Fluid
Microascus ennothomasiorum	Microascales	Biopsy of thumb nodule
Nannizzia perplicata	Onygenales	Skin scrapings
Trichophyton indotineae	Onygenales	Skin scrapings
Wickerhamiella verensis	Saccharomycetales	Blood culture

an important tool for researchers worldwide. The most reliable of these databases is the National Centre for Biotechnology Information (NCBI), followed by Central Bureau voor Schimmelcultures (CBS), Index Fungorum, All Fungal Tree of Life (AFTOL), Q-Bank Fungi Database, Saccharomyces Genome Database (SGD), TrichoKEY, Mycocosm (1000 Fungal

Genome Project), UNITE (http://unite.ut.ee/index.php), Mycobank, and other databases of specific fungi like *Aspergillus*, *Fusarium*, and *Penicillium* (Fusarium-ID, The *Aspergillus* website). This is important for further authentication purposes as well since fungi have many unauthenticated annotated sequences that can hamper precise identification. Mycobank is a comprehensive database that registers all new names, along with a brief description, and thereafter provides a registration number (like EzTaxon for Bacteria). There is a need to publish all the updated valid names in an official journal, as done in the case of bacteria. ISHAM ITS-reference DNA barcoding database is another quality-controlled reference database by an international consortium of medical mycologists, which helps in the reliable identification of clinical as well as animal fungi.

Challenges in Fungal Taxonomy

Owing to the complex evolutionary process and heterogeneity of taxonomical approaches, the precise identification of fungi remains obscure and ambiguous. Despite having simpler body plans, microscopy is imperative for the identification of fungi. The choice of typing method varies with the group under scrutiny. Though conventional microscopic and culture techniques can help to identify clinically relevant fungi, the process is tedious and time-consuming. This paves the path for molecular diagnosis, but the same is not available in all the laboratories and needs standardisation and validation.

Fungal taxonomy is a complex area of research, and mycologists have no consensus on varied aspects like what species is and what is the ideal modality of delineating the same. 'One fungus, one name' has been brought into practice, but the diversity of fungi in the environment makes the process complex. The difference between the 'species concept' and 'determinants for categorising species' needs to be deciphered precisely. A molecular era has revolutionised the determination of phylogeny, but the same is evolving with molecular techniques. Though a unified platform is being created, ambiguity in molecular methods also raises concerns. The genome-based sequencing, when used for the speciation of fungi, will again change the taxonomy. Fitzpatrick et al. have already proposed phylogeny based on 42 complete genomes. specI software (species identification tool; http://www.bork.embl.de/software/specI/) is one of the software based on the WGS of fungi. Besides the molecular methods, matrix-assisted laser desorption/Ionisation time of flight (MALDI-TOF) and Raman spectroscopy are also being used for the identification of clinically significant fungi. Though these techniques are evolving with time and might play a substantial role in the species concept, a considerable amount of data with a reliable database needs to be generated before this can be applied on a large scale.

In conclusion, fungal taxonomy is a very intricate subject that is still evolving with the advent of innovations in molecular modalities. It is of utmost significance in deciphering the evolutionary trends, nomenclature, and classification and understanding the interplay between ecological niches, microbial communities, and characteristics exhibited by fungi in varied environments and climates. Moreover, taxonomical basics help in better understanding the pathogenesis of fungi, which is important in distinguishing pathogenic and non-pathogenic species. The concept has revolutionised from morpho-taxonomic approaches to phylogenetic taxonomical approaches, and with the aid of NGS and metagenomics, more concealed areas of taxonomy are being unveiled that will further aid in the comprehension of controversial aspects.

Case Study

A 48-years old male labourer in construction factories/houses presented with a history of persistent headaches for the past 8 months. The patient was a known case of diabetes mellitus, hypertension, and hyperlipidemia. He had visited a local physician for headache, who prescribed him over-the-counter analgesics and an antibiotic. However, he had no relief and presented to our emergency after the headache worsened and he started with the blurring of vision. Physical examination revealed a Glasgow coma scale of 8. His pupils were reactive to light, he had photophobia, and his patellar reflexes were increased. Kernig's sign was positive with nuchal rigidity. The patient was empirically started on fluconazole. The computerised tomography scan did not show any abnormalities. Lumbar puncture revealed an opening pressure of 170 mm H_2O with 60% neutrophil predominance, 51 mg/dL glucose, and protein of 90 mg/dL. India Ink examination revealed doubtful fungal elements and serum Cryptococcal antigen was 1:10. Culture on Sabouraud's dextrose agar showed growth of white, moist colonies after 48 h of incubation, which was identified as *Cryptococcus laurentii* (new name: *Papiliotrema laurentii*) using matrix-assisted laser desorption ionisation-time of flight (MALDI-TOF MS). The therapy was changed to a combination of liposomal amphotericin B and 5-flucytosine for induction therapy, followed by fluconazole as maintenance therapy. The patient responded to treatment without any adverse effects on follow-up.

1. What are the substantial pointers in the clinical history?
2. How to report the above case, with rarely identified species that have undergone taxonomical changes?
3. What are the risk factors and mode of transmission for *P. laurentii*?
4. How to differentiate *C. neoformans* from *P. laurentii* in conventional settings?
5. What is the recommended management for *P. laurentii*?

Research Questions

1. What is the basis of the current classification of fungi?
2. What is the "barcode' for fungal identification and its limitations?
3. What is the need and status of 'One fungus, one name'?
4. How to overcome the challenges faced in fungal identification in routine diagnostics?

Acknowledgement We gratefully acknowledge the help of Dr. Parakriti Gupta and Dr. Dipti Sharma for the literature search.

Further Reading

Aime MC, Miller AN, Aoki T, Bensch K, Cai L, Crous PW et al (2021) How to publish a new fungal species, or name, version 3.0. IMA Fungus 12(1):1–15

Bates ST, Ahrendt S, Bik HM, Bruns TD, Caporaso JG, Cole J et al (2013) Meeting report: fungal its workshop (October 2012). Stand Genomic Sci 8(1):118–123

Berbee ML, Taylor JW (1992) Detecting morphological convergence in true fungi, using 18S rRNA gene sequence data. Biosystems 28(1–3):117–125

Borman AM, Johnson EM (2021) Name changes for fungi of medical importance, 2018 to 2019. J Clin Microbiol 59(2):1–8

Chen M, Zeng J, De Hoog GS, Stielow B, Gerrits Van Den Ende AHG, Liao W et al (2016) The 'species complex' issue in clinically relevant fungi: a case study in Scedosporium apiospermum. Fungal Biol 120(2):137–146

de Hoog G, Guarro J, Gené J, Ahmed S, Al-Hatmi A, Figueras M et al (eds) (2020) The atlas of clinical fungi, 4th edn. Utrecht, Centraalbureau voor Schimmelcultures

Guarro J, Gené J, Stchigel AM (1999) Developments in fungal taxonomy. Clin Microbiol Rev. 12:454–500 p

Hawksworth DL, Kirk PM, Sutton BC, Pegler DN (1996) Ainsworth & Bisby's dictionary of the fungi. Rev Inst Med Trop Sao Paulo 38:272

Hawksworth DL, Lücking R (2017) Fungal diversity revisited: 2.2 to 3.8 million species. www.indexfungorum.org

Hibbett DS, Binder M, Bischoff JF, Blackwell M, Cannon PF, Eriksson OE et al (2007) A higher-level phylogenetic classification of the fungi. Mycol Res 111(5):509–547

Hibbett DS, Ohman A, Glotzer D, Nuhn M, Kirk P, Nilsson RH (2011) Progress in molecular and morphological taxon discovery in fungi and options for formal classification of environmental sequences. Fungal Biol Rev 25(1):38–47

Karthick B, Williams DM (2012) The International Code for Nomenclature for algae, fungi and plants—Nomenclature. Current 102(4):551–552

Lücking R, Aime MC, Robbertse B, Miller AN, Aoki T, Ariyawansa HA et al (2021) Fungal taxonomy and sequence-based nomenclature. Nat Microbiol [Internet] 6(5):540–548. https://doi.org/10.1038/s41564-021-00888-x

Mende DR, Sunagawa S, Zeller G, Bork P (2013) Accurate and universal delineation of prokaryotic species. Nat Methods [Internet] 10(9):881–884. https://doi.org/10.1038/nmeth.2575

Naranjo-Ortiz MA, Gabaldón T (2019) Fungal evolution: diversity, taxonomy and phylogeny of the fungi. Biol Rev 94(6):2101–2137

Odds EC, Arai T, Disalvo AE, Evans EG V, Hay RJ, Randhawa HS et al (1992) Nomenclature of fungal diseases: a report and recommendations from a Sub-Committee of the International Society for Human and Animal Mycology (ISHAM) [Internet]. J Med Vet Mycol 30. https://academic.oup.com/mmy/article/30/1/1/893063

Raja HA, Miller AN, Pearce CJ, Oberlies NH (2017) Fungal identification using molecular tools: a primer for the natural products research community. J Nat Prod 80(3):756–770

Ruggiero MA, Gordon DP, Orrell TM, Bailly N, Bourgoin T, Brusca RC et al (2015) A higher level classification of all living organisms. PLoS One 10(4):e0130114

Sarwar S, Firdous Q, Nasir Khalid A (2019) Importance of molecular and phylogenetic analyses for identification of basidiomycetes. In: Recent advances in phylogenetics

Sharma R, Polkade AV, Shouche YS (2015) "Species concept" in microbial taxonomy and systematics. Curr Sci 108(10):1804–1814

Host–Fungi Relationship: Pathogenicity and Immunity

2

Shreya Singh and Shivaprakash M. Rudramurthy

Learning Objectives

By the end of the chapter, the readers should be able to:

1. Study the common agents implicated in fungal zoonoses, their ecology, and modes of transmission.
2. Enlist the common virulence attributes of zoonotic fungal pathogens.
3. Describe the components of fungal recognition by the immune system.
4. Understand the roles of innate and adaptive immune mechanisms during fungal infections.

Introduction

Zoonoses are infections that can be naturally transmitted from a vertebrate animal to human beings, and such infections, when caused by fungi, are collectively known as "fungal zoonoses." Fungal zoonoses are transmitted in the natural environment between vertebrate animals and humans against the backdrop of the changing environment of the human–animal ecosystems. Animals can either be reservoirs or vectors for zoonotic fungal pathogens. Many wild and domestic animals that contaminate the surroundings with their feces in soil, burrows, etc., maintain natural infection and can also transmit the infection to humans in varying degrees. An increase in physical contact between humans and animals due to increased industrial development and encroachment of human settlements in wildlife habitats has caused an upsurge in fungal zoonotic infections. The common fungal zoonoses/sapronoses include dermatophytosis, sporotrichosis, histoplasmosis, talaromycosis, and lobomycosis. Infections by other less common zoonotic pathogens include *Emmonsia* spp., *Paracoccidioides brasiliensis*, *Conidiobolus* spp., *Basidiobolus ranarum*, and *Cryptococcus* spp. (Fig. 2.1).

Essentially, pathogenic fungi persist in the environment like saprobes but, in addition, utilize a vertebrate host in their life cycle as a vector. Although humans are not the preferred host, infections still occur due to the adaptation of fungi to the unexpected host. Infection is associated with a robust inflammatory response, which can even be fatal in cases with innate immune defects. Understanding disease pathogenesis provides insight into varied clinical manifestations, factors affecting reactivation, and the development of potential therapeutic targets.

S. Singh
Department of Microbiology, Dr B R Ambedkar State Institute of Medical Sciences (AIMS), Mohali, Punjab, India

S. M. Rudramurthy (✉)
Department of Medical Microbiology, Postgraduate Institute of Medical Education and Research (PGIMER), Chandigarh, India
e-mail: mrshivprakash@yahoo.com

S. C. Parija, S. M. Rudramurthy (eds.), *Textbook of Fungal Zoonoses and Sapronoses*, Microbial Zoonoses, https://doi.org/10.1007/978-981-97-3214-2_2

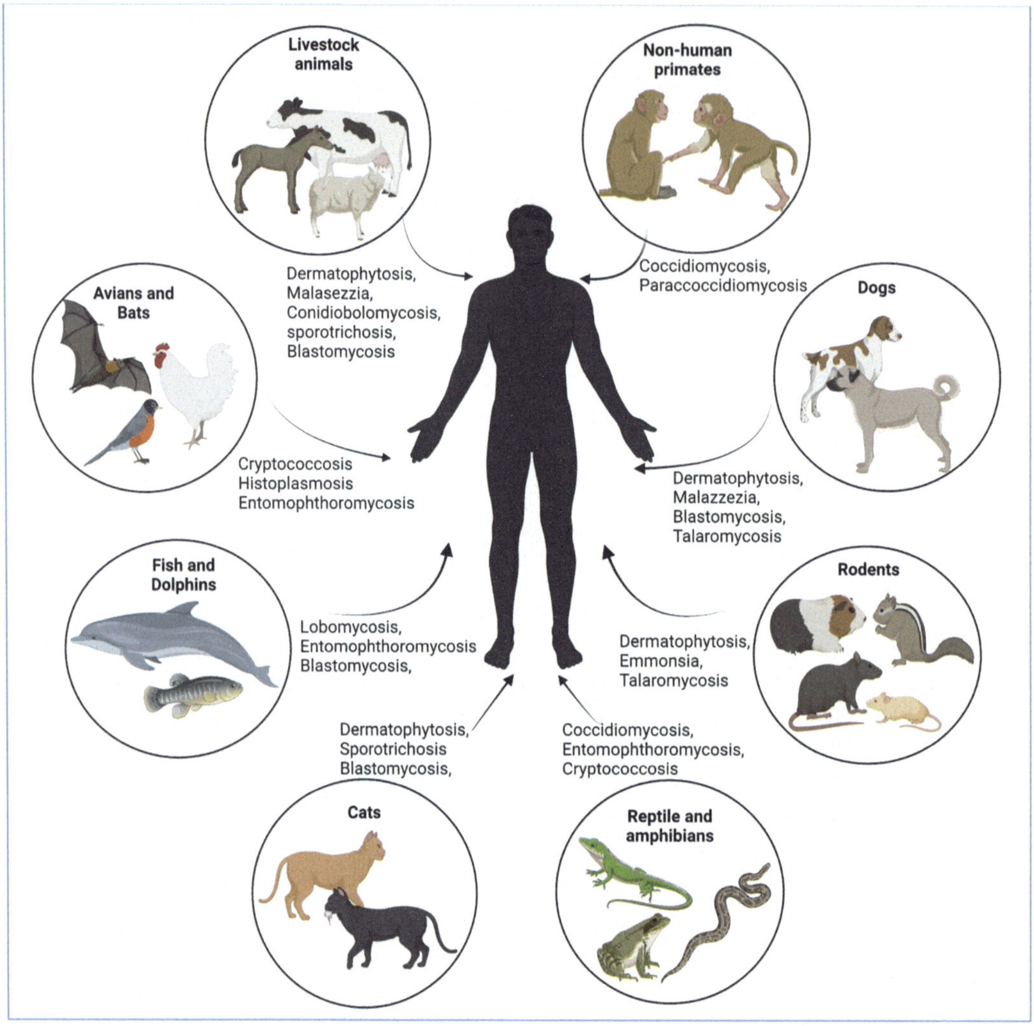

Fig. 2.1 Common fungal zoonotic infections and their animal reservoirs

Sources of Infection and Modes of Transmission

The ecological observations are essential to formulate strategies for the control of zoonotic fungal infections. The outbreak of zoonotic infection due to *Sporothrix brasiliensis* in Brazil, which has been contrary to the fundamental dogma of endemic mycoses, is an interesting case study. Although the condition typically occurs by the environmental spores or conidia from the mycelial pathogen, the yeast of *S. brasiliensis* present in the infected cat can spread infection directly to humans without environmental spores transmitted from the environment. Feline transmission has caused a long-lasting disease outbreak in this region. Animal-derived excrements, carcasses, and organic matter provide substrate and growth factors in the soil, allowing the preferential growth of *Histoplasma capsulatum*, *Coccidioides* spp., and a few other fungal pathogens. The natural habitat and modes of transmission of common fungal zoonoses are summarized in Table 2.1.

Table 2.1 Fungal zoonotic pathogens, ecology/animal source, and transmission

Fungal pathogen	Ecology/animal source	Mode of transmission
Dermatophytes (e.g. *Microsporum canis*, *Trichophyton verrucosum*, *Arthroderma vanbreuseghemii*, *Arthroderma benhamiae*)	Pet animals such as cats, dogs, cattle, Guinea pigs, etc.	On exposure to arthroconidia present in the superficial lesions of the infected animals
Sporothrix spp.	Vegetations and Cats	Cats roaming outdoors spread infection, especially when not vaccinated or neutered or without prophylactic deworming. Also by trauma with vegetations such as thorn prick etc.
Histoplasma. capsulatum sensu stricto *Histoplasma. capsulatum* var. *duboisii* *Histoplasma mississippiense* sp. nov. *Histoplasma ohiense* sp. nov. *Histoplasma suramericanum* sp. nov.	Nitrogen-rich soil, bat guano, chicken feces, pigeon droppings, and excreta of other birds, namely Brewer's blackbirds, grackles, oil birds, red-winged blackbirds, and starlings	Inhalation of conidia by cavers, farmers, miners, gardeners, sweepers, construction, and workers are susceptible
Talaromyces marneffei	Bamboo rats and dogs	Rat-to-human transmission Domestic animals like dogs are possible reservoir in endemic areas
Coccidioides immitis *Coccidioides posadasii*	Soil, especially from desert areas with semi-arid climate, elevated soil salinity, low pH, and extreme temperatures (−40.0 to 60 °C)	*Coccidioides* mould segments into arthroconidia (spores) which are inhaled after aerosolization and transition to spherules in tissue
Lacazia loboi	Soil and vegetation Aquatic mammals like dolphins	Transmission by traumatic inoculation/direct contact with infected animals Direct human-to-human transmission reported under accidental circumstances
Emmonsia crescens *Emmonsia parva* *Emmonsia pasteuriana*	Soil (*Emmonsia crescens*: in Europe and *Emmonsia parva:* in Americas, Central Asia, and Africa) Mammals living in and around soil e.g. rodents, weasels, squirrels, etc	Typical adiaspores are present in infected tissues
Conidiobolus coronatus *Conidiobolus incongruous* *Basidiobolus ranarum*	Plant debris and soil Gastrointestinal tract commensal in bats, fish, amphibians, and reptiles	Transmission dynamics are still under investigation. Infection may follow traumatic inoculation or inhalation of the fungal spores
Paracoccidioides brasiliensis *Paracoccidioides lutzii*	The natural habitat is not precisely known, possibly soil near armadillo burrows Canine infection reported	Inhalational exposure to environmental conidia
Cryptococcus neoformans *Cryptococcus gattii*	Isolated from pet birds and avian excreta Found in tropical and subtropical climates in decaying wood debris under the canopies of eucalyptus, almond, and other trees	Inhalational exposure to aerosolized bird excreta

Clinicopathological Correlations

Attributes of the Fungi

For establishing an infection in a hostile or unusual host, it is essential for the pathogen to display attributes that facilitate early survival and subsequent propagation of the pathogen. These attributes may be general, such as nutrient uptake and metabolism of the host, or may be specific, such as factors involving host immune evasion. Host damage in a fungal infection occurs because of the complex interplay of microbe-mediated damage and inappropriate or unbalanced host responses. Once the fungus enters the host, it must evade the immune response, including intracellular defenses, and find a comfortable niche for subsequent growth and dissemination or latency in the infected host. An overview of the fungal immune response is depicted in Fig. 2.2.

During the initial stage of infection, surface proteins on spores bind to extracellular matrix components like fibrinogen, laminin, and fibronectin and facilitate fungal adhesion and colonization. Fungi exhibit different mechanisms for initiating infections. For example, dermatophytes secrete enzymes like subtilisins and many endo- and exoproteases that cause keratin biodegradation in the infected host tissues. *Malassezia* spp., which lack the main enzymes needed for lipid metabolism, actively carry out enzymatic acquisition of lipids from the host environment. In addition, *Malassezia* spp. excrete various metabolic substances that act as skin irritants, which contribute to pathogenesis. *Coccidioides* produce various enzymes that play an important role in the pathogenesis of the disease caused by *Coccidioides*. It has been observed that there is a reduction in the genes involved in cell wall degradation and an increase in the number of genes required for the enzymatic utilization of carcasses for energy. Thus, the association of *Coccidioides* spp. can be seen with keratin-rich animals in the infection of a living host as well as in carcasses of deceased animals.

Variable virulence levels are even demonstrated in different species of the same genus. For example, pathogenic *S. brasiliensis* is more resistant to oxidative stress, due to mutations in the stress gene, *Hog*1, and has higher melanin production and glycoprotein Gp60 compared to other species. The virulence attributes of pathogenic fungi and their functions are summarized in Table 2.2.

Host Response

Recognition of fungus is mediated by pathogen recognition receptors (PRRs) and is a dynamic and complex process. A broad range of antimicrobial peptides of the defensin family is produced by epithelial cells. Germinating fungal cells are recognized by various PRRs on their surface, subsequently initiating a pro-inflammatory response. In the case of inhaled spores, various soluble factors found in the bronchopulmonary fluid are involved in anti-fungal defense. These include PRRs such as C-type lectins, mannose-binding ligands (MBL), surfactant proteins A and D, and pentraxin (PTX). These soluble factors enhance complement activation and phagocytosis of conidia, thereby contributing to the clearance of the fungus from the site of infection. Surfactant proteins are fungicidal, as seen in *Histoplasma*, where a calcium-dependent, surfactant protein-mediated increase in cellular permeability is observed.

Well-known fungal pathogen-associated molecular patterns (PAMPs) include proteins and polysaccharides (chitin, mannan, α-glucan, and ß-glucan).

The important PRRs implicated in anti-fungal immunity are the following:

C-type lectin receptors (CLRs): The CLR receptor consists of transmembrane receptors as well as soluble receptors with a common carbohydrate-recognition domain and primarily recognize glucan and mannan. Dectin-1 is an important CLR that plays a key role in innate immunity by triggering phagocytosis and in adaptive immunity by balancing the Th1/Th17 response. Pentraxin (PTX3) is another PRR that binds to fungal galactomannan, improving neutrophil uptake.

Toll-like receptors (TLRs): TLRs, including TLR2, TLR4, TLR6, and TLR 9, contribute to

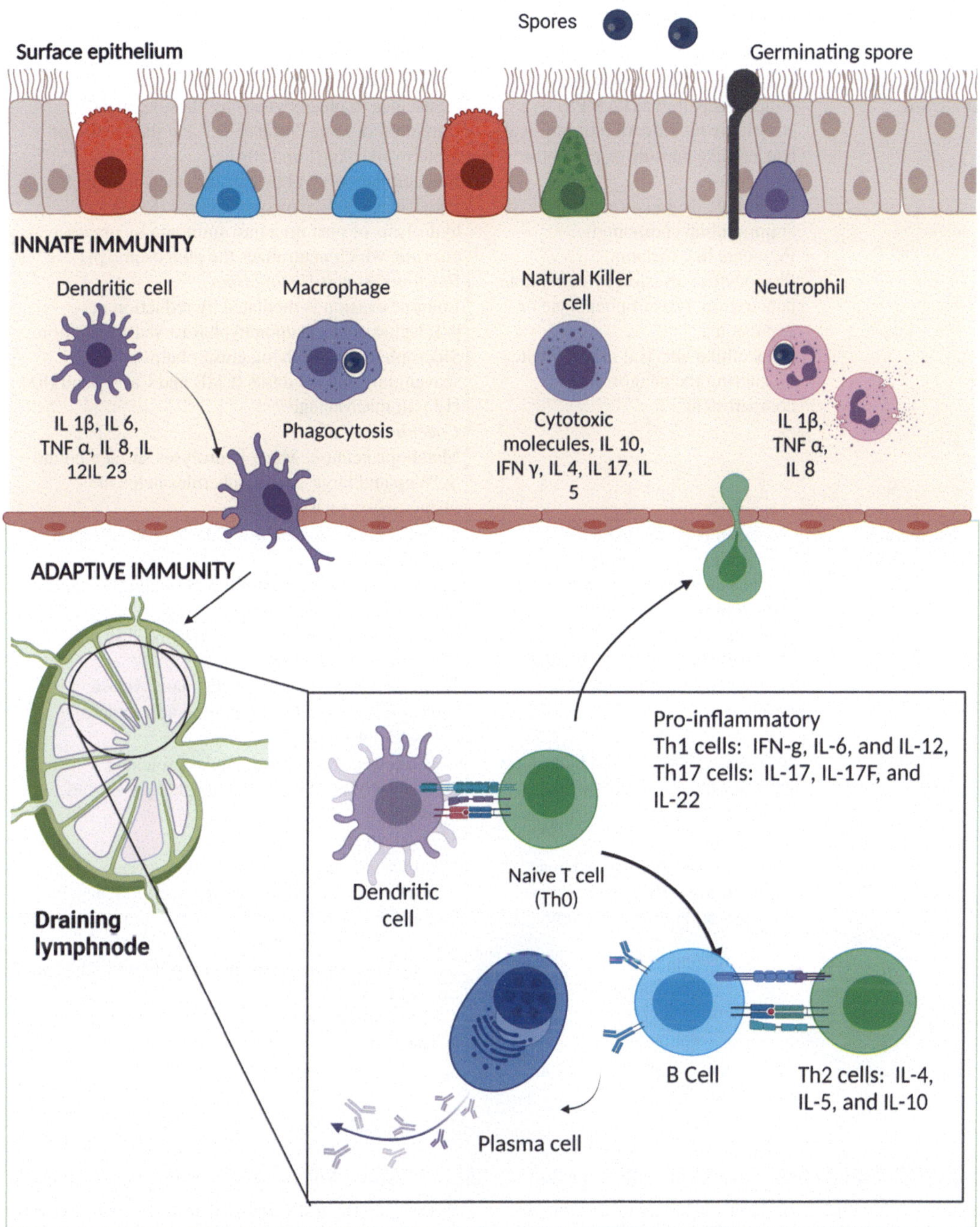

Fig. 2.2 An overview of the interaction between fungal pathogens and the immune system

the recognition of fungi in the pathogenesis of fungal infections.

NOD-like receptors (NLRs): NLRs consisting of NLRs, NLRP1, NLRP3, and NLRC4 are involved in the activation of caspase-1 with subsequent generation of interleukin (IL)-1ß and IL-18. These receptors also trigger an immune response mediated by Th17 on exposure to hyphal forms of fungi.

The pathological changes in the infected host tissue during fungal infections occur because of the host's immune response against these patho-

Table 2.2 Virulence attributes of common zoonotic fungal pathogens

Virulence attribute	Mechanisms	Common examples
Immune evasion strategies	Hiding the pathogen associated molecular patterns (PAMPS) by enzymatic modification or physical barriers like capsule or even host molecules Enzymatic degradation of host antimicrobial compounds Pigments like melanin Phagocytosis evasion by membrane penetration, lysis of phagocyte or expulsion Intracellular survival in phagocytes by altering the phagosomal compartment	*Cryptococcus* Resistance to phagocytosis by capsule, melanin, formation of large "titan cells," and the release of proteins like App1 and Blp1 Intracellular survival by capsule, melanin which inactivates reactive oxygen species (ROS) and hydrolysis of host urea into ammonia by urease enzyme, which neutralizes the phagosome pH *Blastomyces* and *Histoplasma* Immune evasion is mediated by reduction in β-1,3-glucan levels upon hyphal-to-yeast transition Siderophore biosynthetic cluster helps in iron scavenging, and catalases (CatB and CatP) neutralize H_2O_2 in macrophage *Coccidioides* Metalloproteinase, Mep1 hydrolyses surface-bound SOWgp and large size of spherules helps in phagocytosis resistance
Adhesion and invasion factors	Adhesion is prerequisite for colonization and invasion Active penetration by filamentous growth or turgor pressure and passive penetration along necrotic host tissue Host cell invasion to use them as vehicles to overcome immune barriers	*Cryptococcus neoformans:* Cfl1 for cell adhesion and biofilms *Coccidioides immitis*: SOWgp in binding to laminin, fibronectin, and collagen IV *Blastomyces. Dermatitidis:* BAD1/WI-1 in the adhesion to macrophage *Sporothrix* spp.: 3-carboxymuconate cyclase (glycoprotein Gp70 and Gp60) in adhesion to extracellular matrix
Enzymatic host damage	Includes lytic toxins and hydrolytic enzymes like proteases or lipases	*Coccidoides*: Many extracellular proteases Dermatophytes: Proteases, phospholipases, and lipases *Malassezia* spp. Lipases
Morphology variation	Switch in morphology and gene expression profile on exposure to mammalian body temperature	*Histoplasma capsulatum* and *Blastomyces dermatitidis*: yeast form has specific virulence factors for macrophage survival and immune evasion *Sporothrix*: immune evasion of the yeast in macrophages prevents further activation of adaptive immunity *Coccidioides* also switches from arthroconidia to spherule stage
Thermotolerance	Higher thermotolerance may aid dissemination	*Sporothrix schenckii*

gens. Epidermal changes of skin with hyperkeratosis, follicular plugging, and acanthosis are observed in superficial infections. Mono-nuclear cell infiltration in perivascular areas along the dermis surrounding the invading fungal elements may be present. Formation of abscesses in the dermis with neutrophilic exudates and clusters or grains of organisms is also frequent. This is associated with epithelioid histiocytosis, the presence of multinucleate giant cells, or even fibrosis in chronic cases. *Splendore-Hoeppli phenomenon* is a characteristic pathological reaction seen particularly in sporotrichosis, blastomycosis, and entomophthromycosis. In this phenomenon, the fungi are surrounded by retractile eosinophilic material, also called "*Asteroid body*," which represents fungal antigen–host antibody reaction in a sensitized host. In disseminated infections, hematogenous dissemination of fungi to various organs may occur, especially in immune-deficient hosts.

Host Immunity

Innate immunity: Macrophages are important in the innate immune response that expresses a wide range of different receptors, like complement receptors, immunoglobulin receptors, and other PRRs. Following phagocytosis, the fungus is segregated into a phagosome, and endo-lysosomal fusion generates a phagolysosome where fungal killing mediated by reactive oxygen and nitrogen species takes place. Neutrophils form clusters around the spores, restricting them to structures resembling early granulomas. Neutrophil extracellular traps (NETs), reactive oxygen species, and cationic peptides also damage both spores and hyphae. They also produce cytokines to activate other immune cells. Natural killer (NK) cells recognize fungal hyphae, causing damage to activation. Dendritic cells (DCs) link innate immunity with adaptive immunity.

Adaptive immunity: Depending on the antigen presented, DCs promote T helper (Th) cell differentiation. The Th1 produces pro-inflammatory cytokines IFN-g, IL-6, and IL-12, which are required for anti-fungal protection. The Th17 subset of helper T cells expresses IL-17, IL-17F, and IL-22, further promoting pro-inflammatory neutrophil response. Hung et al. have recently demonstrated the role of Th17 responses in protection against coccidioidomycosis. The Th2 response with IL-4, IL-5, and IL-10 production is implicated in impaired anti-fungal host defense. In *Sporothrix* infection, parasite–fungal coinfection enhances disease susceptibility by skewing the host immune profile toward a Th2 response, thereby inhibiting effective fungal clearance. It has even been suggested that the control of helminth infection in cats with sporotrichosis could balance the immune response in favor of a Th1 type, contributing to fungal clearance with a decrease in zoonotic potential. The Th2 response typically not only downregulates the anti-fungal host immunity but induces B cell responses that may protect the host against other pathogens.

Clinical Manifestations

Clinical manifestations of fungal zoonoses vary depending on the causative agent and the organ system affected. Infections are not only confined to the skin and appendages such as hair and nails and subcutaneous tissue but can also cause invasive systemic and disseminated infections.

Spectrum of Clinical Manifestations

Superficial infections: Dermatophytes are fungi that require keratin for growth and can cause superficial infections of the skin, hair, and nails after being directly spread due to contact with infected animals. Infections can be classified as per the site affected, the most common being tinea corporis (body), tinea capitis (scalp), tinea cruris (groin), tinea pedis (feet), and tinea unguium (nail), among others. The condition manifests as a rash associated with itching, scaling, and redness on the skin and white or yellow discoloration of the nail. Sporotrichosis of the skin causes small, painless, reddish bumps, or nodules. These appear over the fingers, hands, or arms at the site of fungal entry and gradually spread subcutaneously along the lymphatics. Cutaneous manifestations of fungal infections after hematogenous dissemination to the skin can also occur in some fungal infections, such as histoplasmosis.

Subcutaneous infections: Sporotrichosis is a typical example of subcutaneous endemic mycosis associated with cutaneous and disseminated cutaneous manifestations. Skin nodules, the typical lesion, suppurate into abscesses and ulcers as the infection spreads via lymphatics. A chain of lymphatics that drains the affected area begins to gradually enlarge, followed by the possible occurrence of bacterial superinfection of ulcerated overlying skin. Lobomycosis is another example in which the auricles and upper and lower limbs are affected, causing keloid-like lesions. Subcutaneous rhino facial conidiobolomycosis is the most common manifestation of *C. coronatus* infection. The infection begins with nasal discharge, epistaxis, unilateral nasal obstruction, sinus tenderness, and extensive and persistent facial swelling that may even result in disfiguration.

Invasive infections: *Histoplasma* infection commonly affects the lungs. The fungus causes acute pulmonary disease, which manifests as pneumonia, including fever, cough, or shortness of breath. Chronic cavitary histoplasmosis can also occur in patients with pre-existing lung disease, while progressive disseminated disease occurs in individuals with poor immunity. *Talaromyces marneffei*, a pulmonary pathogen, causes acute pulmonary infection manifesting as fever, cough, dyspnea, chest pain, and ear, nose, throat, and/or skin lesions. Coccidioidomycosis is a systemic infection associated with fatigue, fever, cough or shortness of breath, headache, myalgia, joint pains, and even rash. Disseminated (extra-pulmonary) coccidioidomycosis manifests as meningitis, a life-threatening condition involving the skin, bones, joints, liver, pancreas, kidneys, and brain. Pulmonary adiaspiromycosis caused by *Emmonsia* is a localized or diffuse pulmonary disease that can present as fever, cough, and progressive dyspnea. Disseminated infections caused by *Emmonsia* spp. have also been reported in immunocompromised adults. *Paracoccidioides* causes lung infection, with presentations of fever, malaise, and cough. *C. neoformans* usually infect the lungs or the central nervous system (the brain and spinal cord) and can also affect other parts of the body. The symptoms of the infection depend on the parts of the body that are affected. In patients with primary lymphocutaneous sporotrichosis, hematogenous spread can cause infections of bones and joints, kidneys, spleen, etc. Chronic pneumonia caused by inhalation of spores is relatively rare, manifests as cavities or infiltrates in the lungs, and is more commonly observed in patients with pre-existing lung diseases.

In conclusion, there is a need for an in-depth study of fungal disease epidemiology and pathophysiology. The factors affecting the emergence of fungal zoonoses/sapronoses are quite complex. Understanding of disease ecology, transmission dynamics, and immunological response is essential to formulating strategies for reducing the spread of fungal zoonoses.

Case Study

A 49-year-old female patient who lived in an urban area of Brazil presented with a 6-month history of symptoms, starting with a pustule over her right-side upper lip, progressing to an inflammatory, intensely painful lesion affecting the entire right cheek and right side of the nose. The patient received treatment without any improvement with systemic antibiotics, oral fluconazole, and acyclovir. She gave a history of kissing a stray cat she had adopted from the street markets. Aspirate from the wound was collected and sent for microbiological studies, which revealed round to elongated budding yeast cells in microscopy of the stained specimen. Eosinophilic reaction with asteroid bodies was observed by histopathology. Culture showed greenish-black mycelial fungi, and conversion to yeast phase was demonstrated. Pathogenicity testing in Swiss albino mice resulted in an infection in the mice with the recovery of *S. schenckii*.

1. What is the possible mode of transmission of this fungal pathogen to the human host?
2. What is the typical phenomenon described in histopathology in such cases?
3. What are the common virulence attributes of *S. schenckii*?

Research Questions

1. What are the immune correlates of protection against zoonotic fungal infections?
2. How can transmission of zoonotic fungal infections be prevented?
3. Which fungal antigens or host immune markers can be used in the diagnosis of fungal zoonoses?

Further Reading

Ameen M, Talhari C, Talhari S (2010) Advances in paracoccidioidomycosis. Clin Exp Dermatol 35:576–580

Baumgardner DJ (2012) Soil-related bacterial and fungal infections. J Am Board Fam Med 25:734–744

Blanco JL, Garcia ME (2008) Immune response to fungal infections. Vet Immunol Immunopathol 125:47–70

Blaschke-Hellmessen R (2000) *Cryptococcus* species-etiological agents of zoonoses or sapronosis? Mycoses 43 Suppl 1:48–60

Borman AM, Simpson VR, Palmer MD, Linton CJ, Johnson EM (2009) Adiaspiromycosis due to *Emmonsia crescens* is widespread in native British mammals. Mycopathologia 168:153–163

Chaiwun B, Vanittanakom N, Jiviriyawat Y, Rojanasthien S, Thorner P (2011) Investigation of dogs as a reservoir of *Penicillium marneffei* in northern Thailand. Int J Infect Dis 15:e236–e239

Chakrabarti A, Singh S (2019) Challenges in invasive fungal disease. In: Advancing frontiers in mycology & mycotechnology. Springer, Singapore, pp 457–478

Dagenais TRT, Keller NP (2009) Pathogenesis of *Aspergillus fumigatus* in invasive aspergillosis. Clin Microbiol Rev 22:447–465

de Lima B, Barros M, de Oliveira Schubach A, Francesconi do Valle AC, Gutierrez Galhardo MC, Conceicao-Silva F, Pacheco Schubach TM, Santos Reis R, Wanke B, Feldman Marzochi KB, Conceicao MJ (2004) Cat-transmitted sporotrichosis epidemic in Rio de Janeiro, Brazil: description of a series of cases. Clin Infect Dis 38:529–535

Del Rocío Reyes-Montes M, Pérez-Huitrón MA, Ocaña-Monroy JL, Frías-De-León MG, Martínez-Herrera E, Arenas R, Duarte-Escalante E (2016) The habitat of *Coccidioides* spp. and the role of animals as reservoirs and disseminators in nature. BMC Infect Dis 16:550

Gibb R, Franklinos LHV, Redding DW, Jones KE (2020) Ecosystem perspectives are needed to manage zoonotic risks in a changing climate. BMJ 371:m3389

Gilbert AS, Wheeler RT, May RC (2015) Fungal pathogens: survival and replication within macrophages. Cold Spring Harb Perspect Med 5:a019661

Greer DL, Friedman L (1966) Studies on the genus *Basidiobolus* with reclassification of the species pathogenic for man. Med Mycol 4:231–241

Karesh WB, Dobson A, Lloyd-Smith JO, Lubroth J, Dixon MA, Bennett M, Aldrich S, Harrington T, Formenty P, Loh EH et al (2012) Ecology of zoonoses: natural and unnatural histories. Lancet 380:1936–1945

Macêdo-Sales PA, Souto SRLS, Destefani CA, Lucena RP, Machado RLD, Pinto MR, Rodrigues AM, Lopes-Bezerra LM, Rocha EMS, Baptista ARS (2018) Domestic feline contribution in the transmission of *Sporothrix* in Rio de Janeiro State, Brazil: a comparison between infected and non-infected populations. BMC Vet Res 14:19

Ricci G, Mota FT, Wakamatsu A, Serafim RC, Borra RC, Franco M (2004) Canine paracoccidioidomycosis. Med Mycol 42:379–383

Rosa PS, Soares CT, de Faria Fernandes Belone A, Vilela R, Ura S, Filho MC, Mendoza L (2009) Accidental Jorge Lobo's disease in a worker dealing with *Lacazia loboi* infected mice: a case report. J Med Case Rep 3:67

Seyedmousavi S, Guillot J, Tolooe A, Verweij PE, de Hoog GS (2015) Neglected fungal zoonoses: hidden threats to man and animals. Clin Microbiol Infect 21:416–425

Theelen B, Cafarchia C, Gaitanis G, Bassukas ID, Boekhout T, Dawson TL (2018) *Malassezia* ecology, pathophysiology, and treatment. Med Mycol 56:S10–S25

Waltzek TB, Cortés-Hinojosa G Jr, Wellehan JFX Jr, Gray GC (2012) Marine mammal zoonoses: a review of disease manifestations. Zoonoses Public Health 59:521–535

Fungal Immunology

3

Pankaj Chandley, Ravikant Ranjan, and Soma Rohatgi

Learning Objectives

By the end of the chapter, the readers should be able to:

1. Understand innate and adaptive immune response during fungal zoonotic infections.
2. Understand antibody-mediated immune responses against fungal zoonotic infections.
3. Learn immune cell-mediated therapy against fungal zoonotic infections.
4. Know the status of vaccine development against fungal zoonotic infections.
5. Evaluate adjunctive/combination therapies against fungal zoonotic infections.

Introduction

The clinical form of fungal disease is associated with increased mortality and morbidity, mostly in immunocompromised hosts. The host defense mechanisms against fungal pathogens range from non-specific protective mechanisms ("innate immunity") to the specific adaptive mechanisms induced during the infection and progression of disease ("adaptive immunity"). Understanding fungal immunology is essential for effective preventive measures against fungal pathogens that affect more than 300 million people globally and are responsible for 1.5 million deaths annually.

Innate Immunity

Despite its lack of specificity, innate immunity plays a crucial role in distinguishing between self and non-self. It acts as the initial defense against pathogens and is responsible for triggering adaptive immune responses through specific signals. The inherent defense mechanisms of innate immunity involve the protective function of body surfaces as well as the mucosal epithelial surfaces in the respiratory, gastrointestinal, and genito-urinary tracts. The innate immunity employs a direct antifungal effector activity to destroy fungal pathogens by either a phagocytic process, usually against fungi residing intracellularly, or indirectly by the secretion of microbicidal components against fungal elements. It also plays an instructive role in the adaptive immune system by producing various pro-inflammatory mediators, including chemokines and cytokines. The mediators are essential for antigen uptake

P. Chandley · S. Rohatgi (✉)
Department of Biosciences and Bioengineering, Indian Institute of Technology, Roorkee, Uttar Pradesh, India
e-mail: soma.rohatgi@gmail.com

R. Ranjan
Chanakya University, Bangalore, Karnataka, India

S. C. Parija, S. M. Rudramurthy (eds.), *Textbook of Fungal Zoonoses and Sapronoses*, Microbial Zoonoses, https://doi.org/10.1007/978-981-97-3214-2_3

and presentation and the induction of co-stimulatory activity by phagocytic cells. The innate immunity in mammals involves various cellular components and humoral factors. Among these, phagocytes play a crucial role, which includes neutrophils, mononuclear leukocytes (monocytes and macrophages), and dendritic cells (DCs) apart from natural killer (NK) cells and non-hematopoietic cells. Macrophages play an important role in antifungal defense in killing fungi by phagocytosis.

Adaptive Immunity

The adaptive cell-mediated immune response is characterized by the generation of antigen-specific T helper effector cells, regulatory T cells, and B cells against fungal pathogens. It is responsible for memory cells, which prevents subsequent infection by the same pathogen. The dendritic cells attack the fungal pathogen at the infection site, transport them to the lymph nodes, and activate different T-cells. Subsequently, different T-cell subsets release a panel of cytokines responsible for activating and inhibitory feedback signals to effector phagocytes, generating a successful immune response against fungi. The T cells contribute to the development of memory antifungal immunity. The plasticity in the T-cell response against fungi causes heterogeneity of the CD4+ and CD8+ T-cell repertoire. These are responsible for the redundancy of effector mechanisms to control and limit fungal infections through T-cell participation. Various activities employed by T cells include direct antifungal activity, lysis of fungus-containing phagocytes, the release of antimicrobial peptides from CD8+ T cells, and other effector functions. The effective tissue response to invasion by fungal pathogens mediated by cell-mediated immunity is characterized by granulomatous inflammation with different patterns of inflammation.

Antibody-Mediated Immunity

Antibodies produced by differentiated B cells are essential tools in the fight against microbial infections. Antibodies are known to have little impact on the effectiveness of treatment for fungal infections. Nevertheless, during the past few decades, the role of antibody-mediated protective responses against fungal infections has been increasingly recognized. Passively immunizing severe immunocompromised patients with monoclonal antibodies to target the fungal pathogen directly or enhancing the patient's remaining anti-fungal immune response has also been evaluated. Antibodies utilize a variety of mechanisms to combat fungal growth and proliferation. These include opsonization, complement activation, neutralization, Antibody-dependent cell-mediated cytotoxicity (ADCC), immunomodulation, direct fungal killing, and biofilm inhibition. The extent of antibody-mediated anti-fungal response varies for different fungal species, and mechanisms of antibody-mediated protection are often species-specific. The development of specific antibody therapies such as monoclonal antibodies variable fragments (scFvs), anti-idiotypic antibodies, antibody-drug conjugates, and synergistic antibodies supplemented with anti-fungal drug therapy may further enhance the antibody-mediated immunity against fungal infections.

Immunology of Zoonotic Fungal Infections

Dermatophytosis

Dermatophytosis is mostly caused by *Trichophyton* and *Microsporum* species. Recently, many dermatophytic virulence factors implicated in the invasion of the keratin network have been identified. The factors include exo-keratinases, peptidases, hydrolases, oxidoreductases, and beta-glucosidase.

Toll-like receptor 2 (TLR2), which is expressed in monocytes, macrophages, and neutrophils, recognizes different ligands in the fungal cell wall, such as mannan and phospholipomannan. Enhanced phagocytosis and production of cytokines are encouraged by TLR2 activation in several fungal infections, such as aspergillosis. TLR2 is critical for human monocyte-mediated phagocytosis of *Trichophyton rubrum* conidia and their generation of pro-inflammatory cytokines. This is a key component of innate response in the initial stage of dermatophyte pathogenesis.

The presence of serum IgE and IgG4 antibodies, specific to *Trichophyton* antigens, is the distinguishing feature of the Th2 response in the initial hypersensitivity response in dermatophytosis. In contrast, delayed-type hypersensitivity mediated by the activated macrophage is the most important effector cell, being mediated by CD4+ T cells (Th1 response) that secrete gamma interferon (IFNγ). Reduced levels of IgG antibodies in response to the *Trichophyton* proteins and the absence of IgE or IgG4 antibodies is a noted feature. This is linked to delayed-type hypersensitivity responses to *Trichophyton* extract. The IgY antibodies against the *T. rubrum* cell wall peptides (anti-trCWP IgY) showed a certain amount of cross-reactivity with several dermatophytic fungi. Anti-trCWP IgY demonstrated a strong dose-dependent growth inhibition of *T. rubrum* in vitro and a substantial dose-dependent therapeutic impact on *T. rubrum* infections in BALB/C mice.

The vaccine studies have been carried out in a variety of animal models, including guinea pigs and rabbits, to evaluate the immunogenicity and protective efficacy. The anti-dermatophyte vaccine, whether given in living or inactivated form, significantly stimulated humoral and cell-mediated immune responses. This also provided adequate protection against infections caused by homologous and heterologous dermatophyte strains. The use of dermatophyte exo-keratinases as a vaccine candidate was found to be immunogenic. They generated severe delayed cutaneous hypersensitivity reactions in animals receiving the vaccination and stimulated high keratinase-specific IgG antibody titter. Table 3.1 provides a list of vaccine candidates and their mechanism of protection against fungal zoonotic infections.

Immunotherapy: Granulocyte macrophage colony-stimulating factor (GM-CSF), a therapeutic cytokine, has been studied as a prospective antifungal agent. *T. rubrum* interactions with keratinocytes trigger the colony-stimulating factor 2 (CSF2) gene implicated in the production of GM-CSF in cases of dermatophytosis. By regulating the superficial infection and reducing the Th1 antifungal response, IL-17 cytokines have a significant effect on the prevention of dermatophytosis in mice infected with *Microsporum canis*. In a mouse model, IL-1 signaling has been found to reduce the growth of *T. rubrum* conidia and regulate the production of IL-17. The effectiveness of the anti-dermatophyte mAB antibody-0014 coupled with the 5-arylhydrazonothiazole derivative led to complete protection in the case of guinea pigs infected with *T. rubrum* and *M. canis*.

Malassezia Infection

The genus *Malassezia*, formally called *Pityrosporum*, contains lipophilic yeasts that are present in the skin's flora. *Malassezia* colonize human skin from birth as commensals. Nevertheless, *Malassezia*-related illnesses, such as head and neck dermatitis, seborrheic dermatitis, pityriasis versicolor, and *Malassezia* folliculitis, have different species distributions on the skin.

TLR2 is linked to innate immune responses that defend against fungus infections. To avoid dermatitis relapses, TLR2 regulation utilizing a topical formulation with TLR2-Regul™ represents a novel therapeutic strategy. Recombinant IFN-omega treatment given in small dosages may be helpful for managing canine atopic dermatitis over the long term. Recombinant canine interferon-gamma (rCaIFN-gamma) therapy has also been shown to be useful in curing the skin lesions associated with atopic dermatitis.

Management of dermatophyte infections by vaccination has yielded promising candidate vaccines for the prevention of *Trichophyton verrucosum* infections in cattle. However, no vaccination studies for the prevention of pityriasis versicolor have been conducted.

Table 3.1 Vaccine candidates and their mechanism of protection against fungal zoonotic infections

S. no.	Fungal zoonotic diseases	Major causative agents	Vaccine candidates	Mechanism of protection
1.	Dermatophytosis	*Trichophyton* and *Microsporum* spp.	Inactivated/live attenuated pathogen, and exo-keratinases	Generated anti-dermatophytic humoral and cellular immune responses in animal models
2.	*Malassezia* infection	*Malassezia furfur*, *Malassezia globosa*	Not investigated	Not investigated
3.	Phaeohyphomycosis	*Exophiala dermatitidis*	Not investigated	Not investigated
4.	Sporotrichosis	*Sporothrix* spp.	Recombinant Enolase and Gp70	Generated humoral immune responses against sporotrichosis in animal models
5.	Coccidioidomycosis	*Coccidioides* spp.	Ag2/PRA, PLB, Pep1, Gel1, Amn1, rHSP60, rURE, and rPep1	Generated cell-mediated immunity against coccidioidomycosis in animal models
6.	Paracoccidioidomycosis (PCM)	*Paracoccidioides* spp.	gp43, P10 peptide, PbHsp60	Generated cell-mediated immunity against PCM in mice model
7.	Blatomycosis	*Blastomyces dermatitidis*	WI-1 and BAD-1	Induced type 1 cytokine responses which conferred resistance to blastomycosis in mice
8.	Histoplasmosis	*Histoplasma capsulatum*	Alkaline extract of *Histoplasma capsulatum* encapsulated in GPs and *H. capsulatum* CFAs	Generated T-cell-mediated immunity against histoplasmosis in mice models
9.	Aspergillosis	*Aspergillus fumigatus*	NXT-2, VesiVax® Af3/9, whole glucan particles (WGP)/WGP-BSA, mannan alone/mannan-BSA conjugates and Lam-CRM	Generated protective immunity against systemic aspergillosis in mice models
10.	Cryptococcosis	*Cryptococcus neoformans*, *Cryptococcus gattii*	*C. gattii* CW and CP proteins, GXM and GalXM	Conferred protective immunity against cryptococcosis via reducing fungal burden in mice models
11.	Adiaspiromycoses	*Emmonsia* spp.	*Emmonsia crescens* UV mutants	Not investigated
12.	Talaromycosis	*Talaromyces marneffei*	Not investigated	Not investigated
13.	Entomophthoromycosis	*Basidiobolus* spp., *Conidiobolus* spp.	Not investigated	Not investigated
14..	Pneumocystosis	*Pneumocystis* spp.	Inactivated whole organisms, gpA (Msg), Kexin, SPD1, P55, and A12	Conferred partial protection against PCP in animal models

Immunotherapy: Allergens-specific immunotherapy is a technique that involves giving an allergic individual a steadily increasing dose of an allergen extract to reduce the symptoms brought on by subsequent exposure to the triggering allergen. Dupilumab, a humanized monoclonal antibody targeting the IL-4R, is the first monoclonal antibody with proven efficacy in preventing atopic dermatitis caused by *Malassezia* species is being reported.

Phaeohyphomycosis

Phaeohyphomycosis is caused by many diverse organisms referred to as "dematiaceous" or "melanized" fungi. The predominance of melanin in the cell walls of these fungi makes them a distinct fungal group. Melanin production also enables the fungus to avoid phagocytosis and acts as a powerful free radical scavenger, shielding the cell from oxidants produced by the immune effector cells.

Exophiala dermatitidis has been observed to stimulate the production of serum IgG antibodies, which were found to relate to an inflammatory response. Since extracellular polysaccharides, particularly acid mucopolysaccharides generated by *E. dermatitidis* yeast cells, aid in evading phagocytes during the invasion of tissue, innate immune cells are ineffective against phaeohyphomycosis. *E. dermatitidis* capsular component helps in adhesion, phagocytosis dysfunction, and resistance to complement-mediated inhibition.

Caspase-recruitment domain 9 (Card-9) signaling plays an important role in antifungal immunity. By knocking out Card-9 signaling in mice, they fail to produce an effective immune response in phaeohyphomycosis. It is suggested that Card9 mediate both the innate and Th17-mediated adaptive immune responses against dematiaceous fungal infections at the initial stage of infection.

No vaccines are available against phaeohyphomycosis. IFN-γ therapy has been shown to improve patients suffering from cystic fibrosis caused by *E. dermatitidis*.

Sporotrichosis

Sporothrix schenckii, a ubiquitous saprophyte of soil, decomposing wood, hay, and sphagnum moss, is the causative agent of chronic granulomatous mycotic illness. LRs, C-type lectin receptors, and complement receptors trigger innate immune cells such as macrophages, monocytes, neutrophils, dendritic cells, and other mediators that act against *S. schenckii* infection. After *S. schenckii* recognition, NLRP3 (nucleotide-binding oligomerization domain-like receptor pyrin domain-containing 3) acts in combating sporotrichosis through T helper cell response and coupling the innate and adaptive immune responses.

Enolase has been widely evaluated as a possible antigenic target for vaccines against sporotrichosis. Recently, multiepitope vaccination against sporotrichosis has been developed. Immunization with *Sporothrix* spp. recombinant enolase (rSsEno) peptides mixture (PeptMix) given with Freund adjuvant (FA) resulted in antigen-specific IgG, IgG1, and IgG2b antibodies that conferred protection of mice infected experimentally with *S. brasiliensis*. Gp70 antigen is also another potential candidate that has been evaluated in experimental animal models.

Immunotherapy: The effectiveness of an antibody against phage-KR, a hybrid phage nanofiber displaying the Gp70 gene, has been evaluated in a mouse model with sporotrichosis. The antibody effectively inhibited the *Sporothrix* spp. infection by reducing progressive fungal colonization, increasing animal survival rates, and alleviating organ inflammation in mice. The number of colony-forming units in the liver and spleen decreased, and the phagocytic index of yeast cells opsonized with mAb P6E7 against Gp70 increased. Given that this antigen is a potential adhesin for fibronectin and laminarin, mAb P6E7 was also able to prevent yeasts from interacting with the subendothelial matrix.

Coccidioidomycosis

"*Valley Fever*," an invasive dimorphic fungal infection, is caused by *Coccidioides immitis* and *Coccidioides posadasii*.

TLRs and CLRs aid in the recognition of spherule wall components of fungi by host cells to initiate innate immune responses and subsequently trigger the development of adaptive immune responses against *Coccidioides*. Activation of IL-1-MyD88 and Decin-1-Card9 signaling pathways are essential for innate immunity and stimulation of Th1 and Th17 mediated adaptive responses against *Coccidioides* infection. Both Th and Tc cells contribute to adaptive immunity against coccidioidomycosis. The T-cell subsets that contribute to a protective immune response against *Coccidioides* infection include Th1, Th2, and Th17 cells and perhaps IL-10-producing cells. No vaccine is available.

Paracoccidioidomycosis

Paracoccidioidomycosis (PCM), caused by the thermodimorphic *Paracoccidioides* species, primarily affects the respiratory tract. In the initial phase of infection, neutrophils, NK cells, and monocytes, supplemented by cytokines and complement systems, act against *Paracoccidioides*. NK cells can be recognized and activated directly by *Paracoccidioides* yeast cells either by directly killing the fungi or by killing *Paracoccidioides*-infected cells. The role of adaptive immunity is not fully understood.

No vaccine is available. *Paracoccidioides* require the glycoprotein gp43 to adhere to epithelial cells and, hence, has been studied as a vaccine candidate antigen. P10 peptides and chitosan nanoparticles have also been evaluated to act on the immunodominant glycoprotein gp43 to protect the mice exposed to *P. brasiliensis* infection.

Immunotherapy: A panel of monoclonal antibodies that recognize gp43 has been studied as protective mAbs by immunotherapy. The mAb 3E was the most effective in lowering the fungus burden in the in vivo studies and encouraging fungal phagocytosis in the in vitro studies. The administration of mAb 3E also increased protection in these studies. IL-12 therapy reduced fungal burdens in the liver and spleen of experimentally infected mice.

Blastomycosis

Blastomycosis is caused by the dimorphic fungus *Blastomyces dermatitidis* by inhalation of aerosolized conidia. Most people with acute infection either show no symptoms at all or manifest a short-lived respiratory infection. Chronic pneumonia is the most prevalent extrapulmonary manifestation. The condition affects the skin and subcutaneous tissue, bones and joints, the prostate gland, and the central nervous system (CNS).

The innate receptor Dectin-3 is required for the development of protective immunity against *Blastomyces* spp. Experimental studies of mice lacking the Dectin-3 receptor do not show Th 17 immune response in the infected animals. CD28 is essential for the activation of optimum, protective T-cell responses to *B. dermatitidis* infection. Impairment in CD28 co-stimulation reduces Th and Tc cell numbers, which results in lower resistance to blastomycosis. Moreover, in hosts deficient in Th cells, Tc17 cells are known to be essential for antifungal vaccination-mediated protection.

Persons who contract *B. dermatitidis* appear to have a robust immunity against the yeast surface adhesin protein, WI-1. To produce an attenuated vaccination strain, mutations were induced in the *WI-1* gene. Vaccination of animal models with mutated strain resulted in type1 cytokine responses and delayed-type hypersensitivity that conferred resistance to infection. In addition, a live-attenuated strain of *B. dermatitidis* has been generated by genetically mutating the key virulence component BAD-1. Vaccination by mutated *Blastomyces* strains successfully protected the mice.

Histoplasmosis

Histoplasmosis caused by *Histoplasma capsulatum* is a soil-based fungus that is acquired by inhalation of airborne conidia. Cytokines like GM-CSF play a crucial role. Treatment by cytokines such as GM-CSF improved survival rates of T-cell-depleted mice infected with *H. capsulatum*. T cell-mediated defense becomes ineffective in the absence of Tumor necrosis factor alpha (TNF-α). The memory T cells decreased by neutralization of this cytokine. Following a lethal or sublethal initial challenge, high mortality and greater fungal load were seen in perforin-deficient knockout mice. The role of perforins in mediating defense against *H. capsulatum* and the ability of CD8+ T cells to mediate protection against *H. capsulatum* has been demonstrated to be caused by both perforin-dependent and/or independent mechanisms.

The IFN-γ and IL-17-dependent immune responses were induced in C57BL/6 mice following immunization with an alkaline extract of *H. capsulatum* encapsulated in glucan particles (GPs). Leukotrienes are administered as adjuvants with significant immunomodulatory properties. In wild-type mice immunization with complete Freund's adjuvant (CFA) from *H. capsulatum*, along with biodegradable leukotriene, enhances the production of IFN-γ recruits memory T cells into the lungs. This combination is linked to effective protection against histoplasmosis.

Immunotherapy: In epidemics of histoplasmosis, protective mAbs could be administered as a preventative measure, especially to high-risk persons with HIV infection or those using TNF inhibitors. For instance, mAbs against Hsp60, when used in combination with other mAbs directed against cell surface antigens, may enhance the protective immunity in histoplasmosis. In a mouse infection model, the survival time was increased when mAbs specific to the histone H2B-like protein were administered prior to *H. capsulatum* infection.

Aspergillosis

Aspergillus niger, *Aspergillus fumigatus*, *Aspergillus terreus*, and *Aspergillus flavus* are the important species that cause aspergillosis. In experimental infection of mice, *A. fumigatus* conidia and hyphae are phagocytosed by murine and human DCs using different recognition receptors or Pattern Recognition Receptors (PRRs). In hematopoietic stem cell transplantation, the adoptive transfer of DCs may improve immunocompetence by assisting the T cell learning process. The ability of adaptive effector CD4+ T cells to fight against invasive aspergillosis is of importance.

A "*pan-fungal*" peptide, NXT-2, derived from homologous sequences of *Aspergillus*, *Cryptococcus*, *Candida*, and *Pneumocystis*, has been explored as a potential vaccine candidate. Animals immunized with NXT-2 show lower mortality and morbidity rates than non-immunized animals. A conjugate vaccine, VesiVax® Af3/9, has been shown to be protective against infection by azole-resistant strains of *A. fumigatus* in mice. Vaccination with whole glucan particles (WGP)/ WGP-BSA, mannan alone/mannan-BSA conjugates, and Lam-CRM has been shown to confer protective immunity against systemic aspergillosis in mice.

Cryptococcosis

Cryptococcosis caused by pathogenic encapsulated yeasts of the genus *Cryptococcus* is distributed throughout the world. *Cryptococcus neoformans* and *Cryptococcus gattii* are the two important species that infect humans.

In experimental infection, it has been demonstrated that macrophages and neutrophils recognize phagocytose and ultimately kill pathogenic *C. gattii* R265 strain in vitro. In the presence of freshly collected mouse serum, B cells and antibodies have been shown to prevent the early spread of *C. neoformans* in experimental studies in mice. The result suggests that IgM may be a significant modulator of early antifungal immunity in the lungs of the mice.

In experimental infection of the mice by the wild-type strain of *C. neoformans*, the TNF-α producing strain has been shown to improve the protective components of various host responses. These responses include preferential T-cell accumulation, improved Th1/Th2 cytokine balance, decreased pulmonary eosinophilia, and substitute activation of lung macrophages during the adaptive phase of infection. The degree of elimination of cryptococcal infection from the CSF is dramatically accelerated in human patients when short-course IFN-γ gamma is added to conventional therapy. Also, effectively stopping IL-10 production prevents *C. neoformans* from spreading to the brain. IL-10 enhances both early and later cellular and molecular mechanisms, such as Th1 and Th17 responses, and causes early fungal clearance in mice.

The vaccine is yet to be available. In one study, following lung challenge with *C. gattii*, mice immunized with *C. gattii* CW (cell wall) and CP (cytoplasmic) protein preparations displayed a substantial reduction in pulmonary fungal loads. *C. neoformans* surface antigens such as glucuronoxylomannans (GXM) and galactoxylomannans (GalXM) are being evaluated as possible potential candidates against cryptococcosis.

Immunotherapy: A pan-fungal murine monoclonal antibody (mAb 2G8) offers considerable protection via targeting β-glucan of *Candida albicans, A. fumigatus*, and *C. neoformans* species. Immune effector responses against *C. neoformans* are suppressed by the programmed cell death protein 1 (PD-1) signaling pathway. Also, enhanced, and persistent levels of PD-1 in lung lymphocytes are linked to continuous *C. neoformans* infection. Anti-PD-1 Ab treatment of infected mice has shown a dramatic increase of lung fungal clearance in mice models.

Talaromycosis

Talaromycosis (penicilliosis) is endemic in tropical and subtropical Asia. Wild bamboo rats that live in mountains and soil in the endemic area serve as a reservoir for *Talaromyces marneffei*. Human infection is acquired possibly via inhalation of *T. marneffei* spores from the environment.

T. marneffei is a highly pathogenic dimorphic fungus that can cause fatal pulmonary talaromycosis in humans. Macrophage activation is crucial for the host's defense against *T. marneffei*. Hence, the innate immune response depends on the polarization of macrophages into fungal-inhibiting M1-like and fungal-promoting M2-like types. The killing ability of macrophages against *T. marneffei* is significantly increased by suppression of the TLR9 pathway, which partially prevents the M2-like polarization caused by *T. marneffei*. In addition, IFN-γ is an essential component of the host's defense mechanism against pathogenic *T. marneffei*; hence, it has a strong therapeutic potential for enhancing the phenotype of M1 polarization. No vaccines are available for talaromycosis infection. To date, no standardized recommendations regarding the appropriate duration of treatment and prophylaxis for *T. marneffei* infection among HIV-uninfected individuals are available.

Pneumocystis

Five distinct *Pneumocystis* species, including *Pneumocystis carinii*, *Pneumocystis wakefieldiae*, *Pneumocystis murina*, *Pneumocystis oryctolagi*, and *Pneumocystis jirovecii* (previously known as *Pneumocystis carinii*) affecting humans, have been identified. *Pneumocystis* pneumonia (PCP), a potentially fatal infection that mostly affects those with impaired immune systems, is caused by *P. jirovecii*.

Immunosuppressed rats experimentally infected with *P. carinii* show significant survival rates when treated with M1 and M2 macrophage cells. Treatment with M2 cells has shown encouraging results, with effective *P. carinii* clearance and a decrease in inflammation when compared to M1 cells. The absence of NK cells during immune suppression causes a significant lung burden of *Pneumocystis* in mice, indicating the protective role of NK cells in mediated protection against *Pneumocystis* infection. Recombinant human IL-7 (rhIL-7) is an effective therapeutic agent for treating murine PCP in CD4-depleted mice. This therapeutic effect is achieved by increased CD8+ T cell recruitment and reduced lung T lymphocyte apoptosis.

No vaccine is available. There are many potential vaccine candidates, including inactivated whole organisms, gpA (Msg), Kexin, SPD1, P55, and A12, which are being evaluated in *Pneumocystis* infections. These vaccines have been shown to offer a partial level of protection against PCP in animal models.

Immunotherapy: Trimethoprim-sulfamethoxazole (TMP-SMX) is used for preventing PCP and treating mild to severe instances of PCP. A combination of anti-*Pneumocystis* antibody and sulfasalazine have been shown to significantly reduce the severity of PCP in experimental infection in mouse.

Entomophthoromycosis

Entomophthoromycosis, a rare fungal infection, is more common in tropical and subtropical areas. Since these fungi were initially detected as parasitic infections of insects, the word "*Entomon*" (which is Greek for "insect") refers to an insect. It is a granulomatous disease of the subcutaneous tissues and skin, characterized by the development of sporadic hard, non-tender swellings, typically on the limbs, trunk, and infrequently other areas of the body. It is suggested that exposure to *Basidiobolus* (*Basidiobolus ranarum*, *Basidiobolus meristosporus*, *Basidiobolus haptosporus*, and *Basidiobolus microsporus*) and *Condiobolous* spp. (*Condiobolous incongruus* and *Condiobolous coronatus*) causes infection in humans upon inhalation of spores, slight trauma to the skin, or insect bites.

The innate immune system plays an important role in recognizing and responding to these fungal infections. The adaptive immune response, involving specific recognition and memory, is likely to be engaged in combating the infection. Vaccine development studies are scanty.

Adiaspiromycoses

Pulmonary disease adiaspiromycosis, caused by the dimorphic fungus *Emmonsia*, primarily infects small wild mammals. Human infection occurs by *Emmonsia screscens*, *Emmonsia sparva*, and *Emmonsia pasteuriana*. A rabbit immunization study was utilized to assess the antigenic activity of three UV mutants of *E. crescens*. Serological reactions revealed the presence of specific IgM and IgG antibodies in the serum of the infected animals. Despite being less potent than virulent strains, all three *E. crescens* mutants induced the production of distinctive antibodies. Currently, there is no report of vaccine studies against adiaspiromycosis.

Table 3.2 summarizes various immunotherapeutic and combination approaches against fungal zoonotic diseases.

Case Study

A 47-year-old HIV-negative male was admitted to the hospital with hoarseness and throat pain. He lost nearly 11 kg in the last 4 months and had periodic fever and loose stools every day for a month. He smoked cigarettes and drank a little alcohol per day. He had no history of substance addiction or extended medication therapy, nor was he diabetic or hypertensive. Mucocutaneous examination showed multiple well-defined, discrete to confluent macerated papuloplaques over the oropharynx involving soft and hard palate, anterior pillar, arytenoids, and epiglottis. He exhibited non-tender moderate hepatosplenomegaly and bilateral submandibular lymphadenopathy. Fine Needle Aspiration Cytology (FNAC) from the submandibular lymph node showed the presence of capsulated organisms suggestive of histoplasmosis. Histopathological analysis of the oral lesions revealed histoplasmosis-related characteristics, which were later validated by *H. capsulatum* culture on Sabouraud dextrose agar (SDA).

Itraconazole (200 mg twice/day) was used as an initial therapy for the treatment because amphotericin B was unavailable. However, his overall health declined (blood pressure: 80/50 mmHg, pulse rate: 103/min), he had aberrant electrolytes (serum sodium: 127 mEq/L, serum potassium: 2.4 mEq/L), his hemoglobin level was low (8.4 mg/dL), and he became delirious within the next 3 days. He had a very low platelet count of 45,000 cells/μL. He was transferred to the intensive care unit. Nevertheless, he left the healthcare facility against the advice of doctors because of financial constraints. When the patient was finally contacted, he had already passed away from his sickness.

Table 3.2 Immunotherapeutic and combination therapy in fungal zoonotic diseases

S. no.	Fungal zoonotic diseases	Antibody approaches	Cellular approaches	Cytokine approaches	Combination approaches
1.	Dermatophytosis	Anti-trCWP IgY led to growth inhibition of *Trichophyton rubrum*	Not investigated	GM-CSF and IL-17 show significant prevention of infection in mice	mAb antibody-0014 coupled with 5-arylhydrazonothiazole provided complete protection in guinea pigs
2.	Malassezia infection	Humanized mAb dupilumab prevented atopic dermatitis	Not investigated	rCaIFN-gamma shows antifungal effects against canine atopic dermatitis	ZnO nanoparticles coupled with ketoconazole show antifungal action against *Malassezia*
3.	Phaeohyphomycosis	Not investigated	Not investigated	IFN-γ therapy improved a CF patient with *E. dermatitidis* infection	Combined use of amphotericin B with flucytosine and triazole is a helpful treatment regimen
4.	Sporotrichosis	Antibody against phage-KR inhibited *Sporothrix* infection, and MAb P6E7 aided in the opsonization of fungal pathogen	Not investigated	Not investigated	Amphotericin B and subsequent itraconazole for maintenance therapy
5.	Coccidioidomycosis	Anti-SOWgp conferred partial protection in mice model	CD4+ T cells are sufficient for the transfer of protection	Exogenous IFNγ or IL-12 are protective in *Coccidioides* strains	Amphotericin-B and fluconazole, itraconazole, voriconazole, and posaconazole along with glucocorticoids
6.	Para coccidioidomycosis	mAbF1.4, mAb 3E, mAbs 7B6, and 4E12 reduced fungal burden in mice model	Adoptive transfer of *Paracoccidioides brasiliensis*-modulated DCs conferred protection in mice model	rIL-12 treatment reduced fungal burden in mice model	Sulfamethoxazole/trimetropim + *Allium sativum* extract, silymarin + cotrimoxazole, and itraconazole + pentoxifylline, showed antifungal activity in mice model
7.	Blatomycosis	Not found protective	CD4+ and CD8+ cells and Tc17 cells are protective in mice model	IL-1, IL-12, and IFN-gamma increase resistance to blastomycosis infection in mice	Oral itraconazole + photodynamic therapy + corticosteroids conferred protection against cutaneous blastomycosis in humans, micafungin + fluconazole protected mice from blastomycosis
8.	Histoplasmosis	Anti-H2B-like protein mAbs and anti-Hsp60 mAbs found protective against histoplasmosis	CD8+ T cells mediate protection against *Histoplasma capsulatum*	TNF-alpha, perforins, and GM-CSF contribute to T-cell immunity against histoplasmosis	Amphotericin B + caspofungin is an effective treatment for severe human histoplasmosis, farnesol + itraconazole showed antifungal activity against *Histoplasma capsulatum*

9.	Aspergillosis	mAbC7, mAbK10, mAbA9, mAb-7, and omalizumab found protective against Aspergillosis	Adoptive transfer of DCs and CD4+ T cells provided protection against invasive aspergillosis	IL-1, IL-18 IL-33, IL-36, and members of the IL-12 family, including IL-12p70, IL-23, IL-27, and IL-35 mediates essential roles during aspergillosis	Caspofungin + azoles, amiodarone + itraconazole or posaconazole and caspofungin + polymixin B showed fungicidal effects against *Aspergillus fumigatus*
10.	Cryptococcosis	mAb 2G8, anti-PD-1 mAb, and anti-idiotypic scFv induced inhibitory effects against cryptococcosis	Neutrophils mediated phagocytosis against *Cryptococcus gattii* R265 strain in vitro, B-cell-mediated reduction in fungal spread in mice	TNF-α and IFN-gamma improve protective cellular immune response against cryptococcal infection	N-acetylcysteine + amphotericin B, everolimus + amphotericin B/azoles and flucytosine + fluconazole may be effective treatments for cryptococcosis
11.	Adiaspiromycoses	Not investigated	Not investigated	Not investigated	Pyrithione, pimaricin, or amphotericin B + corticosteroids may be used
12.	Talaromycosis	Not investigated	M1-like macrophages show fungal-inhibiting properties. NLRP3 inflammasome plays an important role against *Talaromyces marneffei*	IL-1β production is induced by *Talaromyces marneffei*, which may have an antifungal role	Berberine + conventional antifungal drugs (Amphotericin B and itraconazole) show antifungal effects against *Talaromyces marneffei* in vitro
13.	Entomophthoromycosis	Not investigated	Not investigated	Not investigated	Itraconazole + saturated potassium iodide against conidiobolomycosis and potassium iodide along with cotrimoxazole, amphotericin B, ketoconazole, itraconazole, and fluconazole against basidiobolomycosis
14.	Pneumocystosis	mAb M5E12 conferred partial protection against PCP in animal models	Treatment with M2 macrophage cells induced fungal clearance in rat model, adoptive transfer of B cells conferred protection against *Pneumocystis* infection in IFrag(−/−) mice	IL-5, recombinant human IL-7, IL-23, and IL-33 improved antifungal effects against *Pneumocystis* infection in mice models	Trimethoprim-sulfamethoxazole (TMP-SMX), TMP-SMX + caspofungin and caspofungin and clindamycin may be used as alternative therapies against pneumocystosis

1. What could be an ideal initial treatment option for histoplasmosis?
2. What are the prospective combination therapies for the treatment of histoplasmosis?
3. Which cytokines contribute to the host immune response against histoplasmosis?
4. What are the preventive strategies for conferring long-term immunity against histoplasmosis?

Research Questions

1. What are the alternative therapies against histoplasmosis?
2. What is the role of *Histoplasma* in chronic disease?
3. How do the host immune responses prevent histoplasmosis?
4. What is the role of protective and non-protective monoclonal antibodies in histoplasmosis?
5. Which *H. capsulatum* vaccine candidates show protection in animal models?
6. What are the strategies to enhance the immunogenicity of vaccine candidates?

Further Reading

Blanco JL, Garcia ME (2008) Immune response to fungal infections. Vet Immunol Immunopathol [Internet] 125(1–2):47–70

Bonifaz A, Tirado-Sánchez A (2017) Cutaneous disseminated and extracutaneous sporotrichosis: current status of a complex disease. J Fungi (Basel) 3(1):6

Borchers AT, Gershwin ME (2010) The immune response in Coccidioidomycosis. Autoimmun Rev 10(2):94–102

Brummer E, Kethineni N, Stevens DA (2005) Immunological basis for susceptibility and resistance to pulmonary blastomycosis in mouse strains. Cytokine 32(1):12–19

Dias LDS, Silva LBR, Nosanchuk JD, Taborda CP (2021) Neutrophil cells are essential for the efficacy of a therapeutic vaccine against paracoccidioidomycosis. J Fungi (Basel) 7(6):416

El-Shabrawi MHF, Arnaout H, Madkour L, Kamal NM (2014) Entomophthoromycosis: a challenging emerging disease. Mycoses 57:132–137

Eymann C, Wachlin G, Albrecht D, Tiede S, Krummrei U, Jünger M et al (2018) Exoproteome analysis of human pathogenic dermatophyte species and identification of immunoreactive proteins. Proteomics Clin Appl 12(6):e1800007

García Carnero L, Lozoya Pérez N, González Hernández S, Martínez ÁJ (2018) Immunity and treatment of sporotrichosis. J Fungi 4(3):100

Gingerich AD, Norris KA, Mousa JJ (2021) *Pneumocystis* pneumonia: immunity, vaccines, and treatments. Pathogens 10(2):236

Kirchhoff L, Olsowski M, Rath PM, Steinmann J (2019) *Exophiala dermatitidis*: key issues of an opportunistic fungal pathogen. Virulence 10(1):984–998

Lian X, Scott-Thomas A, Lewis JG, Bhatia M, MacPherson SA, Zeng Y et al (2022) Monoclonal antibodies and invasive Aspergillosis: diagnostic and therapeutic perspectives. Int J Mol Sci 23(10):5563

Narayanasamy S, Dat VQ, Thanh NT, Ly VT, Chan JFW, Yuen KY et al (2021) A global call for talaromycosis to be recognised as a neglected tropical disease. Lancet Glob Health 9(11):e1618–e1622

Nosanchuk JD, Steenbergen JN, Shi L, Deepe GS, Casadevall A (2003) Antibodies to a cell surface histone-like protein protect against *Histoplasma capsulatum*. J Clin Invest 112(8):1164–1175

Rayens E, Rabacal W, Willems HME, Kirton GM, Barber JP, Mousa JJ et al (2022) Immunogenicity and protective efficacy of a pan-fungal vaccine in preclinical models of aspergillosis, candidiasis, and pneumocystosis. Wilson I, editor. PNAS Nexus 1(5)

Rohatgi S, Pirofski L (2015) Host immunity to *Cryptococcus neoformans*. Future Microbiol 10(4):565–581

Roth MT, Zamith-Miranda D, Nosanchuk JD (2019) Immunization strategies for the control of histoplasmosis. Curr Trop Med Rep 6(2):35–41

Saunte DML, Gaitanis G, Hay RJ (2020) *Malassezia*-associated skin diseases, the use of diagnostics and treatment. Front Cell Infect Microbiol 10:112

Taborda CP, Urán ME, Nosanchuk JD, Travassos LR (2015) Paracoccidioidomicose: desafios no desenvolvimento de uma vacina contra micose endêmica nas Américas. Rev Inst Med Trop Sao Paulo 57:21–24

Tomsíková A, Stěrba J, Prokopic J, Novácková D (1982) Antifungal activity of pyrithione against *Emmonsia crescens* in vitro and in vivo. Folia Parasitol 29(3):271–278

Werfel T, Biedermann T (2015) Current novel approaches in systemic therapy of atopic dermatitis. Curr Opin Allergy Clin Immunol 15(5):446–452

Fungal Genomics

4

Pankaj Singh, Ritika Harchand, and Hugo Madrid

Learning Objectives

By the end of the chapter, the readers should be able to:

1. Understand various aspects of fungal genomics.
2. Know various resources for fungal genomics.
3. Evaluate contributions of fungal genomics in taxonomy and diagnosis.

Introduction

Fungal genomics includes structural genomics (a structural feature of chromosome and genome mapping), functional genomics (gene characterization and expression), comparative genomics (phylogenetic relationship between different fungi), and epigenomics (gene interactions with each other and environments). The fungal genomics uses a combination of molecular techniques, recombinant DNA technology (RDT), and bioinformatics for the analysis of fungal genomes. Currently, the advancement in genomic technology has immensely contributed to the science of medical mycology, including taxonomy, epidemiology, pathogenesis, and diagnosis, gradually shifting from morphology to molecular-based concepts in mycology.

The fungal kingdom consists of nearly 11–13 million species compared to 2.2–3.5 million different species estimated by different techniques, making it the most diverse group within infectious pathogens. This fungal diversity is reflected by phenotypic, ecological, and genetic diversity. Fungal symbiosis ranges from mutualism, saprophytic, commensalism, entophytic, and parasitic. Approximately, 625 fungal species have been associated with vertebrate infection, 200 of which exist as human microbiota or as pathogens. Globally, the estimated fungal burden is very high, exceeding 1 billion infections, causing the death of >1.5 million people each year. The fungal clinical and ecological importance, with advancement in sequencing technology, has led to the sequencing of thousands of fungal genomes in the past few decades. In this chapter, among others, recent advances in fungal genomics for the understanding of fungal zoonoses will be emphasized.

P. Singh (✉)
Indian Council of Medical Research-National Institute of Virology, Pune, Maharashtra, India
e-mail: rawat.pankaj899@gmail.com

R. Harchand
Department of Medical Microbiology, Post Graduate Institute of Medical Education and Research (PGIMER), Chandigarh, India

H. Madrid
Departamento de Tecnología Médica, Facultad de Ciencias de la Salud, Universidad de Tarapacá, Sede Iquique, Iquique, Chile

S. C. Parija, S. M. Rudramurthy (eds.), *Textbook of Fungal Zoonoses and Sapronoses*, Microbial Zoonoses, https://doi.org/10.1007/978-981-97-3214-2_4

Fungal Genomics Tools and Database

Fungal genomics tools include molecular techniques (Polymerase chain reaction (PCR), Real-Time PCR, microarray, genotyping, Sanger Sequencing, next-generation sequencing), recombinant DNA technology (cloning, Gene silencing, and reverse genetics), and bioinformatics (software and databases for the genomic data analysis). Multiple fungal genomic databases have been created in the past few decades, with advancements in molecular mycology. Among fungal genome databases, *Saccharomyces* Genome Database (SGD) and *Candida* Genome Initiative (CGI) are examples of genus-specific databases. The broader fungal genome database includes FungiDB, Ensembl Fungi CFGP, NCBI, and Mycocosm. Mycocosm database not only offers fungal genome data access but also provides visualization and analysis tools of genomics.

History

In 1980, molecular mycology emerged in synchronous with PCR and resulted in the discovery of new fungal species. Initially, fungal genome sequencing was neglected compared to bacteria. *Saccharomyces cerevisiae* was the first fungal genome published in 1996, followed by *Schizosaccharomyces pombe* (2002) and *Neurospora crassa* (2003). In the past few decades, the use of fungi as a model organism for the eukaryote genome has resulted in the sequencing of many fungi. In 2000, Broad Institute launched the Fungal Genome Initiative (FGI-http://www.broad.mit.edu/annotation/fgi/), targeting annotation of the fungal genome from throughout the kingdom. Over 100 fungi genomes have been sequenced in this program, including pathogenic fungi, till now. The exponential growth of fungal genome data resulted in requirements for new approaches for data integration and analysis. In 2013, MycoCosm portal: 1000 fungal genomes sequencing project was launched to resolve this issue. In 2021, Centers for Disease Control and Prevention (CDC) launched FungiNet, a network for molecular surveillance and genomic epidemiology for fungal diseases. FungiNet is considered the gold standard method for whole genome sequencing (WGS) (Table 4.1).

Taxonomy

Fungal taxonomy has remained controversial for several decades. Initially, the dual nomenclature system based on phenotypic characteristics has been used for fungi. Now, it is shifting from phenotype to genotype-based concepts with the emergence of molecular techniques. The molecular taxonomy has completely revolutionized the fungal nomenclature: many phyla revised, multiple cryptic species or species complexes identified in pathogenic fungi, multiple species shifted to new genera based on DNA barcode similarity, and new genera and species created for novel DNA sequences. Fungal taxonomy dynamics is further expedited following the emergence of high throughput sequencing in diagnosis, leading to the increased discovery of novel species.

Currently, there is no single source for microbiologists and mycologists that capture all the latest nomenclatural modifications of medically important fungi. Diagnostic laboratories, hence, must remain updated with various databases for accurate and reliable identification. The changes hopefully would contribute to a robust and clear fungal taxonomy. Table 4.2 summarizes recent changes in nomenclature for medically important fungi from 2016 to 2019.

Fungal Genomes

Fungal genome sizes range from a few Mb to less than 10 Gb with 1800–35,000 genes, suggesting the high genome plasticity of fungi. Obligate parasitic fungi (*Microsporidia* and *Pneumocystis*) and yeast-forming species (*Candida*, *Schizosaccharomyces*) have smaller genomes. The plant-associated fungi (*Tuber melanosporum*) and insect-killing fungi (*Entomophthora muscae* and *Zoophthora radicans*) have large genomes ranging from ~700 Mb to 1.5 Gb. Genome expansions in these fungi are controlled by a transposable element. Table 4.3 summarizes the characteristics of medically important fungi genomes.

Table 4.1 Databases for fungal genome

Fungi database	Database URL	Year
NCBI GenBank	http://www.ncbi.nlm.nih.gov/GenBank/	1982
Centraalbureau voor Schimmelcultures databases (CBS-KNAW)	http://www.cbs.knaw.nl/collections/	1989
Broad Institute databases	http://www.broadinstitute.org/scientific-community/data/	2003
Unified system for the DNA-based fungal species linked to the classification (UNITE)	https://unite.ut.ee/	2003
Candida Genome Database (CGD)	http://www.candidagenome.org/	2004
MycoBank	http://www.mycobank.org/	2004
Fusarium databases	http://www.cbs.knaw.nl/fusarium/http://isolate.fusariumdb.org/	2005
Multi Locus Sequence Typing (MLST.net)	http://www.mlst.net/databases/	2005
Assembling the Fungal Tree of Life (AFTOL)	https://aftol.umn.edu/	2006
Barcode of Life Data Systems (BOLD)	http://v4.boldsystems.org/	2007
Aspergillus Genome Database (AspGD)	http://www.aspgd.org/	2008
International Society for Human and Animal Mycology (ISHAM-ITS)	http://its.mycologylab.org/	2011
International Society for Human and Animal Mycology (ISHAM-MLST)	http://mlst.mycologylab.org/	2011
NCBI Whole Genome Database	https://www.ncbi.nlm.nih.gov/genome/	2011
FungiDB	http://fungidb.org/fungidb/	2012
The Fungal Genomics Resource (MycoCosm)	http://genome.jgi.doe.gov/programs/fungi/index.jsf	2013
NCBI RefSeq database	http://www.ncbi.nlm.nih.gov/refseq/	2014
Institut Pasteur FungiBank	http://fungibank.pasteur.fr/	2015
Mitofun	http://mitofun.biol.uoa.gr/	2014
FungiNet	https://www.cdc.gov/fungal/outbreaks/wgs.html	2021
Anaerobic Fungi Network	https://anaerobicfungi.org/databases/	
Alternaria Genome Database	http://alternaria.vbi.vt.edu	

Note: NCBI National Center for Biotechnology Information

Table 4.2 List of revised fungal taxa

Previous species name or synonyms	Revised species name
Candida	
Candida infanticola	*Wickerhamiella infanticola*
Candida pararugosa	*Wickerhamiella pararugosa*
Candida haemulonii group II	*Candida duobushaemulonii*
Pichia anomala, Hansenula anomala, Candida pelliculosa	*Wickerhamomyces anomalus*
Candida glabrata	*Nakaseomyces glabrataa*
Candida kefyr, Saccharomyces marxianus, Candida pseudotropicalis, Kluyveromyces fragilis, Kluyveromyces cicerisporus	*Kluyveromyces marxianus*
Candida krusei, Candida acidothermophilum	*Pichia kudriavzevii*
Candida guilliermondii	*Meyerozyma guilliermondii*
Candida utilis, Hansenula jadinii, Pichia jadinii, Candida guilliermondii var. Nitratophila	*Cyberlindnera jadinii*

Table 4.2 (continued)

Previous species name or synonyms	Revised species name
Candida lusitaniae	*Clavispora lusitaniae*
Trichosporon	
Trichosporon cutaneum, Geotrichum cutaneum	*Cutaneotrichosporon cutaneum*
Trichosporon loubieri	*Apiotrichum loubieri*
Trichosporon mucoides	*Cutaneotrichosporon mucoides*
Rhodotorula	
Rhodotorula minuta	*Cystobasidium minutum*
Rhodotorula slooffiae	*Cystobasidium slooffiae*
Dermatophytes	
Microsporum gypseum, Trichophyton gypseum	*Nannizzia gypsea*
Microsporum nanum, Arthroderma obtusum, Nannizzia obtuse	*Nannizzia nana*
Microsporum gallinae, Epidermophyton gallinae	*Lophophyton gallinae*
Microsporum persicolor, Trichophyton persicolor, Arthroderma persicolor	*Nannizzia persicolor*
Microsporum cookie	*Paraphyton cookie*
Fusarium	
Fusarium solani species complex 6 (FSSC6)	*Fusarium metavorans*
Fusarium metavorans	*Neocosmospora metavorans*
Fusarium solani species complex 7 (FSSC7)	*Neocosmospora gamsii*
Fusarium solani species complex 20 (FSSC20)	*Neocosmospora suttoniana*
Fusarium solani species complex 43 (FSSC43)	*Neocosmospora catenata*
Fusarium keratoplasticum	*Neocosmospora keratoplastica*
Fusarium lichenicola	*Neocosmospora lichenicola*
Fusarium petrophila	*Neocosmospora petroliphila*
Cryptococcus	
Cryptococcus neoformans var. grubii (serotype A, molecular types VNI and VNII)	*Cryptococcus neoformans*
Cryptococcus gattii (serotype D, molecular type VNIV)	*Cryptococcus deneoformans*
Cryptococcus gattii (serotype B, molecular type VGI)	*Cryptococcus gattii*
Cryptococcus gattii (serotype C, molecular type VGIII)	*Cryptococcus bacillisporus*
Cryptococcus gattii (serotype B, molecular type VGII)	*Cryptococcus deuterogattii*
Cryptococcus gattii (serotype C, molecular type VGIV)	*Cryptococcus tetragattii*
Cryptococcus gattii (serotype B, molecular types VGIV and VGIIIc)	*Cryptococcus decagattii*
Bipolaris	
Bipolaris australiensis, Drechslera australiensis, Helminthosporium australiense, Pseudocochliobolus australiensis	*Curvularia australiensis*
Bipolaris hawaiiensis, Pseudocochliobolus hawaiiensis, Cochliobolus hawaiiensis	*Curvularia hawaiiensis*
Bipolaris spicifera	*Curvularia spicifera*
Emmonsia	
Emmonsia crescens	*Emergomyces crescens*
Emmonsia soli	*Emergomyces soli*
Emmonsia helica	*Blastomyces helicus*
Emmonsia parva	*Blastomyces parvus*
Emmonsia pasteuriana	*Emergomyces pasteurianus*
Others	
Chaetomium atrobrunneum	*Amesia atrobrunnea*
Diutina (Candida) mesorugosa	*Diutina rugosa*
Allescheria boydii (Pseudallescheria boydii)	*Scedosporium boydii*
Trichosporon capitatum	
Hendersonula toruloidea (Scytalidium hyalinum, Scytalidium dimidiatum, Fusicoccum dimidiatum, Nattrassia mangiferae, Neofusicoccum mangiferae, Neoscytalidium hyalinum)	*Neoscytalidium dimidiatum*

Table 4.3 Characteristics of medically important fungi genomes

S. no.	Fungi	Genome size	No. of chromosome	Ploidy
1.	*Candida*	3.2–4.0 Mb		
	Candida albicans	16 Mb	8	Diploid
	Candida tropicalis	15 Mb	5–6	Diploid
	Candida glabrata	12.3 Mb	13	Haploid
	Candida auris	12–13 Mb	7	Haploid
2.	*Cryptococcus*			
	Cryptococcus neoformans	20 Mb	12	Haploid/diploid
	Cryptococcus gattii	18.7 Mb	14	Haploid/diploid
3.	*Malassezia*	6–14 Mb	6–9	Haploid
4.	*Aspergillus*	29–36 Mb		
	Aspergillus flavus	36 Mb	8	Haploid
	Aspergillus fumigatus	30 Mb	8	Haploid
5.	*Penicillium*	29.5 Mb	3–6	Haploid
6.	Dermatophytes			
	Trichophyton	22–25 Mb	5	Haploid
	Microsporum	23–24 Mb	5	Haploid
7.	Zygomycetes			
	Mucor	39 Mb		
	Rhizopus	42–47 Mb	8–16	Haploid
	Apophysomyces	35–40 Mb		
	Basidiobolus			
	Conidiobolus			
8.	*Sporothrix schenckii*	33–44 Mb	6–8	Haploid
9.	*Trichosporon*	16–42 Mb		
10.	Dimorphic fungi			
	Blastomyces dermatitidis	28 Mb	ND	Haploid/diploid
	Coccidioides immitis	29 Mb	4	Haploid
	Histoplasma capsulatum var. capsulatum	23–25 Mb	7	Haploid/diploid
	Paracoccidioides brasiliensis	29–33 Mb	5	Haploid
11.	Obligate parasite			
	Microsporodia	2–6 Mb	11	Not known
	Pneumocystis	8 Mb	15	Haploid/diploid

Application of Genomic Studies

Genomic studies have made a significant impact in the field of medical mycology. These studies have allowed researchers to gain insights into fungal pathogenesis, taxonomy, evolution, diagnosis, and treatment. Genomic data has revolutionized fungal taxonomy and phylogenetics and allowed researchers to construct more accurate and well-defined taxonomy based on genomics data. The comparative genomics helps in the identification of virulence factors and potential drug targets, leading to the development of more effective antifungal treatments. Metagenomics reveals the diversity of mycobiomes in health and diseases. Sanger sequencing allows researchers to study the effect of single nucleotide polymorphism (SNP), gene deletion, acquisition, and rearrangements in the evolution of pathogenic fungi. Additionally, next-generation sequencing allows the assemblage of many fungal genomes up to the chromosome level and highlights the importance of chromosomal structural variation in pathogenic fungi such as polyploidy and aneuploidy. Genomic studies in knock-out animal models have increasingly contributed to the understanding of antifungal resistance mechanisms, pathogenesis, and efficacy of new drugs.

Fungal Genomics in Epidemiology

Like molecular taxonomy, molecular epidemiology uses molecular techniques, especially nucleic acid-based techniques, for the detection and identification of microbes. Molecular epidemiology aims to identify the host, microbes, and environmental factors associated with the disease transmission cycle in nature. The item to be tracked in a population can be a novel fungal strain or a resistant strain.

Multiple molecular typing schemes used for the identification of pathogenic fungi include Ribotyping, Polymerase chain reaction-Restriction Fragment Length Polymorphism (PCR-RFLP), Pulsed Field Gel Electrophoresis (PFGE), Amplified Fragment Length Polymorphism (AFLP), Random Amplification of Polymorphic DNA (RAPD-PCR), Karyotyping, Inter-Simple Sequence Repeats (ISSR), Minisatellites, Macrosatellites, Multi Locus Satellite Typing (MLST), Variable number tandem repeats (VNTR), Single-nucleotide polymorphism (SNP), Small Insertions or Deletions (InDels), DNA barcoding, and Whole genome sequencing (Table 4.4). The molecular typing in fungal outbreaks is used in tracking patient-to-patient transmission, patient-to-healthcare worker's transmission, patients-to-environment transmission, and source/reservoir identification. Among these typing techniques, MLST and STR-based typing are the most popular methods, with high reproducibility and discriminatory power. WGST emerged as a new typing method but was restricted to apex mycology centers and laboratories.

Table 4.4 Molecular markers used for medically important fungi

S. no.	Fungi	Molecular markers used	Gold standard
1.	*Candida albicans*	MLEE, RAPD, RFLP, AFLP, MLP, MLMT, MLST (*CaACC1*, *CaVPS13*, *CaGLN4*, *CaADP1*, *CaRPN2*, *CaSYA1*), REP-PCR, Southern blot with DNA fingerprinting	MLST (*CaACC1*, *CaVPS13*, *CaGLN4*, *CaADP1*, *CaRPN2*, *CaSYA1*)
2.	*Aspergillus*	PCR (ITS, RPB2, BT2, CaM, aflR), Cytochrome b), RFLP, RAPD (R-108, R-151, and AP12 h), PCR–SSCP, AFLP, STR ($(GTG)^5$ and $(GACA)^4$), MLST	MLST (*Aspergillus fumigatus*) and MST
3.	Zygomycetes	PCR (ITS, 18S, ESI-MS, Microarray)	No consensus typing scheme
4.	*Malassezia*	PCR (26S, 5.8S, ITS-2, PCR-RFLP (26S rDNA), RAPD)	No consensus typing scheme
5.	Dermatophytes	PCR (ITS, D1/D2, BT2, TEF-1α, CaM)	No consensus typing scheme
6.	*Cryptococcus*	Microsatellite- (M13) or minisatellite-specific primers (e.g., $(GACA)_4$ or $(GTG)_5$), RAPD, AFLP, RFLP (*URA5*, *PLB1* genes), MLST and MLMT	MLST *Cryptococcus neoformans*: *MPD1*, *TOP1*, *MP88*, *CAP59*, *URE1*, *PLB1*, *CAP10*, *GPD1*, *TEF1*, *SOD1*, *LAC1*, and *IGS1* *Cryptococcus gattii*: *SXIa or SXIa*, *IGS1*, *TEF1*, *GPD1*, *LAC1*, *CAP10*, *PLB1*, and *MPD1*
7.	*Pneumocystis*	PCR (msg, mt LSU rRNA, mt SSU rRNA), RFLP, SSCP, STR, MLST (mt SSU rRNA and mt LSU rRNA)	MLST (mt LSU rRNA and mt SSU rRNA)
8.	*Sporothrix*	DNA Barcoding (ITS1), RAPD (T3B), RFLP (*CAL*), MLSA (CAL, BT2, EF1α, CHS1, and ITS1), PFGE, MAT (*MAT1-1 and MAT1-2*), AFLP, SSRs, WGS	No consensus typing scheme
9.	*Fusarium*	PCR (TEF1-α,) MLST (*TEF1-α*, *RPB1*, and *RPB2*)	No consensus typing scheme

Fungal Genomics in Host–Parasite Interactions

Fungal pathogenesis is the interplay of host–pathogen–environment. Several genomics techniques have been used to understand the host–pathogen interaction dynamics in fungal pathogenesis; both host and pathogen genetic variation affects the susceptibility to mycoses. The host's genetic susceptibility increases the predisposition to fungal infection. The development and advancement of genomics techniques enhance further the efforts to detect these genetic variations. Several studies reported the link between host genetic variations and increased risk for fungal infection (Table 4.5).

Fungal genomic variation includes small-scale evolution (SNP, transposition, and repetitive sequences editing) and large-scale evolution (chromosomal modification), both of which affect fungal virulence. Comparative genomics has been used to understand the evolution of fungal pathogens such as *Candida* species (diversification of cell adhesion proteins), *Candida auris* (expansion of drug efflux pumps and ERG11 mutations), and *Fusarium oxysporum* (horizontal acquisition of virulence genes) (Table 4.6).

Table 4.5 Human genetic mutation associated with susceptibility to fungal infection

Gene	Pathogen	Disease
CARD9, TLR1, MPO, CYBB, NADPH oxidase	*Candida*	Chronic mucocutaneous candidiasis (CMC), candidemia
PTX3, TLR4, NADPH oxidase, MPO	*Aspergillus*	Invasive aspergillosis (IA)
MBL	*Pneumocystis jirovecii*	Pneumocystis pneumonia (PCP)
IL-6	*Blastomyces*	Blastomycosis
IL-2	*Histoplasma*	Histoplasmosis
STAT1	*Histoplasma* *Coccidiodes*	Histoplasmosis Coccidioidomycosis
NOD2, STAT3, CYP2C19	*Aspergillus*	Aspergillosis
CARD9	*Malassezia*	Pityriasis versicolor
MBL, MASP-2	*Sporothrix*	Sporotrichosis
Ferroxidase	Mucorales	Mucormycosis
HLA-B22	*Histoplasma capsulatum*	Histoplasmosis
Dectin 1, IL-10	*Coccidioides immitis*	Coccidioidomycosis
Immunoglobulin gene defect, IL gene defect, MHC class II gene defect	*Pneumocystis jirovecii* *Candida* *Aspergillus* *Blastomyces* *Coccidioides* *Cryptococcus* *Histoplasma* *Paracoccidioides*	Pneumocystis pneumonia (PCP) Candidiasis Aspergillosis Blastomycosis Coccidioidomycosis, cryptococcosis Histoplasmosis Paracoccidioidomycosis

Table 4.6 Fungal genome mutation associated with fungal pathogenesis

Fungi	Genomic changes
Cryptococcus neoformans	RNAi loss by genetic modification, leading to the emergence of resistant isolates
Candida auris	Azole resistance emergence: mutation in ERG11, TAC1b, Y132F, K143R, and F126L, ATB-binding cassette (ABC)
Cryptococcus	Hypermutation via mobilization of transposon
Candida auris and *Aspergillus fumigatus*	Subtelomeric evolution on virulence
Malassezia and *Microsporidia*	Horizontal gene transfer: acquisition of bacterial genes by *Malassezia* and *Microsporidia*
Malassezia and *Candida*	Chromosome rearrangements
Candida albicans, *Candida tropicalis*, *Cryptococcus*	Polyploidy
Candida albicans and *Candida neoformans*	Aneuploidy
Pneumocystis species	Species variation in the *msg* superfamily

Fungal Genomics in Fungal Pathogen Discovery

A combination of clinical observation, conventional, and molecular approaches are used for the identification of novel fungal pathogens. The molecular level identification has higher sensitivity and specificity and thus is used as a gold standard for fungal species and strain differentiation in medical mycology. Rapid identification of novel strains and species from precious clinical samples is now achievable with sequencing technologies like Sanger sequencing and next-generation sequencing (NGS: illuminate, PacBio, and Oxford Nanopore). Pathogenic fungi's spatial and temporal evolution can be tracked by molecular tools such as phylogenetics. Rapid and reliable novel fungal pathogen detection from clinical samples has been studied using functional genomics (transcriptomics, proteomics, and metabolomics) such as Matrix-assisted laser desorption ionization-time of flight mass spectrometry (MALDI-TOF-MS). The combination of fungal genome data and modern technology such as artificial intelligence (AI) can be used for the identification of novel fungal pathogens by comparative genomics.

Fungal Genomics in Diagnosis

The traditional fungal diagnosis methods are slow, time-consuming, low throughput, and associated with low sensitivity and specificity, leading to the emergence of molecular diagnostic methods. These molecular methods (genomics-proteomics-based methods) increasingly used in diagnostic mycology have high sensitivity and specificity and less turnaround time (Fig. 4.1). Currently, multiple molecular assays are available for the identification of both wild and resistant isolates of pathogenic fungi in a clinical setting worldwide.

Polymerase Chain Reaction (PCR)

Polymerase chain reaction (PCR) targeting ribosomal RNA (rDNA) (conserved and multiple copies) locus is used for fungal identification worldwide. The combination of PCR amplification (rDNA) with sequencing is used as a gold standard for fungal species identification in clinical mycology (Fig. 4.2). Additionally, other markers such as beta-tubulin, Intergenic Spacer (IGS), actin, chitin synthase, and translation elongation factor genes are also used for fungal species identification (Table 4.7).

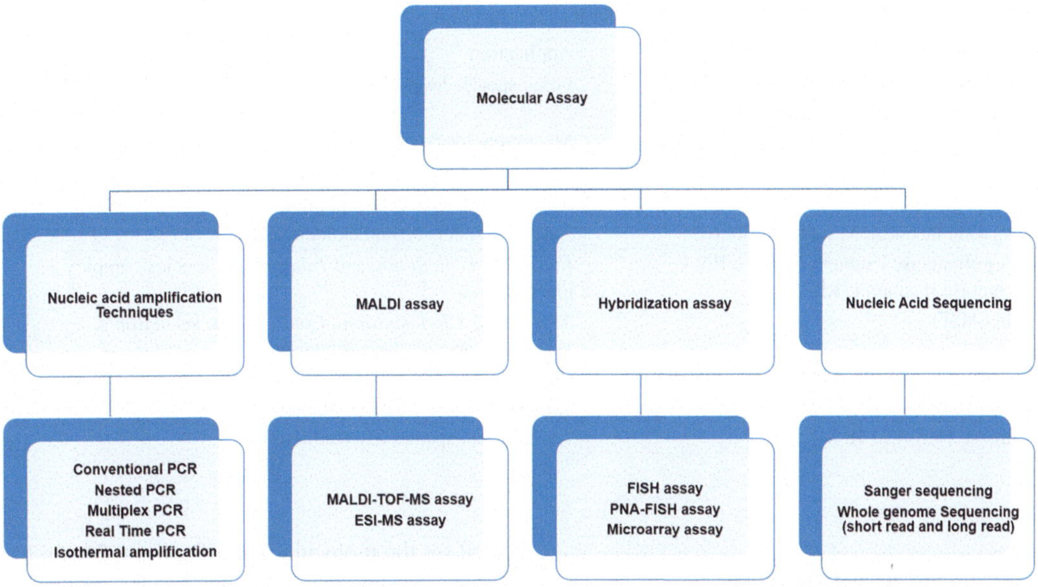

PCR: Polymerase chain reaction, MALDI: Matrix assisted laser desorption ionization, Matrix assisted laser desorption ionization time of flight mass spectrometry, ESI-MS: Electrospray ionization mass spectrometry, FISH: fluorescent *in situ* hybridization, PNA-FISH: Peptide nucleic acid fluorescent *in situ* hybridization

Fig. 4.1 Molecular diagnostic platforms used for fungal diagnosis

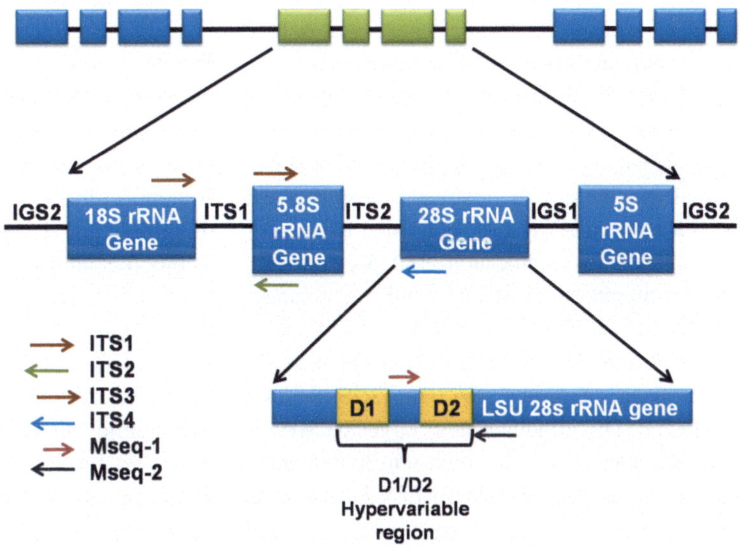

Fig. 4.2 Structural organization of fungal ribosomal gene

Table 4.7 Gene target used for fungal identifications

Molecular target	Application
Internal transcribed spacer regions (ITS1-5.8S-ITS2)	Fungi species level identification
D1/D2 variable domains of the 28S rDNA gene	Mucorales species level identification
Beta tubulin II (β-tubulin)	*Aspergillus* species discrimination
Calmodulin (Cal)	*Alternaria* species discrimination
Elongation factor alpha subunit (EF-1α)	*Fusarium* species complex identification
RNA polymerase I subunit (RPB1), RNA polymerase II subunit (RPB2)	*Fusarium*, *Penicillium*, and *Talaromyces* species complex identification
Actin (ACT)	*Aspergillus*, *Cladosporium*, *Coniochaeta*, *Verticillium*, *Verruconis* species discrimination
Glycerol-3-phosphate dehydrogenase (GPDH)	*Bipolaris*, *Curvularia*, *Verticillium* species discrimination
Chitin synthase (CHS)	*Sporothrix* species discrimination
Chitinase 18-5 (Chi 18-5)	*Trichoderma* species discrimination

Isothermal Amplification

Isothermal amplification, an alternative approach to standard PCR, offers low turnaround time, low cost, and portability. These techniques do not require any expensive equipments such as thermocyclers, making them suitable for resource-limited settings, including primary healthcare centers and remote areas. Currently, isothermal amplification techniques used for fungal detection include nucleic acid sequence-based amplification (NASBA), rolling-circle amplification (RCA), loop-mediated isothermal amplification, strand displacement amplification (SDA), primer-generation RCA, helicase-dependent amplification (HDA), crossing priming amplification (CPA), and recombinase polymerase amplification (RPA). Additionally, CRISPR/Cas9-based molecular diagnostic methods are evolving, including NASBA-CRISPR/Cas9, DETECTR (DNA Endonuclease Targeted CRISPR Trans-Reporter), specific high-sensitivity enzymatic reporter unlocking (SHERLOCK) platform, and opvCRISPR (one-pot visual RT-LAMP-CRISPR).

Sanger Sequencing

Sanger sequencing is the "gold standard" for fungal identification. The fungal rDNA is most targeted for the molecular identification of fungi by Sanger sequencing. The nucleotide sequence databases containing the rDNA sequence of the most clinically important fungi are well-established for fungal identification.

Whole Genome Sequencing

Whole genome sequencing (WGS) is a rapidly emerging technique in molecular diagnosis. The development of third-generation Next Generation Sequencing (NGS), especially Nanopore, has enabled the direct identification of pathogenic fungi from clinical samples such as blood, sputum, BALF, tissue, and stools. The value of WGS lies beyond simple identifications, including molecular epidemiology details about each strain, the potential for drug resistance and identification of novel fungal species, and public surveillance of epidemic or pandemic strains. The first case of fungal infection diagnosed by NGS was reported in 2014: *Aspergillus fumigatus* and *Schizophyllum commune* were detected in an allergic bronchopulmonary mycosis case. Since then, many fungi have been reported by NGS, including *Pneumocystis jirovecii*, *Aspergillus*, *Candida*, *Cryptococcus*, *Rhizopus*, *Fusarium*, *Alternaria*, *Histoplasma capsulatum*, etc.

Mass Spectrometry

Two mass spectrometry-based techniques emerging in molecular diagnosis are matrix-assisted laser desorption ionization-time of flight mass spectrometry (MALDI-TOF MS) and PCR/electrospray ionization-mass spectrometry (ESI-MS). These techniques have high sensitivity, specificity, and less turnaround time (<6 h).

MALDI-TOF MS Biosystems, including MALDI Biotyper (Bruker Daltonics), generates peptide fingerprints and compares them with commercial or in-house databases for fungal identification. MALDI-TOF MS has been approved for clinical diagnosis worldwide for reliable identification of clinically important fungi, including *Aspergillus*, *Candida*, *Cryptococcus*, *Fusarium*, dimorphic fungi, melanized fungi, dermatophytes, and zygomycetes.

PCR/electrospray ionization-mass spectrometry (ESI-MS) biosystem, including IRIDICA, PLEX-ID uses a combination of PCR amplification and mass spectrometry for pathogen identification. In brief, automated multiplex PCR is used for pathogen nucleic acid amplification, followed by automated ESI-MS analysis of each PCR product. PCR/EMI-MS has good diagnostic accuracy in clinical samples for the identification of fungal pathogens.

Hybridization

Fungal hybridization-based assay ranges from blotting techniques, *in situ* hybridization, fluorescence in situ hybridization (FISH), peptide nucleic acid (PNA) FISH, dot hybridization array (DHA), microarray, and macroarray. Numerous commercialized assays based on hybridization exist for fungi, including fluorescence in situ hybridization (FISH), oligonucleotide probe-based, and a cytogenetic assay that targets rDNA sequences. These diagnostic assays can be performed directly on clinical samples with minimum sample processing and have been implemented for multiple fungi, including *Candida*, *Aspergillus*, *Fusarium*, and *Scedosporium* species.

Magnetic Resonance

A single diagnostic assay based on magnetic resonance is available for the diagnosis of fungi: T2Candida (T2 Biosystems, MA, USA). T2Candida is an FDA-approved and fully automated diagnostic test for multiple fungal pathogen detection directly from whole blood. In this biosystem, the fungal DNA is amplified by polymerase chain reaction (ITS region) followed by species-specific T2 magnetic resonance detection. T2 Candida panel can detect five *Candida* species, including *C. albicans*, *C. tropicalis*, *Candida krusei*, *Candida glabrata*, and *Candida parapsilosis*, with high sensitivity (91%), specificity (99%), 1–3 CFU/mL limit of detection (LOD), and 3–5 h turnaround time.

Fungal Genomics and Antifungal Resistance

Fungal infections are associated with high rates of morbidity and mortality due to limited antifungal therapy and the emergence of resistant isolates. The fungi have developed multiple resistance mechanisms, such as gene mutation, over-expression of efflux pump proteins, and biofilm formation. Thus, antifungal susceptibility testing (AFST) became mandatory to guide antifungal therapy in clinical settings. The major drawback of culture-based AFST is that it is labor-intensive and time-consuming. Hence, during the transition from conventional to molecular diagnostics, there are new techniques for concurrent identification and susceptibility testing of fungi. The molecular techniques can also provide rapid identification of both intrinsic and acquired resistant isolates than classical AFST methods such as MALDI-TOF-MS. Table 4.8 provides a list of commercially available molecular methods-based assay kits for the rapid detection of wild-type, mutant strain, and resistant isolates.

Table 4.8 List of commercially available molecular methods-based assay kits for the diagnosis and detection of wild-type and resistant isolates

S. no.	Kits	Company	Target
1.	Affigene *Aspergillus* tracer *Asp. fumigatus* Bio-Evolution artus R Aspergillus diff. RG PCR AsperGenius R Species and AsperGenius R Resistance *Aspergillus* spp. ELITe MGB R Kit	Cepheid, Rolling Meadows, IL, USA Bio-Evolution, Bry-sur-Marne, France Qiagen, Düsseldorf, Germany PathoNostics B.V., Maastricht, Netherlands ELITechGroup S.p.A, Turin, Italy	Target unknown *Aspergillus* spp. ITS1 region *Aspergillus fumigatus* Target unknown *Aspergillus fumigatus*, *Aspergillus terreus*, *Aspergillus flavus* 28*S* rDNA *Aspergillus fumigatus* complex, *Aspergillus terreus*, *Aspergillus* spp. TR34/L98H and TR46/Y121F/T289A mutations 18*S* rDNA *Aspergillus* spp. (*Aspergillus niger*, *Aspergillus nidulans*, *Aspergillus terreus*, *Aspergillus flavus*, *Aspergillus versicolor*, *Aspergillus glaucus*)
2.	FilmArray Blood Culture Identification (BCID) Panel	BioFire Diagnostics, Salt Lake City, Utah, United States	Target unknown *Candida albicans*, *Candida glabrata*, *Candida krusei*, *Candida parapsilosis*, and *Candida tropicalis*
3.	*Asp*ID, *Cand*ID R and *Auris*ID	OlmDiagnostics, Newcastle, United Kingdom	Target unknown *Aspergillus* spp., *Aspergillus terreus*
4.	FungiPlex R Universal FungiPlex R *Aspergillus* and Fungiplex R *Aspergillus* Azole_R FungiPlex R *Candida* FungiPlex R *Candida auris*	Bruker Daltonik GmbH, Bremen, Germany	Target unknown Target unknown *Aspergillus* spp. (*Aspergillus fumigatus*, *Aspergillus flavus*, *Aspergillus niger*), *Aspergillus terreus* TR34 and TR46 mutations Target unknown *Candida* spp. Target unknown *Candida auris*
5.	SepsiTest-UMD	Molzym Molecular Diagnostics, Bremen, Germany	18*S* rDNA All fungal species
6.	T2*Candida*	T2 Biosystems, Lexington, MA, United States	ITS2 region *Candida albicans*/*Candida tropicalis*, *Candida glabrata* complex, Candida *krusei*, and *Candida parapsilosis* complex
7.	LightCycler Septifast	Roche Diagnostics, Mannheim, Germany	ITS region *Aspergillus fumigatus* and *Candida* spp.
8.	Magicplex Sepsis Real-Time Test Magicplex Sepsis Real-Time Test	Seegne, Seoul, South Korea	Target unknown *Aspergillus* and *Candida* spp.
9.	MycoReal *Aspergillus* MycoReal *Candida*	Ingenetix, Vienna, Austria	ITS2 region *Aspergillus* and *Candida* spp.
10.	MycoGENIE R *Aspergillus* species and MycoGENIE R *Aspergillus fumigatus* and resistance TR34/L98H		28*S* rDNA *Aspergillus* spp., *Aspergillus fumigatus* TR34/L98H mutations
11.	*Pneumocysist jiorovecii* Bio-Evolution PneumoID Real Star *Pneumocystis jirovecii* PCR kit 1.0	Bio-Evolution, Bry-sur-Marne, France OlmDiagnostics, Newcastle, UK Altona Diagnostics, Hamburg, Germany	Target unknown *Pneumocystis jirovecii*
12.	MucorGenius	PathoNostics B.V., Maastricht, Netherlands	Target unknown Pan-Mucorales, *Rhizopus* spp., *Mucor* spp., *Lichtheimia* spp., *Cunninghamella* spp., and *Rhizomucor* spp.

Fungal Genomics in Vaccine Development

Fungal vaccines are an emerging research area as compared to bacterial and viral vaccines. Fungal genomics offers insights into the pathogenic fungi genome, virulence factors, pathogenesis, and host interactions, all of which are critical for designing effective vaccines. The vaccine approaches to fungal infection include live-attenuated vaccines (genetically modified strain), subunit vaccines (recombinant protein), conjugate vaccines, DNA vaccines, and pan-fungal vaccines. In fungal vaccine development, fungal genomics aids in the vaccine candidates identification (gene or protein conserved in pathogens), fungal antigens or epitopes identification (fungal components with high antigenicity and immunogenicity), strain variants analysis (pathogenic strains mutant), and virulence factors identification, understanding of the molecular mechanisms of pathogen–immune system interactions, adjuvant selection, selection of vaccine delivery systems, and monitoring vaccine efficacy (genomic surveillance).

Conclusion

Fungal genomics plays a central role in the advancement of medical mycology. The application of fungal genomics in mycology includes species identification, taxonomy, and systematics, understanding pathogenesis, virulence factors identification, antifungal development, phylogenetics, molecular epidemiology, host–pathogen interaction, diagnosis, and vaccine development. In summary, fungal genomics is a powerful tool, which has transformed medical mycology research from a traditional to a molecular approach.

Case Study

A 50-year-old woman presented with a nodule on her right hand for the past 4 years. The lesion fluctuated in size with time. The patient had a history of immunosuppressive chemotherapy. On examination, a soft tissue mass was observed on the left leg. The MRI reported a hard mass with calcification. The patient was managed clinically, and another MRI was performed after 5 months. The MRI examination reported an enlargement of the lesion. Sarcoma was suspected, and a biopsy sample was sent for microbiology and pathology examination. The histopathology examination reported necrotizing granulomatous inflammation with fungal infection. The fungal culture showed a black color colony after 14 days suggesting a dematiaceous fungus. In microscopy, fungal hyphae and spores were also observed, but fungi could not be identified. The isolate was identified as *Curvularia australiensis* by MALDI-TOF MS, ITS sequencing, and NGS. The fungal infection was successfully resolved by surgical treatment.

1. List the name of the Fungal genome database.
2. What is the revised species name for *Bipolaris australiensis*?
3. What is MALDI Biotyper?
4. What is GenBank Blast?
5. What are the advantages of Sanger sequencing over NGS/WGS?

Research Questions

1. How the advancement in fungal genomics can be used for the development of standard universal fungal nomenclature guidelines?
2. What is the importance of fungal molecular epidemiology in the prevention and control of fungal infection?
3. How the novel molecular marker for fungal virulence can be used for fungal identification and vaccine development?
4. What are the novel molecular markers can be used for fungal typing?
5. How the fungal genomics can be used for the development of molecular tests for POC testing?

Further Reading

Borman AM, Johnson EM (2021) Name changes for fungi of medical importance, 2018 to 2019. Kraft CS, editor. J Clin Microbiol [Internet] 59(2). https://journals.asm.org/doi/10.1128/JCM.01811-20

de Vries RP, Grigoriev IV, Tsang A (2018) Introduction: overview of fungal genomics. Methods Mol Biol 1775:1–7

Fisher MC, Alastruey-Izquierdo A, Berman J, Bicanic T, Bignell EM, Bowyer P et al (2022) Tackling the emerging threat of antifungal resistance to human health. Nat Rev Microbiol 20(9):557–571

Hawksworth DL, Lücking R (2017) Fungal diversity revisited: 2.2 to 3.8 million species. In: The fungal kingdom, pp 79–95

Heitman J, Priest SJ, Yadav V (2020) Advances in understanding the evolution of fungal genome architecture. F1000Res 9

Masumi L, Fakhim H, Vaezi A, Pourhassan-Moghaddam M, Ebrahimi-Kalan A, Zarghami N (2021) Strategies for isothermal amplification of nucleic acids: are they ready to be applied in point of care diagnosis of mycosis? Biointerface Res Appl Chem 11(3):10559–10571

McEwen JG, Taylor JW, Carter D, Xu J, Felipe MSS, Vilgalys R et al (2000) Molecular typing of pathogenic fungi. Med Mycol 38:189–197

Mylonakis E, Clancy CJ, Ostrosky-Zeichner L, Garey KW, Alangaden GJ, Vazquez JA et al (2015) T2 magnetic resonance assay for the rapid diagnosis of candidemia in whole blood: a clinical trial. Clin Infect Dis 60:892–899

Naik B, Ahmed SMQ, Laha S, Das SP (2021) Genetic susceptibility to fungal infections and links to human ancestry. Front Genet 12:709315

Paul S, Singh P, Rudramurthy SM, Chakrabarti A, Ghosh AK (2017) Matrix-assisted laser desorption/ionization-time of flight mass spectrometry: protocol standardization and database expansion for rapid identification of clinically important molds. Future Microbiol 12(16):1457–1466

Paul S, Singh P, Shamanth AS, Rudramurthy SM, Chakrabarti A, Ghosh AK (2018) Rapid detection of fluconazole resistance in *Candida tropicalis* by MALDI-TOF MS. Med Mycol 56(2):234–241

Paul S, Singh P, Sharma S, Prasad GS, Rudramurthy SM, Chakrabarti A et al (2019) MALDI-TOF MS-based identification of melanized fungi is faster and reliable after the expansion of in-house database. Proteomics Clin Appl 13(3):1–8

Procop GW (2007) Molecular diagnostics for the detection and characterization of microbial pathogens. Clin Infect Dis 45(Suppl. 2):99–111

Simner PJ, Uhl JR, Hall L, Weber MM, Walchak RC, Buckwalter S et al (2013) Broad-range direct detection and identification of fungi by use of the PLEX-ID PCR-electrospray ionization mass spectrometry (ESI-MS) system. J Clin Microbiol 51(6):1699–1706

Stajich JE (2017) Fungal genomes and insights into the evolution of the kingdom. Heitman J, Stukenbrock EH, editors. Microbiol Spectr [Internet] 5(4):139–48. https://journals.asm.org/doi/10.1128/microbiolspec.FUNK-0055-2016

Tsang CC, Teng JLL, Lau SKP, Woo PCY (2021) Rapid genomic diagnosis of fungal infections in the age of next-generation sequencing. J Fungi (Basel) 7(8):636

Fungal Proteomics

5

Asmita Sarowgi, Manisha Ghosh, and Rupinder Kaur

Learning Objectives
By the end of the chapter, the readers should be able to:

1. Understand various proteomic approaches being employed in medical mycology.
2. Learn the possible application of proteomics to identify antifungal drug resistance and virulence mechanisms in fungal pathogens of public health importance.

A. Sarowgi
Laboratory of Fungal Pathogenesis, Centre for DNA Fingerprinting and Diagnostics, Hyderabad, India

Graduate Studies, Manipal Academy of Higher Education, Manipal, Karnataka, India

M. Ghosh
Laboratory of Fungal Pathogenesis, Centre for DNA Fingerprinting and Diagnostics, Hyderabad, India

Graduate studies, Regional Centre for Biotechnology, Faridabad, India

R. Kaur (✉)
Laboratory of Fungal Pathogenesis, Centre for DNA Fingerprinting and Diagnostics, Hyderabad, India
e-mail: rkaur@cdfd.org.in

Introduction

The word 'proteomics' refers to the global analysis of all proteins present in a biological system, viz., an organelle, cell, tissue, organism, etc., with this analysis yielding key insights into the abundance, localisation, interactions, functions, structures, and post-translational modifications of proteins. Proteomics is a multi-step methodology that starts with the crucial step of sample preparation, followed by protein separation and identification using various techniques. Proteins are cell's molecular machines that perform enzymatic and structural functions and drive several basic cellular processes. These are made up of 20 common amino acids, which vary in mass, charge, and/or molecular formulae. Protein profiling to map the dynamic cellular proteome is largely based on these differences.

In infection biology, three types of proteomic analyses, namely, expression, structural and functional analyses, are carried out. These provide information on protein expression profiles in (a) patient and healthy control samples, (b) structures of protein core complexes in membranes and other cellular organelles, and (c) biological functions of proteins and protein complexes. These analyses play an important role in the early diagnosis, prognosis, development, and progression of an infection, as well as in biomarker discovery, drug resistance mechanism elucidation, and testing the efficacy of new drugs.

S. C. Parija, S. M. Rudramurthy (eds.), *Textbook of Fungal Zoonoses and Sapronoses*, Microbial Zoonoses, https://doi.org/10.1007/978-981-97-3214-2_5

Proteomic methodologies have revolutionised the field of medical mycology in several aspects, including aiding clinicians in accurate detection and identification of the causative fungal agent of infections. These have facilitated a better understanding of the failure of antifungal drug therapy and the identification of novel virulence traits in clinical isolates of the fungi.

Diseases caused by human fungal pathogens range from superficial infections of the skin, nails, hair, and mucosal surfaces to fatal deep-seated and bloodstream infections. Invasive fungal infections are often associated with prolonged hospital stays, high economic costs, and high (20–95%) mortality. The World Health Organization (WHO) has recently released a list of 19 fungal priority pathogens to guide research, development, and public health action for managing fungal diseases. The development of resistance during antifungal therapy, as well as the emergence of multidrug-resistant fungal species such as *Candida auris*, is a matter of serious concern. Among an estimated six million fungal species, nearly 300 cause human diseases. These fungal species, belonging primarily to six genera: *Aspergillus*, *Cryptococcus*, *Candida*, *Pneumocystis*, *Histoplasma* and *Mucor*, account for most life-threatening invasive fungal infections. Secreted hydrolytic activity, biofilm formation, capsule and melanin formation, metabolic flexibility, and morphological plasticity are the major virulence factors of the fungi.

Proteomic Approaches in Medical Mycology

A broad range of proteomic approaches has been applied to fungal pathogens. These include (a) gel-based techniques, viz., two-dimensional polyacrylamide gel electrophoresis (2D-PAGE), two-dimensional differential gel electrophoresis (2D-DIGE), and gel-free high-throughput techniques, and (b) mass spectrometry stable isotope labelling by amino acids in cell culture (SILAC), isotope-coded affinity tag (ICAT), isobaric tagging for relative and absolute quantitation (iTRAQ), label-free quantitative proteomics, multidimensional protein identification technology (MudPIT), electrospray ionisation mass spectrometry (ESI-MS), and liquid chromatography-tandem mass spectrometry (LC-MS/MS). Additionally, the matrix-assisted laser desorption/ionisation time-of-flight (MALDI-TOF) MS-based identification of pathogenic yeasts and moulds is increasingly being employed in hospitals to reduce the turn-around time while ensuring accurate species-level identification. Furthermore, the three-dimensional structural information on proteins is obtained by X-ray crystallography and nuclear magnetic resonance spectroscopy. Finally, targeted proteomics applies to a limited set of peptides of interest and is often used for biomarker research.

2-Dimensional Polyacrylamide Gel Electrophoresis (2D-PAGE)

In 2D-PAGE, proteins are first separated based on their net charge, followed by their molecular mass-based separation. 2D-DIGE is a modified version of 2-DE (2-dimensional electrophoresis; 2D-PAGE) wherein two or three samples are covalently tagged with different fluorescent dyes and run on the same gel. Proteins that are common among samples are seen as spots and display a fixed fluorescent signal intensity ratio, while differentially present proteins exhibit different fluorescence ratios, thereby allowing the detection of expression differences and posttranslational modifications (PTMs). 2D-PAGE and 2D-DIGE are used for many comparisons, including between disease conditions, untreated and drug-treated samples, and mutant/clinical isolates and *wild-type* strains.

Mass Spectrometry (MS)

MS-based proteomic analysis uses the mass-to-charge ratio (m/z) to detect, identify, and quantify a protein, as mass and charge accuracy is pivotal to the process. Mass spectrometry can be used to identify one protein or thousands of proteins and is applicable to samples of varied complexity. It involves amino acid ionisation to generate charged ions that are then detected and analysed by different mass analysers, including TOF, ion trap, and quadrupole spectrometers. Following data acquisition on mass analysers, several computational database search tools, including SEQUEST, Mascot, and SONAR, are employed for the spectral sequence assignment in an automated, high-throughput fashion. The frequently used data analysis software is listed in Table 5.1.

Notably, with the advent of new technologies, proteins in complex biological samples can be identified at subattomolar concentrations. The commonly used ionisation and mass spectrometry techniques are described below.

Ionisation Techniques

Two main ionisation methods, electrospray ionisation (ESI) and matrix-assisted laser desorption/ ionisation (MALDI), are used to convert neutral amino acids in the protein from the liquid phase to the desolvated gaseous ions.

ESI

ESI ionises proteins via soft ionisation by multiple charging into small droplets and usually causes no fragmentation. ESI generates multiple-charged ions with m/z values substantially lower than their masses, thereby permitting detection via a broad range of mass analysers.

MALDI

In this soft ionisation technique, the sample is mixed/coated with an energy-absorbent matrix

Table 5.1 List of commonly used software for data analysis and storage in proteomics

Application	Software	URL
Identification and visualization of MS proteomics data	Peptide Shaker Open MS	http://compomics.github.io/projects/peptide-shaker http://www.openms.de
Visualization of raw data, identification of peptide, validation, data storage and mining, biological inference	Raw Meat TPP	https://rawmeat.software.informer.com/2.0/ http://www.tppms.org/ SPC Proteomics Tools (proteomecenter.org)
De novo sequencing	PEAKS	https://www.bioinfor.com/peaks-studio/
Analysis and identification of PTM sites	delta Masses Scaffold PTM	https://github.com/chhh/deltamass https://www.proteomesoftware.com/products/scaffold-ptm
Open-source software for DIA data, detection of precursor and fragment chromatographic features	DIA-Umpire	https://diaumpire.sourceforge.net/
Metabolomics data analysis	MetaboAnalyst	https://www.metaboanalyst.ca/
Analysis and visualisation of target sites	MultiQuant Skyline	MultiQuant Software (sciex.com) https://skyline.ms/project/home/software/Skyline/begin.view
Visualisation of quantitative proteomics	MS Quant MaxQuant IsobariQ Proteome Discoverer	https://msquant.sourceforge.net/ https://www.maxquant.org/ https://bio.tools/IsobariQ http://www.thermoscientific.com
Graphs and networks	CytoScape Gephi	https://cytoscape.org/ https://gephi.org/
Repositories for visualization of proteome databases	PRIDE PeptideAtlas Global Proteome Machine Database (GPMD) neXtProt Proteomics DB	PRIDE—Proteomics Identification Database (ebi.ac.uk) https://peptideatlas.org/ https://www.re3data.org/repository/r3d100010883 https://www.nextprot.org/ https://www.proteomicsdb.org/

solution and dried, leading to sample and matrix co-crystallisation. Following laser beam-based ionisation of the matrix that transfers charge to analytes in the sample and generates singly-charged ions, the ions are accelerated and analysed based on their m/z ratio via the TOF mass analyser.

LC-MS/MS

This technique is liquid chromatography coupled with tandem mass spectrometry and involves peptide fractionation by high-pressure LC before ionisation and MS analysis.

MS Work-Flows

The MS-based proteomic analysis is carried out using either bottom-up or top-down work-flows, as discussed below.

Bottom-Up Approach

The bottom-up approach analyses peptides that are products of cleaved proteins. In this, mostly trypsin, occasionally other proteases, are used to digest proteins in solution or resolved proteins in SDS-PAGE gel, followed by resultant peptide ionisation, ion separation based on their m/z ratio, and ion detection. The bottom-up MS analysis performed on a mixture of proteins is referred to as shotgun proteomics.

Top-Down Approach

The top-down approach analyses intact undigested proteins, thereby detecting biologically active protein forms and their proteoforms (all different molecular forms of a protein). Due to the difficulty in gas-phase ionisation of an intact protein, the top-down approach is more suited to proteins of $<$70 kDa size.

MS-Based Quantitative Proteomics

MS-based quantification of proteins is based on MS1 and MS/MS spectra and is largely done by three methods, as described below.

Metabolic Labelling (SILAC)

This method involves metabolic labelling wherein isotope (2H, 13-C, 15-N, etc.)-containing amino acids are incorporated into proteins during protein synthesis, followed by mixing of this isotopically-labelled (heavy) sample with the unlabelled-protein containing sample (light) and LC-MS/MS-based relative quantitation of digested/undigested proteins.

Chemical Labelling

ICAT and iTRAQ are chemical labelling methods that label specific chemical groups with isobaric stable isotopes and use isotope labels to derivatise peptides/proteins from various samples. For iTRAQ, digested peptides are mostly used, as these contain a primary amino group, which can be labelled with different tags for quantification. iTRAQ involves labelling proteins with isobaric tags at the N terminus of peptides and at the side chain of the lysine residue. ICAT quantifies only cysteine-containing peptides and proteins, relying on labelling with the cysteine-reactive chemical reagent with a biotin tag and a linker region containing either zero or eight deuterium atoms.

Label-Free Quantification (LFQ)

As the name suggests, this cost-efficient technique does not require any label or mass tag and quantifies proteins, normally at the MS1 level, based on AUC (area under the curve), signal intensity measurements, or spectral counts.

The frequently used proteomic approaches for analysing fungal physiology and virulence are presented in Fig. 5.1.

Proteomic Applications in Medical Mycology Research

Genomes of the common fungal pathogens such as *Candida*, *Aspergillus*, and *Cryptococcus* species contain about 5000 to 12,000 protein-encoding genes. Over the last two decades, several studies have been carried out to profile the global and cell wall proteomes as well as proteomes of extracellular vesicles, drug-resistant isolates, biofilm-forming forms, culture supernatant, etc., of medically important fungi. Additionally, protein profiling in response to

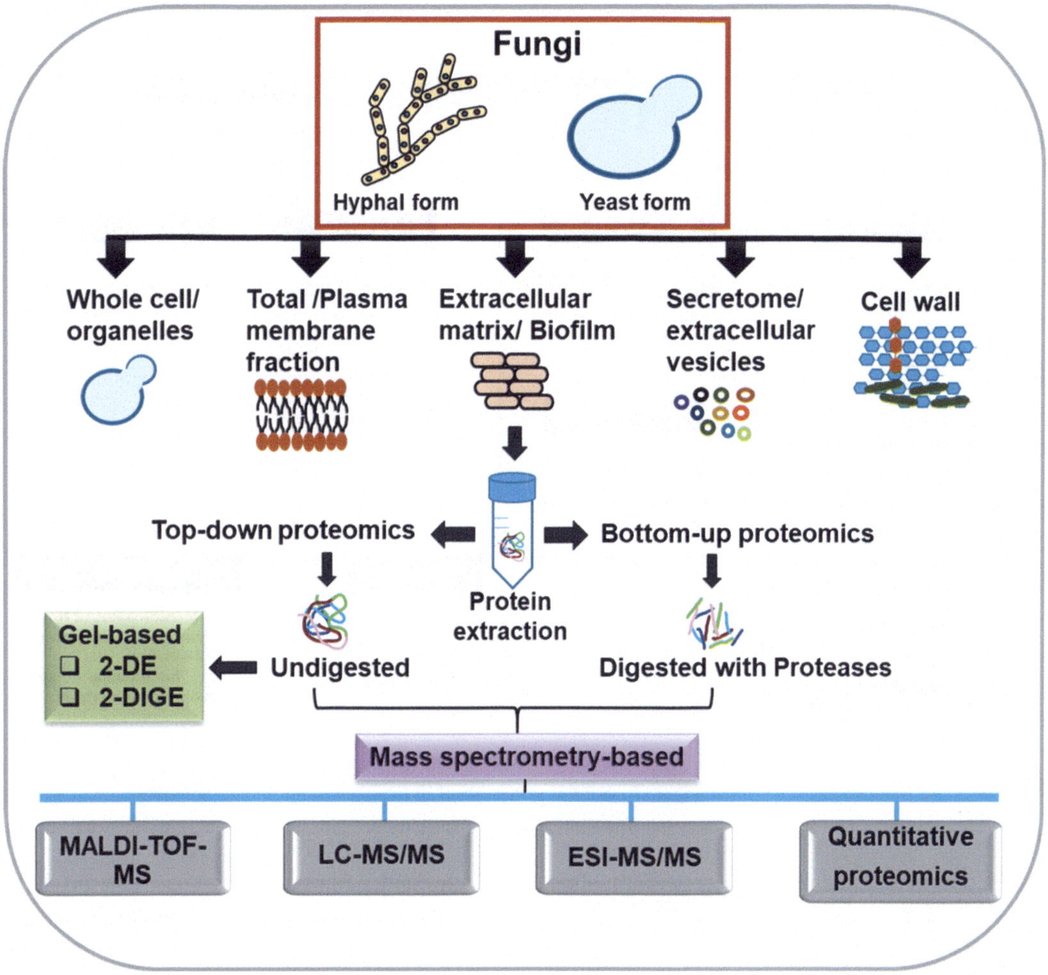

Fig. 5.1 A schematic outlining commonly used proteomic approaches

varied stresses, including oxidative stress, has been studied, which has yielded mechanistic insights into the biology of human fungal pathogens. The key findings of proteomic analyses in major pathogenic species of *Candida*, *Cryptococcus*, and *Aspergillus* are summarised below. These findings are also depicted in Fig. 5.2.

Candida: *Candida albicans*, *Candida glabrata*, *Candida tropicalis*, *Candida parapsilosis*, and *C. auris* are important causative agents of invasive infections. Among these, *C. albicans* and *C. glabrata* proteomes have extensively been profiled under several conditions. The initial LC-MS/MS analysis identified 14 covalently-linked cell wall proteins in the cell-wall proteome analysis of logarithmic-phase *C. albicans*. Furthermore, the proteome analysis of extracellular vesicles and vesicle-free fractions in *C. albicans* identified 75 and 61 proteins, respectively, that included metabolic and cell wall proteins. This analysis also shed light on the classical and non-classical secretion pathways and verified the vaccine potential of one of the secreted immunogenic proteins, Bgl2, against invasive *C. albicans* infections. Notably, the label-free quantitative MS analysis of *C. albicans* response to the osmotic stress highlighted the role of glycerol accumulation in salt adaptation, as upregulated proteins in stressed—*C. albicans* cells included metabolic enzymes linked with glycerol synthesis. SILAC analysis has also revealed

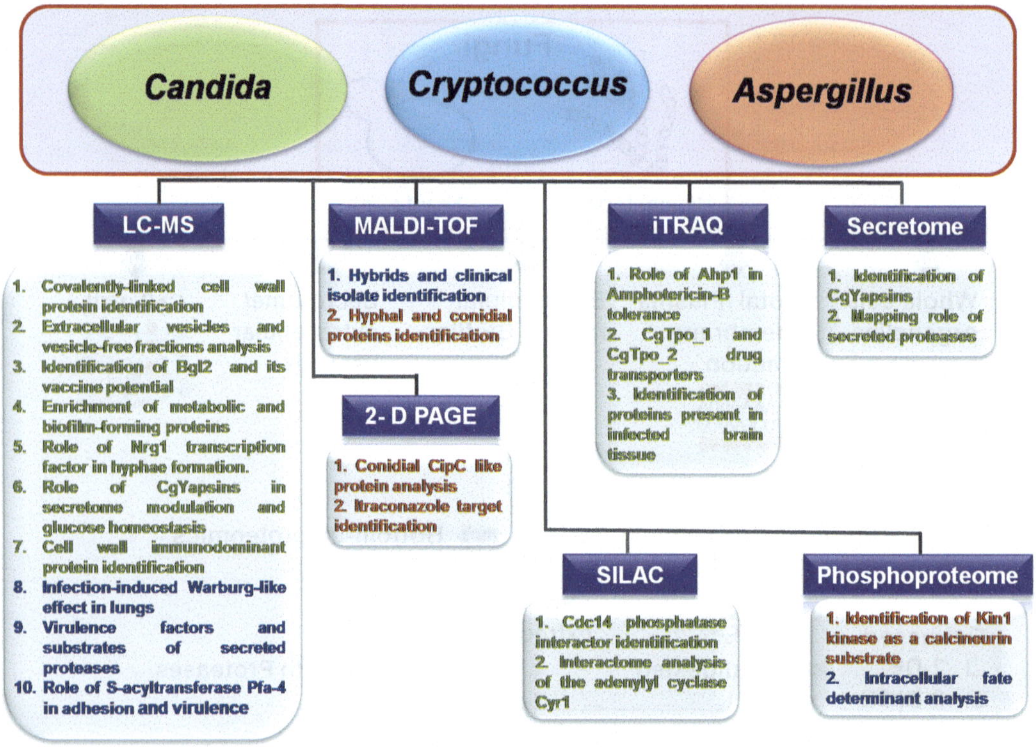

Fig. 5.2 Contributions of proteomic analyses to cellular processes of indicated human fungal pathogens

interactors of the key cell cycle regulator phosphatase Cdc14 and the sole adenylyl cyclase Cyr1. Biofilm proteome profiling of *C. albicans* has revealed cell wall structural and stress-responsive proteins. Similarly, the mixed biofilm analysis of *C. albicans* and *Streptococcus mutans* highlighted the contribution of *C. albicans* to sugar metabolism and environmental acidification.

C. albicans is known to evade macrophages by undergoing morphological switching. The first proteomic analysis of macrophage-internalised *C. albicans* revealed downregulation and upregulation of carbon metabolism and protein degradation processes, respectively. Also, dual proteomic analysis of *C. albicans*–macrophage interaction revealed altered levels of 227 *C. albicans* and 5 macrophage proteins, thereby underscoring active *C. albicans* breakout mechanisms. Similarly, LC-MS/MS analysis revealed enrichment of many proteins, including metabolic enzymes and biofilm-formation proteins, and elevated pathogenicity of *C. albicans* in a serial disseminated candidiasis model. Recently, Far-Western (detects direct interactions among natively structured proteins) and tandem MS analysis revealed *C. albicans*-induced multi-component complex information that is important for *C. albicans* endocytosis by oral epithelial cells and during oropharyngeal candidiasis.

Like *C. albicans*, proteome profiling has elucidated various pathobiological aspects of *C. glabrata* and another *Candida* spp. The proteogenomics (proteomics application to existing genome annotation)-based profiling of *C. glabrata* proteome verified 4421 proteins and revealed novel genes and mRNA isoforms. Proteome profiling of biofilm-forming *C. glabrata* and the biofilm matrix revealed stress-responsive proteins and virulence factors, viz., cell surface adhesins and proteases, in the biofilm-state and biofilm-matrix proteomes, respectively. Similarly, the cell wall protein profiling of hyper-adhesive clinical isolates of *C.*

glabrata underscored the differential incorporation of adhesins in the cell wall. Intriguingly, 2D-LC-MS/MS of non-glucan attached cell surface proteins from biofilm (in human skin-mimicking synthetic sweat medium)-forming and planktonic *C. auris* cells revealed cell surface localisation of the Nrg1 transcription factor, pointing towards dual roles for Nrg1.

Notably, proteomes of different *Candida* species have also simultaneously been profiled and compared under different conditions, including oxidative stress. These had yielded key insights into the functions of cell-wall proteins that provided defence against reactive oxygen species. Furthermore, LC-MS/MS analysis identified 2591 and 780 proteins in cell lysate and conditioned medium, respectively, of *C. tropicalis*, which also included many unique secreted proteins. Similarly, the secretome and the total membrane proteome analysis of *C. glabrata*, via LC-MS/MS and LFQ techniques, unveiled the unanticipated roles of CgYapsins (aspartyl proteases) in the secretion pathway and glucose homeostasis regulation. Furthermore, LC-MS analysis of pathogenic and multidrug-resistant *C. parapsilosis* strains identified 933 cell wall antigenic proteins with immunodominant proteins. These proteins were shown to provide protection to the mice against *C. parapsilosis* infections, thereby posing these proteins as good biomarkers and/or vaccine candidates.

Cryptococcus: *Cryptococcus neoformans* and *Cryptococcus gattii* are two important species that cause cryptococcosis. MALDI-TOF analysis has successfully differentiated various strains of these two species at the intra- and inter-species level. MudPIT (Multidimensional Protein Identification Technology) and 2D-LC-MS/MS-based shotgun proteomics analysis of the lungs of *C. gattii*-infected rats identified several proteins. The study has shown that *C. gattii* infection induces a Warburg-like effect in the lungs, thereby shedding new light on *C. gattii* pathogenesis traits. The bottom-up and targeted proteomics have revealed the potential biomarkers of cryptococcal infection. These secretory proteins were detected and analysed in *C. neoformans* and were also demonstrated in the lysates of *C. neoformans*-infected macrophages and blood and bronchoalveolar lavage of mice.

Similarly, to identify potential therapeutics, substrates of secreted proteases that may be pivotal to *C. neoformans* virulence were identified and cleavage specificities were mapped using secretome samples via MS-based multiplex substrate profiling. Proteome comparison of the planktonic and biofilm-forming *C. neoformans* forms identified 76 proteins. These proteins were either uniquely represented or more abundant in the biofilm form of *C. neoformans*. Some of these identified proteins were involved in oxidative stress, metabolism, and proteolytic activity regulation. The comparative proteome profiling of spore and yeast forms of *C. neoformans* led to the identification of 18 spore-specific proteins. The phosphoproteome analysis of calcineurin (conserved calcium-dependent serine-threonine-specific protein phosphatase)-deficient and calcineurin-activated *wild-type* cells identified 56 targets of calcineurin. All these findings provided mechanistic insights into how calcineurin governs growth, thermal stress survival, sexual cycle, and virulence in *C. neoformans*. MS-based proteomics, in combination with CLICK chemistry and palmitoylome-profiling approaches, identified 72 host substrates of the fungal protein S-acyltransferase Pfa4, which is important for *C. neoformans* adhesion and virulence.

The iTRAQ has been applied to study the brain tissues (frontal lobe) of cryptococcal meningitis patients, which led to the identification of 317 proteins. These proteins exhibited differential abundance in patient samples and may facilitate the pathogen invasion of the host brain. Phosphoproteomic analysis of *C. neoformans*-infected macrophages revealed that phosphorylation of components of the AMP-activated protein kinase-autophagy initiation complex regulates the intracellular fate of *C. neoformans*. LC-MS/MS analysis of lung and brain tissues of *C. neoformans*-infected C57BL/6 mice revealed the enrichment of many pathways, including Fcγ R-mediated phagocytosis, calcium signalling, B-cell receptor signalling, and platelet activation in the lung proteome, thereby shedding light on the host responses to *C. neoformans* infections.

Aspergillus: *Aspergillus fumigatus* and *Aspergillus flavus* are two important *Aspergillus* species. Proteome profiling in *wild-type* and ergothioneine (trimethylated and sulphurised histidine derivative) biosynthesis-deficient mutants of *A. fumigatus*, using LFQ proteomics, has unveiled the anti-oxidant function of ergothioneine. 2-DE and MALDI-TOF analysis of hyphae and resting and germinating conidia showed CipC-like protein to be exclusive to the hyphal form of *A. fumigatus*. The cell wall proteomic analysis of the *A. fumigatus* resting conidia identified a new cell surface protein, CcpA, which was important for conidia recognition by the immune system and virulence of *A. fumigatus*. Furthermore, the phosphoproteome analysis successfully established the Kin1 kinase as a substrate of calcineurin in *A. fumigatus*.

The analysis of host-*A. fumigatus* interaction has revealed an altered abundance of 153 proteins in primary human bronchial epithelial cells upon exposure to *A. fumigatus* conidia, with 73 and 80 proteins showing increased and decreased levels, respectively. Similarly, the simultaneous detection of host and *A. fumigatus* proteins (dual proteome profiling) in the phagolysosomes of *A. fumigatus* conidia-infected macrophages revealed a differential abundance of 637 host and 22 *A. fumigatus* proteins, thereby aiding in mapping host immune cell–*A. fumigatus* conidia interactions. Recently, the proteogenomic analysis in *A. flavus* has been carried out, which led to the profiling of the global proteome, mapping of PTMs, and identification of many novel proteins that may control the production of the major toxin aflatoxin in *A. flavus*.

In essence, these studies emphasise the potential applications of proteomics in the study of the pathogenesis and physiology of fungal infections.

Antifungal Drug Resistance

Several proteomic analysis studies have been carried out to understand the mechanisms of drug resistance in human pathogenic fungi. For example, iTRAQ analysis identified 100 differentially abundant proteins in biofilm-forming haploid and diploid *C. albicans* strains. This study showed that the alkyl hydroperoxide reductase 1 (Ahp1) contributes to maintaining the anti-oxidant capability of the persister population and governs amphotericin B (ergosterol-binding polyene antifungal) tolerance in *C. albicans*. In another study, the iTRAQ analysis of membrane-enriched fraction in clotrimazole (ergosterol biosynthesis-inhibitory azole drug)-treated *C. glabrata* cells led to the identification of two new drug transporters, CgTpo_1 and CgTpo_2. Furthermore, the nano-LC-ESI-MS analysis showed enrichment of cell wall organisation and ribosome proteins in caspofungin (cell wall-targeting echinocandin antifungal)-treated *C. auris* cells, with caspofungin treatment also resulting in elevated production of extracellular vesicles. The 2-DE-MS identified multiple targets of itraconazole (azole drug) in *A. fumigatus*. In contrast, quantitative ESI-MS/MS of naphthofuranquinones-treated fluconazole-resistant *C. tropicalis* strains unveiled oxidative stress protein upregulation, with naphthofuranquinones elevating reactive oxygen species production and causing cell death.

Extracellular vesicles prepared from a triazole-resistant, amphotericin B-susceptible strain promoted the amphotericin B resistance in *C. auris* but not in *C. albicans*, thereby underscoring the species-specific response to the amphotericin B. The quantitative proteomic analysis of the cell walls of 13 human pathogenic fungal species, including clinical strains, revealed 42 of the most abundant proteins. These proteins shared little or no homology with human proteins, with 1,3-β-glucanosyltransferases emerging as potential targets for vaccine development. The comparative cell wall proteome analysis of caspofungin-resistant and sensitive isolates has recently been used to identify biomarkers for diagnosing echinocandin resistance. In essence, the proteomic analyses have significantly enhanced the understanding of how fungal pathogens maintain fitness and proliferate under varied stress conditions, including antifungal drugs and the hostile host environment. In addition, it has also uncovered many potential new drugs, vaccines, and biomarker candidates.

Despite wide applications in fungal pathogenesis, the emerging field of proteomics is not without pitfalls. The proteomic analysis of fungal samples has many limitations. These include (a) the complex nature of samples especially patient sera, (b) insufficient sensitivity to detect low-abundant proteins, (c) incomplete protein separation and/or lack of a broad dynamic range for accurate quantification, (d) unstable nature of some post-translational modifications, such as phosphorylation and SUMOylation during sample processing, (e) under-representation of cell wall proteins in commonly-used cell lysate samples, and (f) loss of protein–protein interactions during sample processing and unavailability of comprehensive reference databases.

Conclusions

Proteomic analyses have positively impacted all key aspects of fungal pathogenesis, (a) unveiling novel pathogenicity mechanisms, (b) identifying new targets for drugs and vaccines, and (c) shedding light on vegetative, sexual, and biofilm-forming forms, etc. The clinical mycology field is currently poised to benefit from recent advances in high-throughput proteomic and bioinformatic technologies. The dual proteome profiling of the fungal pathogen and the host would potentially provide unrivalled insights into spatial and temporal responses during an infection. The simultaneous employment of proteomic, immunologic, and metabolomic methodologies will delineate the role of (a) metabolic reprogramming in shaping the host immune response towards a fungal infection and (b) nutritional immunity in restricting fungal pathogenicity. The advanced mass spectrometry techniques can potentially be applied to detect protein–protein, protein–lipid, protein–carbohydrate, and protein–DNA interaction in clinical samples as well as during the culture of fungi. This would facilitate understanding the multi-pronged interactions among major cellular biomolecules as well as among human commensals.

Case Study

A diabetic female in her 60s was admitted to a hospital as she was tachycardic (108 beats per minute) and in need of oxygen supplementation. Previously, this patient was diagnosed with COVID-19, incompletely vaccinated, and was treated with remdesivir, dexamethasone, and oxygen supplementation via nasal canula at another medical facility. She was discharged after 7 days. However, soon after discharge, she developed a dry cough, dyspnoea, and fatigue, which led to appetite and weight loss. After 10 days of discharge from the initial hospitalisation, she was readmitted to a hospital, as she was tachycardic and required oxygen supplementation. She also had hyperglycaemia and low albumin, with leukocyte count, platelet count, and haemoglobin levels of 6400 cells/cubic mm, 433,000 cells/cubic mm, and 9 g/dL, respectively.

The patient's chest radiograph suggested multifocal pneumonia and a left-sided empyema thoracis. Her blood culture was negative, but Gram staining of the pleural fluid disclosed Gram-positive cocci in pairs, Gram-negative bacilli, and yeast with *Candida*-like morphology, suggesting a polymicrobial infection. Culturing of the pleural fluid revealed extended-spectrum β-lactamase-producing *Klebsiella pneumoniae* and *Enterococcus faecalis*. Initially, the yeast was identified, based on the automated biochemical method (BD Phoenix), as *Candida firmetaria*; however, MALDI-TOF MS analysis identified the yeast isolate to be *C. glabrata*.

Empiric treatment with intravenous vancomycin (1 g every 24 h) and piperacillin-tazobactam (4.5 g every 8 h) was initially provided to the patient. After Gram staining results, intravenous micafungin (100 mg every 24 h) was added to the treatment regimen on day two of admission. After the isolates' drug susceptibility profile analysis, piperacillin-tazobactam was replaced with intravenous meropenem (500 mg every 6 h) on day three. After treating the patient with micafungin for 4 days, antifungal therapy was changed to oral voriconazole tablet (300 mg every 12 h) for two loading doses, followed by 200 mg every

12 h to increase antifungal bioavailability in the pleural space. Although this treatment significantly improved the patient's respiratory symptoms, she expired due to acute cardiac arrest on the tenth of admission.

1. What are the risk factors for *Candida* empyema?
2. Which drugs are recommended for the treatment of *C. glabrata* infections?
3. How can polymicrobial infections be identified?
4. How does MALDI-TOF aid in fungal pathogen detection?

Research Questions

1. What kind of models can be used to study the pharmacokinetics of systemic antifungal drugs for the treatment of *Candida* empyema?
2. What is the optimal treatment for fungal empyema?
3. How can proteomic approaches help in the accurate diagnosis of the infectious agent?
4. Does the antifungal drug regimen impact the prevalence of *Candida* species?

Further Reading

Aebersold R, Mann M (2016) Mass-spectrometric exploration of proteome structure and function. Nature 537:347–355. https://doi.org/10.1038/nature19949

Datta KK, Patil AH, Patel K, Dey G, Madugundu AK, Renuse S, Kaviyil JE, Sekhar R, Arunima A, Daswani B et al (2016) Proteogenomics of *Candida tropicalis*—an opportunistic pathogen with importance for global health. Omics 20:239–247. https://doi.org/10.1089/omi.2015.0197

de Groot PWJ, de Boer AD, Cunningham J, Dekker HL, de Jong L, Hellingwerf KJ, de Koster C, Klis FM (2004) Proteomic analysis of *Candida albicans* cell walls reveals covalently bound carbohydrate-active enzymes and adhesins. Eukaryot Cell 3:955–965. https://doi.org/10.1128/EC.3.4.955-965.2004

Geddes JMH, Croll D, Caza M, Stoynov N, Foster LJ, Kronstad JW (2015) Secretome profiling of *Cryptococcus neoformans* reveals regulation of a subset of virulence-associated proteins and potential biomarkers by protein kinase A. BMC Microbiol 15:206. https://doi.org/10.1186/s12866-015-0532-3

Gil-Bona A, Amador-García A, Gil C, Monteoliva L (2018) The external face of Candida albicans: a proteomic view of the cell surface and the extracellular environment. J. Proteomics 180:70–79. https://doi.org/10.1016/j.jprot.2017.12.002

Gong X, Srivastava V, Naicker P, Khan A, Ahmad A (2023) *Candida parapsilosis* cell wall proteome characterization and effectiveness against hematogenously disseminated candidiasis in a murine model. Vaccines 11:674. https://doi.org/10.3390/vaccines11030674

Graves PR, Haystead TAJ (2002) Molecular biologist's guide to proteomics. Microbiol Mol Biol Rev 66:39–63. https://doi.org/10.1128/MMBR.66.1.39-63.2002

Han X, Aslanian A, Yates JR (2008) Mass spectrometry for proteomics. Curr Opin Chem Biol 12:483–490. https://doi.org/10.1016/j.cbpa.2008.07.024

Meleady P (2018) Two-dimensional gel electrophoresis and 2D-DIGE. In: Ohlendieck K (ed) An automated irrigation system using Arduino microcontroller, vol 1664. Methods in molecular biology. Springer New York, New York, pp 3–14. isbn: 978-1-4939-7267-8. https://doi.org/10.1007/978-1-4939-7268-5_1

Muselius B, Durand SL, Geddes-Mcalister J (2021) Proteomics of *Cryptococcus neoformans*: from the lab to the clinic. Int J Mol Sci 22:12390. https://doi.org/10.3390/ijms222212390

Pais P, Costa C, Pires C, Shimizu K, Chibana H, Teixeira MC (2016) Membrane proteome-wide response to the antifungal drug clotrimazole in *Candida glabrata*: role of the transcription factor CgPdr1 and the drug:H^+ antiporters CgTpo1_1 and CgTpo1_2. Mol Cell Proteomics 15:57–72. https://doi.org/10.1074/mcp.M114.045344

Paudyal A, Vediyappan G (2021) Cell surface expression of Nrg1 protein in *Candida auris*. J Fungi 7:262. https://doi.org/10.3390/jof7040262

Prasad TSK, Harsha HC, Keerthikumar S, Sekhar NR, Selvan LDN, Kumar P, Pinto SM, Muthusamy B, Subbannayya Y, Renuse S et al (2012) Proteogenomic analysis of *Candida glabrata* using high resolution mass spectrometry. J Proteome Res 11:247–260. https://doi.org/10.1021/pr200827k

Ramírez-Quijas MD, López-Romero E, Cuéllar-Cruz M (2015) Proteomic analysis of cell wall in four pathogenic species of *Candida* exposed to oxidative stress. Microb Pathog 87:1–12. https://doi.org/10.1016/j.micpath.2015.07.011

Rasheed M, Kumar N, Kaur R (2020) Global secretome characterization of the pathogenic yeast *Candida glabrata*. J Proteome Res 19:49–63. https://doi.org/10.1021/acs.jproteome.9b00299

Ribeiro GF, Denes E, Heaney H, Childers DS (2021) What omics can tell us about antifungal adaptation. FEMS Yeast Res 21:1–15. https://doi.org/10.1093/femsyr/foab070

Schmidt H, Vlaic S, Krüger T, Schmidt F, Balkenhol J, Dandekar T, Guthke R, Kniemeyer O, Heinekamp T, Brakhage AA (2018) Proteomics of *Aspergillus fumigatus* conidia-containing phagolysosomes identifies processes governing immune evasion. Mol Cell Proteomics 17:1084–1096. https://doi.org/10.1074/mcp.RA117.000069

Smeekens SP, Van De Veerdonk FL, Netea MG (2016) An omics perspective on *Candida* infections: toward next-generation diagnosis and therapy. Front Microbiol 7:1–8. https://doi.org/10.3389/fmicb.2016.00154

Swaminathan N, Anderson K, Nosanchuk JD, Akiyama MJ (2022) *Candida glabrata* empyema thoracis—a post-COVID-19 complication. J Fungi (Basel) 8:923. https://doi.org/10.3390/jof8090923

Yang M, Zhu Z, Zhuang Z, Bai Y, Wang S, Ge F (2021) Proteogenomic characterization of the pathogenic fungus *Aspergillus flavus* reveals novel genes involved in aflatoxin production. Mol Cell Proteomics 20:100013. https://doi.org/10.1074/mcp.RA120.002144

Epidemiology of Fungal Zoonoses

6

Shamanth A. Shankarnarayan and Dipika Shaw

Learning Objectives
By the end of the chapter, the readers should be able to:

1. Understand the epidemiology of common fungal zoonoses.
2. Recognise the ecological niche of zoonotic fungal pathogens and their transmission.
3. Choose the appropriate strategies for the investigation of zoonotic fungal outbreaks, responsible agents, and their prevention.

Introduction

Zoonoses are infections that are naturally transmitted to humans from animals and to animals from humans. Earlier, the disease transmitted from animals to humans was referred to as "anthropozoonoses" and humans to animals as "zooanthroponoses." Due to the interchangeable usage of these terminologies, the expert committee recommended using "zoonoses" as "diseases and infections are naturally transmitted between animals and humans." Zoonoses can be further classified depending on the ecosystem they circulate. Zoonoses with urban (domestic) cycle are known as "synanthropic" zoonoses, where domestic and synanthropic animals (urban rabies, cat scratch disease, and zoonotic ringworm) are involved. Zoonoses outside human habitat (e.g., arboviruses, wildlife rabies, Lyme disease, and tularemia) with sylvatic (feral and wild) life cycles are known as "exoanthropic" zoonoses. Furthermore, some zoonotic infections have both urban and natural cycles (e.g., yellow fever and Chagas disease). Fungal zoonoses, once considered a rare entity, are currently being reported worldwide. Until 1990, fungal infections were primarily observed as plant-infecting agents challenging agricultural yields and food security. Subsequently, fungal infections amongst animals started increasing spatially and temporally and in terms of virulence. Fungi usually survive in the environment for a prolonged duration, but a few may select an appropriate vertebrate host as part of their life cycle. Systemic infections in immunocompetent hosts are usually mild, but they tend to be severe and often fatal in immunocompromised hosts.

Classification of Fungal Parasitism

Fungal parasitism is classified as follows:

Obligate fungal parasite: Obligate fungal parasites are those that cannot survive or complete

S. A. Shankarnarayan (✉)
Department of Physics, University of Alberta, Edmonton, AB, Canada
e-mail: shamanth.adekhandi@gmail.com

D. Shaw
Department of Microbiology, Jawaharlal Nehru Medical College, Datta Meghe Institute of Higher Education and Research, Wardha, Maharashtra, India

S. C. Parija, S. M. Rudramurthy (eds.), *Textbook of Fungal Zoonoses and Sapronoses*, Microbial Zoonoses, https://doi.org/10.1007/978-981-97-3214-2_6

their life cycle without a host. These fungi depend entirely on their host organisms for nutrients and shelter. Obligate fungal parasites typically infect plants and cause various diseases, such as rusts, smuts, and powdery mildews. They penetrate the host tissues and extract nutrients, ultimately causing damage and disease symptoms.

Facultative fungal parasite: Facultative fungal parasites can exist both as parasites and as free-living organisms. These fungi can survive and reproduce outside their hosts and, under certain conditions, cause opportunistic infections in susceptible hosts. For example, certain *Candida* can exist as commensal organisms in the human host body but can cause infection in the same host, following a weakened immune system or disruption of the body's natural microflora.

Accidental/Incidental fungal parasites: Accidental or incidental fungal parasites are those that infect hosts that are not their usual or preferred hosts. These fungi may enter an unusual host and may be able to initiate some degree of development but cannot complete their life cycle or establish a persistent infection. For example, *Aspergillus fumigatus*, which is found in soil and decaying organic substances, can cause opportunistic infections in humans, particularly in individuals with compromised immune systems.

Aberrant fungal parasites: Aberrant or wandering fungal parasites are those that enter a host organism other than their usual host but fail to survive or develop further. These fungi may enter an atypical host and reach a site where they cannot establish a suitable environment for growth or reproduction. *Pneumocystis jirovecii* is an example of such an aberrant fungal parasite that is typically found in the lungs of immunocompromised individuals but cannot thrive in individuals with a healthy immune system.

Opportunistic pathogens: The animal body can be considered as an extreme environment for fungal growth. If the fungus occupies an environmental niche but can survive within the animal's host tissues because of the factors required in that environmental niche happen to enhance survival within the host, the infection is said to be opportunistic. Only a few of the species known in the fungal kingdom can cause infection. *Candida* normally resides on the skin, mucous membranes, and in the gastrointestinal tract of healthy individuals. However, under certain conditions, *Candida* species can overgrow and cause infections. These opportunistic infections can occur in individuals with weakened immune systems, such as those with HIV/AIDS, undergoing chemotherapy, or taking long-term immunosuppressive medications. *Candida* infections can affect various body parts, including the mouth (oral thrush), vagina (vaginal yeast infections), and bloodstream (candidemia).

Although fungi can exhibit parasitic behaviours, not all fungi are parasites. Many fungi are saprophytes that decompose dead organic matter, while others have mutualistic relationships with their hosts, providing benefits to both parties.

Transmission of Fungal Infections

In a meta-analysis study, a total of 307 fungal agents among 1415 pathogens were found to cause infections in humans. Of these pathogens, the majority (61%) were zoonotic infections, of which zoonotic fungal infections comprised 38%. Nevertheless, some of the fungal infections caused by *Aspergillus, Alternaria, Acremonium, Bipolaris, Curvularia, Mucor, Paecilomyces*, and *Rhizopus*, which were classified earlier as zoonoses, are being considered as sapronoses (acquired from the saprophytic environment). Certain fungal infections, such as cryptococcosis and coccidioidomycosis caused by *Cryptococcus* and *Coccidioides*, respectively, can grow on non-living entities by metabolising nitrogen-rich substrates. However, they are also found in pigeon faeces and small-mammal carcasses, which are associated with human infections and hence can be considered "true" zoonoses. Dermatophytosis is a classic example of animal-to-human transmission of fungi from pet animals to humans. Currently, sporotrichosis caused by *Sporothrix brasiliensis* in the feline population has emerged as a true zoonosis that affects humans.

Fungal zoonoses are prevalent worldwide and can occur in both tropical and temperate regions. Animals, including mammals, birds, reptiles, and amphibians, can act as reservoirs for fungal pathogens. The transmission dynamics of fungal zoonoses vary depending on the specific fungal pathogen. Transmission of the infection can be either direct contact or indirect. Some fungi are primarily transmitted through direct contact with infected animals, such as dermatophytes causing ringworm infections. Other fungal pathogens can be transmitted to humans through direct contact with infected animals, inhalation of fungal spores, consumption of contaminated food or water, or bites from infected arthropods. For example, the fungus *Sporothrix schenckii* can be transmitted through scratches or bites from cats or handling contaminated soil, leading to a localised infection known as sporotrichosis. Ringworm infection is another zoonotic fungal infection that can be transmitted from infected animals, such as dogs, cats, or farm animals, to humans by direct contact.

Direct Transmission

Fungal infections can be transmitted through various routes, including direct contact, inhalation of fungal spores, ingestion, or exposure to contaminated environmental sources.

Direct contact: Direct contact with infected persons or contaminated surfaces is a common mode of transmission of many fungal infections. Fungal pathogens can be present on the skin, nails, or mucous membranes of infected individuals. When a healthy person meets these infected areas, either through touch or sharing personal items like towels or clothing, the fungi can be transferred. Dermatophyte fungi, which cause infections like athlete's foot and ringworm, are often spread through direct contact. *Candida albicans* can be transmitted through sexual contact or from mother to newborn during childbirth.

Inhalation of fungal spores: Certain fungi produce microscopic spores, which are present in soil, decaying organic matter, or indoor spaces with mould growth. Inhalation of these air-borne spores causes infections in humans and animals such as aspergillosis, histoplasmosis, or coccidioidomycosis.

Aspergillus species are commonly found in the environment, particularly in soil, compost, and decaying vegetation. Inhalation of *Aspergillus* spores can cause allergic or invasive pulmonary aspergillosis. *Histoplasma capsulatum* and *Coccidioides immitis* are other examples of fungi that also cause respiratory infections by inhalation. Certain occupations, like agriculture, construction, or forestry, where workers are exposed to contaminated environments, may have a higher risk of inhaling fungal spores, thus suffering from these fungal infections.

Exposure to Contaminated Environment

Fungi are widespread in the environment and contaminate various sources, including soil, water, and air. They are present in warm and moist environments, including public showers, swimming pools, and gymnasiums. Inhalation of fungal spores present in indoor environments with poor ventilation and damp conditions, such as basements or bathrooms, can also contribute to fungal infections.

Exposure to hospital environments: Healthcare-associated fungal infections can also occur in hospital settings. Fungi colonise various surfaces, including medical equipment, linens, and air conditioning systems. Immunocompromised patients, individuals with indwelling medical devices (e.g., catheters), and those undergoing invasive procedures are particularly susceptible to these infections. *Candida*, *Aspergillus*, and other opportunistic fungi can cause hospital-acquired infections, including bloodstream infections, surgical site infections, and pneumonia.

Occupational exposure: Certain occupations pose a higher risk of acquiring fungal infection transmission. Workers in agricultural settings may be exposed to fungi present in soil or crops, leading to infections like blastomycosis or sporotrichosis. Healthcare workers may also meet fungal pathogens in hospital settings, particularly in areas with immunocompromised patients or poor infection control practices. Occupational safety measures, including personal protective equipment and proper hygiene protocols, are important for reducing the risk of fungal infection transmission in these settings.

Indirect Transmission

Fungal infections can also be indirectly transmitted by vectors or fomites. Vectors, such as insects or animals, can carry fungal pathogens and transmit them to humans. Fomites, on the other hand, are inanimate objects or surfaces that can harbour fungal pathogens and facilitate transmission. Examples include contaminated medical equipment, shared combs or brushes, or surfaces in public spaces like gymnasiums or swimming pools.

Preventing the transmission of fungal infections involves implementing appropriate hygiene practices, such as regular handwashing, avoiding direct contact with infected individuals or contaminated surfaces, and maintaining a clean and dry environment. In certain cases, antifungal prophylaxis or treatment may be necessary to prevent the spread of infections, particularly in high-risk individuals or healthcare settings.

Geographic Distribution of Fungal Infections

Fungal infections can occur worldwide, but their prevalence and distribution can vary based on geographic factors such as climate, environmental conditions, and regional endemicity. Understanding the geographic distribution of fungal infections is crucial for proper diagnosis, treatment, and public health measures.

Superficial fungal infections: Superficial fungal infections, such as dermatophytosis (ringworm), are prevalent worldwide. The species causing these infections may vary by region. For example, *Trichophyton rubrum* is a common cause of dermatophytosis in many parts of the world, while *Trichophyton tonsurans* is more prevalent in North America. The incidence of these infections can be influenced by climate, with higher rates observed in tropical and subtropical regions due to the warm and humid conditions that favour fungal growth.

Systemic fungal infections: Systemic fungal infections often have distinct geographic distributions, as they are affected by many factors including environmental reservoirs and host susceptibility. Histoplasmosis, caused by *H. capsulatum*, is prevalent in certain regions of the world, known as endemic areas (Fig. 6.1). These areas include parts of North America (Ohio and Mississippi River valleys), Central and South

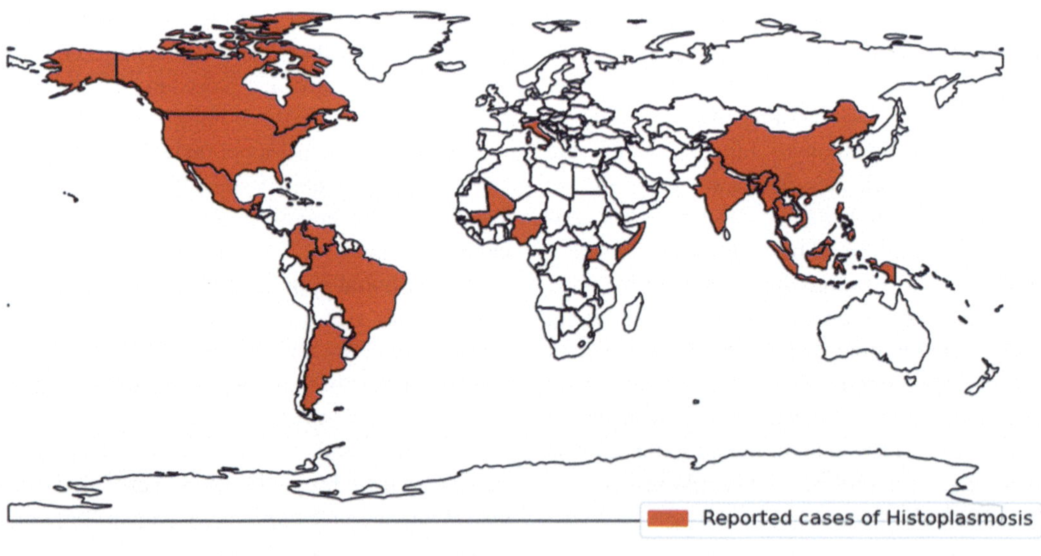

Fig. 6.1 World map showing the distribution of histoplasmosis worldwide

America, and Africa. Coccidioidomycosis, caused by *Coccidioides immitis* and *Coccidioides posadasii*, is primarily found in arid regions of the southwestern United States, Mexico, and parts of Central and South America. Other infections such as paracoccidioidomycosis, caused by *Paracoccidioides brasiliensis*, blastomycosis, caused by *Blastomyces dermatitidis*, and talaromycosis, caused by *Talaromyces marneffei*, are also considered systemic infections.

Opportunistic fungal infections: Opportunistic fungal infections occur primarily in individuals with weakened immune systems, such as those with HIV/AIDS, organ transplant recipients, or individuals undergoing chemotherapy. The distribution of these infections is often influenced by the prevalence of underlying immunosuppressive conditions. For instance, in regions with high HIV prevalence, opportunistic fungal infections like cryptococcosis, caused by *Cryptococcus neoformans*, and pneumocystis pneumonia, caused by *Pneumocystis jirovecii*, may be more common.

Endemic mycoses: Certain fungal infections are endemic to specific regions due to environmental factors. For example, blastomycosis, caused by *B. dermatitidis*, is prevalent in regions of North America, including the Mississippi and Ohio River valleys, the Great Lakes region, and parts of Canada. Paracoccidioidomycosis, caused by *P. brasiliensis*, is predominantly found in Central and South America, particularly in rural areas with agricultural activities. Histoplasmosis, talaromycosis, and coccidioidomycosis are also considered endemic diseases.

Global travel and migration patterns also contribute to the spread and emergence of fungal infections in new geographic areas. Climate change and environmental disturbances may also impact the distribution of fungal infections by altering ecological conditions or expanding the ranges of certain fungal species.

Understanding the pattern of geographic distribution of fungal infections allows healthcare providers to consider specific fungal pathogens when diagnosing patients, especially those with a history of travel or residency in endemic regions. It also enables public health authorities to implement appropriate surveillance measures, develop guidelines for prevention and treatment, and allocate resources effectively to areas with higher disease burdens.

Burden of Fungal Zoonoses

Fungal zoonoses can pose a significant burden on public health. These infections can vary in severity and clinical manifestations, and their burden depends on factors such as the prevalence of the zoonotic fungi, exposure to infected animals, and the ability of the fungi to cause disease in humans.

Sporotrichosis: Sporotrichosis is a fungal infection caused by *Sporothrix* species. *S. brasiliensis* is the most virulent species of this genus that causes infection in cats. Humans acquire infection after meeting cats, or scratches or bites from cats. This species is exclusively present in Brazil and Argentina. Other species of *Sporothrix* that cause infections in humans include *S. shenckii*, *S. globosa*, and *S. mexicana*. The infection due to these species is often acquired by meeting plants, soil, or materials contaminated with the fungus. The infection is not highly contagious for humans, but it can lead to localised skin lesions or more severe disseminated infections in individuals with poor immunity. The burden of sporotrichosis varies geographically, with higher prevalence reported in certain regions of Latin America, Africa, and Asia.

Cryptococcosis: Cryptococcosis, caused by the yeast, *C. neoformans*, or *Cryptococcus gattii*, is commonly associated with exposure to bird droppings or contaminated soil. Although most cases occur in individuals with compromised immune systems, including those with HIV/AIDS, cryptococcosis can also affect immunocompetent individuals. It primarily manifests as a respiratory infection but can also invade the central nervous system, leading to meningitis. The infection has a global distribution and is a significant cause of morbidity and mortality, particularly in regions with high HIV prevalence.

Histoplasmosis: Histoplasmosis, caused by the fungus *H. capsulatum*, is associated with exposure to bird or bat droppings, particularly in areas where it is present in large populations such as in caves. The condition manifests as a mild respiratory illness or progresses to severe disseminated disease, particularly in individuals with compromised immune systems. The condition is endemic in North America (Ohio and Mississippi River valleys), Central and South America, Africa, and Asia. It is a major public health problem in these endemic areas due to both acute outbreaks and chronic infections.

Blastomycosis: Blastomycosis, caused by the fungus *B. dermatitidis*, is primarily acquired by inhalation of spores present in soil or decaying organic matter. The condition manifests initially as respiratory symptoms, subsequently disseminating to other organs, leading to severe infections. The condition is endemic in the Great Lakes region, Mississippi and Ohio River valleys of North America, and parts of Canada. The burden of blastomycosis varies in these endemic areas, with reports of sporadic outbreaks.

Dermatophytosis: Dermatophytosis is a common fungal infection caused by dermatophyte fungi, including *Trichophyton* and *Microsporum* species. The condition is transmitted from infected animals to humans, particularly cats, dogs, and farm animals. The condition manifests as skin lesions, hair loss, itching, and significant discomfort but is not life-threatening but may require prolonged treatment. The burden of the infection depends on many factors, including the prevalence of infected animals and contact with contaminated environments.

Talaromycosis: *Talaromyces marneffei* is a dimorphic fungus that can cause a systemic fungal infection known as talaromycosis. It is endemic to certain regions of Southeast Asia, particularly in Thailand, Vietnam, and parts of China. Talaromycosis primarily affects individuals with weakened immune systems, particularly those with HIV/AIDS. The infection is acquired through inhalation of fungal spores present in the environment, such as in soil or decaying organic matter. Bamboo rats are considered a reservoir or source of this fungi. The condition can manifest as a respiratory infection, but it can also disseminate to other organs, leading to severe systemic disease. Early diagnosis and antifungal treatment are crucial for managing talaromycosis.

Paracoccidioidomycosis ceti (Lobomycosis): *Paracoccidioides ceti*, earlier known as *Lacazia loboi*, is the causative agent of paracoccidioidomycosis ceti or lobomycosis, a chronic fungal infection primarily affecting the skin and subcutaneous tissues. This species is endemic and found in different dolphin species around the coast of Asia, Europe, and North and South America. The exact mode of transmission of this fungus is still not fully understood but is believed to be acquired through contact with contaminated environments or infected animals, such as dolphins or manatees. The condition manifests as nodules or plaques on the skin and can cause significant disfigurement. Treatment typically involves surgical excision of the lesions and antifungal medications.

***Conidiobolus* and *Basidiobolus* infections**: *Conidiobolus* and *Basidiobolus* are fungi belonging to the order Entomophthorales that can cause subcutaneous fungal infections in humans. These infections are commonly associated with rural or agricultural settings, where individuals may meet soil or plant material contaminated with fungal spores. These entomophthoralean fungi can infect both insect and mammalian hosts. They are usually found in droppings of reptiles, insects, and amphibians and in soil. *Basidiobolus* species are found in the gut of amphibians and lizards, which capture and consume insects harbouring propagules of *Basidiobolous*.

Similarly, *Conidiobolus* has been recovered from insects. *Conidiobolus* infections are more prevalent in Africa, whereas *Basidiobolus* infections are more common in Central and South America. They are more common in tropical and subtropical countries. The infection occurs by the bite of the infected insects or open skin contact with the contaminated environment. The clinical manifestations include localised swelling, nodules, or abscesses in the subcutaneous tissues. Surgical excision and antifungal therapy form the basis for managing the condition.

Health education about proper hygiene and handling of animals, particularly those at higher risk of transmitting fungal infections, is important in preventing fungal zoonoses. Diagnosis, treatment, and prevention strategies, such as antifungal medications, vector control, and environmental management, are essential for reducing the impact of these infections on human health.

The geographical distribution of fungal zoonoses is summarised in Table 6.1.

Table 6.1 Epidemiology of zoonotic fungal pathogens

Diseases	Etiological agent	Type of infection	Most common prevalent area	Animals associated with infection	Remark
Dermatophytosis	*Trichophyton* spp., *Microsporum* spp., and *Epidermophyton* spp.	Superficial infection	Worldwide	Cats, dogs, horses, rabbits, rodents, cattle, hedgehogs, ruminant guinea pigs, and other rodents	Rarely cause invasive infection, i.e., deep dermatophytosis Zoophilic dermatophytes cause chronic or asymptomatic infection among primary hosts and cause acute, highly inflammatory among secondary hosts
Sporotrichosis	*Sporothrix schenckii, Sporothrix braziliensis, Sporothrix globosa, Sporothrix mexicana, Sporothrix luriei, Sporothrix chiliensis*, and *Sporothrix stenoceras*	Subcutaneous infection	Worldwide, but *S. braziliensis* only in Brazil and its adjoining countries	Cats, dogs, camels, armadillos, rats, mice, donkeys, horses, and cattle	*Sporothrix* spp. are dimorphic fungi
Cryptococcosis	*Cryptococcus neoformans* and *Cryptococcus gatti*	Subcutaneous, and invasive infection	Worldwide, most common in Brazil, Thailand, Malaysia, and Papua New Guinea and sub-Saharan Africa	Avian, amoeba, mites, amphibians, reptiles, birds, cats, dogs, foxes, ferrets, goats, monkeys, cheetahs, pigs, llamas, alpacas, koalas, mammals, and marine mammals	Cryptococcosis is an opportunistic infection caused by basidiomycetous yeast
Blastomycosis	*Blastomyces dermatitidis, Blastomyces, Blastomyces gilchristii, Blastomyces percursus*	Invasive infection	North America around Great Lakes, Ohio, Mississippi, Missouri, and St. Lawrence rivers	Dog, cat's kinkajou, ferret, red ruffed lemur, rhesus monkey, black bear, and marine mammals such as sea lions, dolphins	*Blastomyces* spp. are dimorphic fungi
Coccidioidomycosis	*Coccidioides immitis*	Invasive and superficial infection	California, Arizona, northwest Mexico, Central and South America	Armadillos, cattle, sheep, pigs, horses, and llamas	Less common among domestic animals like cats and dogs
Paracoccidioidomycosis	*Paracoccidioides brasiliensis*	Invasive and superficial infection	Central and South America and Latin American countries	Dogs, cats, chickens, dairy cattle, pigs, goats, rabbits, and sheep	*Paracoccidioides* spp. are dimorphic fungi
Histoplasmosis	*Histoplasma capsulatum*	Invasive infection	Emerging infections worldwide	Horses, mules, donkeys, hedgehogs, cattle, pigs, cats, dogs, foxes, bears, raccoons, skunks' badgers, and sea otters	*Histoplasma* spp. are dimorphic fungi

Fungal Outbreaks

Fungal outbreaks occur in various settings, including healthcare facilities, communities, and even in the environment. These outbreaks can result in significant morbidity and mortality, particularly in individuals with compromised immune systems or underlying health conditions. Detecting, diagnosing, and preventing fungal outbreaks is essential for timely intervention and effective control.

The examples of some fungal outbreaks mentioned below highlight the potential for fungal outbreaks to occur in various settings, including healthcare facilities and communities. Rapid detection, prompt intervention, and implementation of preventive measures are crucial for controlling and mitigating the impact of fungal outbreaks on public health.

***Candida auris* outbreak**: *C. auris* is an emerging multidrug-resistant fungal pathogen that can cause invasive infections, particularly in healthcare settings. The fungus is known for its ability to persist on environmental surfaces, leading to its transmission between patients in healthcare facilities. Since its first identification in 2009, *C. auris* has been associated with several outbreaks worldwide. These outbreaks have been documented in the United States, the United Kingdom, India, and South Africa. Outbreak control measures include enhanced surveillance, strict infection control practices, and the use of appropriate antifungal agents.

Dermatophytosis outbreak: Dermatophytosis caused by newly identified *Trichophyton indotineae* previously belonging to *Trichophyton mentagrophyte* genotype VIII has been found in near epidemic form in the Indian subcontinent. Many reports suggest that this organism has spread from the Indian subcontinent to other countries across the globe. However, indigenous familial cases of this infection were reported from Iran without any travel history to India. *T. mentagrophytes* is considered a zoophilic fungus and is usually transmitted to humans via infected animals. However, an increase in cases of chronic and recurrent dermatophytosis led to the investigation of the causative agent. Genotypic analysis revealed that the agent belonged to *T. mentagrophytes/T. interdigitale* complex assigning to genotype VIII. They often cause inflammatory and pruritic forms of tinea corporis, tinea cruris, and tines faciei and are resistant to terbinafine. Animal infection due to this species is documented in six cases to date, but most of the cases are transmitted from person to person via direct contact or indirect transmission via inanimate objects such as bathrooms, beds, and linens. Extensive abuse of topical steroids containing antifungal/antibacterial agents is considered the primary reason for this dermatophytosis outbreak. The combination of oral antifungal therapy along with topical antifungals is recommended for clinical resolution of *T. indotineae* infection.

Sporotrichosis outbreaks: Sporotrichosis usually occurs sporadically across the globe. The first reported outbreak of sporotrichosis is from South Africa associated with Rand gold mines caused by *S. schenckii* found on the wooden pit props. Another outbreak of sporotrichosis was reported from Guatemala around Lake Ayarza, caused by injury to fishermen via fish spines. Several outbreaks due to saprophytic sporotrichosis have been reported from Florida, Wisconsin, and China. The first zoonotic sporotrichosis outbreak was reported in Brazil due to cats. It is believed that the spores generated during the mycelial phase of this dimorphic fungi are the infective form. However, the yeasts found in the cats of this region harbour *S. schenckii* sensu lato in the fingernails and mucosal lesions. An outbreak due to *S. braziliensis* has been reported in Brazil, where 96.9% of the cats were found to harbour this fungus. This outbreak spans more than 15 years and has infected more than 4000 patients in Rio de Janeiro. Another zoonotic outbreak of *S. schenckii* was reported among hunters who met the infected nine-banded armadillos.

***Aspergillus* outbreaks in hospitals**: *Aspergillus* species are commonly found in the environment and can cause severe respiratory infections in susceptible hosts with weak immune systems. Outbreaks of invasive aspergillosis have been reported in hospitals, especially in settings where construction or renovation activities are taking place. These outbreaks occur due to the release of *Aspergillus* spores into the air, leading to increased exposure. Prevention and control measures include appropriate air filtration systems, isolation precautions for high-risk patients, and minimising construction-related disturbances in healthcare facilities.

Blastomycosis outbreak in the United States: Blastomycosis is a fungal infection caused by *B. dermatitidis*, primarily found in soil, and decaying organic matter in certain regions of North America. In 2015, an outbreak of blastomycosis occurred in Wisconsin, affecting over 100 individuals. The outbreak was associated with a specific construction site where workers were exposed to the fungus. Investigation and control measures involved identifying the source of exposure, raising awareness among healthcare providers, and implementing appropriate infection control practices.

***Cryptococcus gattii* outbreak in Vancouver Island**: *C. gattii* is a fungus primarily found in tropical and subtropical regions. However, an outbreak of *C. gattii* infections occurred in the mid-2000s on Vancouver Island in British Columbia, Canada. The outbreak affected both humans and animals, including cats, dogs, and marine mammals. The source of the outbreak was traced to the environment, particularly certain tree species. Control measures involved informing the public about the risk factors, enhancing surveillance, and implementing environmental management strategies.

Diagnosis of Fungal Outbreaks

Diagnosing a fungal outbreak involves multiple steps to identify the causative fungus and determine the extent of the outbreak. Diagnosis of the outbreaks typically includes clinical evaluation, laboratory testing, and epidemiological investigations. Clinical evaluation involves assessing patients for symptoms and signs consistent with fungal infections, such as respiratory symptoms, skin lesions, or systemic manifestations. Laboratory testing includes collecting appropriate specimens, such as blood, tissue, or respiratory samples, for microbiological analysis. Diagnostic methods, including fungal culture, microscopy, serology, and molecular diagnostics, aid in identifying the fungi responsible for the outbreak. Serological testing, including antigen detection, is frequently used for the diagnosis of certain fungal infections. The combined efforts of clinicians, laboratory personnel, and epidemiologists are crucial for accurate diagnosis and tracking of fungal outbreaks.

Detection of Fungal Outbreaks

Detecting a fungal outbreak involves active surveillance, monitoring, and reporting systems. Healthcare facilities should have robust infection control programmes in place to identify and investigate unusual clusters of fungal infections. Monitoring patient data, including clinical information and laboratory results, helps identify trends or patterns indicative of an outbreak. Timely communication and collaboration between healthcare providers, laboratory staff, and public health authorities are vital for early detection. In addition to healthcare-associated outbreaks, environmental monitoring may be necessary in settings such as agricultural or construction sites, where exposure to fungal spores can occur.

Prevention of Fungal Outbreaks

Preventing fungal outbreaks involves a multifaceted approach that includes infection control measures, environmental management, and education. Key strategies for prevention are as follows:

1. **Infection control practices**: Implementing and adhering to strict infection control practices, such as hand hygiene, proper disinfection, and appropriate use of personal protective equipment, can minimise the transmission of fungal infections in healthcare settings.
2. **Environmental management**: Identifying and addressing potential sources of fungal contamination, such as water systems, ventilation systems, construction sites, etc., is crucial for preventing outbreaks. Regular monitoring,

maintenance, and appropriate remediation measures can reduce fungal exposure.
3. **Antifungal stewardship**: Rational and judicious use of antifungal medications can help prevent the emergence of drug-resistant fungi and reduce the risk of healthcare-associated fungal infections.
4. **Education and awareness**: Promoting awareness among healthcare providers, patients, and the public about the risk factors, symptoms, and preventive measures for fungal infections can contribute to early detection and prompt management of outbreaks.

Fungal outbreak management requires a coordinated response involving healthcare facilities, public health agencies, and the community. Prompt diagnosis, detection, and prevention efforts can help mitigate the impact of fungal outbreaks, minimise transmission, and protect vulnerable populations. Continued research, surveillance, and collaboration are crucial for improving our understanding of fungal outbreaks and developing effective prevention strategies.

Prevention and Control

Preventing fungal zoonoses requires a multi-pronged approach that involves both animal and human health interventions. Some key preventive measures include:

Education and awareness: Health education campaigns targeted at both animal owners and healthcare professionals can raise awareness about fungal zoonoses, their transmission routes, and preventive measures. Health education promotes early detection, timely treatment, and appropriate management of fungal infections in both animals and humans.

Personal protective measures: Individuals working in close contact with animals or handling contaminated material need to use personal protective equipment (PPE) such as gloves, masks, and protective clothing to minimise the risk of fungal exposure. Proper hand hygiene needs to be followed when handling infected animals or animal products.

Environmental control: Reducing environmental factors that favour fungal growth can help prevent fungal zoonoses. Suggested measures include improving ventilation in animal housing, minimising moisture, and humidity, and controlling the presence of vectors such as arthropods.

Animal management: Proper hygiene practices in animal husbandry, including regular cleaning of animal living spaces and appropriate waste disposal, can help reduce the risk of fungal infections in animals. Veterinary care, including regular check-ups and treatment of infected animals, is equally important (Fig. 6.2).

Case Study

A patient presented with respiratory symptoms, including coughing, wheezing, and shortness of breath. Chest X-rays revealed bilateral infiltrates and a computed tomography (CT) scan showed multifocal nodular lesions in the lungs. The patient had a history of close contact with farm

Fig. 6.2 (a) Lesion above the muzzle of the calf; (b) lesion over the forehead of the cat; (c) tinea corporis in human

animals, specifically poultry, due to their occupation as a poultry farmer.

1. What fungal zoonoses are associated with close contact with farm animals, particularly poultry?
2. What personal protective measures should individuals working near farm animals, such as poultry farmers, undertake to prevent respiratory fungal infections?
3. What are the necessary preventive measures at a community level to control and reduce the transmission of fungal zoonoses associated with occupational exposure to farm animals?

Research Questions

1. What factors contribute to the limited understanding of the source of infection and control measures for uncommon fungal parasites, such as opportunistic moulds in immunocompromised individuals?
2. Should there be a re-evaluation of potential reservoir hosts, such as wildlife or domestic animals, for fungal zoonoses, considering the sporadic cases reported in areas where these diseases were previously believed to be absent?
3. Does evidence support the hypothesis of an anthroponotic nature of specific fungal zoonoses, like the Indian variant of *C. neoformans* infection? Are potential reservoir hosts involved in the transmission dynamics of this fungal pathogen?

Acknowledgement Dr. Abhishek, Department of Microbiology, Indian Veterinary Research Institute, Izatnagar, Uttar Pradesh, 243122, India (for providing animal picture with fungal infection).

Further Reading

Benedict K, Thompson GR III (2014) Fungal contamination in the environment as a cause of healthcare-associated infection. Clin Infect Dis 59(9):1433–1439

Benedict K, Richardson M, Vallabhaneni S, Jackson BR, Chiller T (2017) Emerging issues, challenges, and changing epidemiology of fungal disease outbreaks. Lancet Infect Dis 17(12):e403–e411. https://doi.org/10.1016/S1473-3099(17)30443-7. Epub 2017 Jul 31. PMID: 28774697; PMCID: PMC5712439

Chakrabarti A, Singh R (2010) Fungal rhinosinusitis: a microbiological perspective. Indian J Med Res 132(6):648–658

Fisher MC, Henk DA, Briggs CJ et al (2012) Emerging fungal threats to animal, plant, and ecosystem health. Nature 484(7393):186–194

Francesconi VA, Klein AP, Santos AP, Ramasawmy R, Francesconi F (2014) Lobomycosis: epidemiology, clinical presentation, and management options. Ther Clin Risk Manag 10:851–860. https://doi.org/10.2147/TCRM.S46251. PMID: 25328400; PMCID: PMC4199563

Gugnani HC (2011) Ecology and taxonomy of human-pathogenic yeasts and fungi: a global overview. Crit Rev Microbiol 37(3):211–227

Orofino-Costa R, Macedo PM, Rodrigues AM, Bernardes-Engemann AR (2017) Sporotrichosis: an update on epidemiology, etiopathogenesis, laboratory and clinical therapeutics. An Bras Dermatol 92(5):606–620. https://doi.org/10.1590/abd1806-4841.2017279. PMID: 29166494; PMCID: PMC5674690

Reddy DL, Nel J, Govender NP (2023) Review: emergomycosis. J Mycol Med 33(1):101313. https://doi.org/10.1016/j.mycmed.2022.101313. Epub 2022 Jul 23. PMID: 36272380

Rhodes J, Fisher MC (2019) Global epidemiology of emerging *Candida auris*. Curr Opin Microbiol 52:84–89. https://doi.org/10.1016/j.mib.2019.05.008. Epub 2019 Jul 3. PMID: 31279224

Richardson M, Lass-Flörl C (2008) Changing epidemiology of systemic fungal infections. Clin Microbiol Infect 14(Suppl 4):5–24

Rudramurthy SM, Shaw D (2019) Clinical practice of medical mycology in Asia: epidemiology of superficial fungal infections in Asia. Springer. https://doi.org/10.1007/978-981-13-9459-1

Shaikh N, Hussain KA, Petraitiene R, Schuetz AN, Walsh TJ (2016) Entomophthoramycosis: a neglected tropical mycosis. Clin Microbiol Infect 22(8):688–694. https://doi.org/10.1016/j.cmi.2016.04.005. Epub 2016 Apr 19. PMID: 27109491

Vanittanakom N, Cooper CR Jr, Fisher MC, Sirisanthana T (2006) *Penicillium marneffei* infection and recent advances in the epidemiology and molecular biology aspects. Clin Microbiol Rev 19(1):95–110. https://doi.org/10.1128/CMR.19.1.95-110.2006. PMID: 16418525; PMCID: PMC1360277

Williamson P, Jarvis J, Panackal A et al (2017) Cryptococcal meningitis: epidemiology, immunology, diagnosis, and therapy. Nat Rev Neurol 13:13–24. https://doi.org/10.1038/nrneurol.2016.167

Diagnosis of Fungal Zoonoses

7

Sanchita Mitra, Ritika Harchand, and Shivaprakash M. Rudramurthy

Learning Objectives

By the end of the chapter, the readers should be able to:

1. Understand the proper methods of collection and transport of different samples to the laboratory for diagnosis of fungal zoonoses.
2. Learn various methods of conventional mycology in practice for diagnosing fungal zoonoses.
3. Review the basics of molecular microbiology for the diagnosis of zoonotic fungal infections.

Introduction

Zoonoses are a wide range of infectious diseases in humans that are caused by bacteria, viruses, fungi, and parasites of animal origin. Most of the fungal infections are caused by fungi present in the environment, using animals as vectors or reservoirs for their transmission. There has been an increase in this intensity of transmission due to the settlements and agricultural activities of humans in the natural environmental habitats of wildlife. This has resulted in emergence and re-emergence of many fungal zoonotic infections like *Malassezia* infection, aspergillosis, dermatophytosis, cryptococcosis, phaeohyphomycosis, sporotrichosis, coccidiomycosis, paracoccidiomycosis, blastomycosis, histoplasmosis, adiaspiromycoses, talaromycosis, entomophtoromycosis, and pneumocystosis.

Laboratory diagnosis of fungal infections is challenging. Early and specific diagnosis of fungal infections, especially in immunosuppressed patients, helps in reducing mortality and morbidity. Diagnosis of fungal pathogens is primarily based on conventional methods as well as recent molecular, proteomic, and genomic methods. Conventional mycology techniques require expertise, effort, and time, while recent molecular mycology diagnostics are expensive, and occasionally, data obtained may be difficult to interpret.

The resurgence of zoonotic fungal infections necessitates a need for a correlative diagnostic approach. Multiple factors such as clinical presentation, radiographic features, and laboratory tests are used together to make a diagnosis as proven, probable, or possible fungal infections. Laboratory techniques used include conventional or traditional methods such as microscopy, culture, histopathology, and serology. These conventional technologies have stood the test of time but

S. Mitra (✉)
Jhaveri Microbiology Centre and Brien Holden Eye Research Centre, LV Prasad Eye Institute, Hyderabad, India
e-mail: sanchita034@gmail.com

R. Harchand · S. M. Rudramurthy
Department of Medical Microbiology, Postgraduate Institute of Medical Education and Research (PGIMER), Chandigarh, India

S. C. Parija, S. M. Rudramurthy (eds.), *Textbook of Fungal Zoonoses and Sapronoses*, Microbial Zoonoses, https://doi.org/10.1007/978-981-97-3214-2_7

are time-consuming and require skilled manpower. Molecular approaches have been increasingly used for the diagnosis of fungal zoonoses and have been proven to be an efficient approach.

Collection and Transport of Samples

Collection of appropriate samples for microscopy and culture is most important. Specimens frequently collected for different fungal infections include blood, bone marrow, cerebrospinal fluid, cutaneous specimens (skin, hair, nails), ocular samples (corneal/scleral scrapings, orbital abscess pus, intraocular fluids like aqueous or vitreous, ocular tissues), body fluids or lavage fluids (pleural, peritoneal, joint, bronchoalveolar), sputum, urine, and stool (Table 7.1).

Proper sample collection and transport to the lab without delay is essential to isolate fungal pathogens. Collection of samples in leak-proof containers, labelled with relevant clinical details, is important for the proper handling of pathogens like *Coccidioides immitis* and *Histoplasma capsulatum* that belong to Biosafety level III (BSL III). Tissue and fluid samples, transported in sterile containers or inoculated at the collection point, are the specimens of choice for isolation of pathogens. Specimens transported to the laboratory are processed without any delay. Specimens such as cerebrospinal fluid (CSF) are stored at 4 °C in case of any delay.

Microscopy and Histopathology

Direct examination by microscopy is the most economical and rapid test for the diagnosis of fungal infections. Direct microscopic examination is a rapid method of detection. Based upon the morphology of the fungi, the tissue-specific diagnosis can be done for the detection of *Talaromyces marneffei*, *H. capsulatum*, *Coccidioides immitis*, *Pneumocystis jirovecii*, and *Blastomyces dermatitidis*. Microscopy of CSF, pus, or ocular samples

Table 7.1 Sources of clinical specimens for the diagnosis

	Clinical specimen source								
Suspected pathogen	Blood	Bone marrow	Brain and cerebrospinal fluid	Joint fluid	Eye	Urine	Respiratory specimens	Skin scraping or mucous membranes	Multiple systemic sites
Aspergillus spp.			✓		✓	✓	✓	✓	✓
Histoplasma capsulatum	✓	✓	✓	✓	✓	✓	✓	✓	✓
Blastomyces dermatitidis			✓	✓		✓	✓	✓	✓
Coccidiodes immitis	✓	✓	✓	✓	✓	✓	✓	✓	✓
Paracoccidiodes brasiliensis		✓	✓				✓	✓	✓
Talaromyces marneffei	✓	✓	✓	✓			✓	✓	✓
Pneumocystis jirovecii		✓					✓		✓
Sporothrix schenckii	✓		✓	✓			✓	✓	✓
Cryptococcus neoformans	✓	✓	✓	✓	✓	✓	✓	✓	✓
Dermatophytes								✓	
Malassezia								✓	

✓ Represents the sites from where the sample can be taken
(Adapted from Clinical Mycology/[edited by] Anaissie, E. J., McGinnis, M. R., & Pfaller, M. A.)

is performed by smear preparations followed by staining immediately after preparation. In contrast, harder samples like nails, hair, or tissue specimens are treated with lytic substances, 10–40% potassium hydroxide (KOH), followed by microscopy. Microscopy of KOH wet-mounts, helps in differentiating fungi as yeasts and molds (septate and aseptate hyphae) It can also give information on whether the hyphae are dematiaceous or hyaline (Figs. 7.1 and 7.2). Usually, the addition of a small volume of Chicago blue, trypan blue or chlorazole black to KOH has shown an increased visibility of fungal elements. Chlorazole black has a higher specificity than Chicago blue, but Chicago blue provides more contrast. Calcofluor white (CFW) stain, being a fluorescent stain, specifically binds to the chitin present in the fungal cell wall and is a rapid and more sensitive technique for detection of the fungi in direct smears by fluorescence microscopy (Fig. 7.3). Excitation of CFW stain requires the UV range, hence a fluorescence microscope should have a specific 4′, 6-diamidino-2-fenylindool (DAPI) filter. Both KOH and CFW stained smears are examined under 400x magnification for confirmation of identification of fungal elements by direct microscopy. Gram stain of such direct smears (examined by bright field microscopy, 1000x magnification) helps to rule out bacterial co-infection and also visualization of the associated inflammatory cells.

Gram stain, however, is not suggested for CSF as it cannot differentiate the ultrastructural features of *Cryptococcus*. Wright stain and Giemsa stain help to detect intracellular yeasts of *H. capsulatum* in blood or bone marrow smears. Negative staining of samples like CSF with India ink helps in the detection of the capsule of *Cryptococcus neoformans* as unstained. Negative staining helps in avoiding the distortion of morphology.

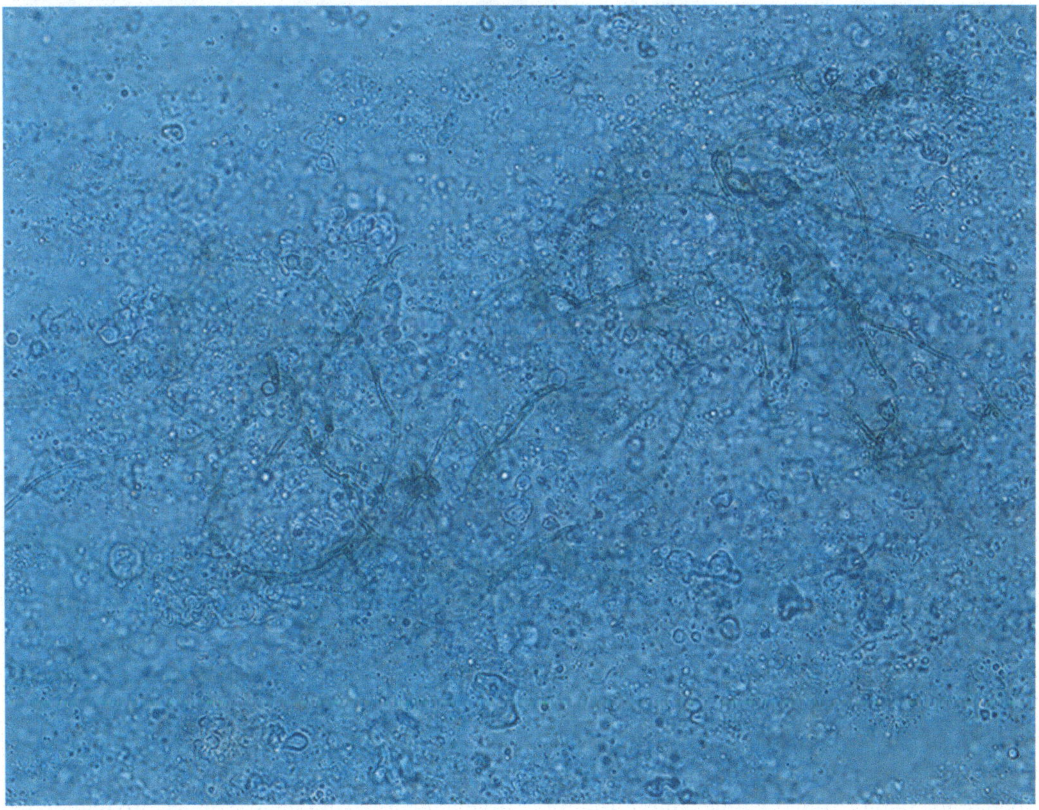

Fig. 7.1 Bright field microscopy shows pigmented septate hyphae on KOH wet mount of brain abscess, 400X, suggestive of dematiaceous or melanized fungus

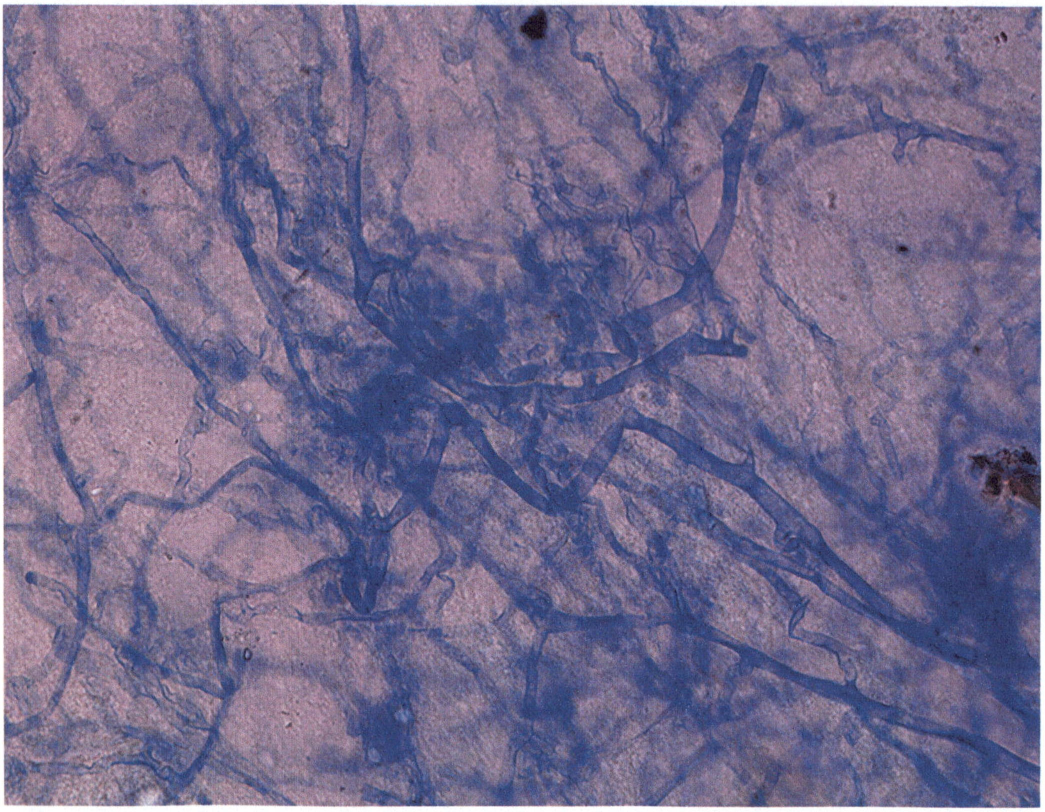

Fig. 7.2 Bright field microscopy shows broad, sparsely septate hyphae on KOH wet mount of sinus tissue, 400X, suggestive of Mucorales spp.

Hematoxylin and eosin (H&E) stains, though not specific for fungi, help to detect the inflammatory response and also detect fungi. Grocott-Gomori methenamine silver (GMS) stain and periodic acid-Schiff (PAS) stain are fungal-specific stains used for the detection of fungi in the histopathological or cytological specimen. In the PAS stain the fungi appear pink (Fig. 7.4) whereas in the GMS stain it appears brownish black (Fig. 7.5). *Pneumocystis jirovecii* can be identified by using the GMS stain and toluidine blue. Direct immunofluorescence stains are also available for the diagnosis of *P. jirovecii* but they are expensive and are slightly better than the GMS stain. Fontana-Masson stain specifically stains the melanin of the fungi and helps to differentiate dematiaceous (melanin-producing) fungi from hyaline fungi (no or minimal melanin producing). A few important stains used in fungal microscopy are listed in Table 7.2.

Isolation and Identification of Fungi

Culture is considered the gold standard for the diagnosis of fungal infection. Recovery of the fungi from the clinical samples is dependent on the type of fungi suspected and the media used for isolation. In routine clinical diagnostic laboratories, the media that supports the growth of the majority of the fungi is generally used. For primary isolation of the fungi, originally described Sabouraud dextrose agar (SDA) with pH of 5.5–5.6 and 4% glucose was predominately used for selective isolation of fungi. Currently, Emmons modification of SDA is increasingly used in the diagnostic laboratory due to its neutral pH with a glucose concentration of 2%, which makes it ideal for the isolation of fungi from human and animal specimens. To inhibit the growth of bacteria in the media, chloramphenicol and gentamicin are added. To prevent the growth of the saprophytic fungi, one set of SDA with

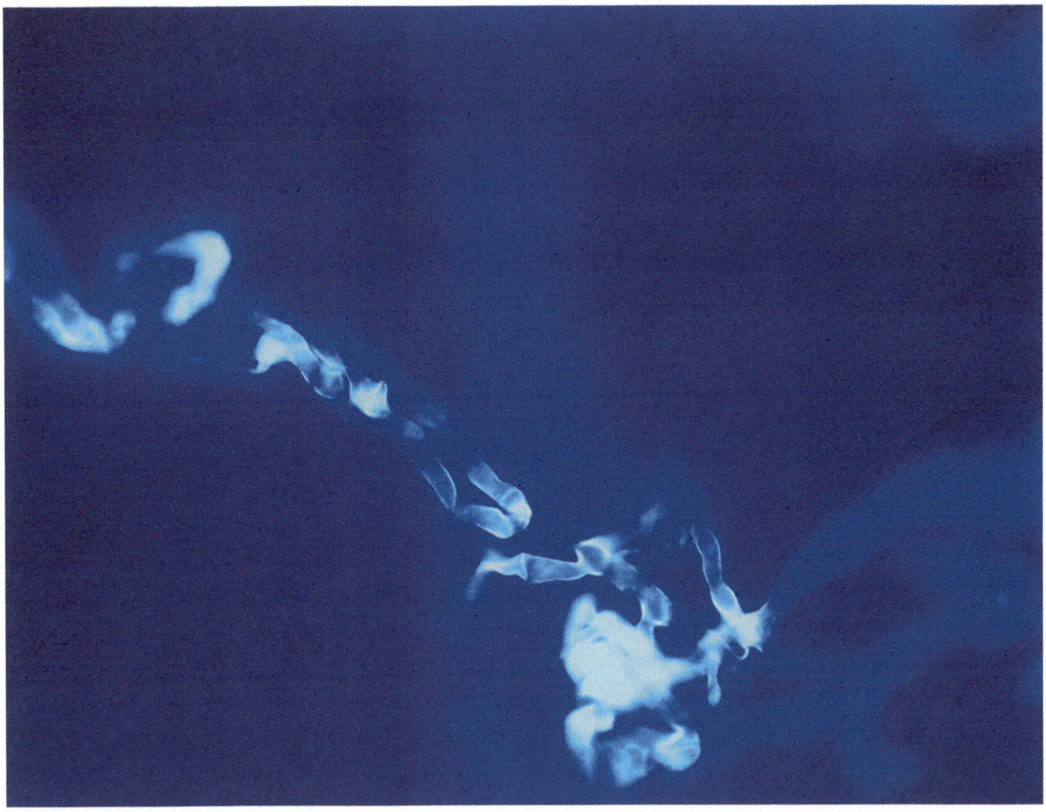

Fig. 7.3 Fluorescence microscopy shows broad, aseptate hyphae in sinus tissue on CFW wet mount, 400X, suggestive of Mucorales spp.

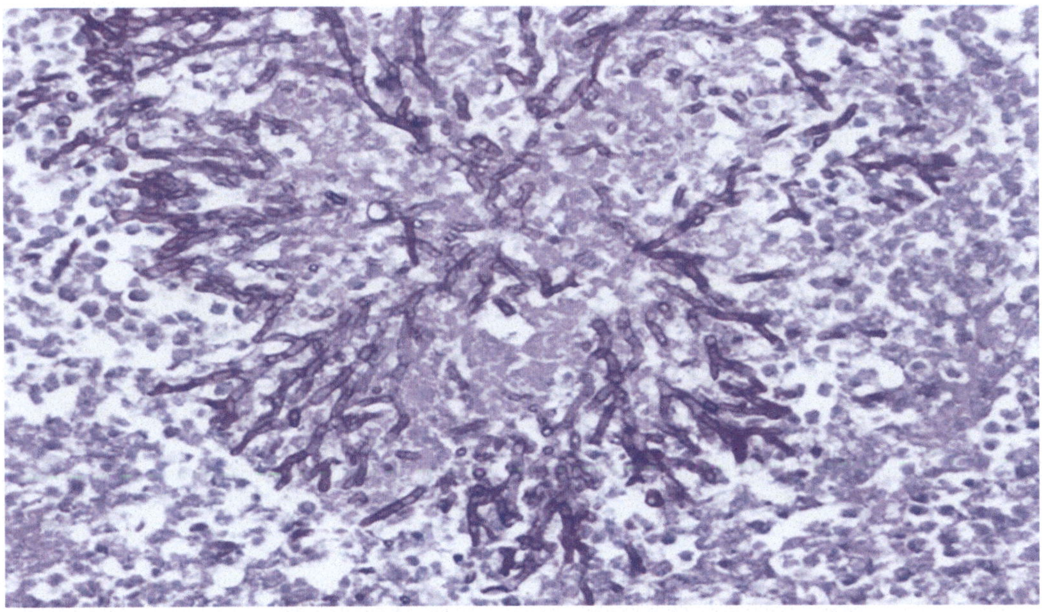

Fig. 7.4 Periodic acid Schiff stained histopathology section of lung tissue showing septate hyphae with acute angle branching suggestive of *Aspergillus* spp, 400X

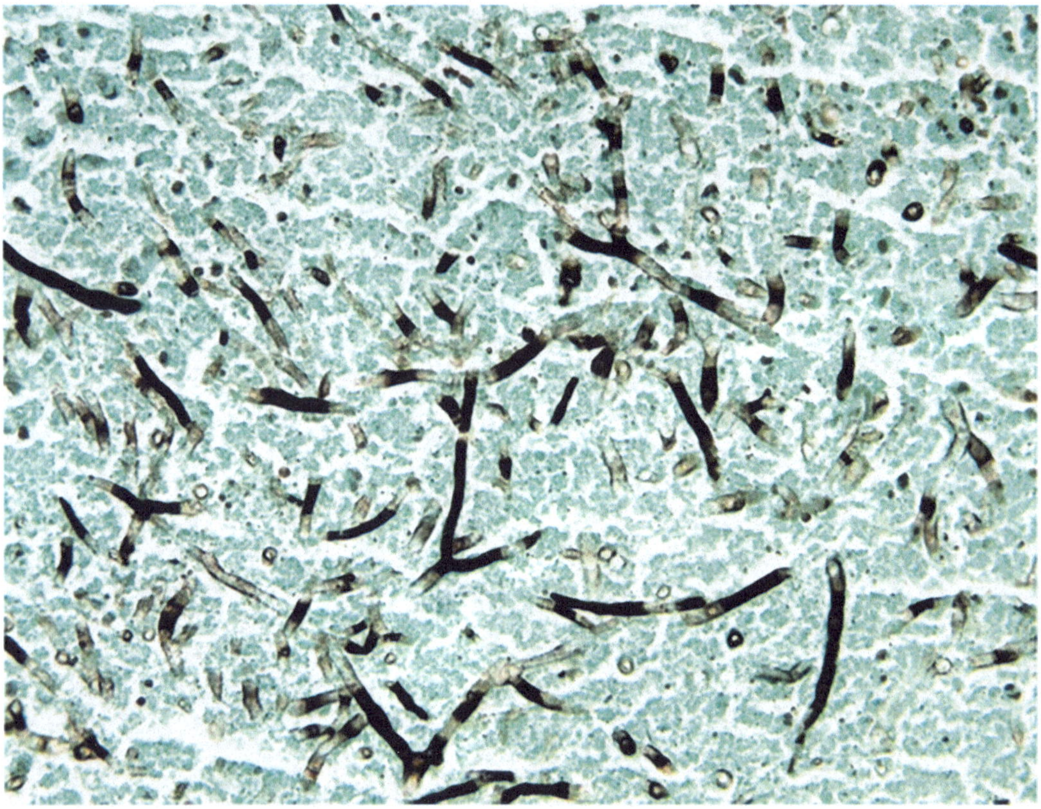

Fig. 7.5 Grocott-Gomori methenamine silver stained histopathology section of lung tissue showing septate hyphae, 400X

Table 7.2 Characteristics of the method and stains involved in fungal microscopy

Methods	Working/principle	Application
India Ink	It's a negative stain used for demonstrating the capsule	*Cryptococcus neoformans*, a capsule forming fungi, is detected
KOH	Being a strong alkali, it digests the tissue protein for visualization of fungal elements	Generally used for direct microscopic demonstration of any fungus in the specimen
Giemsa stain	Attaches itself to the region of DNA having a high content of adenine-thymine bonding	Helpful in the diagnosis of *Pneumocystis jirovecii* and *Histoplasma*
Periodic acid-Schiff stain	Uses the oxidation of fungal cell wall polysaccharides to visualize fungal infections	Helps in staining both yeasts and hyphae Identification of *Histoplasma capsulatum*, *Cryptococcus neoformans*, *Blastomyces*, and *Aspergillus fumigatus*
Hematoxylin and eosin	Cell nuclei are stained by hematoxylin and cytoplasm is stained pink by eosin stains	Helpful in visualizing the yeast cells (nucleus), *Blastomyces* (double refractive cell wall), and *Coccidioides* (spherule)
Mucicarmine stain	Stains mucin and mucopolysaccharide of *Cryptococcus*	Helps in the detection of *Cryptococcus neoformans* in differentiating it from other yeasts, especially in the tissue Also, stains *Blastomyces* cell wall
Fontana-Masson	Detects melanin present in human tissue and fungal cell wall	Detects the presence of melanin in dematiaceous fungi and *Cryptococcus*

Table 7.2 (continued)

Methods	Working/principle	Application
Gomori's methenamine silver stain	Detects fungal polysaccharide	Can identify both degenerated as well as viable fungi *Pneumocystis jirovecii* identification in respiratory samples and other fungi in histologic sections
Toluidine blue	A metachromatic dye; detecting cyst form of *Pneumocystis jirovecii*	Sensitivity of detection in BAL fluid is low
Calcofluor white	A fluorescent dye that stains the chitin present in the cell walls of fungi and plants	All fungi are detected including *Pneumocystis jirovecii*

BAL bronchoalveolar lavage

cycloheximide (inhibits saprophytic fungi, thus allowing the growth of slow-growing fungi) and another set without cycloheximide (helps for the growth of dimorphic fungi and dermatophytes) are used. In addition to SDA, one tube of brain heart infusion (BHI) agar that supports the growth of dimorphic fungi should be included for primary isolation of fungi. Some laboratories use BHI with 5–10% sheep blood (SABHI) instead of two sets of SDA and one set of BHI.

For the diagnosis of fungemia or candidemia, detection of the fungi in the blood by different means is essential. For the isolation of fungi from blood using conventional blood culture, biphasic BHI bottles are ideal. However, the sensitivity of this method is around 50% and there is a delay in the isolation and identification. At present, many automated blood culture identification systems are available and can detect fungi in the blood rapidly compared to the conventional system such as biphasic blood culture bottles. The commonly used systems include BacT/ALERT® (bioMérieux, Marcy-l'Étoile, France) and BD BACTEC™ (Becton Dickinson, East Rutherford, NJ, USA) aerobic systems. These systems are used to isolate both bacterial and fungal pathogens from blood. These systems can also be used to identify the fungi from sterile fluids such as cerebrospinal fluid, peritoneal fluid, and pleural fluid. For specific isolation of fungi from the blood, the culture bottles BACTEC™ Mycosis-IC/F (Mycosis) are available, and the use of this media along with the aerobic culture media improves the sensitivity of blood culture. Also, the use of BACTEC™ Mycosis-IC/F helps in the identification of molds and mixed polymicrobial infections, whereas aerobic media helps in the isolation of the fungi in the presence of antifungals. Many modifications of the systems are available to improve the sensitivity of this technique. One of the modifications uses a lysis centrifugation technique that helps in increasing the sensitivity and isolation of yeasts, molds, and dimorphic fungi. These are specifically helpful for the isolation of intracellular organisms such as *H. capsulatum*. To increase the sensitivity of the blood culture, it is suggested to include more blood volume in the culture media, at least 10 mL, in one set of blood culture.

For selective isolation of dermatophytes, dermatophyte test media (DTM) is used, which also helps in the presumptive identification of dermatophytes. Dermatophytes break down the phyton peptone present in the media and produce a red color surrounding the colony due to the liberation of alkaline metabolites. *Malassezia* spp. are lipophilic yeasts, and for growing this yeast in vitro, the media should be supplemented with lipids. All currently known *Malassezia* species are lipophilic except *M. pachydermatis*, which grows without supplementation of lipids. Lipid supplementation can be given in the form of olive oil, milk, malt extract, or oleic acids. Dixon agar or modified Dixon agar, and Leeming–Notmann agar or modified Leeming–Notmann agar are the selective media for *Malassezia*. Bird seed agar or niger seed agar is a selective media for *Cryptococcus* spp. *Cryptococcus* colonies appear brown due to their ability to utilize creatinine in the media as the sole source of nitrogen. This media especially helps in the differentiation of *Cryptococcus* from *Candida*, especially while isolating the former from the respiratory specimens whereas the latter is seen as commensals.

One set each of inoculated media is incubated in a biological oxygen demand (BOD) incubator at 26–28 °C and 37 °C with humidified air. Lid or cotton plugs are kept slightly loose to allow for aeration but not too loose to avoid contamination. They are examined daily for any significant growth for the first week and twice a week after the first week till the fungus grows or up to 4–6 weeks. For suspected dimorphic fungi, this period of incubation can be extended to 8 weeks, while for suspected yeasts, 7–8 days of incubation is considered adequate. Though all specimens with suspected fungal pathogens are processed in Biosafety level 2 (BSL-II) laboratories with proper precautions like masks and gloves, some dangerous fungi, such as the dimorphic fungi, *Cladophialophora bantiana* and *Rhinocladiella mackenziei*, are handled with utmost caution by the laboratory personnel.

Once the growth occurs, the fungi are identified based on whether the fungi isolated is yeast or molds. Yeast identification is made by a phenotypic identification system such as the morphology of the yeast on the cornmeal agar (CMA). On CMA, we look for the production of terminal chlamydospore, the presence of pseudohyphae, the pattern of pseudohyphae such as match stick appearance (*Candida krusei*), pine tree appearance (*Candida tropicalis*), and giant hyphae production (*Candida parapsilosis*). Other tests include sugar fermentation and sugar assimilation tests. Identification of the molds is generally done on the basis of macroscopic and microscopic characteristics. Some molds, however, require selective media for good sporulation, which is essential for their identification by microscopic features. Media such as potato dextrose agar (PDA) allows the additional advantage of rapid sporulation of molds, compared to SDA. Some fungi may not sporulate even after prolonged incubation for over a month and will thus be labeled as "unidentified" by conventional mycology; such fungi can be identified by molecular methods such as DNA sequencing.

Serology/Biomarkers

Detection of fungal biomarkers provides a means for rapid and non-culture-based diagnosis, making it a better choice. Although culture and histopathology are the foundation of the fungal diagnosis, serology plays an important role in providing much faster results. Detection of specific antigens or antibodies is an important technique for the diagnosis of fungal infections. The different biomarker tests available for the diagnosis of fungal zoonoses are as shown in Table 7.3.

Cryptococcal Antigen Detection

Polysaccharide antigen (glucouronoxylomannan) detection in CSF, Bronchoalveolar lavage (BAL), or serum is a highly sensitive and specific technique for the diagnosis of cryptococcosis. This antigen can be detected by the latex agglutination (LA) method and is considered one of the best antigen detection tests for microbial detection. This test is both qualitative and semi-quantitative. Recently, an immune chromatic test in the form of lateral flow assay (LFA) has been developed and proved its utility as a rapid method with minimal requirements. Hence, LFA provides a better budget-friendly technique for the detection and quantification of cryptococcal antigens. These tests also provide help in monitoring the treatment and prognosis.

Galactomannan Antigen Detection

Detection of galactomannan (GM) is considered a good diagnostic test for the diagnosis of invasive aspergillosis (IA). GM is released from the tip of the hyphae only during invasion into the tissues making it specific for diagnosis of invasive disease. GM also helps in the diagnosis of other fungal infections such as histoplasmosis, talaromycosis, and pneumocystis pneumonia. FDA-approved sandwich Enzyme-linked immunosorbent assay (ELISA) is a highly sensitive method that can detect levels up to one nanogram. GM can be detected from serum, plasma, CSF, and pleural

Table 7.3 Biomarker tests available for the diagnosis of fungal zoonoses

Biomarker test	Fungus detected	Sensitivity	Manufacturer
Cryptococcal antigen lateral flow assay	*Cryptococcus*	99% in CSF, 90–98% in serum	IMMY, StrongStep, Nanjing, Dyamiker
Aspergillus galactomannan lateral flow assay	*Aspergillus*	97% in serum and 91% in BAL	IMMY diagnostics
*Asp*LFD (Aspergillus specific lateral flow device)	*Aspergillus*	80% sensitivity in BAL	OLM diagnostics
ASPERGILLUS IgG lateral flow assay	*Aspergillus*	90.6% for ABPA and 86.7% for CPA	LDBIO diagnostics
Fungitell® (1→3)-β-D glucan colorimetric assay	*Aspergillus* and *Pneumocystis*	100% in serum for *Pneumocystis jivoreci* pneumonia	Associates of Cape Cod, Inc.
Histoplasma urine antigen LFA	*Histoplasma*	96% Sensitivity in urine	MiraVista
Platelia™ Aspergillus galactomannan EIA	*Aspergillus*	40% Sensitivity in serum	Bio-Rad Laboratories
MVista® Coccidioides IgG, IgM antibody EIA	*Coccidioides*	92% Sensitivity in serum	MiraVista

CPA chronic pulmonary aspergillosis, *APBPA* allergic bronchopulmonary aspergillosis, *BAL* bronchoalveolar lavage

fluid samples. For the diagnosis of invasive pulmonary aspergillosis, the BALF (Bronchoalveolar lavage fluid) GM test has better diagnostic utility in comparison to the serum GM. For the faster detection of invasive pulmonary aspergillosis (IPA), lateral flow tests, such as point-of-care tests, have been developed. These include lateral flow assay (LFA) (sōna Aspergillus galactomannan lateral flow assay, IMMY, Norman, OK, USA), which has high sensitivity and specificity.

Glucoprotein Detection

A lateral flow device (LFD) for the detection of the *Aspergillus*-specific glucoprotein has been developed by OLM Diagnostics, Newcastle upon Tyne, UK. This test is developed as a point-of-care test and can diagnose invasive aspergillosis in about 15 min with a sensitivity of 50–100% and specificity of 100%.

1,3 Beta-D-glucan (BDG) Detection

A major biomarker, 1,3 beta-D-glucan (BDG), is present in all the fungi except *Cryptococcus* and Mucorales. It has become an important pan-fungal biomarker for the diagnosis of invasive candidiasis, invasive aspergillosis (IA), and pneumocystis pneumonia. It is mainly used for the diagnosis of invasive candidiasis. BDG detection kit by Fungitell is kinetic ELISA test and approved by the US-FDA. The principle of this test is that BDG present in the fungal cell wall will coagulate the limulus amoebocyte lysate that is present in the horse-crab. The major limitation of Fungitell is that it is available as batch tests and increases the turnaround time. At present, this test is available in a single tube to circumvent the delay of testing and reporting (FungiStat, Associates of Cape Cod, East Falmouth, MA, USA).

Antibody Detection

Antibody detection tests are generally used for the detection of endemic fungi or dimorphic fungi. They have a role in the diagnosis of endemic mycoses such as histoplasmosis and coccidioidomycosis. Antibody detection test are not helpful for the diagnosis of the invasive opportunistic fungal infections. Less sensitive tests like immunodiffusion and complement fixation are still used, as just the presence of antibodies is not diagnostic, especially in the endemic regions. In comparison to complement fixation, immunodiffusion tests are less sensitive. Moreover, indirect immunofluorescence, and

ELISA, are also available but they are less specific. Antibody detection tests also help in the diagnosis of allergic fungal diseases as well as chronic pulmonary aspergillosis. Detection of IgG and IgE by Immunocap reaction is a standard method for allergic and invasive fungal infections such as allergic bronchopulmonary fungal infections, and chronic pulmonary fungal infections reported mainly from humans.

Molecular Diagnosis

Conventional diagnostic and serological methods for diagnosing fungal infections are either less sensitive, delayed, or non-specific or require expertise for rapid identification. In addition, clinical diagnostic procedures, such as biopsies, and (BAL) are not feasible in many critically ill patients. Even histopathology of the tissue cannot provide confirmed identification at the species level. On many occasions, samples are not sent for mycological investigation but instead sent for histopathology and the biopsy samples are fixed in formalin; hence, they cannot be used for mycology workup. Nucleic acid-based molecular markers are now an advanced technique for the diagnosis of fungal infections. Fungal molecular diagnostics involve different methods, such as nucleic acid amplification tests like conventional and real-time polymerase chain reaction (PCR) and nucleic acid sequencing methods. Though many PCR-based techniques are available for the diagnosis of fungal infections, it still needs further standardization. At present, PCR-based diagnostic tests are accepted as one of the criteria for the diagnosis of invasive aspergillosis, and its applicability for the diagnosis of other invasive fungal infections is still evolving. Molecular-based diagnostics are applied for the diagnosis of fungal infections directly from clinical specimens, identification of the fungi cultured from the specimen, or the detection of the antifungal resistance from the isolated fungi or directly from the clinical specimen.

DNA Extraction

Extraction of the DNA from the fungi is difficult due to the tough cell wall as it contains chitin, which is hard. Fungal DNA can be efficiently extracted by using different techniques, such as disruption of the cell wall by using strong enzymes such as lyticase or zymolase, using dilute alkali, or by physical methods such as bead beating, freeze-thawing, or sonication.

Target Selection

Selection of the target genes is important for the efficient detection of the fungal nucleic acid. As the fungal nucleic acid is present in very low amounts in the clinical specimen, it is recommended to use the multicopy gene of the fungi for increasing the sensitivity of detection. Parts of the ribosomal genes are one of the most commonly utilized genes for detection up to the species level. Fungal ribosomal DNA (rDNA) complex, which is present in the multicopy gene (30–50 copies) consisting of 18S (small subunit rDNA), 5.8S, and 28S (large subunit rDNA) is commonly used as gene targets. This gene complex is relatively preserved among members of different fungi and hence serves as the universal or pan-fungal target. These subunits are separated by regions called spacers, internal transcribed spacer 1 (ITS1) between 18S and 5.8S subunits, and internal transcribed spacer 2 (ITS2) between 5.8S and 28S. The ITS1 and ITS2 regions contain variable genes, differing between different species, and thus help in identification at genus and species levels by methods such as sequencing. Other genes used for fungal identification are β-tubulin, calmodulin, and elongation factor *EF1α*. As *ITS* sequences lack the power to discriminate the species of the same complex, calmodulin, beta-tubulin, *RPB1*, and *IGS* are the most commonly used protein-coding genes. These protein-coding genes have the conserved regions for the development of the specific primer for amplification and sequencing of the variable regions of the differentially related species.

PCR-Based Techniques

Many PCR-based techniques are increasingly used for the detection and identification of fungal pathogens. These techniques are not well standardized or developed in-house. At present, many kits are available commercially for the diagnosis of fungal infections, some well-standardized and others less evaluated. Major breakthrough in the molecular diagnosis of fungal infections is the diagnosis of candidemia by T2 magnetic resonance, as it can detect the five major *Candida* species from the blood in less than 3 h of collection of blood. *Candida* nucleic acid is amplified by PCR and detection is done by magnetic resonance with specificity of 99% and sensitivity of >90%. However, it still needs refinement in terms of expanding the number of species and reducing the cost of detection and its wide availability. The major requirement of molecular diagnosis for fungal infections is for the diagnosis of pneumocystis pneumonia as its detection by staining of smear has very low sensitivity, and also it is nonculturable fungi. As *P. jirovecii* is also present as commensal in the respiratory tract, detection of its nucleic acid alone may not always be relevant, and quantitative estimation of its nucleic acid is a specific diagnosis. Real-time PCR kits with cut-off levels are commercially available to target different genes and cut-off levels. Major targets used for the diagnosis of pneumocystis pneumonia (PCP) from BAL or other respiratory samples are major surface glycoprotein, dihydrofolate synthase, mitochondrial large subunit, and beta-tubulin. Kits are also available for the simultaneous detection of PCP and mutations in dihydropterate synthase gene mutation, which is responsible for resistance of *P. jirovecii* to cotrimoxazole, which is used for the prophylaxis and treatment of PCP.

Syndromic-based multiplex PCR-based panels are also available commercially for the detection of fungal etiology in meningitis, pneumonia, and sepsis. These tests are expensive and promising but still need further evaluation.

Due to the non-availability of the sero-diagnostic test or the low sensitivity of conventional techniques for the diagnosis of mucormycosis, there is a need for the development of molecular diagnostic techniques for this disease. Nucleic acid-based tests, such as in-house or commercial tests, are available but still need further evaluation and standardizations. Test to detect Mucorales in the fresh tissue specimen is highly specific and sensitive. The sensitivity of detection from formalin-fixed paraffin-embedded tissue is less due to degradation or damage to the fungal nucleic acid. The use of Mucor-specific 18s rDNA primers is more specific than the pan-fungal primers. Diagnosis of mucormycosis using sterile samples such as serum or plasma is still under evaluation but appears to be promising.

PCR-based techniques always have a limitation that there is a high risk of contamination (as fungal conidia or spores are ubiquitous and present everywhere), and it can be a colonizer (needs to differentiate infection from colonization). Hence, the laboratory performing the PCR for the fungal infection should be as per the standard described.

Matrix-Assisted Laser Desorption Ionization-Time of Flight Mass Spectrometry (MALDI-TOF-MS)

MALDI-TOF-MS is currently one of the standard methods for the identification of yeasts and molds. It is now considered the standard technique for the identification of yeast and has been successfully applied for the identification of molds. MALDI-TOF MS identifies the pathogen using the pattern-based protein profile. Hence, it is important to have a thorough database of protein profiles of the spectrum of organisms to match with it and identify the agent. Though initial investment in setting up MALDI-TOF MS is high, it has less recurring cost and is rapid for identification of fungi. MALDI-TOF MS can also help in the rapid detection of yeasts from the automated blood culture bottle that has flagged as positive. Its utility for the diagnosis of fungal infection directly from the clinical specimen is difficult due to the interference of the protein profiles of the clinical samples. It is also shown to be a promising technique for the detection of antifungal resistance in yeasts.

Case Study

A 70-year-old HIV-negative male was admitted to the hospital because of cough, shortness of breath, respiratory failure, weight loss of 8 kg, and periodic fever episodes with a history of benign prostatic hypertrophy and diabetes mellitus type 2. He was not responding to any antibiotics administered prior to the admission. Chest CT scan disclosed the presence of a number of reticular and nodular infiltrations with distribution in all lobes. Parts of mycelium and numerous oval spores were revealed by the cytopathology of bronchoalveolar lavage. Considering all the clinical suspicions and travel history of the patient, pulmonary histoplasmosis was diagnosed. Liposomal amphotericin B (300 mg i.v. for 14 days) was given to the patient, followed by itraconazole (600 mg daily for 3 days, then 400 mg daily for 3 months). A faster recovery was seen after the administration of the antifungals in relation to the respiratory function and the clinical picture.

1. What is the appropriate diagnostic approach for histoplasmosis?
2. What type of sample is used for the serodiagnosis?
3. Can molecular diagnosis be made for histoplasmosis?
4. What are the conventional approaches used for the diagnosis?

Research Questions

1. What are the commercial platforms available for the diagnosis of fungal zoonoses?
2. What are the advantages and disadvantages of conventional diagnosis?
3. What is the drawback of ELISA-based diagnosis?
4. What type of special culture media is used for the detection of *Malassezia*?

Further Reading

Adebiyi AI, Oluwayelu DO (2018) Zoonotic fungal diseases and animal ownership in Nigeria. Alex J Med 54(4):397–402

Anaissie EJ, McGinnis MR, Pfaller MA (2009) Clinical mycology e-book. Elsevier Health Sciences

Blanco JL, Garcia ME (2010) Animal as reservoir of fungal diseases (zoonoses?). In: Ahmad I, Owais M, Shahid M, Aqil F (eds) Combating fungal infections. Springer, Berlin

Chander J (2018) Textbook of Medical Mycology, 4th edn. Jaypee Brothers Medical Publishers Private (P) Ltd

Clinical and Laboratory Standards Institute (2008) Reference method for broth dilution antifungal susceptibility testing of yeasts, approved standard M27-A3. Clinical and Laboratory Standards Institute, Wayne

Fang W, Wu J, Cheng M, Zhu X, Du M, Chen C, Liao W, Zhi K, Pan W (2023) Diagnosis of invasive fungal infections: challenges and recent developments. J Biomed Sci 30(1):42

Ferrer C, Colom F, Frasés S, Mulet E, Abad JL, Alió JL (2001) Detection and identification of fungal pathogens by PCR and by ITS2 and 5.8S ribosomal DNA typing in ocular infections. J Clin Microbiol 39(8):2873–2879

Friedman DZP, Schwartz IS (2019) Emerging fungal infections: new patients, new patterns, and new pathogens. J Fungi 5(3):67

Kozel TR, Wickes B (2014) Fungal diagnostics. Cold Spring Harbor Perspect Med 4(4):a019299

Larone DH (2011) Medically important fungi: a guide to identification, 5th edn. ASM Press, Washington, DC

Richardson M, Page I (2018) Role of serological tests in the diagnosis of mold infections. Curr Fungal Infect Rep 12(3):127–136

Sangeetha J, Thangadurai D (2013) Staining techniques and biochemical methods for the identification of fungi. In: Gupta V, Tuohy M, Ayyachamy M, Turner K, O'Donovan A (eds) Laboratory protocols in fungal biology. Fungal biology. Springer, New York

Wang Y, Yang M, Xia Y, Yan J, Zou J, Zhang D (2021) Application and evaluation of nucleic acid sequence-based amplification, PCR and cryptococcal antigen test for diagnosis of cryptococcosis. BMC Infect Dis 21(1):1020

Treatment and Management of Fungal Zoonoses/Sapronoses

8

Abhishek, Sonu S. Nair, V. Athira, Prasad Thomas, and Bablu Kumar

Learning Objectives

By the end of the chapter, the readers should be able to:

1. Understand the treatment and management strategies for fungal zoonoses/sapronoses among humans and animals.
2. Learn the advancements in drug development or repurposing of drugs for the management of unresponsive cases of fungal zoonoses/sapronoses.

Introduction

Infections transmitted naturally among animals and humans are called zoonotic infections. Frequent contact with pets and livestock could be the main reason for the human infections. Diverse pathogens, including bacteria, viruses, fungi, parasites, and prions are involved in zoonotic infections. Based on the source of infection and mode of spread, zoonotic diseases can be either anthroponoses (infection originates from human), zoonoses (animals act as the source of infection), or sapronoses (infection originates from non-living substance like soil, water, decaying plants, or animal corpses, excreta, etc.). Fungal zoonoses originate from animals infected with fungi, including molds and/or yeast. Fungi are eukaryotic microorganisms inhabiting the soil and organic matter as saprophytes or living as parasites on plants, animals, or humans. Even though numerous zoonotic fungal diseases exist and have the potential to cause serious public health problems, not good enough attention has been given, which has led to deficient management, prevention, and control measures.

Fungal infections may be either invasive or superficial. Particularly in immunocompromised patients, being an important cause of morbidity and mortality, many invasive mycotic infections have become a serious public health issue. Moreover, the incidence and prevalence of the mycotic diseases have continued to increase over the years. The most common etiological agents of invasive fungal diseases are the infections caused by mucormycetes, *Cryptococcus*, *Histoplasma*, *Blastomyces*, *Aspergillus*, and *Fusarium*. The lack of an effective diagnostics and the emergence of resistance against commonly used antifungal agents have become more challenging for the management of these infections. The selection of a suitable therapeutic agent is risky due to the unknown mechanisms of

Abhishek (✉) · S. S. Nair · V. Athira · P. Thomas
Division of Bacteriology and Mycology, ICAR-Indian Veterinary Research Institute, Bareilly, Uttar Pradesh, India
e-mail: abhivbm@gmail.com

B. Kumar
Division of Biological Products, ICAR-Indian Veterinary Research Institute, Bareilly, Uttar Pradesh, India

S. C. Parija, S. M. Rudramurthy (eds.), *Textbook of Fungal Zoonoses and Sapronoses*, Microbial Zoonoses, https://doi.org/10.1007/978-981-97-3214-2_8

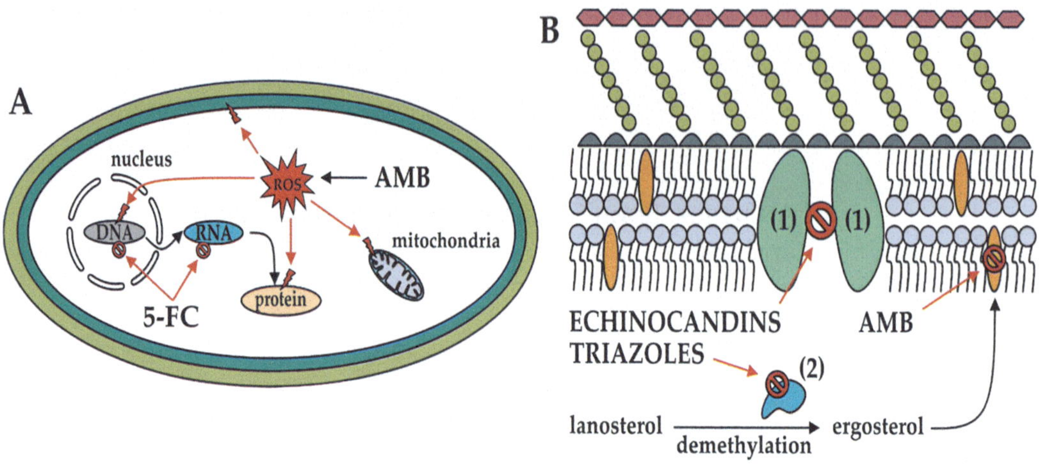

Fig. 8.1 Mechanism of action of a few antifungal drugs (**A**). Amphotericin B (AMB) results in the production of reactive oxygen species and causes damage to cellular structures. 5-flucytosine (5-FC) prevents the synthesis of proteins by acting on DNA and RNA (**B**). 1,3-β-D-glucan synthase enzyme (1) is inhibited by echinocandins. 14-α-Demethylase enzyme (2) is inhibited by triazoles. Sequestration of ergosterol present in the membrane by amphotericin B (AMB) results in pore formation. [Reproduced from Houšť et al. (2020) under the Creative Commons Attribution 4.0 International License (http://creativecommons.org/licenses/by/4.0/)]

drug interactions and adverse effects. The antifungal drugs used for managing mycotic diseases are very limited (Fig. 8.1). Amphotericin B formulations, echinocandins, and the triazole group of drugs are mostly used against invasive fungal infections, whereas flucytosine (5-fluorocytosine) is rarely used for such infections (Table 8.1).

Griseofulvin, azoles, and allylamines are usually recommended drugs for the therapy of superficial mycotic infections. Topical application is effective in the superficial infections, and such formulations may be used in the form of creams, solutions, or powders. Apart from the nature of lesions and the mechanism of action of the drug, the consistency, hydrophobicity, and pH of the drug preparations should be considered for the treatment of superficial or cutaneous infections. Even with effective antifungal therapeutic strategies, there is an increasing trend in incidence and mortality rate, especially due to the systemic fungal infections. It necessitates the boosting of the patient's immune system by immunotherapy, immunoprophylaxis, and or by vaccines.

The present chapter briefly elaborates on the antifungal drugs used for the management of fungal zoonoses/sapronoses in animals, adjunct therapy, development of newer antifungal drugs, and repurposing of drugs for the management of fungal zoonoses and sapronoses.

Table 8.1 Antifungal drugs, their mechanism of action, and indications in animals and humans

Class	Drugs	Mechanism of action	Indications	
			Animals	Humans
Imidazoles	Ketoconazole Miconazole Enilconazole Clotrimoxazole	Inhibits the 14-α-demethylase enzyme involved in the ergosterol synthesis	Dermatophytosis Canine and feline histoplasmosis (itraconazole) Canine blastomycosis (itraconazole) Canine paracoccidioidomycosis (itraconazole) Canine aspergillosis (itraconazole) Basidiobolomycosis (itraconazole) Eumycetoma (itraconazole) Feline chromoblastomycosis (itraconazole) Canine and feline sporotrichosis (ketoconazole, itraconazole)	Dermatophytosis Histoplasmosis (oral itraconzole is indicated after amphotericin B therapy) Mild or moderate forms of blastomycosis Paracoccidioidomycosis (itraconazole) Coccidioidomycosis (fluconazole) Basidiobolomycosis (itraconazole) Eumycetoma (itraconazole) Chromoblastomycosis (itraconazole) Sporotrichosis (itraconazole) Mucormycosis (posaconazole, isavuconazole)
Triazoles	Fluconazole Itraconazole Voriconazole Posaconazole Isavuconazole			
Allylamines	Terbinafine Naftifine	Inhibits squalene epoxidase enzyme involved in the ergosterol synthesis	Dermatophytosis	Dermatophytosis Eumycetoma (combination therapy with itraconazole) Chromoblastomycosis (terbinafine) Sporotrichosis (terbinafine)
Echinocandins	Caspofungin Micafungin Anidulafungin	Inhibits 1,3-β-D-glucan synthase enzyme involved in the synthesis of fungal cell wall polysaccharide β-D-glucan		Invasive aspergillosis
Pyrimidine synthesis inhibitor	5-flucytosine	Enters the fungal cell with the help of permease and deaminates to 5-fluorouracil which is then incorporated in the mRNA leading to poor functioning 5-fluorouracil is converted into 5-fluorodeoxyuridine monophosphate and inhibits thymidine synthase enzyme involved in DNA synthesis		Cryptococcosis (combination therapy with amphotericin B)

(continued)

Table 8.1 (continued)

Class	Drugs	Mechanism of action	Indications	
			Animals	Humans
Polyenes	Amphotericin B Natamycin Nystatin	Binds to ergosterol present in the fungal cell membrane and alters membrane permeability	Canine coccidioidomycosis (amphotericin B) Cryptococcosis in animals exhibiting CNS manifestations (amphotericin B) Mucormycosis (amphotericin B)	Histoplasmosis (amphotericin B) Blastomycosis with severe manifestations (amphotericin B) Severe paracoccidioidomycosis (amphotericin B) Coccidioidal meningitis (amphotericin B) Cryptococcosis (amphotericin B) Invasive aspergillosis (amphotericin B) Mucormycosis (amphotericin B) *Fusarium* keratitis (Natamycin)
Others	Griseofulvin	Disorganizes the spindle microtubules and inhibits mitosis	Dermatophytosis	
	Ciclopirox	Interferes with DNA repair mechanism, cell signaling, and intracellular transport		Dermatophytosis

Antifungal Drugs Used for the Management of Fungal Zoonoses/Sapronoses in Animals

Dermatophytosis

Though spontaneous recovery in animals is common, treatment may check the spread of infection. Local treatment comprises of the removal of crusts by scraping or brushing followed by application of topical agents like tincture iodine, Whitfield ointment, 10% ammoniated mercury ointment, mycoderm powder, and nystatin cream. Systemic treatment, which includes intravenous injection of sodium iodide and oral administration of griseofulvin, may prove useful when lesions are widespread. Topical medications are more frequently used in animals than the systemic formulations. Polyenes, miconazole, and enilconazole are mostly used as topical preparations in the form of ointments or shampoo. Oral preparations of terbinafine, itraconazole, and griseofulvin are also applied in the severe and prolonged conditions. Steroids can be administered along with antifungal drugs in the typical inflammatory conditions such as kerion.

Histoplasmosis

There is no chemotherapeutic agent available that gives consistently satisfactory treatments results in histoplasmosis. Though, Amphotericin B, ketoconazole, and itraconazole are recommended for some clinical cases but they do not have desirable effects. Itraconazole is the drug of choice for the treatment of histoplasmosis in dogs and cats. Relapse is common in animals treated with amphotericin B.

Blastomycosis

The drug of choice for blastomycosis in dogs is itraconazole. A combination of ketoconazole and amphotericin B can be used to treat ocular blastomycosis. Central nervous system (CNS) manifestations in cats can be treated with amphotericin B.

Paracoccidioidomycosis

Itraconazole can be successfully used for the treatment of canine paracoccidioidomycosis, but remissions may be common. Treatment with ketoconazole may result in the recurrence of the disease.

Coccidioidomycosis

Amphotericin B and azole drugs are used in the treatment of canine coccidioidomycosis. Ketoconazole is used to treat canine coccidioidomycosis, but relapses are common. Fluconazole has been successfully used to treat few cases of equine pulmonary coccidioidomycosis.

Cryptococcosis

Amphotericin B is the drug of choice for animals exhibiting CNS manifestations. For pulmonary and osseous forms, itraconazole can be employed. Amphotericin B, along with sodium iodide and surgical interventions, can be used to cure the cryptococcal rhinitis.

Aspergillosis

Itraconazole is used for the systemic antifungal therapy in dogs. Azole infusion therapy or azole topical therapy, or combination therapy, including systemic and topical drugs, are also used in canine aspergillosis. Azole drugs can be used for equine aspergillosis, but the success rates are low.

Entomophthoromycosis

Surgical resection followed by itraconazole therapy is the best treatment for animals. Potassium iodide and sodium iodide, along with antifungal drugs, are used for equine conidiobolomycosis.

Maduramycosis

Itraconazole is used to treat eumycetoma in dogs. In equines, surgical intervention along with thiabendazole topically can be employed for the cure of eumycetoma. Surgical resection, along with sodium iodide and potassium iodide, has been successfully employed for oral eumycetoma in horses.

Chromoblastomycosis

Itraconazole is used for the treatment of feline chromoblastomycosis, but a relapse may occur.

Fusariosis

Surgical resection, along with itraconazole therapy, is used to treat granulomatous pododermatitis in cats. Itraconazole has been used to treat canine onychomycosis caused by *Fusarium* spp.

Sporotrichosis

Ketoconazole, itraconazole, and potassium iodide are commonly used for the management of canine sporotrichosis. Terbinafine has also been used for the successful treatment of canine cutaneous sporotrichosis. Itraconazole and ketoconazole are also used for feline sporotrichosis. Cats treated with ketoconazole have higher resolution of clinical signs than the cats treated with itraconazole. Posaconazole can be used for treating severe systemic sporotrichosis.

Mucormycosis

Amphotericin B is considered the drug of choice for mucormycosis in animals, but the recovery rate is poor. Potassium iodide and mercuric chloride can be used to treat cutaneous lesions.

Adjunct Therapy Used for the Management of Fungal Zoonoses/Sapronoses

The clinical syndrome associated with mycotic disease are manifestations associated with the growth of fungal organisms and the damage in the tissues associated with the physical and biological activities of the agent. Antifungal drugs inhibit or destroy the etiological agents; however, to minimize the damages and associated clinical symptoms, adjunct therapy may be benifical to the infected individual.

Immunotherapeutic/ Immunomodulating Therapeutics

Breaches or defects in the immune defense mechanism are a major contributing factor to the progression of many mycotic diseases. Immunotherapeutic agents along with antifungal drugs are used for better clinical response and survival rates. Monoclonal antibodies, cytokines, and cellular immunotherapy are most commonly employed as immunotherapeutic agents.

Monoclonal antibodies (Mabs) are increasingly used in the treatment of infectious diseases, cancer, and immune-mediated diseases. The utility of Mabs in the management of mycotic diseases has been demonstrated in certain clinical conditions, such as allergic bronchopulmonary aspergillosis. Omalizumab (Mab against immunoglobulin E) and mepolizumab (Mab against interleukin 5) have been used to improve the outcome of allergic bronchopulmonary aspergillosis. Monoclonal antibodies such as AP3, 7B8, 8G4, C7, and A9 have been generated against *Aspergillus* spp. and demonstrated promising results in animal models. Mabs targeted against the CotH3 (involved in the invasion) protein found to enhance the phagocytosis of the etiological agents, reduce mortality in mice models, and were effective against many Mucorales species. These antibodies were also found to increase the therapeutic efficiency of amphotericin B and posaconazole against mucormycosis. Dupilumab, Mab against the IL-4 receptor, has been successfully employed to resolve the clinical syndromes observed in disseminated coccidioidomycosis due to the excess synthesis of IL-4. Similarly, Mab-based therapy can be employed for the management of infections due to *C. neoformans*, *Sporothrix schenkii*, *B. dermatitidis*, etc.

Cytokines are chemical mediators of the immune system that are capable of upregulating or downregulating immune responses. Deficiency of cytokines favor the establishment and enhanced proliferation of opportunistic pathogens in immune-compromised patients. Interferon-gamma (INF-γ) has been used in the management of aspergillosis in patients subjected to solid organ transplantation. Recombinant INF-γ partially stimulates the cell-mediated immune response, resulting in elevated cytokine synthesis and increasing the number of monocytes and lymphocytes. Exogenous INF-γ has also been employed for the successful therapeutic management of the infection due to *Exophilia dermatitidis*. INF-γ therapy has also been observed to increase the clearance rate of *Cryptococcus* from the cerebrospinal fluid (CSF). It has also been employed in augmenting immune response against disseminated coccidioidomycosis. Tumor necrosis factor-alpha and interleukins such as IL-12 have also been experimented as immunotherapeutic agents against fungal pathogens. Similarly, colony-stimulating factors (CSFs) such as macrophage-colony stimulating factors, granulocyte-colony stimulating factors, and granulocyte-macrophage colony-stimulating factors have also been employed as immunotherapeutics for the management of fungal diseases. They can be used prophylactically, and it has been observed that two or more CSFs, in combination with antifungal drugs, exhibited better efficiency in reducing mycotic infections.

T cells play an important role in the defense mechanism against fungal diseases. Adaptive transfer of fungi-specific T cells can be employed in T cell-deficient patients. *Aspergillus*-specific T cells have been used for the management of invasive forms of aspergillosis. A mixture of CD4 and CD8 cells produced by stimulation with *Aspergillus fumigatus* lysate was found to be effective against *Aspergillus* spp., *Fusarium* spp., and *Candida* spp. Another emerging aspect of immunotherapeutics is the use of engineered chimeric antigen receptor T cells or CAR-T cells. These are cytotoxic T cells that are modified to express antigen-specific receptors such as single-chain variable fragments. For tackling invasive fungal infections, these cells express fungal binding regions of receptors such as Dectin-1 and are named D-CART cells. These cells have exhibited promising results in mice models for *A. fumigatus* infections. Granulocyte transfusion for the management of the fungal disease has limited success. Dendritic cell therapy using dendritic cells stimulated with *A. fumigatus* and transduced with IL-12 demonstrated a strong Th1 response. Similarly, NK cell therapy is found to have immunotherapeutic potential against many fungal agents, including *P. brasiliensis*, Mucorales, *C. neoformans*, *A. fumigatus*, etc.

Other immunotherapeutic agents, such as TLR agonists and pentraxin-3, have also been tested for their utility in managing fungal diseases. Checkpoint inhibitors such as anti-PD-1 antibodies have been used in combination with INF-γ for the management of mucormycosis, which was unresponsive to conventional therapy and exhibited a better clinical response.

Neurapheresis

The neurapheresis system was initially developed to remove the blood products from the CSF. It has been investigated for clearing fungal pathogens such as *C. neoformans* from the CSF. It is a closed-loop platform that collects the CSF from the lumbar subarachnoid space, filters and redirects the CSF back through the midthoracic region. This filtration technique removes the pathogen and virulence factors and, thus, can be used as adjunct therapy for the management of fungal-borne meningitis.

Photodynamic Therapy

Photodynamic therapy is a promising approach for the management of cutaneous fungal infections, including onychomycosis, cutaneous sporotrichosis, chromoblastomyosis, etc. Photodynamic therapy utilizes molecular oxygen, a photosensitizer, and a light source. The commonly used photosensitizers include methylene blue, 5-aminolevulinic acid, and methylamino levulinate. The photosensitizer absorbs energy from the light source, which induces alterations in its energy state. The photosensitizer then reacts with molecular oxygen to produce reactive oxygen species, which will damage the injured tissue. The photosensitizer may also direct the synthesis of singlet oxygen by the transfer of energy to molecular oxygen. In effect, photodynamic therapy will result in fungal apoptosis and cellular injury.

Photodynamic therapy is widely employed for the treatment of onychomycosis. A combination of photodynamic therapy and carbon dioxide laser therapy was found to be highly effective for the treatment of recurrent onychomycosis. The photosensitizers exhibit variations in the penetration capability of nails, with methylene blue exhibiting better penetration. Nanotechnology has also been used to deliver photosensitizers in the form of nanoemulsions, which assists in better penetration ability. Photodynamic therapy is commonly practiced along with the administration of antifungal drugs. Photodynamic therapy, along with itraconazole or tetracycline, has been successfully employed in the treatment of cutaneous blastomycosis. Similarly, photodynamic therapy, along with drug isotretinoin and carbon dioxide laser, has also been used for the treatment of chromoblastomycosis. In the case of ulcerated lesions in *Paracoccidioides brasiliensis* infection, photodynamic therapy can be combined with amphotericin B, and cotrimoxazole that may assist in the repairing process and decontamination of the lesion. Methylene blue-based photodynamic therapy, along with itraconazole, has resulted in complete clinical and mycological cures in patients suffering from recalcitrant cutaneous sporotrichosis. Similarly, 5-aminolevulinic acid-based photodynamic therapy, along with oral terbinafine and itraconazole, has been successfully employed for treating phaeohyphomycosis caused by *Exophiala* spp.

Supportive Medical Assistance and Medications Other Than Antifungal Drugs

Supportive medical devices and non-antifungal medications assist in elevating the survival rate of patients suffering from severe fungal diseases. These supportive medications are essential for sustaining life and to lessen the severity of the clinical syndrome associated with mycoses. Most of the fungal diseases are manifested in a range of clinical forms, and thus, the treatment protocol may vary accordingly. The immune status of the patients also influences the treatment regime.

In severe forms of paracoccidioidomycosis, anti-inflammatory drugs such as corticosteroids are incorporated into the treatment regime. Early initiation of oseltamivir therapy in influenza-affected patients avoids the risk of developing invasive pulmonary aspergillosis. Iron chelators such as deferasirox are also used for the clinical management of mucormycosis, as iron overload is a major predisposing factor for the development of mucormycosis. In severe pulmonary blastomycosis with hypoxia and hypoxemia, endotracheal intubation and extracorporeal membrane oxygenation can be employed.

Hyperbaric and hyperoxic conditions negatively affect the growth of Mucorales and have been employed in the management of invasive mucormycosis. Extracorporeal membrane oxygenation (ECMO) utilizes a heart-lung machine to remove the carbon dioxide from the blood pumped outside the body and redirects oxygen-rich blood back to the patient body. ECMO can be employed in patients suffering from acute respiratory distress syndrome associated with blastomycosis. Surgical debridement of devitalized tissue or granulomas can be performed in cutaneous mycotic diseases. Surgical interventions can also be attempted in *Aspergillus*-associated osteomyelitis.

Newer Antifungal Drugs

The major hurdle in the therapeutic management of mycotic diseases is the availability of a limited antifungal drugs. However, both fungi and mammals are eukaryotes and share similarities in biochemical and physiological activities; currently, only a few antifungal targets have been identified. Thus, the development of antifungals against new cellular or biochemical targets in the fungal pathogens or the generation of superior antifungals in the already existing antifungal classes may require strenuous laboratory research and clinical trials. Isavuconazole is the single antifungal drug that has been approved by the US FDA for the treatment of fungal disease over the past few years. However, multiple trials at varying stages are ongoing across the world for the generation of newer and better antifungals.

Rezafungin, a novel echinocandin, has the potential to be used as a prophylactic agent against aspergillosis. It has also been found to be effective in preventing infection due to *Pneumocystis* spp. and *Candida* spp. It is a structural analog of anidulafungin and is considered to have reduced hepatotoxicity. Ibrexafungerp, a derivative of enfumafungin, is an oral glucan synthase inhibitor and is indicated for invasive aspergillosis and invasive candidiasis. Biofungin, a novel echinocandin, has shown promising results in mice models of invasive aspergillosis.

Fosmanogepix belongs to the newer class of antifungal drugs and inhibits the GWT1 (GPI-anchored wall protein transfer) protein required for the inositol acylation associated with the biosynthesis of glycosylphosphatidyl inositol anchors. These drugs subsequently result in the disruption of the cell wall integrity and do not affect the inositol acylation in humans. The drug is found to be effective against several fungal zoonoses/sapronoses agents such as *Aspergillus* spp., Mucorales, *C. immitis*, *C. neoformans*, etc.

Nikkomycin Z is a chitin synthesis inhibitor and has demonstrated good in vitro activity against *B. dermatitidis* and *C. immitis*. Along with echinocandins, they exhibit synergistic action against *A. fumigatus*. Oteseconazole belongs to a next-generation class of azole drugs named tetrazoles.

They have a higher affinity for fungal CYP450 enzyme than that of mammals and exhibited improved selectivity toward lanosterol 14α demethylase. These drugs exhibit potential in vitro activity against *C. neoformans* and Mucorales.

Encochleated amphotericin B is a novel oral formulation of amphotericin B where cochleates act as a platform for the safer delivery of amphotericin B. Cochleates are phospholipid bilayer sheets that are arranged into a spiral structure and consist of phospholipids such as phosphatidyl serine and divalent cations such as calcium. Cochleates prevent the denaturation of amphotericin B in the gastrointestinal tract and exhibit reduced toxicity compared to the conventional amphotericin B formulations. Aureobasidin A is an antibiotic compound isolated from *Aureobasidium pullulans*. This compound inhibits the inositol phosphoryl ceramide synthase enzyme involved in the synthesis of sphingolipids, which are the components of the fungal cell membrane. It is highly specific for fungi as the enzyme is exclusively detected in fungal cell membranes, resulting in reduced toxicity in mammals. The drug exhibits potential *in vitro* activity against *Cryptococcus* spp. and *Aspergillus* spp.

Orotomides are a novel class of antifungal drugs act by inhibiting the pyrimidine synthesis in fungi. They inhibit the enzyme dihydroorotate dehydrogenase (DHODH), which is involved in the conversion of dihydroorotate to orotate. Olorofim belongs to this class of drugs and has no activity against the corresponding enzymes in humans, ensuring a good safety profile. These drugs are found to be effective against triazole-resistant fungi and can be effective against invasive aspergillosis and infections due to *Lomentospora* spp. Certain drugs, such as 4-methoxynaphthalene-*N*-acylhydrazone, have promising activity against both paracoccidioidomycosis and tuberculosis.

Repurposing of Drugs

Drug repurposing is an alternative approach to the development of newer antifungal drugs. The commercially available non-antifungal drugs are tested for their antifungal activity, and those exhibiting satisfactory antifungal activity are repositioned as antifungal drugs. Drugs can be repositioned using in silico/computational approaches or experimental approaches. Experimental approaches include (a) dilution methods based on standard guidelines such as CLSI or EUCAST, (b) the effect on biofilm formation, or (c) phenotypic variations in the fungi, and the use of suitable animal models.

Members from different classes of drugs, such as immunosuppressants, anti-arrhythmic drugs, statins, antipsychotic drugs, antidepressant drugs, and NSAIDs, have been repurposed for their antifungal activities. Haloperidol, an antipsychotic drug, was observed to inhibit melanin production and biofilm formation in *C. neoformans*. Mefloquine, an antimalarial drug, can inhibit melanin synthesis and capsule formation in *C. neoformans*. Ebselen, an organoselenium compound, has shown promising results in the mice model of invasive aspergillosis. Drug repositioning has indicated promising anti-cryptococcal activity of benzimidazole anthelmintics like fenbendazole, mebendazole, albendazole, and flubendazole. The antidiabetic drug pioglitazone can be combined with standard antifungal therapy for cryptococcosis. The combination acts as an antidote for amphotericin toxicity, thereby increasing the survival rate of the patients. Miltefosine is an anti-leishmanial drug, which inhibits the filamentous/yeast phase of *H. capsulatum* and the filamentous phase of *C. posadasii*. Robedine, an anti-coccidial drug, demonstrated inhibitory activity against *A. fumigatus*. Raltegravir, a retroviral integrase enzyme inhibitor, has reduced the fungal burden of *Paracoccidioides* spp. in a mouse model. *N*-acetyl cysteine administration resulted in increased survival and reduced fungal burden in the brain and lungs of mice infected with *C. gattii*. Minocycline, an antibiotic of class tetracycline, has in vitro activity against *A. flavus*, *A. fumigatus*, and *Fusarium* spp. Similarly, other antibiotic drugs, such as gentamicin, streptomycin, azithromycin, ciprofloxacin, polymixin B, linezolid, etc., demonstrated antifungal activities.

Repurposed drugs are used in combination with conventional antifungal drugs, and synergism can be observed in certain combinations of drugs. In many cases, synergism can occur when the repurposed drug potentiates the activity of the antifungal drugs. Lovastatin, a cholesterol-lowering drug, has a synergistic action with itraconazole or amphotericin B against several fungal pathogens, including *Rhizopus oryzae*, *T. mentagrophytes*, *Mucor circinelloides*, *A. fumigatus*, etc. Tamoxifen, an estrogen receptor modulator, exhibits synergism with amphotericin B against *C. neoformans*. Antibiotics also exhibit synergic effects when used in combination with antifungals. Tobramycin, in combination with voriconazole and amphotericin B, has synergistic action against *Fusarium* spp. Similarly, linezolid exhibits synergistic action with amphotericin B against *C. neoformans*.

Conclusion

The selection of appropriate antifungal drugs and supportive medications at an early stage is the key to the successful treatment of fungal zoonoses/sapronoses. A suitable antifungal drug has to be chosen from the limited collection of the drugs available. The route of administration needs to be taken into consideration while selecting the drug for the treatment. Some formulations may have more side effects depending on the physiological conditions of the patients. The strain variation in the antifungal susceptibility may also create further hurdles in the treatment strategies. The probable emergence of resistance and its mechanisms are also to be considered with utmost importance. Repositioning or repurposing of other medications may also be a viable option for combating fungal diseases. Apart from the selection of appropriate antifungal drugs, physical interventions, and supportive medications also play a crucial role in the recovery process. A common treatment regime may be formulated for the most frequently encountered fungal diseases after conducting a number of trials on people from different physiological statuses and ages. This will help to resist the fungal diseases at the initial stage, thereby increasing the recovery rate. Along with currently available treatment options, there is need to focus on other diverse strategies for proper management of fungal infections. Particularly against invasive fungal infections. These include prophylactic therapy, immunotherapy, and the use of vaccines. The means to restrict exposure of fungal spores in the case of high-risk patients need to be strictly followed. Direct or indirect contact with any objects that can predispose to the fungal infection should be avoided. The prophylactic therapy is practiced with an azole group of drugs and interferon-γ for successful management of the infection. Timely initiation of the prophylactic antifungal therapy, to avoid the development of resistance. Immunotherapy has shown promising results in the management of invasive aspergillosis in both humans and animals, may be explored for the effective management of fungal diseases.

Case Study

A 2-year-old Persian cat was found to have nodules on the forehead. The animal was also suffering from alopecia, itching, and crust formation in the face, limbs, and forehead. The physiological parameters were within the normal range, and the animal was properly vaccinated and dewormed. Skin scraping was collected from the regions of alopecia and crust formation. Intact hair, along with follicles, was plucked using sterile forceps, and fluid was aspirated from the subcutaneous nodules. The hair, skin, and scab materials were subjected to direct microscopic examination (DME) after treatment with 10% KOH. An ectothrix pattern of spores was observed along the hair shaft. Spores were also found to be dispersed in bunches in various fields. The fluid material was collected from the nodules using hypodermic needles. Furthermore, the clinical samples were inoculated in abouraud dextrose agar (SDA) plates supplemented with chloramphenicol and cycloheximide and incubated at 25–30 °C. After 5 days of incubation, multiple yellow-pigmented colonies were detected. On microscopic examination, after Lactophenol cotton blue (LPCB) staining spindle-shaped, rough-walled macroconidia could be detected, which is suggestive of

Microsporum canis infection. The animal was treated with itraconazole for 3 months, along with topical medications. The animal was completely recovered without any recurrence.

1. What are the phenotypic and molecular methods used to characterize *M. canis*?
2. What are the predisposing factors associated with subcutaneous infection by dermatophytes?
3. What is the difference between kerion and pseudomycetoma?
4. What are the components of lactophenol cotton blue stain?

Research Questions

1. Optimum concentration of the antifungal drugs in deeper tissues like bones, brain, and lungs is a major barrier in eliminating the etiological agents. What new formulations or delivery systems can be developed for successful drug delivery and attaining the required concentration?
2. Toxicity of antifungal drugs is a major setback and results in adverse health effects in the patients. How can we reduce the toxic effects of the various drugs without altering the efficacy?
3. Resistance in fungal agents, including dermatophytes, results in reduced success rate and recurrence. What may be the possible selection pressure for developing antifungal resistance apart from overuse of the drug?
4. What may be the possible mechanism of transfer of drug resistance within and among various species?

Further Reading

Ademe M (2020) Immunomodulation for the treatment of fungal infections: opportunities and challenges. Front Cell Infect Microbiol 10:469

Boniche-Alfaro C, Kischkel B, Thomaz L, Carvalho-Gomes MM, Lopes-Bezerra LM, Nosanchuk JD, Taborda CP (2021) Antibody-based immunotherapy combined with antimycotic drug TMP-SMX to treat infection with *Paracoccidioides brasiliensis*. Front Immunol 12:725882

Brandolt TM, Madrid IM, Poester VR, Sanchotene KO, Basso RP, Klafke GB, Rodrigues MDL, Xavier MO (2019) Human sporotrichosis: a zoonotic outbreak in southern Brazil, 2012–2017. Med Mycol 57(5):527–533

Chapman SW, Dismukes WE, Proia LA, Bradsher RW, Pappas PG, Threlkeld MG, Kauffman CA (2008) Clinical practice guidelines for the management of blastomycosis: 2008 update by the Infectious Diseases Society of America. Clin Infect Dis 46(12):1801–1812

Espinel-Ingroff A (2009) Novel antifungal agents, targets or therapeutic strategies for the treatment of invasive fungal diseases: a review of the literature (2005–2009). Revista Iberoamericana de Micología 26(1):15–22

Houšť, J., Spížek, J. and Havlíček, V., 2020. Antifungal drugs. Metabolites, 10(3), p.106.

Kim JH, Cheng LW, Chan KL, Tam CC, Mahoney N, Friedman M, Shilman MM, Land KM (2020) Antifungal drug repurposing. Antibiotics 9(11):812

Rauseo AM, Coler-Reilly A, Larson L, Spec A (2020) Hope on the horizon: novel fungal treatments in development. In: Open forum infectious diseases, vol 7. Oxford University Press, p ofaa016

Williams TJ, Harvey S, Armstrong-James D (2020) Immunotherapeutic approaches for fungal infections. Curr Opin Microbiol 58:130–137

Zhang Q, Liu F, Zeng M, Mao Y, Song Z (2021) Drug repurposing strategies in the development of potential antifungal agents. Appl Microbiol Biotechnol 105:5259–5279

9 Antifungal Agents and Resistance

Rimjhim Kanaujia and Gagandeep Singh

Learning Objectives

By the end of the chapter, the readers should be able to:

1. Classify antifungal agents based on the mechanism of their actions.
2. Understand the mechanisms of resistance of antifungal agents.
3. Review antifungal agents used for the treatment of fungal diseases.
4. Learn methods of antifungal resistance detection.

Introduction

Fungal infections constitute a significant burden in veterinary medicine. The spectrum of fungi causing fungal infections is increasing, while the treatment options are limited. The currently available antifungal agents include azoles, polyenes, echinocandins, pyrimidine analogs, allylamine, thiocarbamates, and morpholines. Other agents like iodine and hydroxystilbamidine isethionate (HSB) are also used in veterinary medicine for the treatment of fungal infections. Antifungal resistance is defined as the growth of fungi beyond their minimum inhibitory concentration (MIC). Some fungal species have intrinsic resistance due to the presence of efflux pumps or their inability to bind to the antifungal agent, e.g., the resistance of *Cryptococcus* spp. against amphotericin B as it lacks beta-D-glucan in the cell membrane. Acquired resistance refers to the acquisition of resistance mechanisms during the therapy or from the environment, which is enabled either by the alteration/overexpression of the drug-target site or inactivation of the conversion of pro-drug to the active form. Understanding the underlying mechanisms contributing to antifungal resistance helps physicians to provide antifungal susceptibility-guided therapy and formulate institutional guidelines for the treatment of fungal infections. The chapter will review available antifungals along with underlying molecular resistance mechanisms and various methods to detect antifungal susceptibility.

Azoles

Ergosterol serves as a crucial component for the maintenance of integrity and membrane fluidity of fungal cell walls.

R. Kanaujia
Department of Medical Microbiology, PGIMER, Chandigarh, India

Department of Microbiology, Punjab Institute of Liver and Biliary Sciences, Mohali, Punjab, India

G. Singh (✉)
Department of Microbiology, AIIMS, New Delhi, India
e-mail: drgagandeep@gmail.com

S. C. Parija, S. M. Rudramurthy (eds.), *Textbook of Fungal Zoonoses and Sapronoses*, Microbial Zoonoses, https://doi.org/10.1007/978-981-97-3214-2_9

Mechanism of Action

Azoles are fungistatic drugs, primarily targeting the 14′-lanosterol demethylase (LAD) enzyme of ergosterol biosynthesis. The abrogation of LAD causes inhibition of cytochrome P450-dependent conversion of ergosterol from lanosterol. The lack of ergosterol hampers membrane integrity and cell growth (Fig. 9.1). Additionally, ergosterol depletion also disrupts vacuolar ATPase functions, i.e., acidification of vacuoles and ion homeostasis. Finally, LAD inhibition leads to the accumulation of toxic methylated sterols produced by ergosterol, leading to membrane stress in the fungal cell wall. This mechanism is more evident with certain azoles like itraconazole/fluconazole, where obtusifolione accumulates in the cell, leading to stress in the yeast membrane. In *Candida* spp. and *Cryptococcus* spp., ERG11 gene encodes the LAD. In *A. fumigatus*, LAD is encoded by *cyp*51A and *cyp*51B genes.

Azoles are classified depending on the presence of nitrogen atoms in the cyclic azole ring. Imidazoles (miconazole, econazole, oxiconazole, tioconazole, ketoconazole, and clotrimazole) contains two nitrogen atoms while triazoles (itraconazole, fluconazole, voriconazole, posaconazole, ravuconazole, albaconazole, isavuconazole, and efinaconazole) contains three nitrogen atoms. Due to broad-spectrum activity, bioavailability, and safety profile, azoles have remained a widely used class of antifungal. Except for ketoconazole, all imidazoles have limited utility in the treatment of superficial fungal infections. Triazoles have a broader spectrum of applications in the treatment of both superficial and systemic fungal infections. Triazoles also have a better safety profile than imidazoles as they have an affinity for mammalian cytochrome P-450 enzymes compared to fungal enzymes (Table 9.1).

Mechanisms of Azole Resistance

Azoles have been used widely as a prophylactic agent. Also, the fungistatic mechanism allows the selection and survival of resistant fungal cells. These factors have led to the emergence of resistance. Broadly, two routes, long-term azole prophylaxis or treatment and use of azole fungicides in agriculture, i.e., environmental route, have been described. Continued monitoring and surveillance of azole resistance are necessary, espe-

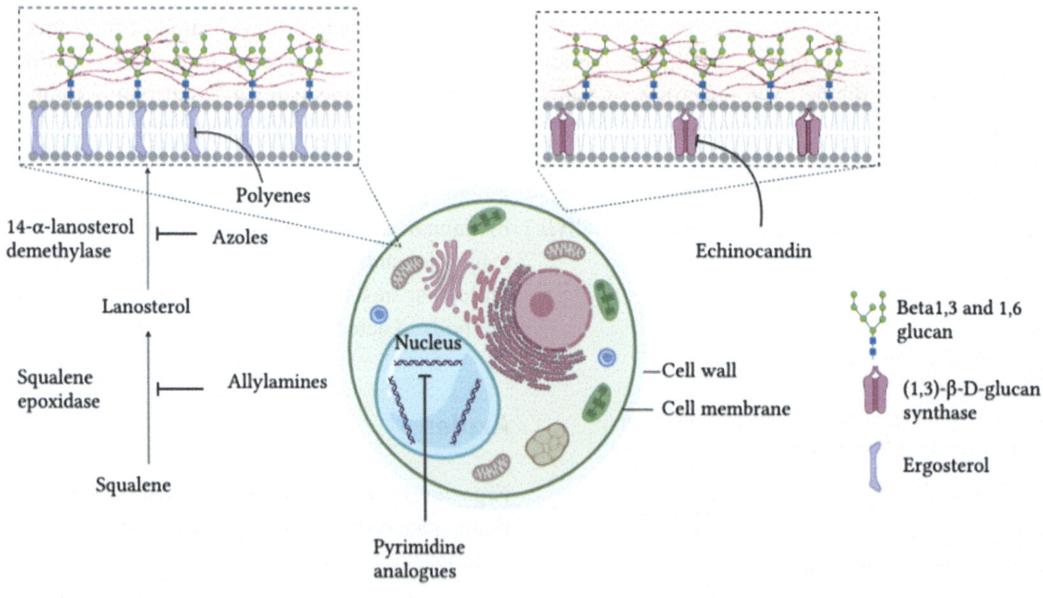

Fig. 9.1 Anti-fungal agents and their mechanism of resistance

Table 9.1 Mechanisms of action and development of resistance in common antifungal agents

Antifungal groups	Antifungal agents	Mechanism of action	Mechanism of development of resistance	Comments
Azoles	*Imidazoles*: Miconazole, oxiconazole, econazole, ketoconazole, tioconazole, and clotrimazole *Triazoles*: Itraconazole, fluconazole, voriconazole, posaconazole, ravuconazole, albaconazole, isavuconazole, and efinaconazole	Azoles bind to 14′-lanosterol demethylase (LAD) enzyme of ergosterol biosynthesis, which leads to inhibition of cytochrome P450-dependent conversion of ergosterol from lanosterol. Lack of ergosterol hampers membrane integrity and cell growth Ergosterol depletion disrupts vacuolar ATPase functions, i.e., acidification of vacuoles and ion homeostasis The inhibition of LAD leads to accumulation of toxic methylated sterols produced by ergosterol and responsible for membrane stress	Active efflux of drug by overexpression of CDR-1/2 and MDR-1 Overexpression/point-mutation of drug targets, i.e., *ERG11* and *Cyp51A* Mutation in ERG3 leads to tolerance to methylated sterols	Active against *Candida* spp. like *Candida albicans*, *Candida glabrata*, *Candida krusei*, and common molds such as *Aspergillus* and *Fusarium* spp.
Polyenes	Nystatin Natamycin *Amphotericin B formulations*: conventional AmB, Lipososmal AmB, deoxycholate AmB	Binds to ergosterol present in the fungal cell-membrane and increases membrane permeability Alteration of the membrane function by binding of ergosterol (depleting cells of ergosterol)	Defects in the ERG3 gene leads to decreased ergosterol content in cell membrane Mutation in ERG11, ERG6, ERG5, ERG3, and ERG2—formation of non-ergosterol cytoplasmic membrane sterols like lanosterol and eburicol	Active against Mucorales Intrinsic resistance has been observed in *Candida lusitanea*, *Candida guilliermondii*, *Aspergillus terreus*, *Scedosporium* spp., and *Fusarium* spp.
Echinocandins	Caspofungin, micafungin, and anidulafungin	Noncompetitive inhibition of 1,3-b-D-glucan synthase (and, to a lesser extent, 1,6-b-D-glucan synthase)	Mutation in FKS (FKS1 and FKS2 gene) subunits of glucan synthase decreases the affinity of drugs for the glucan synthase	Active against *Candida* spp. and *Aspergillus* spp. less active against *Candida parapsilosis* and *Candida guillermondii* Intrinsic resistance against species of *Cryptococcus* and Mucorales

(continued)

Table 9.1 (continued)

Antifungal groups	Antifungal agents	Mechanism of action	Mechanism of development of resistance	Comments
Pyrimidine analogs	5-Fluorocytosine (5-FC)	The 5-FC enters the fungal cell by permeases and is subsequently converted to 5-fluorouracil (5-FU) by cytosine deaminase. The 5-FU inhibits pyrimidine metabolism in the fungal cell	Mutation in cytosine deaminase, i.e., *FCY1* and *FCA1* genes Mutation gene encoding uracil phosphori-bosyltransferase (*FUR1*) decreases the conversion of 5-FU into toxic metabolite (5-FC monophosphate)	Active against *Candida* spp. and *Cryptococcus* spp. Intrinsic resistance against *Aspergillus* spp. Rapid emergence of resistance has been observed in *Cryptococcus* spp. and *Candida* spp. restricts use of 5-FC as monotherapy
Allylamine	Terbinafine and naftifine	Inhibit ERG1 gene and inhibit squalene epoxidation—causes accumulation of squalene and toxic sterols	Mutation in drug target, i.e., ERG1	Active against dermatophytes

cially with the increasing use in hospitals and agriculture sectors.

1. **Reducing intracellular drug accumulation/ decreased drug concentration**
   (a) **Active efflux**: Active efflux pumps at the site of action result in decreased drug concentration. Two major families of transporters, i.e., the ATP-binding cassette superfamily (CDR genes) and major facilitators class (MDR genes), encode the efflux pumps in *Candida* species. The upregulation of CDR-encoded efflux pumps (CDR1 and CDR2) has been associated with resistance to all azoles. In *Candida albicans*, the transcriptional activator Tac1 regulates the ATP-binding cassette (ABC) transporters Cdr1 and Cdr2. The loss of heterozygosity at the TAC1 locus confers a higher level of azole resistance. Upregulation of several other ABC transporters like CgCDR1, CgCDR2, and CgSNQ2 genes in *Candida glabrata* and AFR1 from *Cryptococcus neoformans* have also been implicated in conferring azole resistance. In *Aspergillus fumigatus*, AtrF is known to have a direct role in azole resistance.

   The upregulation of MFS genes (MDR1) confers resistance in *C. albicans* and *Candida dubliniensis*. The MDR gene upregulation has been observed to confer resistance specifically for fluconazole. The FLU1 (for Fluconazole resistance), an MFS gene, is upregulated in *C. albicans*, serving as a fluconazole efflux transporter. *A. fumigatus* encodes four Mdr-like (Mdr1–4) efflux pumps. However, the role of AfuMDR3 in clinical isolates remains elusive, as it has been observed to confer itraconazole resistance in laboratory mutants.

   (b) **Overexpression of drug target**: Increasing the target for drug attachment is another mechanism by which azoles acquire resistance in fungi. Thus, an increased concentration of the effective drug is required to saturate all drug targets. The resistance in *C. albicans* and *A. fumigatus* has been associated with overexpression of drug targets, i.e., ERG11 and Cyp51A, respectively. The duplication of 34- and 42-bp elements in the Cyp51A promoter region leads to specific mutations (Y121F/T289A, L98H) in *A. fumigatus* isolates.

(c) **Drug sequestration in extra- or intracellular compartments:** The fungi can form biofilms in specific growth conditions and sequester drugs within extracellular compartments and matrix. Recent studies have shown that the ability of fungi to sequester antifungals has led to resistance in azoles and other agents like polyenes and pyrimidine analogs.

2. **Target site alteration**

   The LAD is encoded by Cyp51A/Cyp51B in *A. fumigatus* and ERG11 in *C. albicans*. The point mutations in the ERG11 gene decrease the binding affinity of azoles to the enzymatic site. Various non-synonymous amino acid substitutions due to ERG11 mutations have been associated with azole resistance, which is mostly expressed in fluconazole and less in posaconazole. The high MIC values in many azole-resistant isolates have been observed due to simultaneous ERG11 mutations on the same ERG11 allele, leading to drug transport modifications. Point mutations in LAD genes have been reported in Cyp51A from azole-resistant *A. fumigatus*. Single Cyp51A mutations can confer high-level resistance to azoles in *A. fumigatus*.

3. **Upregulation of the target enzyme**

   Various mechanisms like increased rate of transcription, gene amplification, or decreased degradation of the gene product led to target enzyme upregulation. Higher intracellular concentrations of ERG11p lead to an overwhelming of the enzyme. Thus, azoles do not inhibit ergosterol synthesis effectively, leading to the failure of therapy.

4. **Development of bypass pathways**

   The formation of metabolic pathways is a compensatory mechanism, which is exerted due to the toxic effects of some antifungal agents. Exposure to azole leads to ergosterol depletion and growth arrest due to the accumulation of toxic products (14a-methyl-3,6-diol). The ERG3 gene mutation in fungi prevents the formation of toxic 14a-methyl-3,6-diol and contributes to resistance.

5. **Cross-resistance**

   Substitution of amino acid at the M220 and 54 positions of Cyp51A is associated with cross-resistance to itraconazole and posaconazole, respectively.

Polyenes

The polyenes are fungicidal, amphipathic, and naturally produced by *Streptomyces*.

Mechanism of Action

Polyenes bind to ergosterol present in the fungal cell membranes and increase permeability (Fig. 9.1). The presence of close structural similarity among mammalian membrane sterol cholesterol and fungal ergosterol limits its utility in animals due to severe toxicity. Nystatin, natamycin, and amphotericin B (AmB) belong to this class of antifungals. The common side-effect, i.e., nephrotoxicity of conventional AmB, is partially mitigated by alternative amphotericin B preparations with lipid-based drug delivery systems (Table 9.1).

Mechanisms of Polyene Resistance

The ERG3 gene plays a pivotal role in ergosterol synthesis, and mutations in this gene lead to the accumulation of various sterols in the cell membrane.

1. **Reducing intracellular drug accumulation/decreased drug concentration**

   Owing to the large molecular volume of polyenes, a mechanism like the efflux of the drug is not described in fungus. Recent data have shown that matrix in biofilm sequester AmB and contribute to the development of resistance in *C. albicans*.

2. **Target site alteration**
   Alterations in the composition of sterol are the most common mechanism of acquired resistance in polyenes. Various mutations in ergosterol synthesis pathway (ERG) genes have been implicated with this mechanism. The loss of function of ERG11, ERG6, ERG5, ERG3, and ERG2 genes leads to the formation of alternative sterols like lanosterol and eburicol.
   The increase in the 1,3-α-glucan production on exposure to polyenes has also been implicated as another mechanism contributing to resistance. The glucan molecules inhibit the polyene penetration through the fungal cell wall.
3. **Oxidative stress**
   Amphotericin B (AmB) induces oxidative stress in the cell membrane and is responsible for the fungicidal effect. The presence of increased catalase levels has been observed in species intrinsically resistant to AmB, like *A. terreus*. The catalase enzyme breaks the harmful reactive oxygen species produced during AmB treatment and thus protects cells from the toxic effects of polyenes.
4. **Cross-resistance**
   The increased fitness cost of the ergosterol synthesis pathway rarely causes cross-resistance. However, the inactivation of ERG3 and ERG11 causes inhibition of the synthesis of the toxic sterol 14-methyl-3,6-diol and ergosterol synthesis, respectively, and leads to cross-resistance toward azoles.

Echinocandins

The echinocandins were first approved for clinical use in 2001. They are synthetically modified non-ribosomal water-soluble cyclic lipopeptides, originally derived from fermentation broths of various fungi.

Mechanism of Action

The 1,3-b-D-glucan forms an important cell wall component of the fungus. The echinocandins noncompetitively inhibit 1,3-b-D-glucan synthase (and 1,6-b-D-glucan synthase) and exhibit a fungicidal effect (Fig. 9.1). This enzyme complex consists of two subunits: Fks1p (encoded by FKS1/FKS2/FKS3 genes) and Rho1p. FKS1 transcription helps remodel the cell wall, and FKS2 transcription is calcineurin-dependent. The key regulatory protein is Rho1p, which upregulates the arrest of 1,3-b-D-glucan synthesis. The echinocandin activity is dependent on the proportion of 1,3-b-D-glucan. For example, echinocandins have no activity against Zygomycetes and *Cryptococcus*, which lack 1,3-b-D-glucan in the cell walls (Table 9.1).

The commonly available echinocandins include micafungin, caspofungin, and anidulafungin. The lack of 1,3-b-D-glucan in animal cells makes this drug an agent of choice for systemic fungal infections, and less toxicity has been reported in comparison to polyenes.

Mechanism of Resistance

The overall prevalence of resistance among echinocandins remains low. However, a paradoxical phenomenon called the "eagle effect," i.e., the ability of echinocandin-susceptible strains to grow at higher in vitro concentrations, has been observed. This phenomenon is strain-dependent and has been implicated in upregulating the chitin synthesis in the cell wall of fungal species.

1. **Target site alteration**. The amino acid substitution at the Fks subunits of glucan synthase is the major mechanism of resistance. This substitution decreases the affinity of drugs for the glucan synthase. Hot spot 1 (HS1) and Hot spot 2 (HS2) are two major regions for amino acid changes. In *C. albicans*, mutation at the HS1 region (640–650) and HS2 region (1345–1365) has been reported in the FKS1 subunit. Few *Candida* species, such as *Candida parapsilosis*, *Candida metapsilosis*, and *Candida orthopsilosis*, naturally harbor FKS1 gene polymorphism and exhibit low levels of resistance to echinocandins.
2. **Adaptive stress responses**. The 1,3-β-D-glucan polymer crosslinking is important for

the survival of fungi. In response to stress induced by echinocandins, adaptation mechanisms like protein kinase C (PKC), calcineurin, and heat shock protein (Hsp90) are upregulated. The Hsp90 organizes a cellular stress response circuit and manifests as resistance to echinocandin. Alternatively, chitin production is upregulated in response to stress. The chitin replaces 1,3-β-D-glucan and contributes to increased resistance among fungi.

Pyrimidine Analogs

Pyrimidine analogs are synthetic structural analogs of nucleotide cytosine.

Mechanism of Action

These include 5-fluorocytosine (5-FC) and 5-fluorouracil (5-FU) (Fig. 9.1). The permeases allow 5-FC to enter the fungal cell, which is converted to 5-FU by cytosine deaminase. The 5-FU inhibits cellular functioning by interfering with pyrimidine metabolism and RNA, DNA, and protein synthesis. These drugs act against various *Candida* and *Cryptococcus* species. Pyrimidine analogs are rapidly absorbed and exhibit good bioavailability. However, due to toxic effects and the rapid development of tolerance, it is not frequently used. 5-FC is generally preferred in combination therapy, especially in cryptococcal disease management (Table 9.1).

Mechanism of Resistance

1. **Reducing intracellular drug accumulation/decreased drug concentration**. Poor metabolism of the drug is also observed when 5-FC resistance occurs. Mutations in cytosine deaminase in *Candida glabrata* (FCY1) and *C. albicans* (FCA1) result in 5FC resistance.
2. **Development of bypass pathways**. A mutation in the gene FUR1, which encodes uracil phosphoribosyltransferase, decreases the conversion of 5-FU into a toxic metabolite, i.e., 5-FC monophosphate, preventing its action on the fungus.

Allylamine, Thiocarbamates, and Morpholines

Allylamine is a fungicidal drug that causes squalene accumulation by inhibiting squalene epoxidation (Fig. 9.1). These high levels of squalene may increase membrane permeability and defective ergosterol formation, leading to the disruption of cell membranes. Terbinafine and naftifine are commonly used allylamine formulations (Table 9.1). Thiocarbamates include tolnaftate, and morpholines include drugs like fenpropimorph and amorolfine. The thiocarbamates and allylamines target the ERG1 gene, while morpholines inhibit the ERG2 genes to inhibit the ergosterol synthesis. Missense mutation or substitution in the ERG1 gene leads to an altered drug target and subsequent development of resistance. These drugs are used to treat dermatophyte infections among veterinary animals (Table 9.1).

Other Agents

Iodine

Iodine exhibits a fungicidal effect by rapidly penetrating the fungal cell wall and oxidizing nucleotides, proteins, and fatty acids, which are essential molecules for fungal growth. PVP-I is used for entomophthoromycosis and has a broad antimicrobial and antifungal spectrum.

2-Hydroxystilbamidine Isethionate (HSB)

It is a member of the diamidines family. Along with antifungal activity, it also exhibits anti-trypanosomal and anti-malarial activities. It is a carcinostatic, immunosuppressant drug used for blastomycosis treatment in veterinary animals. HSB inhibits extracellular DNA activity by binding to extranuclear DNA, causing numerous mutations.

Copper Sulfate

Various exudates like hydroxy and amino acids are released by germinating fungal spores. The "insoluble" copper fungicides react with these exudates and form toxic soluble copper complexes, which exert direct fungicidal action. Copper sulfate (1:2000) in drinking water has been regarded as an effective treatment for avian candidiasis in veterinary medicine.

Techniques for Detection of Antifungal Susceptibility Testing

With the emergence of antifungal resistance, especially in azoles and echinocandin, targeted treatment based on susceptibility is the need of the hour. For routine testing, the phenotypic method, i.e., broth microdilution (BMD) in accordance with Clinical and Laboratory Standards Institute (CLSI) and the European Committee for Antimicrobial Susceptibility Testing (EUCAST) guidelines, is performed in most of the laboratories (Fig. 9.2). Additionally, CLSI recommends disk-diffusion assay for the detection of resistance in yeast isolates. For rapid detection of azole resistance in *A. fumigatus*, the agar assay has been standardized by EUCAST. These conventional methods are time-consuming and laborious procedures. Therefore, several commercial platforms like Sensititre, YeastOne, and Vitek2 are available for the detection of antifungal susceptibility for yeast isolates. Other assays like AsperGenius, Fungiplex Aspergillus, and MycoGenie are suitable for the detection of resistance among molds. Gradient diffusion strips, i.e., Etest and Liofilchem, are available commercially. They can be placed on the inoculated agar plates and, after incubation of 24–48 h, provide the MIC of the strain. Since yeast and mold are slow-growing, the turn-around time ranges from 24 to 72 h. Also, these techniques are operator-dependent, and slight changes in the inoculum and incubation conditions can significantly influence the results. The eagles' phenomenon (paradoxical growth exceeding the MIC in echinocandins) and trailing growth (partial inhibition of growth over a wide concentration range exceeding the MIC in azoles) can lead to an overestimation of resistance.

Recently, newer techniques like Matrix-Assisted Laser Desorption/Ionization Time-of-Flight (MALDI-TOF) mass spectrometry (MS) have been used for the rapid detection of antifungal resistance. This technology detects the highly conserved mass-to-charge ratios of ribosomal proteins (Fig. 9.2). A second semi-quantitative method, MALDI-TOF BioTyper Antibiotic Susceptibility Test Rapid Assay (MBT ASTRA), analyses the peak intensity and areas under the curve (AUC), which display a spectrum that is correlated with the growth. In a recent meta-analysis by Knoll et al. MALDI-TOF had a pooled sensitivity and specificity of 91% and 95%, respectively, while MBT ASTRA showed a higher sensitivity of 96%. These techniques are rapid, reliable, easy to perform, and cost-effective. The turn-around time (4–14 h) is also less in comparison to other conventional culture-based methods.

Molecular assays like polymerase chain reaction targeting hot spots of antifungal resistance markers, e.g., Cyp15 for azole and *FKS* mutation in echinocandins, are being used. A commercially available kit, i.e., AsperGenius multiplex real-time PCR assay (PathoNostics), allows direct detection of mutations conferring azole resistance (TR34, Y121F, L98H, and T289A) from the sample and offers sensitivity and specificity of 89% in the hematology group. Another assay, MycoGENIE, targets TR34 and L98H mutations in the cyp51A gene of *A. fumigatus* and provides good diagnostic accuracy with respiratory samples (sensitivity, 92.9%; specificity, 90.1%) and serum samples (sensitivity, 100%; specificity, 84.6%). Advances in sequencing techniques, especially next-generation sequencing platforms, allow the detection of known antifungal resistance mechanisms and elucidation of newer potential mutations that can confer resistance to the isolate. Castanheira et al. analyzed polymor-

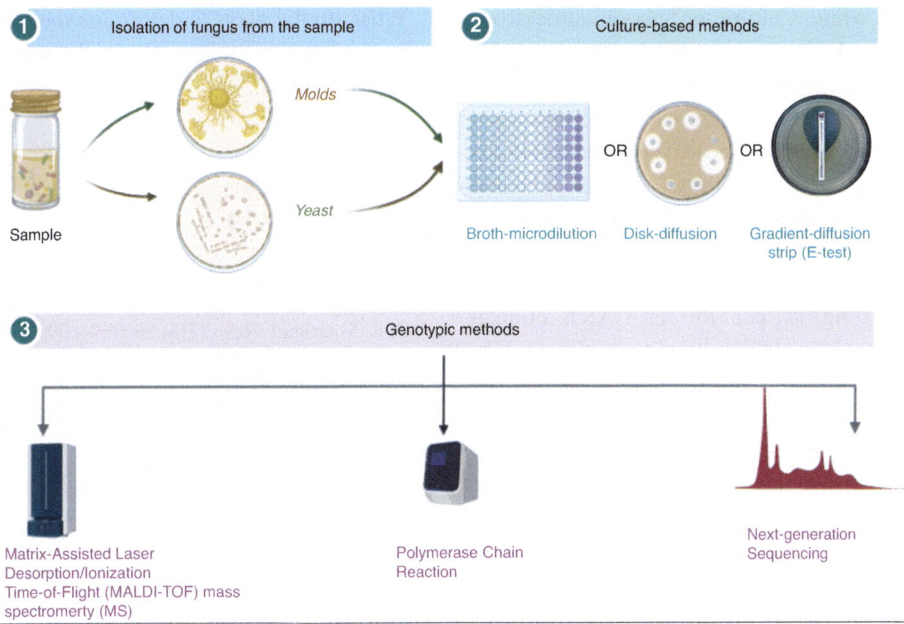

Fig. 9.2 Methods for the detection of antifungal resistance by culture-based (phenotypic) and non-culture-based (molecular)

phisms in genes associated with azole resistance in *C. albicans* via whole-genome sequencing. The whole-genome sequencing also helps in understanding the emergence of resistance in isolates along with its detection.

Conclusion

In conclusion, the emergence of antifungal resistance is an inevitable consequence of using these agents for the treatment of fungal infections in humans and animals. The indiscriminate use of antifungals in agriculture has also led to the development of resistance to important agents. The knowledge and understanding of resistance mechanisms are important components that can help strategize the guidelines for the proper use of these agents. Antifungal susceptibility testing can help us to detect and understand the current status of resistance in the country. Based on these findings, antifungal stewardship policies can be formulated, which can guide the judicious use of these agents and, finally, limit the development of resistance.

Case Study

A 36-year-old female laborer with a history of diabetes mellitus was admitted to the intensive care unit for chief complaints of nonproductive cough, progressively increasing breathlessness, and chest pain from the past 5 days. In view of bacterial pneumonia, She received azithromycin for 5 days from a local practitioner. On examination, she was febrile (>38.3 °C), the heart rate was 96 beats/min, respiratory rate 42 breaths/min, blood pressure 140/90 mmHg, and oxygen saturation 94% on high flow nasal oxygenation. Her Sepsis-related Organ Failure Assessment (SOFA) score at ICU admission was 7. The patient was also treated with the broad-spectrum antibiotic regimen, including piperacillin–tazobactam. On day 3, her condition worsened, and her was intubated with vasopressor support. The chest computed tomography (CT) revealed a halo sign. In view of the fungal infection, voriconazole was initiated, and tracheal aspirate was sent for microbiological examination. Tracheal aspirate showed acute branching hyphae on direct examination with potassium hydroxide and calcofluor white. Culture on Sabouraud dextrose agar revealed blue-green growth with a suede-like surface colony. Lactophenol cotton blue (LPCB) stain revealed the presence of uniseriate and columnar conidial heads with the phialides limited to the upper two-thirds of the vesicle and curving to be roughly parallel to each other. The morphology and colony characteristics were consistent with *A. fumigatus*. The antifungal susceptibility testing as per the EUCAST clinical breakpoints showed the following minimal inhibitory concentrations (MIC) of posaconazole (0.5 μg/mL, resistant MIC >0.25), itraconazole (>8 μg/mL, resistant MIC >1), voriconazole (2 μg/mL, resistant MIC >1), and isavuconazole (4 μg/mL, resistant MIC >1).

1. What are the significant points in the above clinical vignette?
2. How do you determine antifungal resistance in *A. fumigatus* isolate?
3. What are the common mechanisms of resistance in *A. fumigatus*?
4. What is the current status of azole resistance in *A. fumigatus?*
5. In cases of azole resistance in *Aspergillus* species, what other antifungals can be used?

Research Questions

1. How do we determine the burden of antifungal resistance among the veterinary population?
2. How to develop effective therapeutic strategies to combat antifungal resistance.
3. What are the ways to determine the effectiveness of newer antifungals in the pipeline in veterinary use?

Further Reading

Berkow EL, Lockhart SR, Ostrosky-Zeichner L (2020) Antifungal susceptibility testing: current approaches. Clin Microbiol Rev 33(3):e00069–e00019. https://doi.org/10.1128/CMR.00069-19 PMID: 32349998; PMCID: PMC7194854

Cowen LE, Sanglard D, Howard SJ, Rogers PD, Perlin DS (2014) Mechanisms of antifungal drug resistance. Cold Spring Harb Perspect Med 5(7):a019752. https://doi.org/10.1101/cshperspect.a019752 PMID: 25384768; PMCID: PMC4484955

Durand C, Maubon D, Cornet M, Wang Y, Aldebert D, Garnaud C (2021) Can we improve antifungal susceptibility testing? Front Cell Infect Microbiol 11:720609. https://doi.org/10.3389/fcimb.2021.720609 PMID: 34568095; PMCID: PMC8461061

Perfect JR (2017) The antifungal pipeline: a reality check. Nat Rev Drug Discov 16(9):603–616. https://doi.org/10.1038/nrd.2017.46 Epub 2017 May 12. PMID: 28496146; PMCID: PMC5760994

Revie NM, Iyer KR, Robbins N, Cowen LE (2018) Antifungal drug resistance: evolution, mechanisms, and impact. Curr Opin Microbiol 45:70–76. https://doi.org/10.1016/j.mib.2018.02.005 Epub 2018 Mar 13. PMID: 29547801; PMCID: PMC6135714

Robbins N, Caplan T, Cowen LE (2017) Molecular evolution of antifungal drug resistance. Ann Rev Microbiol 71:753–775. https://doi.org/10.1146/annurev-micro-030117-020345

Sanglard D (2016) Emerging threats in antifungal-resistant fungal pathogens. Front Med (Lausanne) 3:11. https://doi.org/10.3389/fmed.2016.00011 PMID: 27014694; PMCID: PMC4791369

10 Prevention and Control of Fungal Zoonoses and Sapronoses

Rajendra Guidsa and Parakriti Gupta

Learning Objectives

By the end of the chapter, the readers should be able to:

1. Decipher various approaches for the prevention and control of fungal zoonoses.
2. Understand the importance of One Health approach.
3. Learn the importance of preparedness and stratification while dealing with pandemics.

Introduction

Fungal infections contribute substantially to the burden of morbidity and mortality, affecting approximately one billion individuals globally, with a recent rise in the incidence of these infections. The recent rise has been attributed to be multifactorial and includes increasing immunosuppression owing to an increased number of transplants, advancements in chemotherapy, usage of indwelling devices, climate changes, and human activities related to urbanisation and deforestation. Even though underlying factors are varied, recent pandemics have underlined the role of disturbed ecosystems in certain emerging infections. Of these emerging infections, zoonoses having an animal reservoir contribute substantially, to the extent of ~75%. Dermatophytosis, cryptococcosis, histoplasmosis, talaromycosis, sporotrichosis, adiaspiromycosis, and emergomycosis are examples of frequently encountered fungal zoonoses.

Fungal zoonoses include the fungal infections transmitted between vertebrate animals and humans. Zoonotic agents are dependent on the host for transmission as well as feeding, which mostly occurs via direct route. Environmental fungal pathogens follow a specialised phase following the invasion of the host, have the propensity to be transmitted via the host, and can escape from the host upon the death of the body.

Opportunistic fungal agents usually inhabit the environment and have the ability to survive in a host as well. However, these fungi are non-transmissible and die after the death of the host. These opportunistic fungi, when derived from non-living sources, such as decaying plant material or soil, are termed sapronosis. Sapronotic agents are fungi with a propensity to affect humans or animals without any specialised adaptation to the host and are non-transmissible. *Sprorthrix brasiliensis* is an example of a zoonotic agent, while *Sporothrix globosa* is a sapronotic/opportunistic agent. Environmental fungal agents include *Histoplasma capsulatum* and *Coccidoides immitis*.

R. Guidsa
Department of Medical Microbiology, Postgraduate Institute of Medical Education, and Research (PGIMER), Chandigarh, India

P. Gupta (✉)
Department of Microbiology, Government Medical College and Hospital, Chandigarh, India
e-mail: parakritii@gmail.com

S. C. Parija, S. M. Rudramurthy (eds.), *Textbook of Fungal Zoonoses and Sapronoses*, Microbial Zoonoses, https://doi.org/10.1007/978-981-97-3214-2_10

Understanding Transmission of Fungal Zoonoses

Transmission of fungal zoonoses can be either direct or indirect. Direct transmission involves contact via skin, bites, or scratches from animals, while indirect transmission is vehicle-borne, utilising air, water, or some vectors for dispersal. Of these modes of transmission, the indirect mode plays a pivotal role in the dissemination of the agent. Animal vectors can disseminate the fungal spores to distant places, culminating in extensive environmental distribution. Zoonotic transmission is further facilitated by the persistence of fungal spores in the environment. In addition, several fungi follow both modes of transmission; for example, dermatophytes are mainly transmitted via direct contact but also have the propensity to survive on keratinised fomites, facilitating indirect transmission.

Host Shift and Concept of Ecological Fitting

Zoonotic transmission of fungi is a consequence of spillover of the agent, which possesses the ability to survive in newer host species. The old concept of 'species barrier' defines the host specificity of fungal agents implicated in human infections and is mostly linked to phylogenetic relatedness. There are possibilities of transmission of infection from non-primate mammals to humans rather than phylogenetically distant species. Fungi can modulate their receptor interactions to colonise the phylogenetically related host for multiplication in the host. However, all fungi do not tend to follow this barrier. For example, *Microsporum canis* does not succumb to this species restriction and has the propensity to be transmitted between canines and humans. This paved the path for more extensive research to delineate the precise transmission interplay, and a newer concept of '*ecological fitting*' has been introduced.

'*Ecological fitting*' is the mechanism by which the agent establishes itself in the new host by means of survival advantages adopted in the previous/novel host. Two hypotheses, 'resource tracking' or 'sloppy fitness space' have been put forward that provide the ecological fitting to the agent; in resource tracking, the agent perceives a newer host in a similar way as the previous host, owing to phylogenetic relatedness. However, in a 'sloppy fitness space,' the agent has a propensity to associate and expand its potential host range, utilising the fitness space. These mechanisms enable the agent to adapt and survive in a broader range of species following spillover episodes.

Prevention of Fungal Zoonoses

Prevention of fungal infections is an undermapped area, with guarded results of varied modalities. Some of the researchers have focused on delimiting the transmission from wildlife by targeted interventions, as was successfully achieved for chytridiomycosis on Mallorca Island. However, the same cannot be extrapolated to other fungal infections in humans, owing to the multifactorial dynamic interplay. Other areas that are being targeted for the prevention of fungal zoonoses include (a) improved diagnostics with increased accessibility, (b) increasing awareness pertaining to zoonotic transmission, (c) stringent trade surveillance, (d) strengthening the preparedness to tackle emerging agents, (e) focusing on immunisation, and (f) immunotherapy against fungal agents and the One Health approach (Fig. 10.1).

Accessibility to Newer Diagnostics

The major roadblock in the management and prevention of zoonotic fungal infections is the availability of diagnostics. Diagnosis of common fungal infections is also limited owing to the lower sensitivity of commonly employed microscopy and culture techniques and the non-availability of advanced newer methods. Though non-culture-based diagnostics, viz. antigen testing, polymerase chain reaction, biomarkers, proteomic-based diagnosis, and whole-genome

Fig. 10.1 Different approaches for the prevention of fungal zoonoses

sequencing, have been developed, their availability in resource-limited regions remains a matter of concern. Moreover, routine diagnostics is prone to subjectivity and precision, thereby leading to a diagnostic dilemma, highlighting the need to integrate newer tests, such as molecular tests, in routine diagnostics to detect emerging zoonotic agents at the earliest.

Mounting Awareness

Increasing public awareness about fungal zoonoses is crucial in preventing infections. Educational campaigns targeting both animal owners and the public can promote understanding of risks, modes of transmission, and associated preventive measures. Providing information on proper hygiene practices, such as handwashing after handling animals, can significantly reduce the risk of transmission. It is necessary to sensitise the public and animal owners as well as clinicians regarding potential threats posed by zoonotic fungi. The clinicians need to be trained to augment their clinical diagnosis, which needs to be supported by the laboratory diagnosis to reduce the delay in treatment, thereby reducing morbidity and preventing mortality and further transmission.

Trade and Travel Surveillance

The transmission of zoonotic fungal infections currently has been escalated due to the well-established travel network across the globe. Hence, there is a need to impose travel restrictions or to have a stringent surveillance system in place during ongoing outbreaks/epidemics/pandemics, failure of which might lead to a large number of cases. The same was observed during the outbreak of *Trichophyton indotineae* infection in India, which was a species that was more virulent and more resistant. The emergence of this novel species was attributed to the host-shifting adopted by geophilic *Trichophyton mentagrophytes*, initially from wild animals to domestic animals and subsequently emerging as an '*anthropophilic clonal offshoot*'. The cases initially restricted to India were soon reported from distant countries as well.

The rise in fungal infections has also been attributed to a certain extent to unguarded and illegal trade marketing. Though global trade is an essential part of development, the rise in infections necessitates the initiation of well-defined preventive actions by transnational committees and organisations for the prevention of zoonoses. World Organization for Animal Health (WOAH) and the World Health Organisation (WHO) are striving hard for the same. WOAH focuses on listing and controlling the movements of infected wild stock. There are many concerns, such as (a) several fungal agents are currently not on their priority list, and new fungi are emerging simultaneously, (b) easily traversing the trans boundaries, and (c) delaying the stipulated response to be addressed. These concerns necessitate further strengthening of biosecurity against zoonotic fungal infections.

Preparedness and Stratification of Mitigation Levels

Viral pandemics such as COVID-19 have exposed the infrastructure of healthcare setup across the globe, underlining the necessity to have preparedness for combating such events in the future. In addition to the diagnostic and management strategies, there is a need to decipher more refined epidemiological tools, mathematical models, and models based on experimentally derived variables to predict future events and escalate the facilities of healthcare to combat the same. With the emergence of viral pandemics and the evolution of fungal zoonoses, toolkits for combating emerging infections are available; however, knowledge to put that to practical use is still very scanty.

Four categories for the wildlife include the (a) pre-arrival stage, (b) invasion stage, (c) epizootic stage, and (d) the established stage. The pre-arrival stage aims to prevent the entry of fungal agents into the population. This is possible by strengthening biosecurity and trade restrictions and implementing stringent scrutiny of vectors to ensure pathogen-free trade. The invasion-front strategy targets eradicating the agent and delimiting its transmission. This is usually achieved either by disinfection of the reservoir, killing of fungal agents, or deactivation. In the epizootic and established phases, the main concern is to reduce the impact of ongoing transmission on the target population. In the established phase, this can be achieved either by immunisation or preventing contact with the agent, in the case of humans. Gene-editing and silencing tools have been employed to mitigate the virulence of fungal agents in the case of wildlife, and the same needs to be studied in humans. In essence, there is a need to formulate and implement the invasion-based management categories available for mitigation at the wildlife level and for zoonotic infections.

Immunotherapy

Immunotherapy is based on the role of antibodies/immunoglobulins in recognising varied antigens via somatic hypermutation and genetic rearrangement. The same mechanism has been exploited to develop monoclonal antibodies against *C. neoformans*. Two hybridoma lines were derived from the same B cell, and based on somatic hypermutations, mAbs varied in variable regions by 11 residues, thereby identifying different epitopes on *C. neoformans*. Of these, two mAbs showed interesting effects; one could penetrate throughout the capsule and recognise epitopes present across the length, while the other one recognised epitopes present merely on the outer surface. Though the results appeared to be encouraging, they need an extensive evaluation, taking into the concerns of the possibility of being prone to resistance and inconsistent protection. Moreover, some mAbs require to be recognised by the human immune system, which will be rendered ineffective in several immunodeficient patients who are at risk of fungal zoonoses.

Dendritic Cell Immunotherapy

This method involves ‘pulsing’ the dendritic cells (DC) ex vivo with selected antigens, after which DCs are returned to the host body. DC immunotherapy provides protection against a specific agent with which it is incubated ex vivo after the diagnosis has been made. Based on the same grounds, DC vaccination, which involves pulsing of DCs before exposure, is also under trial. Peptide P10 is one such example, mapped with the T-cell epitope of *Paracoccidioides brasiliensis.* Mice studies revealed ~2-log reduction in fungal burden with the pre-exposure DC vaccination modality when bone marrow DCs were incubated/pulsed with P10, and subsequently, after returning to the body, mice were exposed to *P. brasiliensis* yeasts. Similar experiments were performed to note the therapeutic effects after the exposure, as well, wherein DCs were treated with P10 following the exposure and gave satisfactory results.

Immunisation

Immunisation is one of the major approaches to prevent the morbidity and mortality caused by infectious agents; however, currently, no fungal vaccines have been approved for use in humans. Several vaccine candidates are presently under research and evaluation. One such example is the heat-killed *Saccharomyces cerevisiae* yeasts (HKY)-based vaccine against *Coccidioides posadasii.* The vaccine helps in the initiation of immune response based on the epitope-sharing (polysaccharide) with *S. cerevisiae*. Another vaccine against *C. posadasii* makes use of formalin-killed *Coccidioides immitis* spherules (FKS) as the target. Mice models have revealed that HKY protected 70% of CD1 mice, whereas FKS protected all of them. Another attenuated vaccine against *C. posadasii* is also under trial, wherein the target candidate is a strain with the absence of two chitinase genes, thereby disabling it to endosporulate. The vaccine was reported to protect 75–100% of mice after two doses (subcutaneous) 14 days apart.

Several vaccine trials are ongoing against *C. neoformans.* One such vaccine uses a genetically engineered strain that produces murine IFN-γ (H99γ), and studies have shown a 100% protection rate in CD4+ T cell-deficient mice. Though this murine strain cannot be used in humans, it provides a platform for testing a humanised version of H99γ. Another example of anti-*C. neoformans* vaccine is the conjugate vaccine, wherein glucuronoxylomannan (GXM) has been linked to tetanus toxoid and monophosphoryl lipid A (adjuvant). The vaccine was able to elicit IgG and IgA at high levels, imparting 70% protection to the mice. However, extensive studies revealed that protection could also be derived from the adoptive transfer of antisera from immunised animals. This strategy can be broadened to incorporate saccharide epitopes, viz. β-glucans commonly present in all fungi, creating the 'common vaccine for a broad range of pathogenic fungi.' These conjugate vaccines might prove to be a boon to immunosuppressed individuals, who are prone to frequent fungal infections and are at risk of infection if attenuated vaccines are administered.

One Health Approach

The concept of One Health incorporates the integrated approaches to 'optimise the health of people, animals and ecosystem.' On account of the close association of zoonotic fungal agents with the environment, wildlife, and human health, it is imperative for regional, national, and international bodies to work in coordination to prevent the emergence of fungal zoonoses.

In conclusion, with an increase in human activities and environmental changes, the spillover of zoonotic fungal infections is witnessing an ongoing rise. There is a requisite for interdisciplinary monitoring of the risks posed by these fungi. Collaborative efforts of public health authorities, veterinary professionals, clinicians, microbiologists, researchers, and the public are crucial in preventing the impact of fungal zoonoses and protecting both human and animal health. Extensive research is required to decipher the habitats and spillover mechanisms, which needs to be inter-sectoral between ecologists, epidemiologists, and mycologists.

Case Study

The A family, residing in a suburban area, enjoyed gardening as a family activity and were avid pet lovers. They maintained a garden with various plants, shrubs, and flowers. They owned several cats, dogs, and rabbits, and their child often played with these animals in the backyard. This summer, the family members began developing skin lesions, initially dismissed as minor cuts and insect bites due to their gardening activities. However, the lesions worsened over time, leading to concern and a visit to their family physician.

Case Description:

Mr. and Mrs. A:

Mr. A, the father, and Mrs. A, the mother, were the first to notice small, painless bumps on their hands and arms. Over the next few weeks, these bumps turned into ulcerative sores. Both experienced discomfort, itching, and occasional joint pain. Initially, they attributed it to regular wear and tear from gardening.

B (14 years old, daughter):

B, their teenage daughter, developed similar skin lesions on her face, particularly around her nose and cheeks. The lesions caused her distress, affecting her social life and self-esteem.

Upon visiting their family physician, the family underwent multiple tests, including skin biopsies. Sporotrichosis was diagnosed based on fungal culture results. The possibility of a zoonotic transmission route was explored due to the family's close contact with animals.

1. What is the importance of understanding the mechanisms of transfer of disease in this case?
2. Discuss the advantages of non-culture-based methods for diagnosis of the disease and its advantages over conventional methods.
3. How would One Health approach prevent the above condition?
4. What prevention strategies should be advised for the family?
5. Does vaccination exist for the prevention of fungal zoonoses? If yes, explain with an example.

Research Questions

1. What are the most effective strategies for the prevention of fungal zoonoses in both domestic and wild animal populations, with a focus on reducing the transmission to humans?
2. How can surveillance and early detection systems be improved to identify emerging fungal zoonotic threats and facilitate rapid response and control measures?
3. What are the key factors influencing the emergence and spread of fungal zoonoses, and how can we develop targeted interventions to mitigate these factors?
4. What are the best practices for educating and raising awareness among healthcare professionals, veterinarians, and the public about the risks associated with fungal zoonoses and the importance of prevention and early detection?

Further Reading

Agosta SJ, Klemens JA (2008) Ecological fitting by phenotypically flexible genotypes: implications for species associations, community assembly and evolution. Ecol Lett 11(11):1123–1134

Bauerfeind R, Von Graevenitz A, Kimmig P, Schiefer HG, Schwarz T, Slenczka W, Zahner H (2020) Infectious diseases transmissible from animals to humans. Wiley, Washington, DC

Beenhouwer DO, Yoo EM, Lai CW, Rocha MA, Morrison SL (2007) Human immunoglobulin G2 (IgG2) and IgG4, but not IgG1 or IgG3, protect mice against *Cryptococcus neoformans* infection. Infect Immun 75(3):1424–1435

Bosch J, Sanchez-Tomé E, Fernández-Loras A, Oliver JA, Fisher MC, Garner TW (2015) Successful elimination of a lethal wildlife infectious disease in nature. Biol Lett 11(11):20150874

Capilla J, Clemons KV, Liu M, Levine HB, Stevens DA (2009) *Saccharomyces cerevisiae* as a vaccine against coccidioidomycosis. Vaccine 27(27):3662–3668

Carpouron JE, de Hoog S, Gentekaki E, Hyde KD (2022) Emerging animal-associated fungal diseases. J Fungi (Basel, Switzerland) 8(6):611

Devi SJ (1996) Preclinical efficacy of a glucuronoxylomannan-tetanus toxoid conjugate vaccine of *Cryptococcus neoformans* in a murine model. Vaccine 14(9):841–844

Fisher MC, Gow NA, Gurr SJ (2016) Tackling emerging fungal threats to animal health, food security and ecosystem resilience. Philos Trans R Soc Lond Ser B Biol Sci 371(1709):20160332

Fisher MC, Gurr SJ, Cuomo CA, Blehert DS, Jin H, Stukenbrock EH, Stajich JE, Kahmann R, Boone C, Denning DW, Gow NAR, Klein BS, Kronstad JW, Sheppard DC, Taylor JW, Wright GD, Heitman J, Casadevall A, Cowen LE (2020) Threats posed by the fungal kingdom to humans, wildlife, and agriculture. mBio 11(3):e00449–e00420

Garner TW, Schmidt BR, Martel A, Pasmans F, Muths E, Cunningham AA, Weldon C, Fisher MC, Bosch J (2016) Mitigating amphibian chytridiomycoses in nature. Philos Trans R Soc Lond Ser B Biol Sci 371(1709):20160207

Hubálek Z, Rudolf I (2010) Microbial zoonoses and sapronoses. Springer, Dordrecht

Mesquita JR (2021) Emerging and re-emerging diseases: novel challenges in today's world or more of the same? Animals 11:2382

Mukherjee J, Nussbaum G, Scharff MD, Casadevall A (1995) Protective and nonprotective monoclonal antibodies to *Cryptococcus neoformans* originating from one B cell. J Exp Med 181(1):405–409

Rohde K (2013) The balance of nature and human impact. Cambridge University Press, Cambridge

Taborda CP, Juliano MA, Puccia R, Franco M, Travassos LR (1998) Mapping of the T-cell epitope in the major 43-kilodalton glycoprotein of *Paracoccidioides brasiliensis* which induces a Th-1 response protective against fungal infection in BALB/c mice. Infect Immun 66(2):786–793

Tang C, Kong X, Ahmed SA, Thakur R, Chowdhary A, Nenoff P, Uhrlass S, Verma SB, Meis JF, Kandemir H, Kang Y, de Hoog GS (2021) Taxonomy of the *Trichophyton mentagrophytes/T. interdigitale* species complex harboring the highly virulent, multiresistant genotype *T. indotineae*. Mycopathologia 186(3):315–326

Voyles J, Kilpatrick AM, Collins JP, Fisher MC, Frick WF, McCallum H, Willis CK, Blehert DS, Murray KA, Puschendorf R, Rosenblum EB, Bolker BM, Cheng TL, Langwig KE, Lindner DL, Toothman M, Wilber MQ, Briggs CJ (2015) Moving beyond too little, too late: managing emerging infectious diseases in wild populations requires international policy and partnerships. EcoHealth 12(3):404–407

Wolfe ND, Dunavan CP, Diamond J (2012) Origins of major human infectious diseases. In: Improving food safety through a one health approach: workshop summary. National Academies Press, Washington, DC

Wozniak KL, Young ML, Wormley FL Jr (2011) Protective immunity against experimental pulmonary cryptococcosis in T cell-depleted mice. Clin Vaccine Immunol CVI 18(5):717–723

Xue J, Chen X, Selby D, Hung CY, Yu JJ, Cole GT (2009) A genetically engineered live attenuated vaccine of *Coccidioides posadasii* protects BALB/c mice against coccidioidomycosis. Infect Immun 77(8):3196–3132

Part II

Superficial Fungal Zoonotic Infections

11 Dermatophytoses

Anupma Jyoti Kindo and Mary Kiran Danni

Learning Objectives

By the end of the chapter, the readers should be able to:

1. Understand the pathogenesis and clinical manifestations of dermatophytosis among animals and birds.
2. Understand available diagnostic modalities for the diagnosis of dermatophytes.
3. Describe the microscopic morphology of zoophilic dermatophytes.
4. Understand the need for rational use of antifungal drugs and their implications.

Introduction

Animal fungal pathogens known as zoophilic dermatophytes exhibit a high level of host specificity as well as potential for zoonotic transmission. Zoophilic dermatophyte species include *Microsporum canis* (pet animals, particularly cats and dogs), *Nannizzia nana*, *Nannizzia persicolor*, *Trichophyton vanbreuseghemii* (cats and dogs), *Trichophyton benhamiae* (guinea-pigs), *Trichophyton simii*, *Trichophyton quinckeanum*, *Trichophyton verrucosum* (usually from cattle), *Trichophyton erinacei* (hedgehogs), *Lophophyton gallinae*, and *Trichophyton equinum* (horses). They cause infections in humans, which are highly contagious and require systemic and long-lasting treatment. *Lophophyton gallinae* causes fowl favus, which affects the wattles and comb in chickens and fowls, producing "white comb" lesions.

Sources of infection include farm animals, laboratory animals, wild animals, and pets. Close contact with domestic animals such as cats and guinea pigs, who are frequently asymptomatic carriers, transmits infections to children and adolescents. Direct contact through infected animals and indirect contact through the contaminated environment or inanimate objects are the two important modes of transmission of infections. Rarely, insects have been implicated in transmission of disease from animals. For instance, *Trichophyton equinum* is transmitted from horses (there are two varieties: *T. equinum var. equinum* and *T. equinum var. autotrophicum*), *M. canis* from cats, *T. verrucosum* from cows, and *Trichophyton mentagrophytes* from rodents. *Microsporum gypseum* and *Trichophyton bullosum* are other rare fungi that cause zoonotic infections.

History

David Gruby, a Hungarian physician, first described dermatophytosis in 1841. Prior to Gruby, several scientists reported lesions that

A. J. Kindo (✉) · M. K. Danni
Department of Microbiology, Sri Ramachandra Institute of Higher Education and Research, Chennai, Tamilnadu, India
e-mail: anupamajyoti@sriramachandra.edu.in

S. C. Parija, S. M. Rudramurthy (eds.), *Textbook of Fungal Zoonoses and Sapronoses*, Microbial Zoonoses, https://doi.org/10.1007/978-981-97-3214-2_11

were ring-like and were perceived to be infective. Raimond Sabouraud in 1890 made a significant contribution through his research on the taxonomy, morphology, and therapy of dermatophytes. He classified the fungi into four genera of which three are still being used. During the early 1900s, ringworms were treated with chemicals containing mercury, sulfur, or even iodine.

Taxonomy

Dermatophytes are members of the family Arthrodermataceae and order Onycogenales, and they were previously included in the genera, *Trichophyton*, *Epidermophyton*, and *Microsporum*. Ascomycota's teleomorphic genus *Arthroderma* includes species that can reproduce sexually.

The nomenclature and taxonomy of the dermatophytes, of late, has been revised and has undergone revision on the basis of a "one fungus—one name" policy, mandating that the teleomorphs (sexual stages) and anamorphs (asexual stages) must share a single designation and molecular analysis. Two species, namely *T. vanbreuseghemii* and *T. benhamiae* constitute the zoophilic *T. mentagrophytes* species. Previously, isolates of zoophilic *T. mentagrophytes* may have erroneously been identified as *T. vanbreuseghemii* isolates from other mammals or *T. benhamiae* from rodents and lagomorphs. Since *T. benhamiae* is primarily associated with rodents, mainly guinea pigs, when kept in large groups. It is plausible, that the infection was caused by *T. benhamiae* when guinea pigs owners became infected with *T. mentagrophytes*. Hence, the binomium "*T. mentagrophytes*" is referred to as to animal isolates rather than anthropophilic species.

Genomics and Proteomics

The taxonomy of the dermatophytes has been modified based on recent multilocus phylogenetic analysis studies. These include 21 species of *Arthroderma*, 1 species each of *Ctenomyces*, *Epidermophyton*, and *Lophophyton*, 3 species of *Microsporum*, 9 species of *Nannizzia*, and 16 species of *Trichophyton*. In addition, new genus *Paraphyton* consisting of three species and the genus *Guarromyces* consisting of one species has been described recently. The genus *Arthroderma* frequently occurs in soil and cave environments but seldom in human and animal cutaneous infections. These are geophilic dermatophyte species that are important in the ecological process of decomposing the keratin in animal skin and hair.

Based on morphology, Rippon (1988) recognized 22 species and four variants in the genus *Trichophyton*, which has undergone major revisions. Currently, the genus *Trichophyton* has 16 recognized species based on DNA sequence studies being increasingly used in determining phylogenetic relationships among species. The genus *Microsporum* has three recognized species *M. canis*, *Microsporum ferrugineum*, and *Microsporum audouinii*. Previously classified as *Microsporum* species, the remaining geophilic and zoophilic species are included in different genera such as *Nannizzia*, *Lophophyton*, and *Paraphyton*.

The Fungus Morphology

The macroscopic characteristics of the growth of the fungus supplemented with microscopy facilitate the identification of the fungi (Figs. 11.1, 11.2, 11.3, 11.4, 11.5, 11.6 and 11.7).

Culture Characteristics

Dermatophytes are slow-growers. These grow in Sabouraud's dextrose agar (SDA) and other media such as dermatophyte test medium or oatmeal agar with cycloheximide. Oatmeal agar enhances better fungus sporulation. The culture media after inoculation are incubated at 37 °C and the other at 25 °C. Dermatophytes take a longer time to grow and, hence, are observed for the growth for at least 3 weeks. The macroscopic morphology of dermatophytes is summarized in Table 11.1.

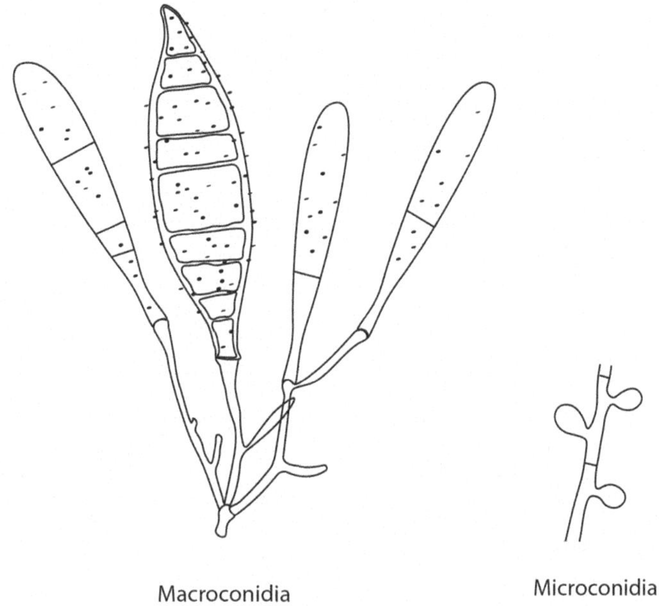

Fig. 11.1 Schematic diagram of *Microsporum canis*

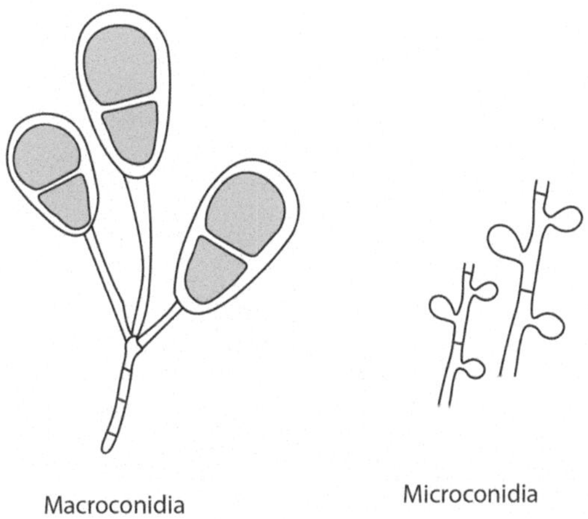

Fig. 11.2 Schematic diagram of *Microsporum nanum*

Animal Models

An experimental animal model has been carried out to study the pathogenesis of dermatophyte infections. The varying skin structure, hair volume, and horny cell layer thickness of experimental ani mals are the recognized issues for establishing dermatophyte infections with animal models. Experimental infection of *Tinea corporis* has been carried out in mice, cockscomb, naked rats, rabbits, germ-free guinea pigs, chickens, and guinea pigs. Guinea pigs are used for experimental infection of *Tinea pedis*, and rabbits for *T. unguuium*, *T. mentagrophytes*, and *M. canis* infections. These models are established by occlusion and abrasion techniques in the laboratory animals.

Pathogenesis and Pathology

Dermatophytes colonize the dead keratinized tissue of the stratum corneum of the human skin, followed by attachment and keratinolysis. Arthrospore, the infective stage, is produced by the disintegration of fungal hyphae. These adhere to keratinocytes, subsequently germinate, proliferate, and cause typical "ringworm" lesions in the infected host. Several molecules play an important role in the attachment sage. Dipeptidyl peptidase (Dpp-IV) and subtilisin protease (Sub3), both of which are involved in adhesion, are produced by *M. canis*. Mannose and galactose in *Trichophyton* residues are recognized by the arthrospore surface to establish infection. It also extends fibrils morphologically into the stratum corneum.

Keratinolysis is an enzymatic process that causes lysis of keratin with the production of monomers of amino acid. Dermatophytes release sulfite using efflux pumps and disrupts

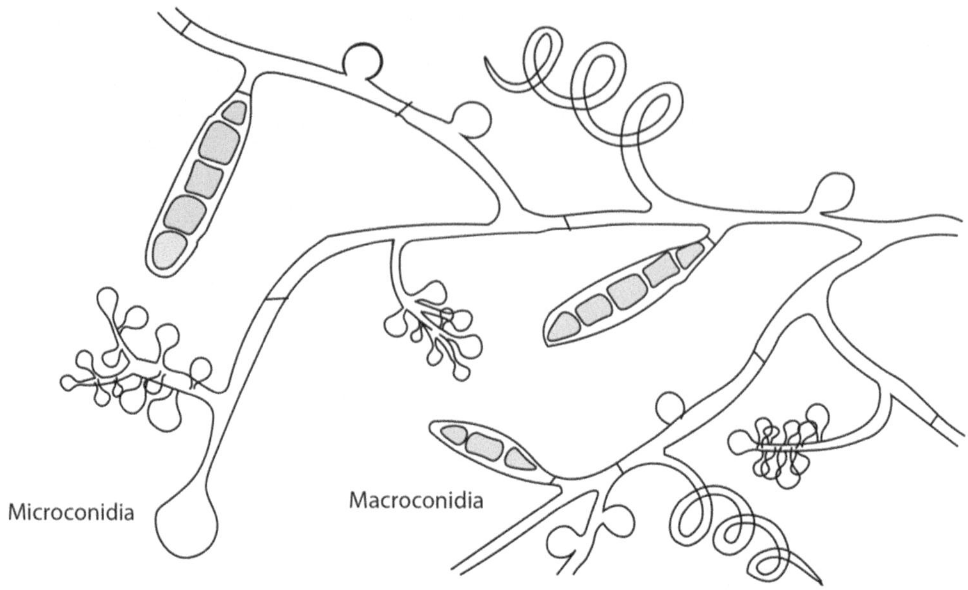

Fig. 11.3 Schematic diagram of *Trichophyton mentagrophytes*

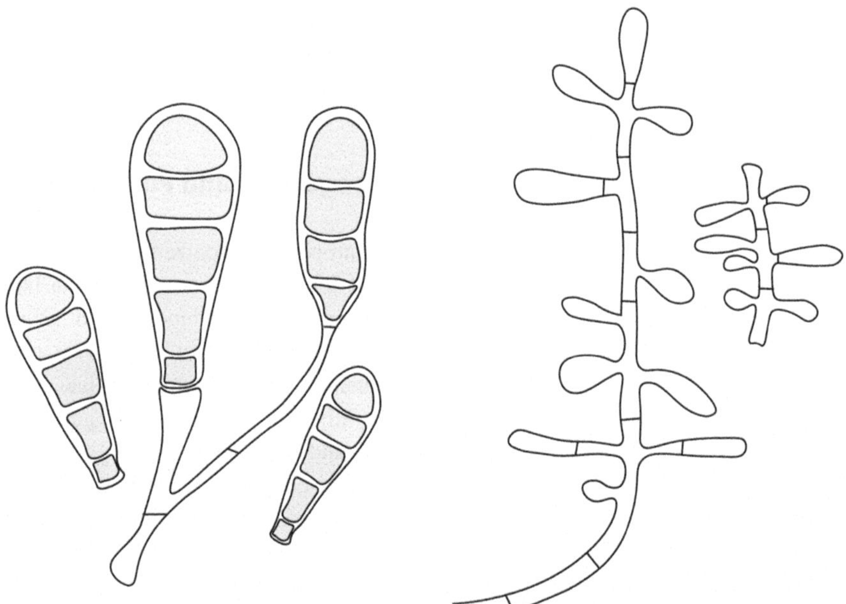

Fig. 11.4 Schematic diagram of *Trichophyton equinum*

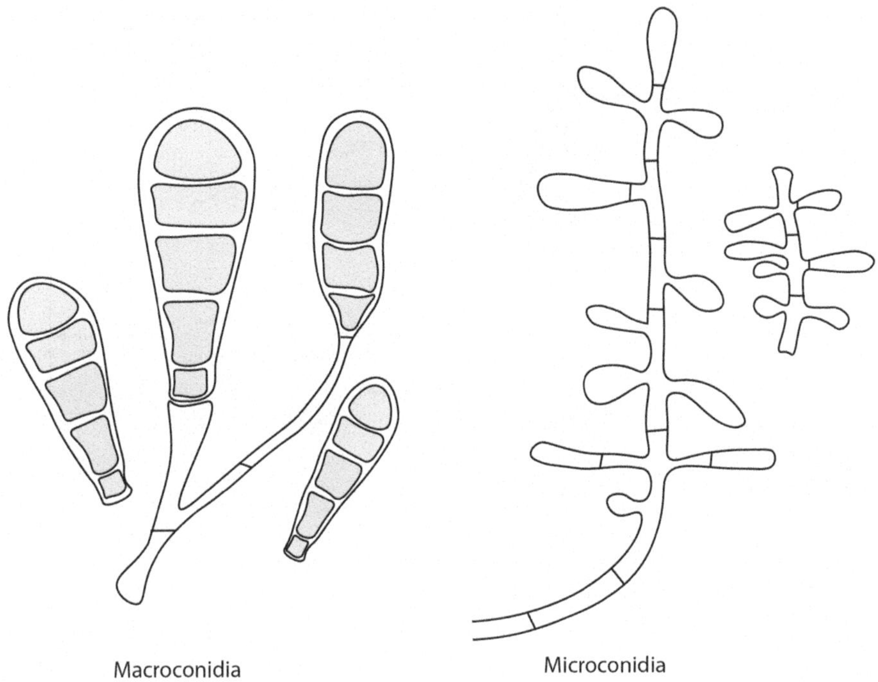

Fig. 11.5 Schematic diagram of *Trichophyton verrucosum*

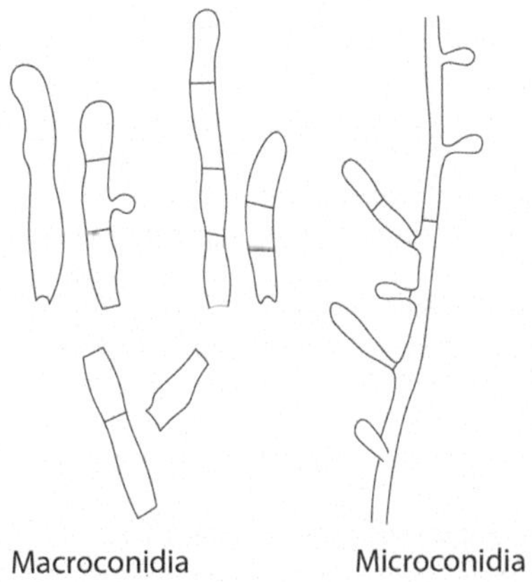

Fig. 11.6 Schematic diagram of *Trichophyton erinacei*

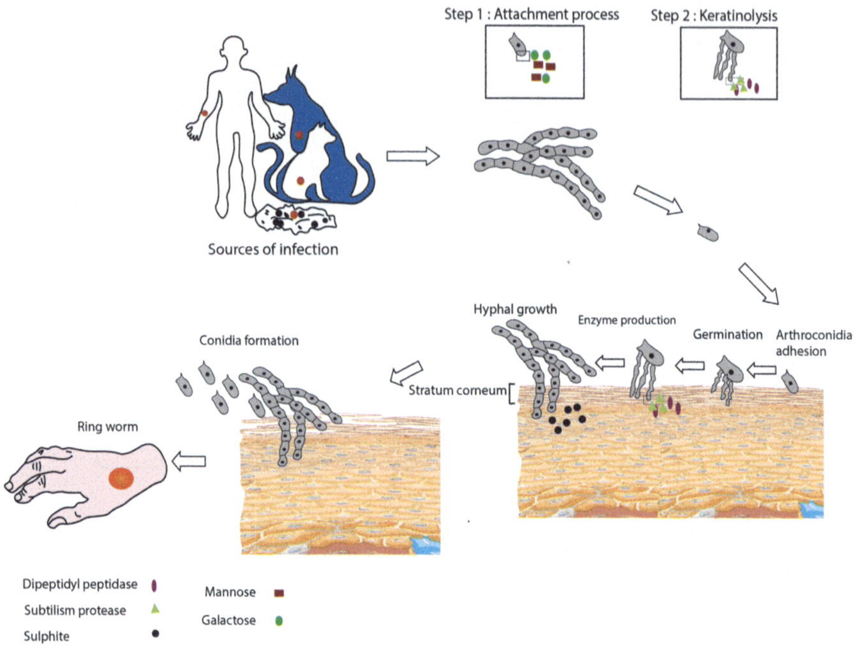

Fig. 11.7 Pathogenesis of dermatophyte infections. (Drawing by Mrs. Anitha S)

Table 11.1 Morphology of important zoonotic dermatophytes

Dermatophyte species	Macroscopic features	Microscopic features
Microsporum canis	Growth woolly to cottony and flat texture Color on the obverse: White to yellowish, color on the reverse: Deep yellow to yellow-orange	Septate hyphae seen Macroconidia: Spindle shaped with an asymmetrical apical knob, 6–15-celled, rough, long, have thick outer cell walls and thin septal walls Microconidia (rare): Clavate to pyriform in shape and unicellular Teleomorph: *Arthroderma otae* *Microsporum canis* differs from *Microsporum audouinii* by growing on polished rice grains and perforating hair Deep yellow pigment produced when grown on rice grains
Microsporum nanum	Moderately rapid growth on Sabouraud's dextrose agar Growth is powdery, cottony, thin, spreading, velvety, or flat texture and frequently has some radial superficial furrows. From the front, the color is white to dark beige and reddish-brown from the reverse	Presence of septate hyphae, club-shaped microconidia with varying abundance, and thin-walled macroconidia Macroconidia contain 1–4 cells usually two. The shape is oval to elliptical
Trichophyton mentagrophytes	White to cream in color, flat, with powdery to granular surface colonies. Certain cultures develop raised central tufts or pleomorphic suede-like to downy areas or exhibit central folding Generally, the reverse pigmentation is yellow-brown to reddish-brown in color	Microconidia: It is single-celled in dense clusters, smooth-walled, hyaline, spherical to subspherical in shape, and occasionally clavate to pyriform Macroconidia: It is smooth, thin-walled, clavate-shaped, multicelled Special forms: They are varying numbers of spherical chlamydospores and spiral hyphae can be seen

(continued)

Table 11.1 (continued)

Dermatophyte species	Macroscopic features	Microscopic features
Trichophyton equinum	Colonies: These are mostly flat, sometimes developing radial grooves or gentle folds, white to buff in color, suede-like to downy in texture, like *Trichophyton mentagrophytes* Pigmentation is deep-yellow with a submerged fringe and a reverse that later becomes dark red in the center	Microconidia: These are abundant, forms laterally along the hyphae, shape may be clavate to pyriform and sessile or spherical and stalked Macroconidia: These are rare, clavate, smooth, thin-walled, and of variable size Special forms: These are abundant chlamydospores in old cultures that undergo a transformation as nodular organs
Trichophyton verrucosum	Colonies: These are small, slow-growing, button or disk-shaped, white to cream-colored, with a suede-like to the velvety surface, a raised center, and a flat periphery with some submerged growth. Reverse pigment: It may differ from non-pigmented to yellow It is the only dermatophyte that grows at 37 °C	Hyphae: These are broad, irregular with many terminal and intercalary chlamydospores in chains Antler effect: The tips of some hyphae are club-shaped and broad and occasionally divided, giving the so-called "antler" effect Microconidia: These are produced by some strains when grown on thiamine-enriched media. They are clavate to pyriform, borne singly along the hyphae Macroconidia: They are rarely produced. They have a string bean shape or characteristic rat-tail Teleomorph: This is not found
Trichophyton erinacei	Colonies: These are flat, white, powdery, and sometimes downy to fluffy with a brilliant lemon-yellow reverse	Microconidia: These are numerous, large clavate and borne on the sides of hyphae Macroconidia: These are two- to six-celled, smooth-walled, clavate, variable in size, and may have terminal appendages. In comparison to what is seen in *Trichophyton mentagrophytes*, macroconidia are much shorter Differentiating points from *T. Mentagrophytes*: (a) Numerous large slender clavate microconidia are borne on the sides of hyphae, and its smooth, thin-walled clavate macroconidia are shown by microscopic morphology (b) On Sabourauds dextrose agar and lactritmel agar, brilliant lemon-yellow reverse pigment (c) On Sabouraud's salt agar, there is a lack of reverse pigment (d) Negative hydrolysis of urea

disulfide bonds, which weakens tightly connected keratinized tissue and leaves it open for enzymes to act. Subtilisins (Sub4) and fungalysins (Mep3 and Mep4) break down large proteins to be degraded by Dpp-V and IV. The ability of dermatophytes to invade the hosts is influenced by several host inflammatory factors. Increased β-defensin expression causes epidermal proliferation in *Trichophyton* infections. The phagocytes fail to eliminate dermatophytes from the stratum corneum but may facilitate permeability and keratin integrity loss. Dermatophyte serine proteases are responsible for severe pruritis associated with infection. Micro-abrasions on the skin also aid in the pathogenesis of infection (Fig. 11.7).

Immunology

Keratinases and glycopeptides are two important dermatophyte antigens recognized by the immune system of the infected human host. Keratinases, when injected intradermally into an animal skin, elicit delayed-type hypersensitivity (DTH). The polysaccharide of the glycopeptides enhances humoral immunity, but the protein component specifically stimulates cell-mediated immunity (CMI). CMI is responsible for removing the fungus from the stratum corneum of the skin, thereby facilitating clinical cure.

Anthropophilic fungi such as *T. rubrum* produce a milder inflammatory response and hence elicit mild DTH response. Nevertheless, geophilic or zoophilic dermatophytes elicit a strong inflammatory response. The DTH response to the trichophytin glycopeptides results in activated lymphocytes and macrophages, which cause inflammation. Mannan glycoprotein, present in the fungal cell wall, plays an important role in controlling the inflammatory response in atopic patients. In experimental infection study, *T. rubrum* mannan (TRM) was purified and incubated with lymphocytes, which resulted in reduced proliferation response to mitogens and other antigenic stimuli. TRM also inhibits the proliferation of keratinocytes, hence affecting epidermal turnover and facilitating a longer chronic infection.

The presence of dermatophyte antigens in the stratum corneum facilitates keratinocytes in producing IL-8, which recruits neutrophils. Macrophages identify dermatophytes using the Dectin-1 and Dectin-2 lectin receptors. Lesion resolution is connected to IFN-γ-producing T cells that cause a delayed-type hypersensitivity reaction. Animals with athymia are unable to recover from an infection.

Infection in Humans

Based on the site of infection, dermatophytosis in humans is designated by various names such as tinea capitis, tinea corporis, tinea barbae, tinea faciei, tinea manuum, tinea cruris, and tinea pedis that denotes dermatophyte infections of the scalp, trunk, beard, face, hands, groin, and feet, respectively.

Tinea corporis or tinea capitis in humans are caused by zoophilic dermatophytes. The ringworm that affects animals is highly contagious to humans. Farmers, veterinarians, experts in artificial insemination, and others that come in contact with sick animals are at high risk. The dermatophytes and host interaction vary depending on the fungus species.

The zoophilic dermatophytes that have less coevolution with humans make individuals more sick. For example, *T. erinacei*, a zoonotic fungus transmitted by hedgehogs, causes a severe inflammatory and pruritic eruption in infected humans. *T. mentagrophytes*, a zoophilic fungus, has global distribution with a wide range of animal hosts that includes mice, cats, guinea pigs, horses, kangaroos, rabbits, and sheep. The species produces inflammatory lesions in the skin and scalp, especially among rural populations. It causes kerion of the beard and scalp, while invaded hair exhibits ectothrix infection.

Anthropophilic dermatophytes are at times pathogenic to humans or animals that principally grow on decomposing keratinous material.

Infection in Animals

Contaminated fomites such as clippers, brushes, and gloved hands are known to transmit infection by spores. Cat whose skin is actively itchy and is exposed to a contaminated environment is another source of infection. *N. nana* causes infections in high-intensity livestock settings, more commonly in pigs, dogs, and ruminants. *L. gallinae* causes infection in chickens and *N. persicolor* causes infection in sylvatic rodents. *T. vanbreuseghemii* affects pets, agricultural, and laboratory animals, while *T. erinacei* commonly infects hedgehogs. Table 11.2 summarizes the common zoophilic dermatophytes and the infections they cause.

Epidemiology and Public Health

Dermatophytes based on their primary habitat have been classified as anthropophilic, zoophilic, and geophilic species. Anthropophilic dermatophytes cause infection in humans and rarely in animals. Zoophilic dermatophytes primarily infect animals and rarely infect humans. Geophilic dermatophytes are often found in soil. Some dermatophytes overlap in ecology, for

Table 11.2 Most common dermatophyte infections

Animal	Fungi	Lesion
Rabbit	*Trichophyton mentagrophytes var. mentagrophytes*	Crusting dermatitis on the head (eyelids, face, and ears), back, legs, and nails
Cattle	*Trichophyton verrucosum (var. album, discoides, ochraceum)*	With non-fluorescent mega-spore endo-ectothrix type of hair infection
	Trichophyton mentagrophytes var. mentagrophytes	Alopecia ringworm lesions are scaly and crusted in cattle
Horses	*Trichophyton equinum, Microsporum canis* and *Trichophyton mentagrophytes, Microsporum gypseum*	Alopecic areas with an underlying dry integument are covered by flaky, grayish scales, and sometimes kerion
Dogs	*Microsporum canis, Microsporum nanum Microsporum gypseum, Trichophyton verrucosum, Trichophyton mentagrophytes*	Brittle hairs and alopecic lesions. The scaly crusted lesions, kerion, onychomycosis, and/or paronychia
Cats	*Microsporum canis, Microsporum gypseum, Microsporum persicolor*, and *Trichophyton mentagrophytes*	Non-inflammatory alopecic areas, with central desquamation, dermatitis that resembles miliary and lesions in the shape of rings that have fur regrowth in the center and inflammation or papules around the periphery
Wild animals	*Trichophyton mentagrophytes* and its perfect state *Arthroderma benhamiae, Microsporum gypseum*, and *Microsporum canis*	May be asymptomatic, otherwise the same as domestic animals

instance, *M. gypseum*. The disseminated arthroconidia from dermatophytes can persist as a contagion in the environment and repeatedly cause outbreaks in certain areas.

Anthropophilic species are highly endemic, but a few of them such as *T. rubrum* and *T. tonsurans* are cosmopolitan due to their transportation by humans. They are also distinctive in body site preference. *T. tonsurans* and *M. canis* are the predominant causative agents of tinea capitis in urban North America. *M. canis* is the important agent in the rural areas and in Europe, the eastern Mediterranean areas, and South America. *T. rubrum*, followed by *T. mentagrophytes*, causes most dermatophyte infections in India, especially in Tamil Nadu.

Diagnosis

Microscopy

A skin biopsy or direct examination of the hair and scales from the lesions can confirm the diagnosis. Both in humans and animals, the examination of hair for culture and/or direct inspection can be made using a Wood's light. Positive hair will glow/fluoresce apple green. To locate aberrant hair for immediate inspection, a dermoscope is used. Using this handheld device, the hair on the skin is examined and used for culture. When determining whether spores are present on the hair coat, fungi culture is carried out in conjunction with the results of a clinical examination. Table 11.3 summarizes the diagnostic approaches for dermatophytosis.

Histopathology

Different histological features are observed in the histopathology of biopsy specimens in dermatophyte infections. Neutrophils, compact orthokeratosis, and the "sandwich sign" are three distinct features in the stratum corneum that are related to dermatophyte infections. The presence of hyphae "sandwiched in" between a lower layer of recently formed stratum corneum is characteristic. It is form of compact orthokeratotic or parakeratotic, with an upper layer that is normal and has a basket-weave structure. Neutrophils in the stratum corneum are not believed to be a reliable indicator of dermatophytosis. Grocott-Gomori's (or Gömöri) methenamine silver stain is useful

Table 11.3 Diagnostic methods available for dermatophytosis

Diagnostic approaches	Methods	Comments
Microscopy	Potassium hydroxide (KOH) wet mount from sample	Hyphal elements visualized
	Lactophenol cotton blue (LPCB) wet mount from culture	Septate hyphae, microconidia, and macroconidia in various shapes and patterns based on the species (summarized in Table 11.1)
	Grocott-Gomori's (or Gömöri) methenamine silver stain for histopathology	Neutrophils, compact orthokeratosis, and the "sandwich sign"
Culture	On SDA, oatmeal agar or dermatophyte test medium	Various colony morphologies are appreciated, as already mentioned in Table 11.1
Molecular methods	PCR-based ITS region amplification	
	Matrix-assisted laser desorption ionization–time of flight mass spectrometry (MALDI-TOF MS)	
	Sequencing techniques	

for demonstration of the fungi in the biopsy specimen.

In Vitro Culture

Diagnosis by culture is a routine test for identification of dermatophytes based on macroscopic examinations and microscopy of fungal culture growth (Figs. 11.1, 11.2, 11.3, 11.4, 11.5, and 11.7 and Table 11.1).

Serology

No serological tests are available for diagnosis of dermatophyte infections.

Molecular Diagnosis

PCR-based ITS region amplification using universal primers is carried out in the detection and identification of a few dermatophytes. The internal transcribed spacer region (ITS) of the rDNA needs to be sequenced to validate the identity of fungal species. Nevertheless, due to non-standardized procedure, non-validated databases, and ongoing changes in renaming and reclassification of dermatophytes, the results of nucleic acid sequences need to be interpreted with caution.

Automated Tests

Matrix-assisted laser desorption ionization–time of flight mass spectrometry (MALDI-TOF MS) can be used effectively for species-level identification of zoophilic dermatophytes if the reference library is well equipped with the different species, including uncommon ones.

Antifungal Drug Susceptibility Testing

Antifungal drug susceptibility testing is not routinely done in laboratories worldwide. However, the recent epidemic-like scenario of anti-fungal drug resistance in the Indian subcontinent and isolated case reports of recalcitrant tinea from Europe have shown the increasing need for the procedure. Standard guidelines for antifungal susceptibility testing method of dermatophytes have been issued by the Clinical Laboratory Standards Institute (CLSI), the European Committee on Antimicrobial Susceptibility Testing (EUCAST), and the British Society of Antimicrobial Chemotherapy (BSAC). The broth (liquid media) microdilution assay, according to the Clinical and Laboratory Standards Institute M38-A2 document, is used largely. This technique requires 4 days of incubation at 35 °C. The minimum inhibitory concentration (MIC) is defined as the concentration at which the growth of the organism will be inhibited by 80% compared with the growth in the control (MIC80).

Treatment

Topical or systemic administration of allylamines, azoles, and griseofulvin is the mainstay of treatment of dermatophytoses. Tinea pedis, tinea corporis, tinea cruris, and tinea manuum are treated effectively with topical antifungals in most cases. However, tinea capitis, tinea barbae, and onychomycosis require long-term treatment by systemic antifungals. The emergence of antifungal drug resistance due to the use of over-the-counter combination treatments containing steroids in dermatophytes is of concern.

Treatment in Humans

Topical medications do not penetrate the hair follicle; hence, dermatophyte infections of the scalp and hair need systemic treatment using oral antifungal agents. Griseofulvin given for 4–8 weeks and terbinafine given for 6 weeks are frequently used drugs with similar efficacy and safety profiles, but griseofulvin is more cost-effective.

Topical antifungal shampoos like 2% ketoconazole are effective in clearing the infection from the skin. It is applied as a 5-min application two times per week for at least 2 weeks or until the infection has cleared.

Oral antifungals are the first-line treatment for tinea barbae, with adjunctive topical medications such as terbinafine and ketoconazole. Tinea corporis treatment is primarily topical. Clotrimazole, miconazole, or terbinafine is usually recommended. However, widespread infections require oral antifungal therapy. Management of tinea cruris is by topical terbinafine and azoles supplemented by in usual cases oral antifungal therapy for extensive disease.

Tinea pedis and onychomycosis are prone to cause cellulitis and hence need immediate treatment with topical azoles such as miconazole, clotrimazole or topical terbinafine. Systemic treatment is needed when the dorsum of the foot, heel, or sole is involved. A refractory infection or one with blisters also must be treated systemically.

Treatment in Animals

Itraconazole along with a miconazole bath and chlorhexidine shampoo is recommended for the treatment of pet cats and dogs. Sodium hypochlorite rinse is also effective. Terbinafine and miconazole are effective for dermatophyte infections in cattle. Itraconazole and griseofulvin are used but fluconazole is not effective.

Prevention and Control

Sanitation, hygienic practices, and treatment are essential for the prevention and control of dermatophytosis in humans. Mass immunization campaigns against the disease in animals have proven useful to control and eradicate infections in cattle and other fur-bearing animals. *T. verrucosum* vaccines have been used in the prevention of cattle ringworm.

Case Study

A 35-year-old lady presented with a 7-month history of chronic rashes on the scalp. She was a known diabetic for 5 years, previously identified as having seborrheic dermatitis and psoriasis; the condition was unsuccessfully treated with a combination of triamcinolone acetonide cream and compound econazole nitrate. HbA1c was 7. Her other blood investigations were within normal range. Multiple scaly erythematous spots with a greyish-yellow crust, patches of alopecia, and broken hair were discovered during the local examination. Under Wood's bright lights, bright-green fluorescence was especially noticeable. Direct microscopic examinations of broken hair using 10% KOH showed ectothrix hyphae and spores. Fluorescence stain also revealed the same. History elicited found interaction with a stray cat that she used to feed frequently. She received daily doses of 250 mg of oral terbinafine, topical imidazole solution, and ketoconazole cream.

1. What is the clinical condition?
2. What are the various fungi that cause this infection?
3. What culture media are used for this group of fungi?
4. What are zoonotic fungi?

Research Questions

1. With the availability of efficacious vaccines for the prevention of dermatophyte infections in animals, what is the direction of research needed to translate it toward human infections?
2. Are rapid, low-cost, sensitive serological tools available to diagnose dermatophyte infections?
3. While the complex fight between dermatophyte virulence mechanisms and the host immune system has been well documented, what are the gaps that need to be filled in understanding the role of immunopathogenesis in host susceptibility and invasive disease?

Acknowledgments The contribution of Mrs. Anitha S. in the preparation of the figures in this chapter is acknowledged.

Further Reading

Baert F, Stubbe D, D'hooge E, Packeu A, Hendrickx M (2020) Updating the taxonomy of dermatophytes of the BCCM/IHEM collection according to the new standard: a phylogenetic approach. Mycopathologia 185(1):161–168

Baumbach CM, Müller S, Reuschel M, Uhrlaß S, Nenoff P, Baums CG, Schrödl W (2021) Identification of zoophilic dermatophytes using MALDI-TOF mass spectrometry. Front Cell Infect Microbiol 11:365

Berlin M, Kupsch C, Ritter L, Stoelcker B, Heusinger A, Gräser Y (2020) German-wide analysis of the prevalence and the propagation factors of the zoonotic dermatophyte *Trichophyton benhamiae*. J Fungi 6(3):161

Chermette R, Ferreiro L, Guillot J (2008) Dermatophytoses in animals. Mycopathologia 166(5):385–405

de Hoog GS, Dukik K, Monod M, Packeu A, Stubbe D, Hendrickx M, Kupsch C, Stielow JB, Freeke J, Göker M, Rezaei-Matehkolaei A (2017) Toward a novel multilocus phylogenetic taxonomy for the dermatophytes. Mycopathologia 182(1):5–31

Drouot S, Mignon B, Fratti M, Roosje P, Monod M (2009) Pets as the main source of two zoonotic species of the *Trichophyton mentagrophytes* complex in Switzerland, *Arthroderma vanbreuseghemii* and *Arthroderma benhamiae*. Vet Dermatol 20(1):13–18

Kane J (1997) Laboratory handbook of dermatophytes: a clinical guide and laboratory handbook of dermatophytes and other filamentous fungi from skin, hair, and nails. Star Publishing Company

Mariat F, Chatelain J, Rouffaud MA (1976) Study of the contamination by dermatophytes of a population of small wild mammals in Alsace. Mycopathologia 58(2):71–78

Mazur M, Lodyga M, Łańczak A, Adamski Z (2018) Majocchi's granuloma (granuloma trichophyticum) in a Guinea pig owner: a case report and literature review. J Mycol Médicale 28(3):523–526

Mignon B, Monod M (2011) Zoonotic infections with dermatophyte fungi. In: Oxford Textbook of Zoonoses: biology, clinical practice, and public health control. New York University Press

Palmer SR, Soulsby L, Torgerson P, Brown DWG (2011) Oxford Textbook of Zoonoses: biology, clinical practice, and public heath control, 2nd edn. New York University Press

Segal E, Elad D (2021) Human and zoonotic dermatophytoses: epidemiological aspects. Front Microbiol 12:713532

Weitzman I, Summerbell RC (1995) The dermatophytes. Clin Microbiol Rev 8(2):240–259

Malassezia Infections

12

Prasanna Honnavar
and Shivaprakash M. Rudramurthy

Learning Objectives
By the end of the chapter, the readers should be able to:

1. Characterize *Malassezia* species based on taxonomy, genomics, and proteomics.
2. Differentiate *Malassezia* infection based on their clinical presentation.
3. Identify the predisposing factors.
4. Recognize the normal reservoir of *Malassezia* and their modes of transmission.
5. Correlate the virulence factors of *Malassezia* with the pathophysiology and immunopathology of the host.
6. Choose the appropriate laboratory investigation to confirm the diagnosis of *Malassezia* infections.
7. Select the appropriate antifungal agent for the treatment of *Malassezia* infections.

P. Honnavar (✉)
Department of Microbiology and Immunology, American University of Antigua College of Medicine, St. Johns, Antigua and Barbuda
e-mail: prasannna1@gmail.com

S. M. Rudramurthy
Department of Medical Microbiology, Postgraduate Institute of Medical Education and Research (PGIMER), Chandigarh, India

Introduction

Malassezia species (spp.) are lipid-dependent commensal yeasts found on all humans, a variety of animals and birds, and in diverse ecosystems such as deep marine environments. The human skin microbiome consists of bacteria, viruses, fungi, and archaea. Over the last 30 years, much importance has been given to bacteria. Since the last decade, studies on mycobiome (fungal communities) have been on the rise. A single predominant eukaryotic genus, *Malassezia*, represents 5–10% metagenomics sequences on the skin and less than 1% in mucosal epithelia. *Malassezia* constitutes 90% of the eukaryotes on the skin. *Malassezia* are unable to synthesize lipid, hence predominantly colonize the sebum-rich areas such as scalp, chest, face, and upper back than low sebum areas such as the arm and feet.

The diversity of *Malassezia* spp. may vary with geography and ethnicity. *Malassezia* has been associated with skin diseases such as pityriasis versicolor (PV), seborrheic dermatitis/dandruff (SD/D), psoriasis, atopic dermatitis (AD), and folliculitis. *Malassezia* has become an emerging concern in fungemia among immunocompromised patients and premature neonates receiving parenteral nutrition. Among the animals, several species of *Malassezia* are present on the healthy and diseased skin of dogs, cats, horses, goats, pigs, rabbits, hibernating bats, and

S. C. Parija, S. M. Rudramurthy (eds.), *Textbook of Fungal Zoonoses and Sapronoses*, Microbial Zoonoses, https://doi.org/10.1007/978-981-97-3214-2_12

pet parrots. Out of 118 *Malassezia* fungemia published cases, only a couple of studies found the potential for zoonotic transmission from dogs to their owners leading to intensive care nursery outbreaks. The inability to diagnose *Malassezia* bloodstream infection (BSI) due to its fastidious nature would have underestimated the role of *Malassezia* in BSI than candidemia. Future studies are requisite to further explore transmission routes such as animal to human, and human to human (pet owner, healthcare worker, family member, etc.). Other than humans and warm-blooded animals, culture-independent studies have shown that *Malassezia* are present in terrestrial and marine ecosystems such as soils, deep-sea sediments, sponges, corals, cone snails, and nematodes.

History

The nomenclature and terminology used for the *Malassezia* are quite complex. Back in 1846, Karl Ferdinand Eichstedt observed the yeasts and hyphal forms in PV patients' scrapings for the first time. Later in 1853, Robin named the microorganism *Microsporon furfur*. Rivolta, in 1873, observed double-contoured yeasts in psoriasis lesional flakes and placed them within the genus *Cryptococcus*. In 1884, based on the shape of the spores observed in the scalp scales, Bizzozero named them *Saccharomyces ovalis* (oval) and *Saccharomyces sphaericus* (spherical). After creating the genus *Malassezia* (Baillon) 133 years ago, it was recently realized that the genus *Malassezia* consists of multiple species. *Malassezia furfur* was isolated from the human cutaneous lesion in 1889. In 1925, Weidman isolated *Malassezia pachydermatis* from the Indian rhinoceros' exfoliative dermatitis. In the 1950s, *Malassezia* was renamed *Pityrosporum*, consisting of three species *Pityrosporum orbiculare* (round shape), *Pityrosporum ovale* (oval shape), and *Pityrosporum pachydermatis* (isolated from Indian rhinoceros). In 2005, *M. furfur*, *Malassezia globosa*, and *Malassezia restricta* were assigned as *M. furfur* serovars A, B, and C, respectively. Currently, the *Malassezia* genus comprises 1 non-lipid-dependent species (*M. pachydermatis*) and 17 lipid-dependent species.

Taxonomy

Malassezia are basidiomycetous fungi that belong to the subphylum Ustilaginomycotina and class Malasseziomycetes. The phylogenetic niche of *Malassezia* suggests that their ancestors comprised almost exclusively of plant pathogens or soil residents. They successfully adapted to animal and bird's cutaneous ecosystem. After the creation of the genus *Malassezia* (Baillon) 133 years ago, in the year 1996, it was realized that the genus *Malassezia* consists of multiple species. *M. furfur* was isolated from the human cutaneous lesion in 1889. In 1925, Weidman isolated *M. pachydermatis* from Indian rhinoceros' exfoliative dermatitis. In the 1950s, *Malassezia* was renamed as *Pityrosporum,* consisting of three species *Pityrosporum orbiculare* (round shape), *Pityrosporum ovale* (oval shape), and *Pityrosporum pachydermatis* (isolated from Indian rhinoceros). *M. furfur, Malassezia globosa*, and *Malassezia restricta* were designated as *M. furfur* serovars A, B, and C, respectively. Before 1996, only three species were recognized: *M. furfur*, *M. pachydermatis*, and *Malassezia sympodialis*. Further advancement in molecular tools helped scientists to discover novel species and currently consists of 18 species (Table 12.1). The new species were named after the Latin names of animals from which they were initially isolated such as *M. pachydermatis* (pachydermata-rhinoceros); *Malassezia equina* (horses); *Malassezia caprae* (goats); *Malassezia cuniculi* (rabbits); *Malassezia pstittaci* (parrots); and *Malassezia vespertilionis* (bats). As the presence of *Malassezia* in a broader range of wild animals or birds is investigated, there is a high probability that the number of species in the near future may increase.

Table 12.1 Summary of taxonomy, ecology, and physiological characteristics of all 18 currently accepted *Malassezia* species

Species	Taxonomic cluster	Presence in humans	Presence in animals	Utilization					Catalase	β-Glucosidase	Growth at 37 °C	Growth at 40 °C
				Tween 20	Tween 40	Tween 60	Tween 80	Cremophor EL				
Malassezia furfur	A	HS, PV, fungemia	HS (dog, cat)	+	+	+	+	+	+, (−)	−, (w)	+	+
Malassezia obtusa	A	HS, LS	–	–	–	–	–	–	+	+	−, (w)	–
Malassezia yamatoensis	A	HS, SD	–	+	+	+	+	?	+	?	+	–
Malassezia japonica	A	HS, AD, SD	–	–	w	+	–	?	+	?	+	–
Malassezia vespertilionis	A		HS (bats)	w	+	+	w	–	–	–	w	w
Malassezia psittaci	A	–	HS (parrot)	+	+	+	+	+	+	–	–	–
Malassezia brasiliensis	A	–	HS (parrot)	+	+	+	+	+	+	–	+	+
Malassezia pachydermatis	B1	HS, fungemia	HS, LS (dog, cat)	+	+	+	+	+	+, (w)	+, (−)	+	+
Malassezia sympodialis	B1	HS, AD, SD	HS, OT (cats)	–	+	+	+	–	+	+	+	+
Malassezia caprae	B1	–	HS (goat)	–	+	+	+, (−)	–	+	+, (−)	−, (w)	–
Malassezia equina	B1	–	HS, LS (horse)	w	+	+	+	–	+	–	w	–
Malassezia dermatitis	B1	HS, AD	–	+	+	+	+	W	+	?	+	+
Malassezia nana	B1	–	HS (cat, horse), OT (cat, cattle)	v	+	+	w	–	+	–	+	v
Malassezia restricta	B2	HS, SD		–	–	–	–	–	–	–	v	–
Malassezia globosa	B2	HS, PV, SD, AD	HS, OT (cat)	–	–	–	–	–	+	–	−, (w)	–
Malassezia arunalokei	B2	HS, SD	–	–	–	–	v	–	–	–	+	–
Malassezia slooffiae	C	HS, LS	HS (pig, cat)	+	+	+	–	–	+	–	+	+
Malassezia cuniculi	C	–	HS (rabbit)	–	–	–	–	–	+	+	+	+

AD atopic dermatitis, *HS* healthy skin, *LS* lesioned skin, *PV* pityriasis versicolor, *SD* seborrhoeic dermatitis, *OT* otitis, + positive, – negative, *W* weak, *v* variable, no information available for this specific test

Genomics and Proteomics

Phylogenetic analyses of Basidiomycota phylum have revealed that *Malassezia* is distantly related to the human pathogen *Cryptococcus neoformans* and surprisingly close to the plant pathogen *Ustilago maydis*. Although the *Malassezia* genome has similarities with the plant pathogen *Ustilago*, *Malassezia* enzymes closely resemble that of other skin yeasts such as *Candida*. *Malassezia* are even more divergent than other skin mycobiome such as *Candida albicans* and dermatophytes. Laniri et al. examined 45 genome assemblies of 15 known *Malassezia* species (GenBank search on March 7, 2020). The authors tried to resolve the pathogenesis and niche adaption with respect to evolutionary trajectory. Three clades have been identified in the *Malassezia* genus (A, B, and C). Clade B has two subclades B1 and B2 (Fig. 12.1). Clade A: *M. obtusa*, *M. furfur*, *M. brasilienses*, *M. japónica*, *M. yamatoensis*, *M. psittaci*, *M. vespertilionis*; Clade B1: *M. dermatis*, *M. equina*, *M. sympodialis*, *M. caprae*, *M. nana*, *M. pachydermatis*; Clade B2: *M. globosa*, *M. restricta*, *M. arunalokei*; Clade C: *M. cuniculi*, *M. slooffiae*. Sankaranarayanan et al. elucidated the karyotype diversity in the *Malassezia*. The difficult-to-grow *M. globosa* and *M. restricta* have nine ancestral chromosomes. Readily cultivable *M. sympodialis* and *M. furfur* have eight and seven chromosomes, respectively, which would have formed from losing centromeres. Being an outlier, *M. pachydermatis* has six chromosomes (Fig. 12.1).

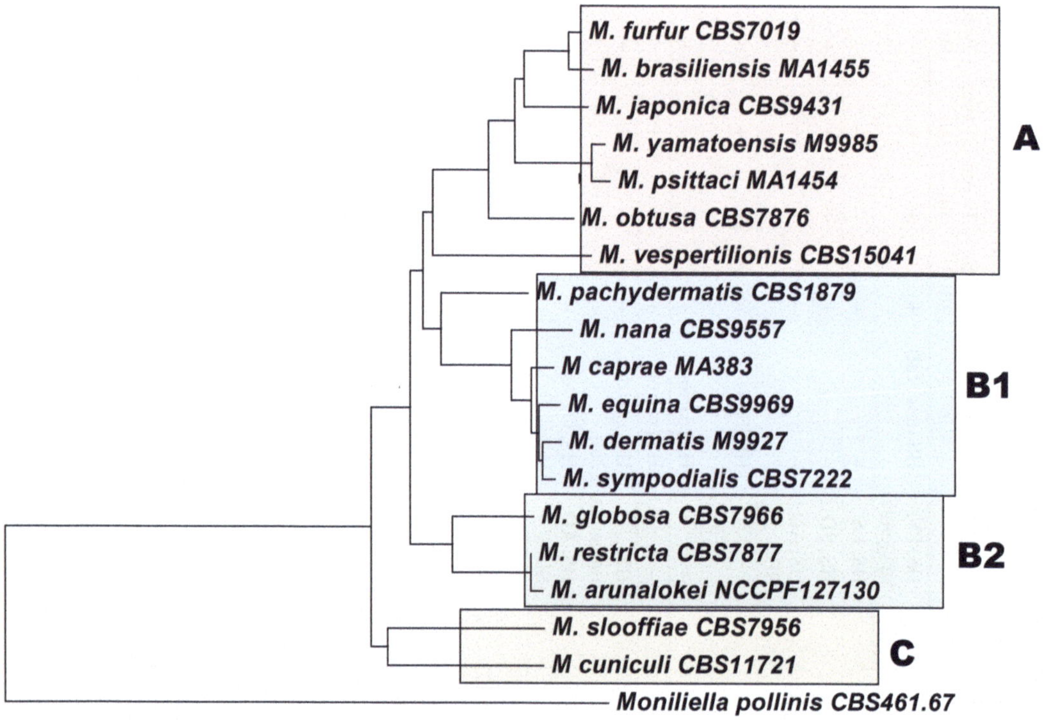

Fig. 12.1 Phylogenetic tree of all 18 currently accepted *Malassezia* species. The evolutionary history was inferred using the Neighbor-Joining method. The optimal tree with the sum of branch length = 0.79824279 is shown. The percentage of replicate trees in which the associated taxa clustered together in the bootstrap test (1000 replicates) are shown next to the branches. The tree is drawn to scale, with branch lengths in the same units as the evolutionary distances used to infer the phylogenetic tree. The evolutionary distances were computed using the Kimura 2-parameter method and are in the units of the number of base substitutions per site. The analysis involved 19 nucleotide sequences. All positions containing gaps and missing data were eliminated. There was a total of 516 positions in the final dataset. Evolutionary analyses were conducted in MEGA6

As the whole genome sequences of many *Malassezia* spp. are yet not available, there is still a long way to understand the phylogeny and evolution. These efforts are hampered by challenges in cultivating the *Malassezia* and understanding the function of unknown genes. *Malassezia* have compact (genes arranged closely with small introns) and smallest (7–9 Mb) haploid genome among free living fungi. *Malassezia* have an abundance of genes that code for lipases, phospholipases, acid sphingomyelinase, proteases, and peptidases that facilitate them to adapt to the host environment. In addition, they secrete many allergens that may potentially cross-react with immune cells leading to inflammation of skin. On the other hand, they have lost the genes that code for fatty acid synthase, $\Delta^{2,3}$-enoyl-CoA isomerase, and Δ^{9}-desaturase, which explains their inability to synthesize lipids and thrive in the oil-rich sebaceous area.

The results of many recent studies are relatively new and need further evidence. *M. sympodialis*, *M. obtusa*, *M. globosa*, *M. restricta*, *M. yamatoensis*, and *M. pachydermatis* harbor dsRNA mycoviruses of the Totiviridae family. These mycoviruses resulted in higher pathogenicity. More than 30 bacterial genes (catalase, flavohemoglobin, septicolysin-like, oxidoreductase, etc.) acquired through horizontal gene transfer have been identified in *Malassezia*.

The Fungus Morphology

Malassezia cells are monopolar budding yeast with narrow to broad bases. The size and shape of yeast cells varies among different *Malassezia* species. The cell shape may be globose, ellipsoidal, cylindrical, or ovoid. The size of yeast cells ranges from 2–8 to 2–3.5 μm. The colony morphology also varies among the species such as dull to shiny, pale cream to yellowish-brown, flat to convex, entire or slightly folded margin, wrinkled to cerebriform, butyrous to brittle, smooth to rough, 2–8 mm in diameter.

Cultivation of Fungi

Cultivating *Malassezia* in a laboratory setting presents a unique set of challenges. This is primarily because of their requirement for lipid supplements in the growth medium, without which they cannot proliferate. Modified Dixon agar (MDA) or Leeming and Notman agar (LNA) are the selective media for *Malassezia* culture. In resource-deficient settings, Sabouraud dextrose agar medium overlaid with olive oil can also be used. CHROM agar *Malassezia* and MalaSelect are the newer selective culture mediums. These are agar-based media supplemented with olive oil, ox bile, and other ingredients that provide a rich source of the necessary fatty acids for *Malassezia* growth. The process of cultivation typically begins with the collection of samples, which can be obtained from skin scrapings, swabs, or biopsy material. Once collected, the sample is placed on the lipid-enriched agar plate and then incubated at a temperature ranging from 32–37 °C. This temperature range is chosen to mimic the physiological temperature of human skin, thereby creating a conducive environment for *Malassezia* growth. The growth rate varies among the *Malassezia* species from 2 days to several weeks. *Malassezia* cultures do not survive at room temperature for a long period. Cultures can be maintained in 30% glycerol kept at −80 °C or lyophilization.

Animal Models

Implementing a host model for *Malassezia* infection is challenging due to its weak virulence. The animal model should be susceptible to repetitive exposure to develop *Malassezia* infection effectively. The guinea pig was the most commonly used animal model for establishing SD and evaluating antifungal activity. Severe inflammatory lesions resembling SD were observed after repeated inoculation (up to 7 days) of *M. furfur* and *M. restricta* cultures. Ketoconazole and luliconazole showed better clinical outcomes. Noted drawbacks of using guinea pigs as animal models are many and include their complex social structure, expense, lack of congenic strains, and their susceptibility to stress. Dogs were used as a model for *M. pachydermatis* external otitis to evaluate the antifungal activity. Plastic film occlusions were necessary to establish *Malassezia* skin infection in rabbits; spontaneous healing happened once the occlusions were discontinued. Transgenic T-cell model was developed to mimic SD in mice, and antifungals resolved the clinical symptoms. Similar to *Malassezia*-induced lesions in humans, *Malassezia* triggered the release of the cytokines TNF and IL-6 in Mincle-deficient mice. Recently, it has been demonstrated that, as observed in AD, *Malassezia* induced the release of IL-17 in skin barrier-disrupted (dorsal skin of the ear) mice.

Very few invertebrates as animal models for *Malassezia* infection have been studied. *Caenorhabditis elegans* larva mortality was higher after exposure to *M. pachydermatis*. Drosophila melanogaster Toll-mutant flies' survival was significantly reduced compared to wild type after being challenged with *M. pachydermatis*. *Galleria mellonella* larva has been standardized and proven reliable infection models for *M. pachydermatis* (CBS1879) and *M. furfur* (CBS1878). *G. mellonella* produced similar results as those of murine models.

Life Cycle of the Fungus

Malassezia yeasts have a unique and somewhat elusive life cycle due to their inability to form sexual reproductive structures, such as spores or fruiting bodies, which are commonly found in many other fungi. They reproduce asexually via a process called budding, where a smaller cell, or bud, grows directly out of an existing yeast cell. As the bud matures, it becomes a fully formed yeast cell, eventually separating from the parent cell. Once separated, the new cell can continue this cycle, forming its own buds. In addition to budding, *Malassezia* can also exist as mycelial forms (filamentous growth) under certain conditions, such as pityriasis versicolor or in the lab when cultured in certain media. Despite this general understanding, there is still much to learn about the full life cycle of *Malassezia*, as the genus has a diversity of species with potentially different growth characteristics and life cycles that are not fully understood.

Pathogenesis and Pathology

Malassezia can either stimulate the immune system leading to a disease state or limit the immune stimulation resulting in commensal carriage. The complexity of *Malassezia*-induced diseases and their prevention may be explained to some extent by deciphering the factors that switch *Malassezia* between pathogenic and commensal states. Multiple *Malassezia* species or genotypes may cause unique or similar pathologies. Before the unified *Malassezia* classification in 1996, the immunological studies conducted remained obscure due to the *Malassezia* species/strains used in those studies. The in vitro studies on cellular and humoral immunity have used various components of *Malassezia*, such as cell surface proteins, cytoplasmic antigens, and whole or lysed yeast cells. In humans, it is suggested that humoral immunity may be related to commensal carriage, and higher cellular immune responses may be linked to *Malassezia*-related diseases.

Histopathological studies have contributed significantly to understanding the pathogenesis and pathology of cutaneous *Malassezia* lesions. The inflammatory cells, such as mature T helper cells and Langerhans cells, are commonly found in PV lesions. Compared to other fungal inflammatory conditions such as tinea corporis, PV has fewer inflammatory infiltrates. This might be due to the thick lipid layer around *Malassezia* that downregulates the cytokine responses. The SD lesions infiltrate consists of macrophages, monocytes, granulocytes, CD4+ Th cells, ICAM1 (intercellular adhesion molecules 1), NK1+ cells, CD16+ cells, and C1q positive cells. *Malassezia* folliculitis (MF) may have lymphocytes in non-raptured follicular infiltrates, whereas neutrophils, macrophages, plasma cells, and eosinophils predominate in raptured follicular infiltrates.

Inflammation with superficial interstitial or focal accumulations of neutrophils, eosinophils, and mast cells and perivascular infiltrate of mononuclear cells (MNC) are demonstrated in dogs with *M. pachydermatis*-associated dermatitis. In some cases, the dermo-epidermal junction may contain linearly aligned mast cells. In cats, the dermo-epidermal junction may comprise a prominent mixed dermal infiltrate or sparse with pancreatic paraneoplastic alopecia combined with an atrophic follicular pattern.

Immunology

The fungal factors such as cell wall constituents, secreted components, species/strain variation along with host factors including genetics, skin condition, and systemic predisposing conditions influence the immune response to *Malassezia*. Immune response to *Malassezia* is largely unknown due to the fact that it exists both as commensal and in skin disorders. IL-17/IL-23 activation is the key factor in inhibiting fungal growth as well as mediating immunopathology. IL-17 has been linked to psoriasis and AD. *Malassezia* dermatitis case reports due to a defect in the IL-17 pathway have not been documented yet. C-type lectin receptors are thought to play a prominent role in sensing the lipid-rich outer layer of *Malassezia*. In vivo and in vitro studies have shown conflicting roles of Dectin-2 and Mincle in antifungal defense. An increase in the *Malassezia*-specific serum IgE level has been correlated with the severity of human AD and canine AD. It is still unclear how the Th2/Th17 response develops in allergic conditions and how the Th cell subsets are related to each other. T-cell independent mechanisms may also be involved in the pathogenesis of AD and other inflammatory skin diseases. The course and severity of *Malassezia*-associated skin disorders may be influenced by neurotransmitters and hormones. Future research may shed more light on the immune response to *Malassezia* on the skin.

Infection in Humans

The change in the microenvironment such as barrier defect and/or host genetic susceptibility predisposes commensal *Malassezia* to become pathogenic. *Malassezia* metabolites can directly cause irritation or indirectly induce inflammation. *Malassezia* has been associated with numerous skin conditions in humans (Fig. 12.2).

In infants, *Malassezia* colonization has been linked to sebaceous miliaria or neonatal acne characterized by hypersensitive reaction in scalp, face, and neck skin with non-follicular papulopustular eruptions. They resolve within weeks. However, due to BSI, the consequences will be lethal in prematurely born and immunocompromised babies. The risk factors include nosocomial and iatrogenic sources, an incomplete fragile skin barrier, and weak immune system.

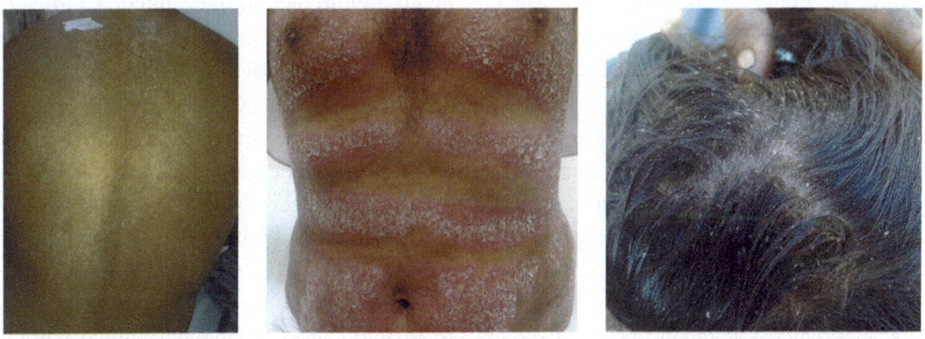

Fig. 12.2 Clinical manifestation of *Malassezia* infection in humans

PV, a disease more prevalent in hot and humid climates, presents with mild itch and scaling in the affected area. It is characterized by mild, chronic, hyper-, or hypo-pigmented patches in the sebum-rich areas of the trunk, shoulder, neck, and upper back. *M. globosa*, *M. furfur*, and *M. sympodialis* are more frequently isolated from PV cases. Numerous mycelial forms can be seen in the microscope from the samples collected from the lesions. PV relapse rate is up to 80% despite effective antifungal treatment.

SD is a common skin disorder characterized by the appearance of flaking over the greasy areas of skin most commonly on the scalp, nasolabial folds, ears, postauricular area, eyebrows, and chest. Dandruff (D) is a mild non-inflammatory form of SD restricted to the scalp. The SD/D may be embarrassing, especially because of the scaling scalp, which may cause uneasiness associated with uncleanliness. Seasonal temperature and humidity variation, use of cleansing agents, oil application, and hormonal imbalances are believed to trigger SD/D. Host susceptibility, sebum (microenvironment), and *Malassezia* are three basic etiological factors for SD/D. Itching is induced by irritants released from *Malassezia* such as oleic acid, malassezin, arachidonic acid, and indole-3-carbaldehyde. Host inflammatory response is triggered by lipase, phospholipase, and protease activity on the defective epidermal barrier. *M. globosa*, *M. restricta*, and *M. furfur* are more frequently isolated species from SD/D. Elimination of *Malassezia* using antifungals improves SD/D.

Psoriasis is characterized by epidermal hyperproliferation and hyper-keratinization that affects the skin, nails, and joints. Psoriasis is an autoimmune disease resulted from genetic and environmental factors. Symptomatic improvement of psoriasis after topical and systemic antifungal treatment suggests the role of *Malassezia* as a strong microbial component at least in scalp psoriasis. *M. globosa* is the predominant species in scalp psoriasis, followed by *M. furfur* and *M. sympodialis.*

AD is a chronic skin inflammatory disorder resulting from the interplay among immune, genetic, and environmental factors. The mutation in the filaggrin gene/tight junction and increased Th2 cytokines together disrupt the skin barrier leading to increased trans-epidermal water loss, altered microbiota, and high pH resulting in itching, dryness, and erythema. The exacerbation of AD lesions after the application of *Malassezia* extracts/antigens and the resolution of severity of symptoms after antifungal treatment suggests the role of *Malassezia* in AD. *M. globosa* and *M. restricta* are predominantly isolated from AD lesions followed by *M. sympodialis* and *M. furfur.*

Fungemia due to *Malassezia* spp. has been increasingly reported in immunocompromised patients and premature neonates receiving parenteral nutrition. BSI from *Malassezia* spp. has received less attention than its dermatological manifestation. Slow, fastidious, and lipid dependence for the growth in the laboratory condition has made isolation of *Malassezia* less common than *Candida* from BSI. *M. furfur*, *M. pachydermatis*, and *M. sympodialis* are more frequently isolated from BSI. Interestingly, only one genotype of *M. pachydermatis* or *M. furfur* has been associated with BSI in immunocompromised patients. Human skin mycobiome may be a potential reservoir for *Malassezia* causing BSI. In 1981, the first case of *Malassezia* BSI and sepsis was reported. As of 2020, 118 cases with only 3 surveillance studies have been reported. *M. furfur* (n = 82) was the predominant species isolated from BSI followed by *M. pachydematis* (n = 33) and *M. sympodialis* (n = 3). BSI from *Malassezia* is more common in neonates (n = 82) than adults (n = 23) and children (n = 17). *M. sympodialis* has caused a case of post-appendectomy fungemia in a child on long-term inhaled corticosteroid for asthma and post-gastrectomy continuous ambulatory peritoneal dialysis (CAPD) associated fungemia in an elderly man. *M. restricta* has caused splenic lesions in a child with neutropenia and acute appendicitis. Recently, in France, *M. restricta* has been isolated from infective endocarditis with a blood culture-negative sample. Two cases of *M. restricta* pneumonia in solid organ transplant recipients have been reported from Italy. Most diagnostic clinical laboratories won't have the facility to culture and identify the *Malassezia* spp. If you look for it, you will find it.

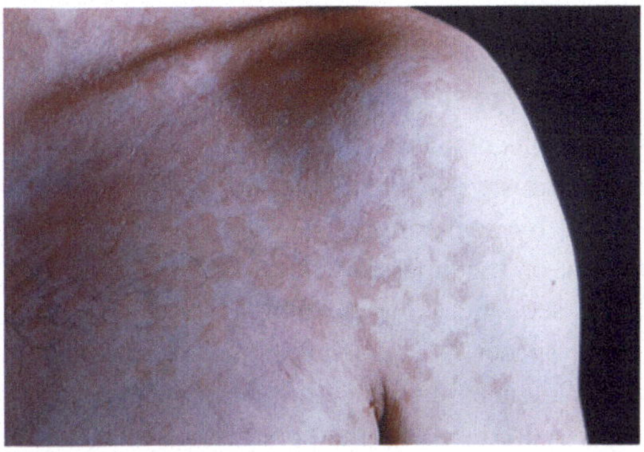

Fig. 12.3 Photograph depicts an anterior view of a patient's left shoulder, revealing a plaque-like, blotchy, erythematous rash, that was diagnosed as a case of tinea versicolor, caused by the fungal organism, *Malassezia furfur*, formerly known as *Pityrosporum ovale*. (Courtesy CDC)

Other than well-established skin manifestation (Fig. 12.3) and BSI, relatively high prevalence and abundance in other anatomical sites have raised the concern of involvement of *Malassezia* in chronic diseases like Crohn's disease, colorectal cancer, pancreatic adenocarcinoma, chronic rhinosinusitis-associated inflammation, cystic fibrosis pulmonary exacerbation, Alzheimer, and Parkinson's disease. However, more evidence is warranted to associate *Malassezia* with these chronic diseases.

Infection in Animals

A panel of veterinary dermatologists and mycologists prepared a comprehensive publication literature review (up to mid-2018) of ecology, pathophysiology, diagnosis, therapy, and prevention of *Malassezia* dermatitis in dogs and cats. The documents were presented at an international veterinary meeting. The World Association of Veterinary Dermatology (WAVD) led the way to come up with a consensus on the diagnosis and treatment of skin diseases associated with *Malassezia* yeasts in dogs and cats.

Switching of *Malassezia* from commensal to pathogen is well appreciated in otitis externa and dermatitis cases among dogs and cats. Few cases have been reported in large animals such as SD from goats, dermatitis from horses, otitis externa from ferrets, pigs, fennec foxes, and camels.

In dogs, *M. pachydermatis* dermatitis is presented with erythema and varying degrees of traumatic alopecia. Scaling is very common. Involvement of inter-digital and inter-triginous areas accompanied with greasy exudate and malodor. Abdominal lesions are erythematous, scaly, well-demarcated, and symmetrical. In a few instances, there may be scattered papules. Skin fold lesions are erythematous with sticky exudate. Otitis externa cases are presented with erythema, lichenification, and yellow/brownish ceruminous discharge over the vertical ear canal and pinnae. The signs of *Malassezia* dermatitis may mimic canine AD. The diagnosis of *Malassezia* dermatitis is mainly by clinical signs, organism load in the affected area, and response to antifungals.

Epidemiology and Public Health

Human factors such as age, gender, and immune status, along with environmental factors like geographical location and seasonal variation, determine the *Malassezia*-induced disease's outcome. Recently developed molecular tools for *Malassezia* identification and typing have revolutionized epidemiology. Immediately after the birth of a human, *Malassezia* colonizes the skin. *Malassezia* microbiota changes with age in a healthy individual. *M. restricta* and *M. globosa* predominate in all age groups irrespective of gender. Most *Malassezia*-induced cutaneous diseases (PV, SD, MF) are prevalent in adult populations when sebaceous gland activity is at its peak. The incidence of PV is higher in summer and is commonly found in individuals in their 20s and 30s. Many studies indicate that *M. globosa* plays a significant role in PV pathogenesis. SD affects 1–3% of the general population and 30–33% of patients with AIDS. SD is more prevalent in men. In contrast to PV, SD is more common, severe, and chronic during winter. *M. restricta* plays a significant role in SD pathogenesis. There are three forms of AD: infantile, childhood, and adult. AD usually onsets during the first 6 months of life, then the incidence decreases, whereas severity increases in adults. Quantitatively, *Malassezia* colonized the adults 2.4-fold higher than children. AD exacerbation may be due to *M. restricta* and *M. globosa*. MF is more common in tropical countries and is associated with leukemia, pregnancy, and Hodgkin's disease. *M. restricta* was predominantly isolated from MF, followed by *M. globosa* and *M. sympodialis*. *M. restricta* plays a major role in psoriasis, especially scalp psoriasis. *Malassezia* has been isolated from all the domestic animals and a few wild animals. *M. pachydermatis*-related cutaneous diseases have been extensively studied in dogs and cats. The age and gender predisposition of dogs to *Malassezia* dermatitis or otitis has been excluded by numerous studies. Future research may be focused on isolating *Malassezia* yeasts from various wild animals.

Potential for Zoonotic Transmission

Chang et al. conducted the first investigation of *M. pachydermatis* outbreak in an intensive care nursery. *M. pachydermatis* was transferred from pet dogs to the hands of healthcare workers and made its way to the intensive care nursery resulting in BSI ($n = 8$), urinary tract infection ($n = 2$), meningitis ($n = 1$), and asymptomatic colonization ($n = 4$). *Malassezia* can persist in the hospital environment for a prolonged period, leading to patient-to-patient transmission. Later sporadic cases of *M. pachydermatis* fungemia and facial granuloma associated with transmission from pet dogs to their owners were reported. The healthcare workers having pet animals should practice rigorous hand hygiene before handling immunocompromised patients and neonates.

Diagnosis

The traditional culture-dependent phenotypic identification of *Malassezia* spp. comprises microscopic features, assimilation of different tweens (20, 40, 60, and 80) and cremophor EL, catalase reaction, and growth at different temperatures (32 °C, 37 °C, and 40 °C) (Table 12.1). The results of phenotypic methods are sometimes ambiguous. The identification should be confirmed by analyzing D1/D2 domain of the 26S rDNA or the internal transcribed spacer (ITS) DNA sequences. For taxonomic purposes, intergenic spacer (IGS), Β-tubulin, and chitin synthase-2 genes can be utilized. The researchers have used Amplified Fragment Length Polymorphism (AFLP) and Multi Locus Sequence Typing (MLST) for strain typing. Recently, MALDI-TOF MS has also been utilized for rapid and accurate identification of *Malassezia* species.

Microscopy

Malassezia can typically be observed using a simple procedure called a potassium hydroxide (KOH) preparation, often used for diagnosing fungal infections. The sample, usually a skin scraping or swab, is placed on a glass slide with a drop of 10–20% KOH solution, which dissolves keratin and other cellular material, allowing for clearer visualization of the yeast under a microscope.

Malassezia yeasts are typically oval or round and vary in size from 2 to 6 μm (Fig. 12.4). The presence of clusters of round to oval budding yeasts and occasionally, short hyphae, is suggestive of *Malassezia* infection. They exhibit a characteristic "spaghetti and meatballs" appearance on microscopy, with yeast cells (the "meatballs") and hyphae (the "spaghetti") visible in the sample obtained from PV skin. Only clusters of round to oval budding yeasts (absence of hyphae) can be observed from other skin conditions or bloodstream infection samples. However, it's worth noting that this morphological examination can not differentiate between the various species of *Malassezia*. For precise identification, molecular methods are required (Table 12.2).

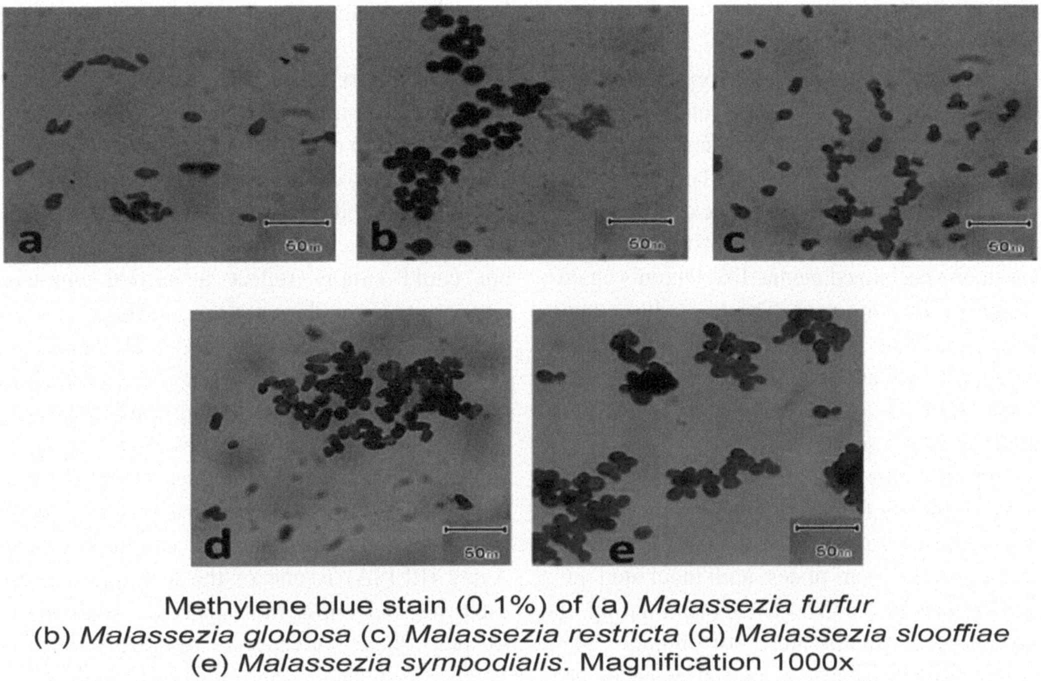

Fig. 12.4 Methylene blue staining of different *Malassezia* species isolated [adapted from Rudramurthy et al., Ind J Med Res. 2014. 139 (3). 431–7]

Table 12.2 Laboratory diagnosis of *Malassezia* infections

Disease	Direct microscopy	Immunodiagnostics	Culture	Remarks
Pityriasis versicolor	Unipolar budding yeast and hyphae (spaghetti and meatballs)			Pale yellow fluorescence scales under ultraviolet (Wood's) light
Seborrhoeic dermatitis				Diagnosis is mostly clinical; biopsy in difficult-to-diagnose cases
Malassezia folliculitis	Unipolar budding yeast (rarely hyphae)			Diagnosis is mostly clinical; histopathology in complicated cases
Atopic dermatitis		*Malassezia* spp. specific IgE (atopy patch/skin prick test) and commercial and standardized kit (ImmunoCAP m70, Phadia)		Diagnosis is mostly clinical
Invasive infection	Unipolar budding yeast (rarely hyphae)		Culture on modified Dixon agar/Leeming and Notman agar/ CHROMagar *Malassezia*	

Note: *Malassezia* species identification from clinical samples by culture or molecular methods, except for *Malassezia* fungemia, is of no diagnostic value

In vitro Culture

In vitro cultures of *Malassezia* spp. are unique in that they require lipid supplementation for their growth, which differentiates them from most other fungi. Consequently, standard fungal culture media, such as Sabouraud dextrose agar, are not effective for *Malassezia*. Instead, they are grown on specialized media, like Dixon's agar or Leeming–Notman agar. These media contain ingredients that provide the necessary lipids, including sources like olive oil, ox-bile, and tweens that act as fatty acid sources for these lipid-dependent yeasts.

To initiate the culture, skin samples are collected, typically from scrapings, swabs, or biopsies. These samples are placed on the lipid-enriched agar plates and incubated at a temperature of 32–37 °C, approximating the physiological temperature of human skin. Visible growth usually occurs after an incubation period of 2 days to several weeks (Fig. 12.5). However, growing *Malassezia* in vitro presents a challenging task due to its fastidious nature and slow growth rate. When more than one species is present in a clinical sample, the fast growers such as *M. furfur* outgrow the slow growers such as *M. globosa*, *M. restricta*, or *M. arunalokei*.

Serodiagnosis

Serodiagnosis of *Malassezia* infection is somewhat challenging, mainly because these yeasts are a part of the normal skin flora in healthy individuals. Therefore, the presence of *Malassezia*-specific antibodies in the serum does not necessarily indicate an infection or disease state, but could simply reflect a normal immune response to these ubiquitous organisms.

However, elevated antibody levels may suggest an overgrowth of *Malassezia* or hypersensitivity to it, potentially indicating conditions such as AD or MF. Some studies have reported increased levels of IgE (Immunoglobulin E) antibodies against *Malassezia* in patients with these conditions. Enzyme-Linked Immunosorbent Assay (ELISA) is one of the techniques often used for detecting these specific antibodies. Despite these advances, serodiagnosis for *Malassezia*-related conditions is still not widely used or accepted due to its ambiguity (Table 12.3).

Molecular Diagnosis

Molecular diagnostics of *Malassezia* infections has dramatically improved the ability to accurately identify and differentiate the various spe-

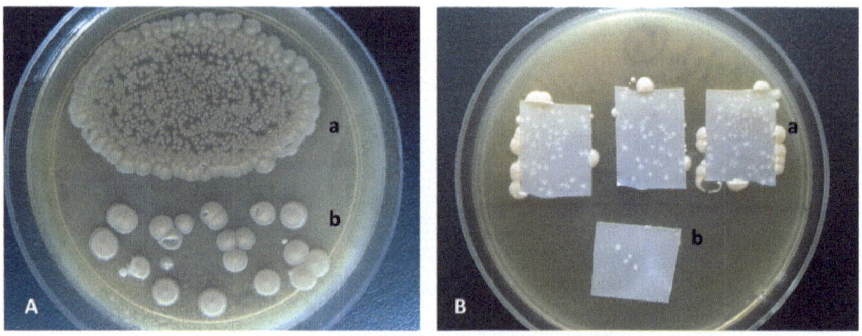

Fig. 12.5 *Malassezia* colony isolated from (**a**) lesional area, (**b**) non-lesional area of human skin using (**a**) cotton swab and (**b**) Scotch tape sampling

Table 12.3 Distribution of *Malassezia* infection in the global population

Malassezia species	Host	Site of infection	Distribution
Malassezia furfur	Dog, cat, human	Skin, blood	Worldwide
Malassezia obtusa	Human	Skin	Worldwide
Malassezia yamatoensis	Human	Skin	Japan
Malassezia japonica	Human	Skin	Japan, India
Malassezia vespertilionis	Bats	Skin	USA
Malassezia psittaci	Parrot	Skin	Brazil
Malassezia brasiliensis	Parrot	Skin	Brazil
Malassezia pachydermatis	Dog, cat, human	Skin, blood	Worldwide
Malassezia sympodialis	Cat, human	Skin	Worldwide
Malassezia caprae	Goat	Skin	Spain
Malassezia equina	Horse	Skin	Spain
Malassezia dermatitis	Human	Skin	Worldwide
Malassezia nana	Cat, horse, cattle, human	Skin	Japan, Brazil
Malassezia restricta	Human	Skin	Worldwide
Malassezia globosa	Cat, human	Skin	Worldwide
Malassezia arunalokei	Human	Skin	India, South Korea
Malassezia slooffiae	Pig, cat, human	Skin	Worldwide
Malassezia cuniculi	Rabbit	Skin	Spain

cies within this genus. Traditional diagnostic methods based on morphological and biochemical characteristics often fall short, especially when it comes to differentiating between the different *Malassezia* species. However, molecular biology techniques can now identify and distinguish amongst the species more accurately. Polymerase chain reaction (PCR) is one of the most commonly used molecular techniques for diagnosing *Malassezia*. It involves the amplification of a specific region of DNA that is unique to the species being tested. By using primers designed to target specific *Malassezia* species, PCR can not only confirm the presence of *Malassezia* but also differentiate between the species. The commonly targeted region is the 26S rDNA or the ITS gene, which has been shown to have sufficient variation between species to allow for differentiation.

Despite the effectiveness of PCR, one drawback is that it requires a laboratory setup and trained personnel to conduct the test. To overcome this limitation, newer technologies are being developed that are simpler, faster, and can be used directly in a clinical setting. For instance, real-time PCR and next-generation sequencing (NGS) are two such advanced techniques that are paving the way for the future of *Malassezia* diagnostics. These methods not only offer a high degree of accuracy, but they also provide quantitative data, which can be valuable in assessing the severity of an infection. However, these advanced techniques are still being refined and are not yet widely implemented in clinical settings.

Disparity Between Cultural and Molecular Studies

To decipher the role of *Malassezia* in various dermatoses, it is imperative to address the disparity between recently developed non-culture-based molecular techniques and conventional culture-based studies. The association of particular species of *Malassezia* with different dermatitis in humans and animals depends on the selection of sampling methods, culture media, and species identification techniques, as reported in different studies. A clear epidemiological picture may evolve depending on whether the mere presence of *Malassezia* or its quantity decides its commensal nature or association with cutaneous lesions. The fastidious and slow-growing *Malassezia* can be easily picked up by culture-independent methods such as next-generation sequencing and qPCR. However, one can argue that it quantifies the organism's dead or alive DNA.

Treatment

The complex interplay between hosts' immune status and *Malassezia* spp. involvement has made it difficult to draw a definitive line between the commensal and pathogenic nature of *Malassezia* leading to uncertainty over treatment options. The reduction of *Malassezia* load or clinical improvement after the usage of antifungals strongly suggests the pathogenic role of *Malassezia* in dermatitis. *Malassezia* dermatitis is usually treated with antifungals, anti-inflammatory therapy, or sometimes in combination. Some antifungals act as anti-inflammatory agents and vice-versa. Relapse of *Malassezia* dermatitis is very common. Daily application of itraconazole (1–2 months) followed by twice weekly improved the clinical severity score of head and neck AD. The itraconazole has also been shown to reduce *Malassezia*-specific IgE and eosinophil count in AD. The topical application of calcineurin inhibitors such as tacrolimus and pimecrolimus has shown promising results against long-term SD. These drugs have antifungal activity as well. The topical antifungal application of ketoconazole or ciclopirox have beneficial role in SD. Some salts such as lithium succinate and selenium sulfide are shown to be efficient in SD. Severe, resistant, and rapidly relapsing AD and SD cases can be treated with systemic antifungals. The PV lesions can be treated by application of azole antifungals for 2–3 weeks. Alternatively, 1% terbinafine or 2.5% selenium sulfide shampoo application appears to clear the PV infections. Extensive and recalcitrant PV cases can be treated with oral itraconazole 100 mg daily for 10 days. Currently, there is no internationally approved treatment guideline for the management of MF. Topical application of azoles/selenium sulfide (once daily for 3 days and then weekly)/propylene glycol (50% twice daily) or 100–200 mg daily systemic itraconazole/fluconazole for 1–4 weeks have proven useful for MF. Photodynamic therapy has also been used for MF.

Recently, the European Confederation for Medical Mycology (ECMM) in association with the American Society for Microbiology (ASM) and International Society for Human and Animal Mycology (ISHAM) prepared a global guideline for the diagnosis and management of rare bloodstream yeast infections such as *Malassezia*. We have sufficient knowledge regarding candidiasis and cryptococcosis in hospitalized patients. However, we lack information about *Malassezia* susceptibility profiles, clinical breakpoints, and comparative treatment efficacy trials. Routine susceptibility testing for epidemiological knowledge is moderately supported, whereas it is weekly recommended to guide treatment. The first-line recommended therapy is liposomal amphotericin B (moderately) and first-line recommended alternative therapy is amphotericin B deoxycholate (moderately). The central venous access device removal is strongly recommended.

Twice weekly use of 2% miconazole and 2% chlorhexidine shampoo showed the "strong" treatment evidence of canine *Malassezia* dermatitis. Three percent chlorhexidine shampoo treatment showed "moderate" evidence. Daily once/twice ketoconazole (5–10 mg/kg orally) and itraconazole (5 mg/kg orally) once daily or two consecutive days per week administration showed "moderate" evidence. A recent randomized, blinded, and placebo-controlled trial of a novel non-azole (leave-on spray containing zinc, laureth-9, ethyl lauroyl arginate, panthenol, urea, glycerin, and butylene glycol) for the treatment of dogs with pododermatitis showed a reduction in yeast load and clinical improvement.

Prevention and Control

Malassezia infections can be prevented and controlled through good hygiene practices, such as washing your skin regularly with soap and water, avoiding tight-fitting clothing, and keeping skin dry. Additionally, avoiding the use of oily or greasy skin products and maintaining a healthy diet can help prevent the overgrowth of the yeast. In cases where an infection has already developed, antifungal medications may be prescribed by a healthcare professional to help control the spread of the yeast and alleviate symptoms. It is important to seek medical attention for a suspected case of

Malassezia infection, as untreated infections can lead to more serious health issues.

Case Study

A 53-year-old male with idiopathic pulmonary fibrosis underwent a double-lung transplant (LT). His immunosuppressive regimen included prednisone, mycophenolate mofetil, and tacrolimus, and 24 months after LT, he required admission for fever and bilateral pneumonia. A chest computed tomography (CT) showed persistent ground glass opacities, with tree-in-bud appearance, of the right middle and lower lobes. Laboratory results showed an elevated C-reactive protein of 78.8 mg/L (normal range 0–5 mg/L) and a normal white blood cell count. Blood cultures turned negative. The forced expiratory volume in 1 s (FEV1) was 1.73 L and the forced vital capacity (FVC) of 2.16 L. Empirical antibiotic and antifungal therapy with meropenem, linezolid, and liposomal amphotericin B was started with mild clinical improvement. A bronchoalveolar lavage (BAL) showed no growth on bacterial and fungal cultures, and galactomannan on BAL was negative. Lung transbronchial biopsy (TBB) showed no signs of acute rejection but was notable for acute bronchitis with aggregates of round and oval elements suggestive for fungi by the PAS staining. PCR 26S-sequencing was performed on the fresh tissue and identified *M. restricta*. The patient was then started on voriconazole, with a significant progressive improvement in respiratory function, with an increase in both FEV 1 (2 L) and FVC (2.55 L). After 12 months, a new lung-TBB (36 months after LT) showed the persistence of *M. restricta* on PCR 18S-RNA molecular identification. At an 18-month follow-up (42 months after LT), both lipid-enriched culture and 26S-RNA on biopsy were negative for fungal pathogens, and the chest CT scan showed a notable improvement of radiographic findings. At a 2-year follow-up (4 years after LT), voriconazole was discontinued and the patient maintained a stable respiratory function. He had no infection recurrence at his 3-year follow-up.

1. What are the significant points in the above clinical vignette?
2. How to identify the pathogen by conventional and molecular methods?
3. What are the risk factors for *Malassezia* systemic infection?
4. What may be the mode of infection transmission?

Research Questions

1. How to investigate the presence of *Malassezia* in various vertebrates, invertebrates, and different ecological niches?
2. What factors trigger commensal *Malassezia* to be pathogenic?
3. How do the host immune factors interplay with mycobiome?
4. Is it high time to start a surveillance study on *Malassezia* BSI?
5. What is the role of *Malassezia* in chronic diseases?

Further Reading

Abdillah A, Ranque S (2021) Chronic diseases associated with *Malassezia* yeast. J Fungi (Basel) 7(10)

Batra R, Boekhout T, Gueho E, Cabanes FJ, Dawson TL Jr, Gupta AK (2005) *Malassezia* Baillon, emerging clinical yeasts. FEMS Yeast Res 5(12):1101–1113

Bond R, Morris DO, Guillot J, Bensignor EJ, Robson D, Mason KV et al (2020) Biology, diagnosis and treatment of *Malassezia* dermatitis in dogs and cats clinical consensus guidelines of the world Association for Veterinary Dermatology. Vet Dermatol 31(1):28–74

Borda LJ, Wikramanayake TC (2015) Seborrhoeic dermatitis and dandruff: a comprehensive review. J Clin Investig Dermatol 3(2)

Chang HJ, Miller HL, Watkins N, Arduino MJ, Ashford DA, Midgley G et al (1998) An epidemic of *Malassezia pachydermatis* in an intensive care nursery associated with colonisation of health care workers' pet dogs. N Engl J Med 338(11):706–711

Chen SC, Perfect J, Colombo AL, Cornely OA, Groll AH, Seidel D et al (2021) Global guideline for the diagnosis and management of rare yeast infections: an initiative of the ECMM in cooperation with ISHAM and ASM. Lancet Infect Dis 21(12):e375–ee86

Gaitanis G, Magiatis P, Hantschke M, Bassukas ID, Velegraki A (2012) The *Malassezia* genus in skin and systemic diseases. Clin Microbiol Rev 25(1):106–141

Gaitanis G, Velegraki A, Mayser P, Bassukas ID (2013) Skin diseases associated with *Malassezia* yeasts: facts and controversies. Clin Dermatol 31(4):455–463

Guillot J, Bond R (2020) *Malassezia* yeasts in veterinary dermatology: an updated overview. Front Cell Infect Microbiol 10:79

Honnavar P, Ghosh AK, Paul S, Shankarnarayan SA, Singh P, Dogra S et al (2018) Identification of *Malassezia* species by MALDI-TOF MS after expansion of database. Diagn Microbiol Infect Dis 92(2):118–123

Honnavar P, Chakrabarti A, Dhaliwal M, Dogra S, Handa S, Lakshmi PVM et al (2021) Sociodemographic characteristics and spectrum of *Malassezia* species in individuals with and without seborrhoeic dermatitis/dandruff: a comparison of residents of the urban and rural populations. Med Mycol 59(3):259–265

Ianiri G, Heitman J (2020) Approaches for genetic discoveries in the skin commensal and pathogenic *Malassezia* yeasts. Front Cell Infect Microbiol 10:393

Kuttin ES, Glas I (1985) Mycotic otitis externa in animals. Mykosen 28(2):61–68

Oh J, Byrd AL, Deming C, Conlan S, Program NCS, Kong HH et al (2014) Biogeography and individuality shape function in the human skin metagenome. Nature 514(7520):59–64

Pinter L, Anthony RM, Glumac N, Hajsig D, Pogacnik M, Drobnic-Kosorok M (2002) Apparent cross-infection with a single strain of *Malassezia pachydermatis* on a pig farm. Acta Vet Hung 50(2):151–156

Rhimi W, Theelen B, Boekhout T, Otranto D, Cafarchia C, Malassezia spp. (2020) Yeasts of emerging concern in Fungaemia. Front Cell Infect Microbiol 10:370

Sankaranarayanan SR, Ianiri G, Coelho MA, Reza MH, Thimmappa BC, Ganguly P et al (2020) Loss of centromere function drives karyotype evolution in closely related *Malassezia* species. elife 9:9

Saunte DML, Gaitanis G, Hay RJ (2020) *Malassezia*-associated skin diseases, the use of diagnostics and treatment. Front Cell Infect Microbiol 10:112

Sparber F, Ruchti F, LeibundGut-Landmann S (2020) Host immunity to *Malassezia* in health and disease. Front Cell Infect Microbiol 10:198

Vijaya Chandra SH, Srinivas R, Dawson TL Jr, Common JE (2020) Cutaneous *Malassezia*: commensal, pathogen, or protector? Front Cell Infect Microbiol 10:614446

Part III

Subcutaneous Fungal Zoonotic Infections?

Phaeohyphomycosis

13

Satwika Rompally, Kalaiselvi Vinayagamoorthy, and Hariprasath Prakash

Learning Objectives

By the end of the chapter, the readers should be able to:

1. Understand the taxonomy and phylogenetics of causative agents of phaeohyphomycosis.
2. Learn the pathogenesis and clinical manifestations of phaeohyphomycosis in humans and animals.
3. Evaluate diagnostic modalities available for phaeohyphomycosis.
4. Study antifungal susceptibility of clinically relevant fungi causing phaeohyphomycosis.

Introduction

Phaeohyphomycosis is an emerging fungal infection caused by a group of fungi that contains melanin in their cell walls. Due to the presence of melanin, the yeasts and the filamentous fungal structures appear darker in colour; hence, terms such as "melanised, phaeoid, dematiaceous or black fungi" have been alternatively used. Black yeasts reproduce by unicellular budding and produce yeast-like cells, and their filamentous relatives produce conidial structures and hyphae during their growth and developmental stages. As these fungal species are ubiquitous, humans acquire the infections mainly by inhalation or after traumatic inoculation. Dematiaceous fungi are oligotrophic, and they can survive in extreme environmental conditions. These dematiaceous fungi are slow growers, and the other common saprophytic fungi often overgrow them in culture media; hence, they are less frequently isolated from the environment. Nevertheless, the culture-independent techniques showed the abundant presence of dematiaceous fungi such as *Cladophialophora*, *Exophiala*, *Fonsecaea*, *Rhinocladiella*, and *Veronaea* species in the environmental samples. The data on phaeohyphomycosis epidemiology is scarce because species implicated in the disease are diverse. Less epidemiological information further complicates the diagnosis and management of the disease. There is no standardised treatment regime for phaeohyphomycosis. Treatment plan and management varies according to the species implicated. In most cases, antifungal treatment remains unsuccessful, as these agents are resistant to different classes of antifungal agents.

S. Rompally
Clinical Microbiology Registrar, Australian Clinical Labs, Osborne Park, Western Australia

K. Vinayagamoorthy
Centre for Public Health (U.I.E.A.S.T.), Panjab University, Chandigarh, India

H. Prakash (✉)
Department of Microbiology, PES Institute of Medical Sciences and Research, Kuppam, Andhra Pradesh, India
e-mail: harisath2003@gmail.com

S. C. Parija, S. M. Rudramurthy (eds.), *Textbook of Fungal Zoonoses and Sapronoses*, Microbial Zoonoses, https://doi.org/10.1007/978-981-97-3214-2_13

History

In mycology, the term "phaeohyphomycosis" emerged due to the dedicated research conducted by Ajello and his team in 1974. This term was coined to encompass a broad spectrum of infections, ranging from superficial to cutaneous, subcutaneous, and even systemic, all attributed to a specific group of phaeoid fungi. This group exhibited filamentous nature and pigmentation in their invasive forms. Although the first instance of cutaneous infection attributed to *Alternaria alternata* was noted in 1933 by Borsook, this case was identified by culture and microscopy, but lacked proper histological documentation. In 1952, Binford et al. reported a case of cerebral phaeohyphomycosis due to *Cladophialophora bantiana* and they have provided a comprehensive description of this ailment. In 1974, Ajello et al. provided a provisional definition of phaeohyphomycosis. They defined phaeohyphomycosis as infections that presents as cutaneous, subcutaneous or systemic disease, with the presence of dark-walled hyphae within the tissue and in culture. The crucial aspect of these infections was the presence of their filamentous phaeoid tissue forms. The term "phaeohyphomycosis" itself originates from the Greek word "phaeo," which translates to "dark." As Ajello and his colleagues proposed, this name aptly reflects the core nature of infections caused by these fungi with dark-walled structures.

Taxonomy

The taxonomy of black yeasts and their filamentous relatives is progressively evolving with the advancement of molecular techniques. Many new species have been reported as causative agents of phaeohyphomycosis over the last few decades, even though frequent changes in the genus and species names hamper understanding phaeohyphomycosis epidemiology. Most infections are caused by fungi from the phylum Ascomycota and the subphylum Pezizomycotina. The fungal species are phylogenetically placed in different classes, such as Dothideomycetes, Eurotiomycetes, Leotiomycetes, and Sordariomycetes. Multiple gene targets such as internal transcribed spacer region (ITS), 28S large ribosomal subunit (LSU), 18S small ribosomal subunit (SSU), translation elongation factor 1 (*TEF1*), RNA polymerase B I and II second largest subunit (*RPB1* and *RPB2*), mitochondrial small ribosomal subunit (mtSSU), minichromosome maintenance proteins (*MCM7*), actin (*ACT*), glyceraldehyde-3-phosphate dehydrogenase (*GAPDH*), and beta-tubulin were used to infer phylogenetic delineation in these ascomycete fungi. Based on the multi-gene analysis, Dothideomycetes comprises 32 orders, 191 families, and 1495 genera; Eurotiomycetes consist of 6 subclasses, 10 orders, and 34 families; Leotiomycetes includes 11 orders, 43 families, and 592 genera; and Sordariomycetes encompass 45 orders, 167 families, and 1499 genera. Numerous fungal species have been identified under each class; however, only a few genera are implicated in human and animal infections.

The causative agents of phaeohyphomycosis in humans and animals are listed in Table 13.1. Most pathogenic species are placed in the order Pleosporales (Class: Dothideomycetes) and Chaetothyriales (Class: Eurotiomycetes). However, the most clinically important species are placed in the family Herpotrichiellaceae (Order: Chaetothyriales). *Cladophialophora*, *Exophiala*, *Fonsecaea*, *Phialophora*, *Rhinocladiella*, and *Veronaea* are the major genera in the family Herpotrichiellaceae that cause phaeohyphomycosis. The fungal genera in Herpotrichiellaceae are associated with fatal infections in humans and animals, including cerebral and disseminated infections. The fungal genera in the order Pleosporales, such as *Alternaria*, *Bipolaris*, *Curvularia*, and *Exserohilum*, are implicated in allergic infections, cutaneous and subcutaneous infections, and ocular infections. However, these agents rarely cause invasive diseases such as central nervous system (CNS) infections and disseminated disease. Sordariomycetes fungi can also cause human and animal infections. The genera such as *Chaetomium*, *Phialemonium*, *Acrophialophora* and *Phaeoacrmo-*

Table 13.1 Taxonomical classification of causative agents of phaeohyphomycosis in humans and animals

Taxonomy classification	Species isolated from humans	Clinical disease in humans	Species isolated from animals	Clinical disease in animals
Ascomycota, Pezizomycotina, Dothideomycetes				
Neoscytalidium (order: Botryosphaeriales)	*Neoscytalidium dimidiatum* (Syn. *Scytalidium dimidiatum*), *Neoscytalidium hyalinum*	Fungal rhinosinusitis, endophthalmitis, subcutaneous, onychomycosis, pulmonary, systemic, cerebral, mycetoma	NA	NA
Cladosporium (order: Cladosporiales)	*Cladosporium cladosporioides*, *Cladosporium oxysporum*, *Cladosporium sphaerospermum*	Cutaneous, endophthalmitis, rhinosinusitis, pneumonitis, pneumonia, cerebral	*C. cladosporioides* (canine, feline, mammals, reptiles), *Cladosporium herbarum* (fish)	Dermatitis, encephalitis, nephritis, cerebral, systemic
Aureobasidium (order: Dothideales)	*Aureobasidium melanogenum*, *Aureobasidium pullulans*, *Aureobasidium proteae*, *Aureobasidium mansonii*	Disseminated, fungemia, keratitis, cutaneous, pneumonia, peritonitis, cerebral	*Aureobasidium pullulans* (canine, rodents)	Cutaneous
Rhytidhysteron (order: Hysteriales)	*Rhytidhysteron rufulum*	Subcutaneous	NA	NA
Verruconis (order: Venturiales)	*Verruconis gallopava* (Syn. *Ochroconis gallopava*)	Cutaneous, systemic, pulmonary, cerebral,	*Verruconis gallopava* (canine, fish)	Systemic
Scolecobasidium (Syn. Ochroconis) (Venturiales)	*Scolecobasidium tshawytschae (Syn. Ochroconis tshawytschae)*	Subcutaneous	*Scolecobasidium tshawytschae* (fish)	NA
Ascomycota, Pezizomycotina, Dothideomycetes, Pleosporales (order)				
Alternaria	*Alternaria alternata*, *Alternaria chlamydospora*, *Alternaria dennisii*, *Alternaria infectoria*, *Alternaria longipes*, *Alternaria malorum*, *Alternaria rosae*, *Alternaria tenuissima*, *Alternaria arborescens*, *Alternaria chartarum*, *Alternaria dianthicola*, *Alternaria longipes*	Ocular infections, allergy, rhinosinusitis, onychomycosis, cutaneous and subcutaneous, cerebral, osteoarticular, pulmonary, peritonitis	*Alternaria alternata* (equine, canine, feline), *Alternaria arborescens* (reptile), *Alternaria infectoria* (feline)	Cutaneous, systemic
Bipolaris	*Bipolaris cynodontis*	Subcutaneous	*Bipolaris* sp. (canine)	Systemic

(continued)

Table 13.1 (continued)

Taxonomy classification	Species isolated from humans	Clinical disease in humans	Species isolated from animals	Clinical disease in animals
Curvularia	*Curvularia hawaiiensis* (Syn. *Bipolaris hawaiiensis*), *Curvularia spicifera* (Syn. *Bipolaris spicifera*/*Drechslera spicifera*), *Curvularia clavata*, *Curvularia lunata* (*Cochliobolus lunatus*), *Curvularia pallescens*, *Curvularia tuberculata*, *Curvularia tsudae* (Syn. *Cochliobolus australiensis*), *Curvularia alcornii*, *Curvularia tuberculata*, *Curvularia inaequalis*, *Curvularia brachyspora*, *Curvularia senegalensis*	Ocular infections, cutaneous and subcutaneous, arthritis, rhinosinusitis, cerebral, systemic, peritonitis,	*C. hawaiiensis* (aquatic mammal, equine), *C. spicifera* (mammals, equine, feline), *C. lunata* (canine)	Cutaneous, rhinitis, cerebral, systemic
Exserohilum	*Exserohilum mcginnisii*, *Exserohilum longirostratum*, *Exserohilum rostratum*	Subcutaneous and cutaneous, invasive sinusitis, rhinosinusitis, systemic, cerebral, ocular	*Exserohilum rostratum* (equine)	Rhinitis and sinusitis
Ascomycota, Pezizomycotina, Eurotiomycetes, Chaetothyriales (order)				
Cladophialophora	*Cladophialophora arxii*, *Cladophialophora bantiana*, *Cladophialophora carrionii*, *Cladophialophora devriesii*, *Cladophialophora modesta*, *Cladophialophora emmonsii*, *Cladophialophora boppii*	Cerebral, cutaneous, and subcutaneous, osteomyelitis, endocarditis, pulmonary, systemic, onychomycosis	*Cladophialophora bantiana* (alpaca, canine, equine, feline)	Cerebral, cutaneous, systemic, endometritis, pulmonary, spinal cord

Table 13.1 (continued)

Taxonomy classification	Species isolated from humans	Clinical disease in humans	Species isolated from animals	Clinical disease in animals
Exophiala	*Exophiala arunalokei, Exophiala attenuata, Exophiala bergeri, Exophiala campbellii, Exophiala cancerae, Exophiala dermatitidis, Exophiala equina, Exophiala hongkongensis, Exophiala jeanselmei, Exophiala lecanii-corni, Exophiala mesophila, Exophiala moniliae, Exophiala oligosperma, Exophiala phaeomuriformis, Exophiala pisciphila, Exophiala polymorpha, Exophiala salmonis, Exophiala spinifera, Exophiala xenobiotica*	Onychomycosis, cutaneous and subcutaneous, peritonitis, pneumonia, ocular infections, pulmonary, cerebral, systemic, endocarditis, arthritis	*Exophiala angulospora* (fish), *Exophiala attenuata* (feline), *Exophiala jeanselmei* (annelids, feline, reptiles), *Exophiala lecanii-corni* (fish), *Exophiala oligosperma* (reptiles), *Exophiala pisciphila* (fish), *Exophiala spinifera* (fish, felines), *Exophiala xenobiotica* (fish), *Exophiala salmonis* (fish)	Systemic, cutaneous, cerebral
Fonsecaea	*Fonsecaea monophora, Fonsecaea pedrosoi*	Subcutaneous, cerebral, osteomyelitis, systemic, pulmonary	*Fonsecaea pedrosoi* (feline, fish)	Cutaneous, systemic
Phialophora	*Phialophora americana, Phialophora verrucosa*	Cutaneous, ocular infections, sinusitis	*Phialophora verrucosa* (feline)	Cutaneous
Rhinocladiella	*Rhinocladiella basitona, Rhinocladiella mackenziei* (Syn. *Ramichloridium mackenziei*)	Systemic, cerebral, ocular infections, subcutaneous	NA	NA
Veronaea	*Veronaea botryosa*	Systemic, cerebral, ocular infections, cutaneous	*Veronaea botryosa* (amphibians, fish, reptiles)	Systemic, obstructive tracheitis
Ascomycota, Pezizomycotina, Sordariomycetes				
Phialemonium (order: Cephalothecales)	*Phialemonium obovatum*	Cutaneous	*Phialemonium obovatum* (canine)	Osteomyelitis
Chaetomium (order: Sordariales)	*Chaetomium globosum, Chaetomium strumarium*	Cerebral, cutaneous, and subcutaneous	*Chaetomium globosum* (fish), *Chaetomium strumarium* (equine)	Systemic, cerebral
Acrophialophora (Sordariales)	*Acrophialophora fusispora, Acrophialophora levis*	Cerebral, pulmonary, ocular infections	*Acrophialophora fusispora* (equine), *Acrophialophora levis* (equine)	Cerebral

nium (order: Togniniales) are identified as agents of phaeohyphomycosis in Sordariomycetes. Other than ascomycetes, *Wallemia sebi* and *Moniliella* species, and the fungi belonging to the phylum Basidiomycota, are reported to cause phaeohyphomycosis.

Genomics and Proteomics

The fungal species in the order Chaetothyriales are the most important pathogenic agents causing phaeohyphomycosis in humans and animals. The genome analysis of these fungal species has revealed unique features contributing to their virulence and pathogenic mechanisms. The genome size varies from 26 Mb (*Capronia coronata*) to 43.03 Mb (*Cladophialophora immunda*). Genomes in Chaetothyriales have fewer repetitive elements, contributing to the moderate sizes of genomes. Black yeasts can survive harsh environmental conditions. The genomes of black yeasts are equipped with genes such as aldehyde dehydrogenases and trichothecene efflux pumps. Aldehyde dehydrogenases play essential roles in stress response and pathogenesis of phaeohyphomycosis. The expansion of aldehyde dehydrogenase may help *Fonsecaea* species survive extreme environmental conditions and host tissue invasion. Trichothecene efflux pumps may help the black fungi colonise in harsh environments, morphogenesis, and virulence.

The gene family expansion analysis has shown that the cytochrome p450 (CYPs) genes are higher in Chaetothyriales than other ascomycete fungi. The essential genes of sterol biosynthesis, such as lanosterol 14-alpha-demethylase (CYP51F) and sterol delta22-desaturase (CYP61A), are conserved in Chaetothyriales. Mutations in CYP51 genes are responsible for azole resistance in fungi; point mutations in this gene may render azole resistance in Chaetothyriales. The cell wall-associated genes, such as chitin synthase and 1,3-α-glucan synthase, were conserved in Chaetothyriales. However, chitosanase, chitinase, and chitin deacetylase genes are not uniformly distributed in Chaetothyriales. The mating locus genes (MAT) and other sexual reproduction-associated genes are conserved in Chaetothyriales. However, in Chaetothyriales, most fungal species show asexual reproduction. For example, heterothallic species contain a single copy of the MAT locus in their genome (e.g. *Fonsecaea*, *Cladophialophora*, *Exophiala*). In contrast, homothallic fungi contain both MAT locus genes in their genome (e.g. *Capronia*).

The gene clusters for biosynthetic pathways such as polyketide synthase (PKS), non-ribosomal peptide synthase (NRPS), and terpenes are clustered in Chaetothyriales. The fungal species such as *Alternaria* and *Curvularia* in *Pleosporales* also encode biosynthetic pathways such as PKS and NRPS in their genomes. Melanin is an important cell wall component and virulence factor in the fungal pathogens. In Chaetothyriales, 1,8-dihydroxy naphthalene (DHN) melanin biosynthesis is mediated by polyketide synthase pathway genes. The genes responsible for the dihydroxyphenylalanine (DOPA)-melanin pathways, such as tyrosinase and laccase, are conserved in Herpotrichiellaceae. However, the other causative fungi, such as *Verruconis gallopava*, lack multicopper oxidases, tyrosinase, and laccase for melanin synthesis. In Pleosporales, melanin production is mediated by the DHN-melanin pathway. The gene associated with melanin synthesis, such as multicopper oxidases, also plays a role in iron acquisition.

Iron acquisition from the host is an important virulence trait of medically important fungi. In *Cladophialophora bantiana*, genes responsible for iron acquisition, such as L-ornithine-N5-monooxygenase (SidA) and other NRPS genes responsible for ferricrocin synthesis (SidC) and extracellular fusarinine (SidD and SidF) have been identified. In addition, the non-siderophore reductive iron assimilation genes such as metalloreductase, iron permease, and ferroxidases are conserved in *C. bantiana*. Iron acquisition genes, such as siderophore iron transporters and ferric triacetylfusarinine C transport genes in *Fonsecaea* are conserved, which may play a role in their virulence. Transcriptome analysis of *Exophiala dermatitidis* during the skin infection model has revealed up-regulation of reductive pathway iron uptake genes such as iron permease (Ftr1) and ferroxidase (Fth1), and siderophore synthesis genes (SidA, SidD, SidF, and SidC responsible for triacetylfusarinine synthesis) have substantiated their role in their virulence. The iron acquisition genes, such as siderophore transporters and oxidoreductase, are also conserved among members of Pleosporales.

The genome analysis of *Phialophora* species has revealed virulence and pathogenic genes such as actin-binding cyclase-associated signalling proteins, transcription factors of Myb-like DNA binding domain (morphogenesis), and SANT/Myb-like domain of Damage-Associated Molecular Patterns (DAMP) (increased virulence), and all these genes are conserved in this fungi. The genome analysis of *Exophiala* species has shown virulence genes belonging to capsular polysaccharide synthesis, adenylate cyclase associated (CAP) genes, heat shock proteins (HSP70), and ubiquitin family. CAP genes play an important role in lipid and sterol metabolism. In addition, the extracellular capsule-like material seen in *Exophiala* species may play a role in virulence. However, this property varies from species to species; *Exophiala dermatitidis* capsular material is intact at 37 °C during host invasion, whereas *Exophiala spinifera* loses the capsule at 37 °C. Hence, the *Exophiala* species exhibiting thermotolerance can cause invasive infections in the host. Similarly, HSP70 proteins are molecular chaperones that may play a role in morphogenesis, growth and development, and stress response. Ubiquitin genes are essential for protein degradation. Transcription factors of the Myb-like DNA binding domain and SANT/Myb-like domain of DAMP are conserved in *Exophiala* species.

The secreted proteases are another important virulence factor of *Chaetothyriales*. The genomes of *Chaetothyriales* encode abundant serine and metalloproteases. These proteases play a role in nutrient acquisition from the environment and in pathogenesis, mediating tissue invasion in the host. Candidapepsin, a secreted aspartyl protease (SAP) gene, in the *Chaetothyriales bantiana* genome has been identified. SAPs also play an important role in the virulence of *Candida* species by mediating adhesion to the host tissue, immune evasion, and tissue invasion by hyphal formation. The function of SAPs in *C. bantiana* is yet to be studied in detail. *E. dermatitidis* during the skin infection model revealed up-regulation of cysteine proteases (C26) and serine proteases (S16). The agents of phaeohyphomycosis use these virulence factors and establish infections in the host.

Fungus Morphology

The identification of fungi causing phaeohyphomycosis is based on their distinctive fungal morphology, both macroscopic and microscopic, as well as selected physiological traits. The colour, growth rate, and cultural characteristic features in standardised media are important for their identification. These also include microscopic features like hyphae, conidiogenous cells, and conidia, along with physiological attributes such as cycloheximide tolerance, nitrate assimilation, urea hydrolysis, and salt concentration preferences. Phaeohyphomycosis-causing agents are characterised by their septate hyphae and yeasts with yellow to brown pigmentation in their cell walls. Slide cultures are carried out to reveal these microscopic traits effectively. The distinctive colony and microscopic features of clinically relevant phaeohyphomycosis agents are provided in Table 13.2, supplemented with illustrative depictions in Fig. 13.1.

Pleomorphism or pleoanamorphism is an intriguing phenomenon observed in particular fungi, exhibiting varying forms. This phenomenon is characterised by an initial yeast-like growth followed by more filamentous growth, observed in black yeasts such as *Exophiala*. Pleoanamorphism can also show distinct types of anamorphic structures (synanamorphs) within the same fungus. For instance, *Pseudallescheria* may exhibit the Graphium state, and *Pleurostomophora richardsiae* may feature variably shaped conidia.

Table 13.2 Colony morphology and microscopic characteristics of clinically relevant phaeohyphomycosis agents

Fungus	Macroscopic features	Microscopic features
Alternaria alternata	Rapidly growing black to olivaceous-black or greyish, suede-like to floccose colonies	Smooth, mostly unbranched conidiophores Pale-brown, obclavate to ellipsoidal conidia, with muriform septation and cylindrical beak
Bipolaris	Moderately fast growing grey to blackish colonies with powdery texture	Bipolar germination, zigzag-shaped conidiophores with smooth to thick walled cylindrical conidia *Bipolaris spicifera*: Predominantly 4-celled conidia with 3 distosepta *Bipolaris hawaiiensis*: Mostly 6-celled conidia with 5 distosepta
Cladophialophora	Cladophialophora bantiana: modertaely fast growing, olivaceous grey colonies with black reverse, velvety to powdery texture. *Cladophialophora carrioni:* moderately growing olivaceous-black colonies with powdery texture	*Cladophialophora bantiana*: Conidia develop from undifferentiated conidiophores, either at their side or tip; form long, connected chains; and vary in shapes from ellipsoidal to fusiform to nearly cylindrical *Cladophialophora carrioni*: Olivaceous-green apically branched elongated conidiophores that produce branched acropetal chains of conidia; pale-olivaceous, smooth-walled or slightly verrucose, and limoniform to fusiform conidia
Curvularia	Rapidly growing, hairy, brown to blackish-brown colonies with black reverse	Geniculate, unbranched, septate conidiophores *Curvularia geniculata*: Dark brown, 4-septate curved conidia with swollen central cell *Curvularia lunata*: Olivaceous brown, 3-septate conidia with swollen subterminal cell
Exophiala	*Exophiala jeanselmei*: initially moist velvety, greenish-grey to black, mucoid yeast-like colonies, later with age aerial mycelium develops. *Exophiala spinifera*: olivaceous-black, initially mucoid and yeast-like in appearance, become raised once aerial mycelium develops.	*Exophiala jeanselmei*: Yeast-like budding cells, along with larger inflated cells transform into unswollen hyphae; conidia form on lateral, annellate conidiogenous pegs, originating at various angles from undifferentiated hyphae or detached conidia; conidia are clear, smooth, thin-walled, broadly ellipsoidal, and have small basal scars *Exophiala spinifera*: Abundant yeast cells; erect, cylindrical, multi-celled conidiophores with narrow, cylindrical annelated zones at apical and intercalary positions; subglobose to ellipsoidal conidia
Exserohilum rostratum	Rapidly growing grey to blackish-brown, floccose colonies with an olivaceous-black reverse	Geniculate conidiophores Bipolar germination Straight, curved, or slightly bent, ellipsoidal to fusiform conidia with a protruding truncated hilum and thickened and dark septum above the hilum
Fonsecaea complex	Slow growing, olivaceous to black, lanose, cottony to velvety restricted colonies with black reverse	Pale olivaceous conidiogenous cells arranged in dense clusters, with prominent denticles bearing pale olivaceous, clavate to ellipsoidal, subhyaline, smooth and thin-walled conidia of variable sizes in short chains
Verruconis (previously *Ochroconis*) *gallopava*	Smooth to suede-like, dry, flat, tobacco-brown to brownish-black colonies often with a dark-brown diffusible pigment	Cylindrical to acicular conidiophores bearing a few conidia at the tip Two-celled, subhyaline to pale-brown, smooth-walled to verrucose, cylindrical to clavate conidia constricted at the septum with apical cell wider than basal cell
Phialophora verrucosa	Slow growing, smooth to woolly, olivaceous black colonies, with black reverse	Flask-shaped or elliptical phialides with distinctive funnel-shaped, darkly pigmented collarettes Ellipsoidal, hyaline conidia with smooth walls aggregate in slimy clusters at the apices of the phialide

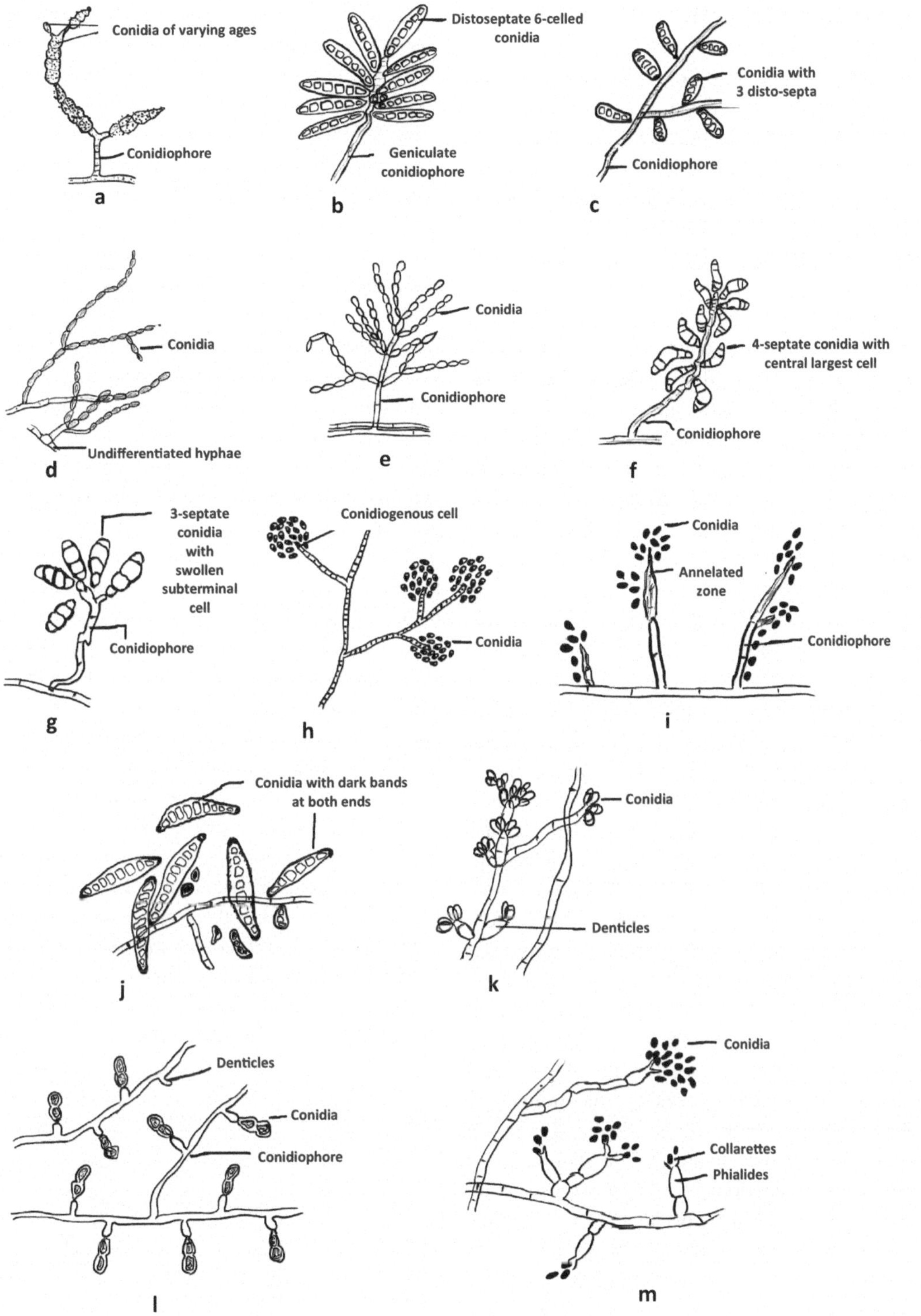

Fig. 13.1 Diagrammatic representation of microscopic features of clinically relevant phaeohyphomycosis agents. (**a**) *Alternaria alternata*; (**b**) *Bipolaris hawaiiensis*; (**c**) *Bipolaris spicifera*; (**d**) *Cladophialophora bantiana*; (**e**) *Cladophialophora carrionii*; (**f**) *Curvularia geniculata*; (**g**) *Curvualaria lunata*; (**h**) *Exophiala jeanselmei*; (**i**) *Exophiala spinifera*; (**j**) *Exserohilum rostratum*; (**k**) *Fonsecaea monophora*; (**l**) *Verruconis gallopava*; (**m**) *Phialophora verrucosa*

Cultivation of Fungi

The cultivation of agents of phaeohyphomycosis requires various approaches to ensure successful isolation and identification. On Sabouraud dextrose agar (SDA), the fungi, during their initial growth, often present as light or pale colonies, later they turn grey to black, making their identification as "pheoid" organisms. Employing plant-based media such as potato dextrose agar (PDA), malt extract agar, and V-8 juice agar is recommended to enhance recognition. The PDA provides colony colours that closely resemble the original description and support the conidiation of the fungi. Phaeoid moulds exhibit diverse colony colours influenced by environmental conditions, aiding in initial differentiation. Most species range from pale grey to dark grey to black, with exceptions exhibiting shades of brown, purple, or even reddish hues.

The selection of medium significantly influences accuracy in the identification of fungi. For accurate recovery of dematiaceous fungi from clinical specimens, a combination of nonselective (commonly SDA) and selective media (brain heart infusion agar (BHI) and inhibitory mould agar (IMA)) containing cycloheximide, media enriched for fastidious organisms. Transferring isolates from SDA to plant-based media leads to more accurate colouration. Initiating culture on a plant-based medium such as PDA is highly recommended, especially in immunocompromised individuals, to avoid potential negative consequences of extended subcultures. Incubation at 30 °C or room temperature around 25 °C, with regular monitoring during the initial days and subsequently twice a week, is recommended. Most phaeoid moulds can be identified within a week, while some isolates might require more prolonged incubation for diagnostic structure development. Storage of inoculated cultures beyond 3 weeks is rarely necessary if no growth is observed. It is important to interpret the clinical relevance of positive cultures, particularly when isolated from non-sterile sites.

Most importantly, the process and examinations of filamentous organisms should be carried out in the certified biological safety cabinets to ensure both safety and precision in the cultivation process. For example, sporulating *Cladophialophora bantiana* cultures require a laboratory biosafety level 3 containment facility. Guidelines for handling potentially infectious specimens need to be strictly followed.

Pathogenesis and Pathology

Inoculation is the usual mode of entry in localised cutaneous and subcutaneous lesions. Disseminated cerebral and pulmonary infections are more commonly seen secondary to inhaling organisms from the environment. The fungal species associated with phaeohyphomycosis are neurotrophic, and these cause a variety of infections, including CNS infections in immunocompetent hosts. The infections in other vertebrate hosts mimic the disease pathology of humans. The genomic characteristics of phaeohyphomycosis agents differ based on the site of infections. The neurotrophic fungi have more thermotolerance genes than the agents causing cutaneous infections. Hence, the pathogenesis of the disease is mediated by genes responsible for many virulence factors, such as thermotolerance, cell wall components (melanin), and proteases. Virulence factors that include structural and functional elements are listed in Table 13.3. Structural virulence factors are further elaborated in Table 13.4.

Functional virulence factors such as yeast to hyphal transition (morphogenesis) is crucial for tissue invasion. A few phaeoid fungi forms muri-

Table 13.3 Virulence factors of agents of phaeohyphomycosis

Structural virulence factors	Functional virulence factors
Melanin Carotenoids Chitin synthases	Muriform cells formation Prevalence of yeast-like phases Thermotolerance Hydrophobicity and adhesion Aromatic hydrocarbon assimilation Extracellular, acidic, or alkaline secondary metabolites and siderophores production

Table 13.4 Description of structural virulence factors of agents of phaeohyphomycosis

Virulence factor	Composition and examples	Mechanism of action
Melanin	a) Dihydroxyphenylalanine melanin b) Catechol melanin c) γ-Glutaminyl-4-hydroxybenzene melanin d) 1,8-Di-hydroxynaphthalene (DHN) melanin	Prevents the action of hydrolytic enzymes on cell wall Prevents oxiative burst by neutralising the free radicals generated by phagocytic cells Inhibits receptor-mediated phagocytosis Prevents the cells from nitric oxide mediated killing
Carotenoids	a) Torulene (3,4-didehydro-γ-carotene) b) Torularhodin (3,4-didehydro-γ-carotene-16-oic acid)	Act as an antioxidant and may play a role in neutralisation of harmful oxidants
Chitin synthases	Class V chitin synthase	Plays a role in cell growth at elevated temperature and virulence in case of *Exophiala dermatiditis*

form cells through meristematic growth, resulting in clustered cells with melanin-enriched walls. These cells help them to survive in the adverse environmental conditions such as extreme temperatures, UV exposure, water scarcity, nutritional deficiencies, acidity, and high salinity. Muriform cells evade host-mediated killing and resist phagocytosis. Morphogenesis of yeast to hyphal conversion in *E. dermatitidis* is mediated by the APSES trancription factor (WdStuAp), the deletion of this gene resulted in decreased hyphal growth and conidiation. Cdc42 gene is essential in conversion of yeast-like cells to sclerotic bodies formation in *E. dermatitidis*. Among black yeasts, *Exophiala* species shows unique trait of unicellular growth, with presence of only yeast-like phase, which may helps in causing disseminated infections.

Thermotolerance (ability to grow ≥ 37 °C) is another significant virulence factor, to cause systemic disease. In Chaetothyriales thermotolerance is particularly evident in species from the bantiana (*C. bantiana*, *F. pedrosoi*), dermatitidis (*E. dermatitidis*), and jeanselmi (*E. jeanselmei*, *E. spinifera*) clades, which grow at temperatures above 37 °C and cause systemic infections. In contrast, the carrionii (*C. carrionii*, *P. verrucosa*) and europea (*Cyphellophora europaea*) clades typically result in superficial and cutaneous infections and exhibit maximal growth at temperatures ranging from 35 to 37 °C.

Cell wall adhesion and hydrophobicity also play pivotal roles in chronic infections. Hydrophilic extracellular polysaccharides contribute to cellular hydrophobicity. They help them in adherence to the epithelial cells, and later converts to muriform cell and resist host-mediated killing. HSP70 family proteins on cell surface play a crucial role in endocytosis of the epithelial cells. Less common virulence factors such as aromatic hydrocarbons, secondary metabolites, siderophores, capsule, enzymes (proteases and urease) are identified. Structural resemblances between monoaromatic hydrocarbons and neurotransmitters, explain the neurological predilections exhibited by black yeasts. Additionally, the production of secondary metabolites and siderophores plays a role in infection dynamics, specifically in iron homeostasis and melanin synthesis. Polysaccharide capsules seen in *E. spinifera* and *E. dermatitidis,* may play a role in virulence.

Fungal cell wall components, including melanin and mannose, are pathogen-associated molecular patterns (PAMPs) that activate host innate immune receptors such as Toll-like receptors and mannose receptors, respectively. Other cell wall associated genes such as chitin synthase, chitin degradation enzymes (chitinase), 1,3-beta-glucan synthesis and ergosterol biosynthesis pathway genes are also involved in virulence.

Pathogenesis in Animals

Melanised fungi can infect invertebrates such as molluscs and crustaceans. The vertebrate hosts such as amphibians, aquatic mammals, fishes, reptiles are severely affected compared to birds and mammals. Skin is a primary site of infection that comes in contact with agents of phaeohyphomycosis. In cold-blooded animals, the fungi inhabitating the water source can cause microbial disintegration of skin, tissue invasion and death. Birds and mammals have highly protected intact skin and mucosal membranes; hence, black yeast infections are sporadic and primarily neurological. Systemic phaeohyphomycoses frequently occur in otherwise healthy animal hosts.

Immunology

To date, information on the pathogenesis of phaeohyphomycosis is limited. The innate immunity and cell-mediated immune response to phaeoid fungi is important. The humoral immune responses do not have any protective role against agents of phaeohyphomycosis.

Innate Immunity

The innate immunity serves as the first line of defence. It comprises the skin, epithelial and mucosal barriers, microbial antagonism by normal flora, complement activation, and opsonisation of invading pathogens. Upon entry of the pathogens, the phagocytic cells, neutrophils, macrophages and dendritic cells (DCs) are activated to contain the infection. Neutrophils are one of the primary phagocytes responsible for killing melanised fungi and prevent tissue invasion. Neutrophils on interaction with *Curvularia lunata* conidia stimulates myeloperoxidase and hydrogen peroxide to induce oxidative burst mediated killing. However, they escape phagocytic killing by production of antioxidants. Neutrophil extracellular traps (NETs) was induced by neutrophils on exposure to conidia of *Phialophora. verrucosa.* NETs mediates the extracellular killing of fungi and act as an effective innate immune mechanism against invading fungi. These factors may contribute black fungi to cause invasive disease. Macrophages play a significant role in the host's defence mechanisms against melanised fungi. Their contribution to phagocytosis, like that of polymorphonuclear neutrophils, is mediated by CD-40 and BL-7. They also stimulate the production of various first- and second-order cytokines, activating humoral and cell-mediated immune responses. Macrophage mediated killing is not specific, as most phaeoid fungi are not killed by resident macrophages. They escape phagocytosis by inhibiting the nitric oxide production.

DCs, as part of the skin and mucosal barrier, play an important role in the cutaneous and subcutaneous modes of infection. In addition to mediating phagocytosis, they are important antigen-presenting cells that induce and modulate adaptive T-cell responses by stimulating cytokines such as IL-12 and other chemokines. IL-12 stimulates the production of IFN-γ- by natural killer cells and T cells, resulting in cell-mediated immunity. Various factors, including hormones, cytokines, and microbial products, can modulate DCs' ability to produce IL-12 and prime Th1 responses. The schematic representation of host immune responses to agents of phaeohyphomycosis is depicted in Fig. 13.2.

Cell-Mediated Immunity

The cell-mediated immunity plays an important role in determining the severity of infection. CD4+ T-helper cells are critical defence components. Patients with defective T-cell-mediated immunity such as HIV/AIDS, are at increased risk for phaeohyphomycosis. A robust T-cell-mediated immune response and adequate IFN-γ production are required for protective immunity. In patients with mild form of infections, the phenomenon of increased IFN-γ and low levels of IL-10 is observed. Whereas, disease severity is

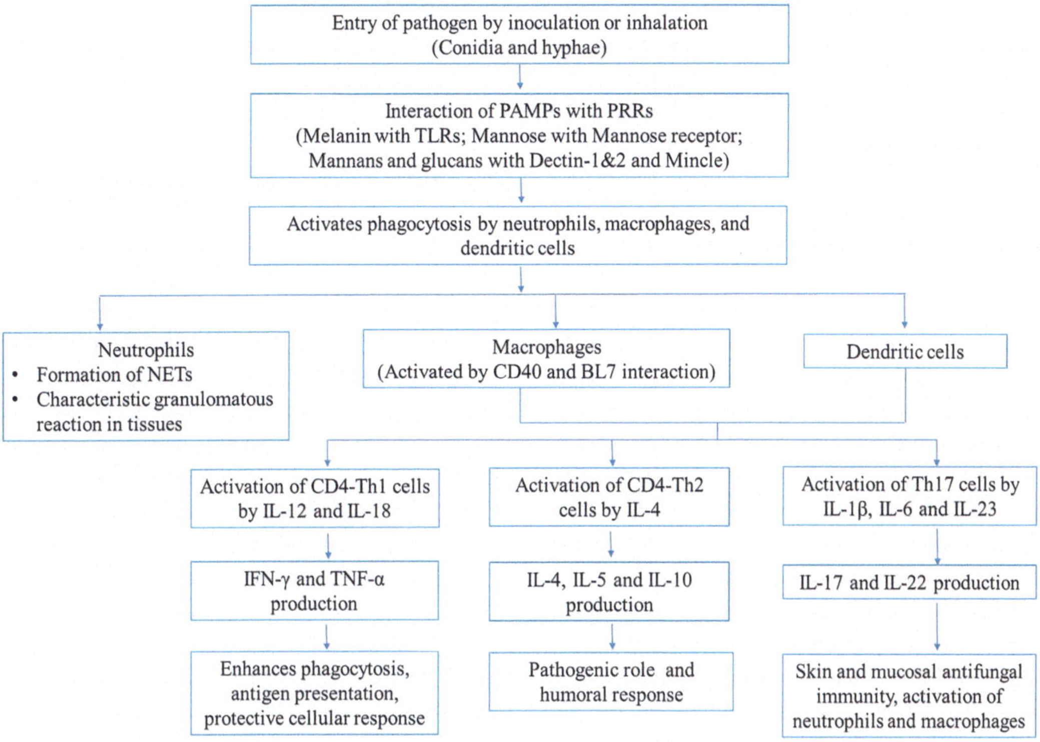

Fig. 13.2 Overview of host immune responses to agents of phaeohyphomycosis. *PAMP* pathogen-associated molecular pattern, *PRR* pathogen recognition receptor, *TLRs* toll-like receptors, *IL-12* interleukin-12, *IL-18* interleukin-18, *IL-1β* interleukin-1-beta, *IL-6* interleukin-6, *IL-23* interleukin-23, *Th1* T-helper 1, Th2 T-helper 2, *Th17* T-helper 17, *IFNγ* interferon-gamma, *TNFα* tumor necrosis factor-alpha, *IL-17* interleukin-17, *IL-22* interleukin-22

associated with increased IL-10 and low levels of IFN-γ. The another important arm in antifungal immunity is Th17 response, the release of cytokines IL-17 and IL-22, are crucial for mucosal and skin antifungal defences. PAMPs such as mannan and glucans activates the cell-mediated response via Dectin 1/2 and mincle receptors, they activate tyrosine protein kinase (Syk) and caspase recruitment domain family member 9 (CARD9) pathway. Syk-CARD9-BCL10-MALT1 pathway activates the NF-kB and stimulates the production of IL-17, which triggers neutrophils and macrophages activation. Patients with CARD9 mutations are at increased risk of acquiring phaeohyphomycosis with severe disease, due to impairment in Th17 response.

Infection in Humans

Clinical manifestations of phaeohyphomycosis in humans are classified as follows:

CNS Phaeohyphomycosis

C. bantiana is the causative agent for more than 50% of cases of cerebral phaeohyphomycosis, being more common in the immunocompetent host. The disease is cosmopolitan; however, most cases are seen in India and the USA. Malignancies, neutropenia, chemotherapy solid-organ transplants, and corticosteroid use predispose to *C. bantiana* CNS infections. Few CNS infections have been recorded in patients with intravenous drug abuse. *Rhinocladiella mackenziei* is another important causative agent of CNS infections, and most cases are documented in the Middle East. A nosocomial outbreak of CNS caused by *Exserohilum rostratum* has been recorded following injection of contaminated methylprednisolone injections. *V. gallopava*, *Chaetomium strumarium*, *Exophiala. dermatitidis*, *Fonsecaea monophora*, *Curvularia spicifera*, and *C. lunata* are less frequent pathogens associated with CNS infections.

CNS infections can present as brain abscess, encephalitis, myelitis, arachnoiditis, and meningoencephalitis. Clinical manifestations include fever, headache, increased intracranial pressure, neck stiffness, blurred vision, hemiparesis, tonic spasm, altered mental status, and coma. Ventricular obstruction, hydrocephalus, and cerebral hernia can be seen during disease progression. A high degree of clinical suspicion is needed to diagnose the disease. Radiological findings show ring-enhanced signs with hyper-dense and hypo-dense lesions. A definitive diagnosis based on histopathology shows dark-pigmented yeast-like cells and hyphal elements in the infected tissue specimens. Despite surgery and antifungal treatment, the mortality rate associated with CNS infections is as high as 65–96%.

Pulmonary Phaeohyphomycosis

Pulmonary phaeohyphomycosis manifests as pneumonia, pulmonary nodules, and endobronchial lesions. The condition presents as fever, cough with sputum, dyspnoea, and haemoptysis, and if not adequately treated, may progress to pneumonia and respiratory failure. Transplant Associated Infection Surveillance Network (TRANSNET) study showed that pulmonary phaeohyphomycosis was more common in haematopoietic stem cell transplant (HSCT) patients; however, solid organ transplant (SOT) recipients are also at considerable risk of acquiring pulmonary infections. Chest radiological findings show nodular and lobar abnormalities, pulmonary infiltrates, pleural effusion, and cavity lesions. Crescent or halo signs are also noted in rare cases of pulmonary phaeohyphomycosis. The mortality rate in pulmonary phaeohyphomycosis cases varies from 45% to 50%.

Disseminated Phaeohyphomycosis

Disseminated infections are common in immunocompromised hosts with haematological malignancies, HSCT and SOT recipients, neutropenia, corticosteroid use, and heart valve replacement recipients. TRANSNET study showed that more than half of the phaeohyphomycosis cases in HSCT and SOT recipients are disseminated infections. Lungs and skin were the frequent sites of infection, followed by the brain, sinuses, heart, and blood. Disseminated infections affecting the liver, spleen, lymph nodes, bones, and joints are relatively less common. The mortality rate varied from as low as 16% to as high as 72%. Multiple fungal species are associated with disseminated infections; no species predilection is observed.

Subcutaneous Phaeohyphomycosis

Most subcutaneous infections are caused by *Exophiala* species, followed by *Phialophora. verrucosa*, *V. botryosa*, *Phaeoacremonium* species, and others (Table 13.1).

Approximately, one-third of subcutaneous infections are observed in immunocompetent hosts. Rheumatoid arthritis, haematological and solid-organ malignancies, corticosteroid therapy, immunosuppressive drugs, diabetes mellitus, and SOT recipients are numerous risk factors for subcutaneous phaeohyphomycosis. Trauma is another known risk factor. In recent years, phaeohyphomycosis has been increasingly reported in CARD9 deficiency patients in China who are at increased risk of developing severe or disseminated infections.

Subcutaneous phaeohyphomycosis initially manifests as mild plaques or indolent, localised lesions. As the disease progresses, lesions may infiltrate and form verrucous plaque and solid or cystic nodules. In a few, the lesions may appear as dry crusts and ulcers with purulent discharge.

The self-limiting lesions often appear in the arms and legs. In severe cases, the head, face, and upper limbs are affected, spreading to the adjacent nose, eyes, pharynx, and bones. Due to the severity of the lesions and the host's immune system, antifungal therapy plays an important role in the prognosis of the disease.

Phaeohyphomycotic Keratitis

Nearly 4.4–44.2% of fungal keratitis cases are caused by dematiaceous fungi. *Curvularia*, *Alternaria*, *Exserohilum*, *Cladosporium*, and *Lasiodiplodia* species are most isolated from the cases of dematiaceous fungal keratitis. *Curvularia* is the most common isolate in 23–72% of cases.

Trauma is the foremost risk factor, followed by topical steroid use, previous surgery, and diabetes mellitus. Clinical manifestations include brown pigmentation of ulcers, epithelial plaque, corneal infiltrate, feathery edges, hypopyon, and raised lesions. Rarely corneal perforations are seen. Most of the patients had lesser visual acuity at the time of presentation. However, the visual acuity depends on the area of infiltrates, lesion depth, presence of hypopyon, and location of the ulcer. Antifungal therapy and keratoplasty are necessary to improve visual acuity.

Deep Local Infections

Allergic fungal sinusitis (AFS) is the most common deep local infection. *Aspergillus* species are the common cause of AFS, followed by dematiaceous fungi such as *Curvularia*, *Bipolaris*, *Alternaria*, and *Exserohilum*. Diagnosis of the condition is made by Bent and Kuhn criteria that include type I hypersensitivity, presence of nasal polyps, computed tomography (CT) findings, presence of fungi by microscopy or culture, and allergic mucin with fungal elements without tissue invasion. AFS involves unilateral or bilateral sinuses; CT shows hyperattenuating areas in sinuses, sinus expansion, and bone erosion. In a few cases, the infraorbital spread of the disease is observed. Intracranial spread from sinuses has rarely been reported. Surgical removal of the nasal polyps and the allergic mucin, along with antifungal treatment, provides a better outcome in AFS patients.

The other deep local infections caused by black fungi include endophthalmitis, infections of bones and joints, hepatic and splenic infections, endocarditis, and peritonitis in peritoneal dialysis patients.

Superficial Infections

Dematiaceous fungi are associated with superficial infections, such as tinea nigra, black Piedra, onychomycosis, and cutaneous infections.

Hortaea werneckii is the most common cause of tinea nigra. The condition presents as superficial, scaling, and brown or black patched skin lesions commonly affecting the palms and soles of the feet. Lesions can also occur in the neck and thorax. The lesions, in most cases, remain asymptomatic.

Black piedra is a superficial infection caused by *Piedraia hortae*. Infection is characterised by the formation of brown-black nodules in the hair shaft, leading to weakness and breakage of the affected hair and, in severe cases, associated with hair loss. The infection is seen predominantly in the scalp, rarely in the beard, moustache, and pubic hair. Infected hair produces a metallic sound due to the hard and gritty nature of the nodules.

Dematiaceous fungi such as *Chaetomium globosum*, *E. jeanselmei*, and *L. theobromae* are associated with onychomycosis, seen mostly among agricultural workers. Toenails are most affected; the infected nails are thickened, cracked, exfoliated, and destroyed. Superficial infections can manifest as cutaneous lesions with dark-brown macules and papules on the skin. The condition often manifests as a skin nodule or cyst, and disfiguring lesions with bony erosions.

The causative agents of infections in humans are summarised in Table 13.1.

Infection in Animals

Most phaeohyphomycosis cases are observed in vertebrate hosts such as amphibians, reptiles, fishes, aquatic mammals, canines, felines, and equines. Few cases, nevertheless, are also reported among invertebrates, such as crustaceans, molluscs, and annelids.

Phaeohyphomycosis in the Vertebrate Hosts

Phaeohyphomycosisis reported in diverse hosts such as amphibians (frogs and toads), aquatic mammals (sea cow), canines (dogs), felines (cats, snow leopard and tiger), equines (horses), fishes (rainbow trout, sea dragons, swell shark, chinook salmon, saffron cod, Atlantic halibut, split nose rockfish, cardinal tetra, lumpfish, sturgeon), reptiles (turtles, tortoise, and green iguana), rodents (red squirrels), and other mammals (sheep, alpaca, cattle, and deer). A few cases of systemic phaeohyphomycosis caused by *Veronaea botryosa* and *Exophiala* species are reported in frogs and toads; the disease manifests as cutaneous ulcers or nodules, lesions in the internal organs such as lungs, heart, spleen, liver, and kidney, leading to tissue necrosis. Rarely, phaeohyphomycosis has been reported in reptiles such as tortoises and turtles due to *V. botryosa*, *Exophiala*, *Ochroconis*, and *Oidiodenron* species. Localised and systemic phaeohyphomycosis has been reported in multiple fish species. The fungal species associated with phaeohyphomycosis in fish are listed in Table 13.1. *Exophiala* species are most reported in fish; *E. salmonis* is frequently associated with cerebral infections. Infected fish show lethargy and disoriented swimming behaviour due to cerebral infections.

Phaeohyphomycosis is commonly seen in mammalian hosts such as bovines, canines, felines, and equines. Clinical manifestations include ocular, cutaneous, pulmonary, cerebral, and disseminated infections. Cutaneous infections manifest as ulcerative lesions, multiple granulomas, and eumycetoma. Rhinitis caused by *Toxicocladosporium irritans*, *Exserohilum rostratum*, *Curvularia lunata*, *Curvularia hawaiiensis*, and *Alternaria alternata* have been reported in horses and Chinese gorals. Pulmonary phaeohyphomycosis caused by a rare agent, *Phialemonium curvatum*, has been reported in dogs. Cerebral infections mostly are caused by *Cladophialophora bantiana*. The condition is more common in cats, dogs, sheep, cattle, and other mammals such as Huacaya alpaca. *Acrophialophora fusispora*, *Acrophialophora levis*, *Cladosporium cladosporioides*, *Exophiala*, and *Bipolaris* species are the other species that cause cerebral infections. Disseminated infections are associated with high mortality. Histopathology of phaeohyphomycosis in vertebrate hosts are characterised by granulomatous inflammation with infiltration of neutrophils, macrophages, multinucleated giant cells, dark-pigmented yeast-like cells, and hyphae.

Phaeohyphomycosis in the Invertebrate Hosts

Epizootic systemic phaeohyphomycosis infections, namely lethargic crab disease (LCD), are seen in edible mangrove land crab *Ucides cordatus* (Brachyura: *Ocypodidae*) in Brazil and are caused by *Exophiala cancerae* and *Fonsecaea brasiliensis*. LCD is a systemic infection; the infected crab presents with weak motor skills, lethargy, and poor balance, leading to death. Histopathology shows disseminated infection, tissue necrosis and degeneration, haemocyte infiltration, agglutination and encapsulation, and phagocytic cells with yeast-like cells. Epizootic infections in mussels (*Bathymodiolus brevior*) due to *Capronia moravica* are reported in the Fiji Basin. The disease is associated with tissue destruction and intense haemocyte infiltration due to enhanced host immune response. *Exophiala jeanselmei* infection has also been reported in earthworm species such as *Octolasium tyrtaeum* and *Eisenia foetida*.

Epidemiology and Public Health

Many dematiaceous moulds are associated with invasive infections in humans and animals, and the cases are reported worldwide. These infections are seen predominantly in tropical and subtropical countries such as South and Central America, Africa, Southeast Asia, and Middle Eastern countries. Inhalation or trauma to the host from the environment is the primary mode of transmission of infection. Phaeohyphomycotic outbreaks have been reported in hospital settings, transmitted from contaminated saline solutions.

Phaeohyphomycosis is seen in both immunocompetent and immunocompromised human hosts. Patients with immunosuppressed conditions such as haematological malignancies, transplants, corticosteroid therapy, acquired immunodeficiencies, and genetic mutations such as CARD9 deficiency are at higher risk. Phaeohyphomycosis in diverse groups of animals manifests with similar disease pathology in humans.

Unlike other invasive fungal infections, the studies describing the epidemiology of phaeohyphomycosis are scanty. The prevalence/incidence of phaeohyphomycosis is not known due to the lack of population-based studies. A study from the USA reported an incidence rate of 1.0–3.1 cases per 100,000 patient days during the study period of two decades (1989–2008). TRANSNET studied invasive fungal infections (IFIs) in transplant patients. Of the IFIs reported, 2.6% of cases were phaeohyphomycosis. A study from China (1987–2021) reported that the number of cases due to phaeohyphomycosis increased 6.6 times since 2000. These data show that there has been an increase in the case burden of the infection in recent years. However, many studies are needed to estimate the exact prevalence/incidence of the infection worldwide.

The underlying diseases and risk factors associated with invasive phaeohyphomycosis cases include haematologic malignancies, chemotherapy, neutropenia, corticosteroid therapy, HSCT and SOT patients, graft versus host disease, renal insufficiency, diabetes mellitus, malnutrition, and patients receiving prophylactic or empirical antifungal therapy. Trauma is the major risk factor for cutaneous and ocular infections. In recent years, more subcutaneous and cutaneous phaeohyphomycosis cases have been observed in patients with CARD9 deficiency. Nevertheless, many cases are seen in patients without any known underlying disease and risk factors.

Diagnosis

Invasive phaeohyphomycosis is associated with severe mortality and morbidity; hence, early diagnosis is essential. Various clinical specimens are obtained using proper aseptic techniques based on the site of infection. Proper collection and processing of the sample and clinical correlation are essential to establish the diagnosis of the condition.

Direct microscopy, histopathology, culture, serological methods, and molecular techniques have been employed to diagnose phaeohyphomycosis.

Microscopy

Direct microscopy of potassium hydroxide (KOH) wet mount from tissues reveals yeast cells or thick branching septate hyphae with dark pigment in their cell walls. The histopathology of infected tissues shows granulomatous inflammatory tissue reactions, infiltrating lymphocytes, neutrophils, macrophages, and multinucleated giant cells with dark-pigmented yeast cells and hyphal elements. Haematoxylin & Eosin (H&E), Fontana-Masson, and Gomori methenamine silver (GMS) (Fig. 13.3) stains are used to diagnose phaeohyphomycosis. H&E and GMS stains show the dark pigmented fungal hyphae; however, visualisation of the darkly pigmented hyphae is not certain. In these cases, the Fontana-Masson stain can be used, which helps to demonstrate fungal melanin in their cell walls. Direct microscopy or histopathology showing the darkly pigmented hyphae with tissue invasion confirms the diagnosis.

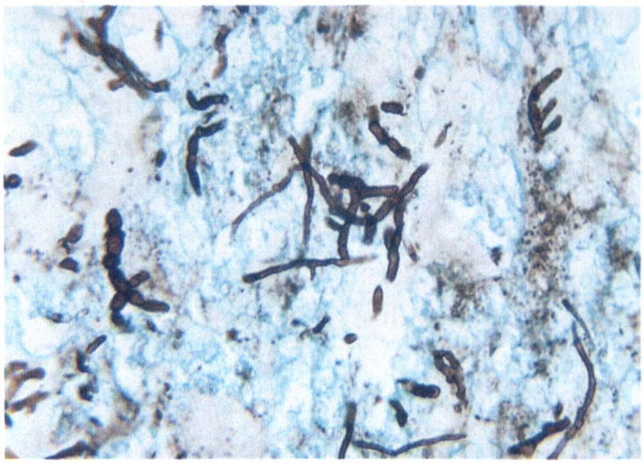

Fig. 13.3 This photomicrograph of a methenamine silver-stained tissue specimen reveals some of the histopathologic details associated with a case of phaeohyphomycosis due to the fungal organism, *Phialophora parasitica*. Note the numerous phaeoid hyphae dispersed throughout the tissue sample. (Courtesy CDC)

In Vitro Culture

Sabouraud dextrose agar and potato dextrose agar are used frequently to isolate dematiaceous fungi. The fungi are identified based on their phenotypic and molecular characteristics.

Phenotypic identification relies on macroscopic and microscopic features such as conidiation and fruiting bodies. In addition, a few dematiaceous fungi produce only sterile hyphae; in such cases, the species is confirmed by molecular methods. Culture techniques have limitations and take a high turnaround time, as the dematiaceous fungi are slow growers, and the identification is non-reliable. Phenotypic identification needs mycological and taxonomical expertise, as multiple genera are implicated in the disease. Hence, accurate species identification is necessary to initiate appropriate therapy. Isolation of dematiaceous fungi in clinical settings is often considered laboratory contaminants. Hence, interpretation of the findings supported by clinical relevance is essential.

Serodiagnosis

The role of serodiagnosis in phaeohyphomycosis is still in its infancy. A few studies showed 1,3-β-D-glucan (BDG) use in diagnosing phaeohyphomycosis, especially in *Exserohilum* meningitis, disseminated *E. dermatitidis* infection, and cutaneous alternosis. BDG analysis of cerebrospinal fluid (CSF) showed >96% sensitivity in proven cases of *Exserohilum* meningitis and is of prognostic value. The decline in CSF BDG levels after treatment correlated with a good prognosis, whereas unaltered or increased CSF BDG levels after the treatment had a poor prognosis. Nevertheless, the utility of BDG in diagnosing phaeohyphomycosis needs to be evaluated in clinical settings with many cases.

Molecular Diagnosis

DNA sequencing of the ITS and LSU regions is increasingly used for species confirmation. The ITS region is used as a gene target in fungal barcoding. SSU, *ACT*, *GAPDH*, *RPB1*, *RPB2*, and beta-tubulin are the other genes used in the species identification. The pan-fungal gene targets such as ITS, 28S (D1/D2 region), and 18S are used to diagnose invasive fungal diseases such as aspergillosis and mucormycosis. Molecular methods to identify the fungi directly from tissue specimens in culture-negative cases have also been evaluated.

Species-specific ITS primers are used to identify *C. carrionii*, *F. pedrosoi*, and *F. compacta*, differentiating them from other dematiaceous fungi. The loop-mediated isothermal amplification (LAMP) technique is used to identify *Fonsecaea* species using the ITS region. ITS gene sequencing of tissue specimens of 174 phaeohyphomycosis cases showed 54% positivity and matched 100% with culture identification. The limitation of the test is that it is not feasible to design genus- or species-specific gene targets because of multiple genera and species causing phaeohyphomycosis. Moreover, the sensitivity and specificity of these gene targets change with each laboratory; hence, more extensive studies are needed for proper validation.

In recent years, protein-based techniques such as matrix-associated laser desorption/ionisation (MALDI)-time of flight (ToF) mass spectroscopy (MS) have been widely used in the identification of fungal pathogens. The most crucial step in MALDI-ToF-MS is to generate species-specific protein spectra (libraries/database) from the control isolates identified by molecular methods (DNA sequencing). MALDI provides accurate, faster, and reliable results within minutes. The accuracy of MALDI depends on the protein extraction methods and the spectra generated from the test isolate. Direct spot method with formic acid (on the plate) or ethanol-formic acid extraction (off plate) methods are used for protein extraction. The in-house database generated using the control isolates belonging to multiple genera and species has shown 100% sensitivity. However, the non-availability of the spectra for the test isolates in the instruments database is the noted disadvantage in identifying dematiaceous fungi using MALDI-ToF MS, making the identification less robust, reliable, and faster.

Treatment

Treatment strategies for phaeohyphomycosis depend on the specific site of infection and the type of organism involved. The susceptibility profile of the relevant species guides the choice of antifungal therapy, as outlined in Table 13.5. For treating localized cutaneous infection or subcutaneous nodule, surgical intervention supplemented by cryotherapy, laser, heat, and photodynamic therapy has proven effective. In case of multiple subcutaneous nodules, antifungal therapy with Itraconazole or voriconazole is effective. In immunocompromised patients, to prevent dissemination, antifungals such as azolescan be used as an adjunctive therapy. Patients with keratitis are treated topically using natamycin alone, or in combination with other topical antifungals such as amphotericin B and azoles for several weeks to months. Refractory keratitis cases may treated with oral azoles and surgical interventions such as keratoplasty. Intravenous

Table 13.5 Antifungal susceptibility profile of clinically relevant agents of phaeohyphomycosis

Fungus	Susceptibility profile
Alternaria	Amphotericin B, itraconazole, posaconazole, and voriconazole are effective against most of the species Echinocandins, fluconazole, and flucytosine have high MIC values. Terbinafine is beneficial in treating cutaneous alternariosis
Acrophialophora	Varying degree of susceptibility to amphotericin B, itraconazole, posaconazole, and voriconazole. Echinocandins resistance is seen.
Aureobasidium	Amphotericin B and voriconazole can be used (limited data)
Bipolaris	Itraconazole, voriconazole, posaconazole, and amphotericin B were effective against *Bipolaris* species Fluconazole and flucytosine is not recommended
Chaetomium	Itraconazole, voriconazole, and posaconazole are effective against *Chaetomium* species. Caspofungin showed high MIC Amphotericin B exhibited variable susceptibility patterns against different *Chaetomium* species.
Cladophialophora	Itraconazole and posaconazole are highly effective Voriconazole penetrates the central nervous system more deeply and has higher bioavailability. In vitro activity of echinocandins and amphotericin B is satisfactory *Cladophialophora carrionii* has low MICs for isavuconazole
Curvularia	Amphotericin B is highly effective. Echinocandins and triazoles are less active.

(continued)

Table 13.5 (continued)

Fungus	Susceptibility profile
Exophiala	Itraconazole, voriconazole, posaconazole, and amphotericin B have variable activity Itraconazole and terbinafine alone or in combination are preferred in cutaneous infections
Exserohilum	Itraconazole, posaconazole, and amphotericin is highly effective, followed by voriconazole.
Fonsecaea	Posaconazole is highly effective; followed by other triazoles, terbinafine, amphotericin B and caspofungin
Hortaea werneckii	Itraconazole, posaconazole, voriconazole, isavuconazole, and amphotericin B have variable MICs
Neoscytalidium dimidiatum	Amphotericin B was most effective, followed by terbinafine. Posaconazole and voriconazole are less active
Ochroconis	Posaconazole and itraconazole are the most effective treatments for *Ochroconis* infection The preferred alternatives are amphotericin B and voriconazole The least effective are flucytosine and fluconazole
Phaeoacremonium	voriconazole, posaconazole, and itraconazole are most effective. Amphotericin B has lesser activity.
Phoma	Itraconazole, voriconazole, amphotericin B can be used to treat cutaneous infections.
Rhinocladiella	Amphotericin B and echinocandins are less effective (have high MICs) Itraconazole, posaconazole, and voriconazole are highly effective (have low MICs)
Veronaea	Itraconazole and posaconazole are the most effective agents High MICs to all other antifungals.

liposomal amphotericin B or azoles (itraconazole, voriconazole, posaconazole), is recommended to treat pulmonary infections over an extended period. Surgical excision is frequently recommended to manage solitary pulmonary nodules in immunocompetent patients.

Complete excision is recommended for the management of brain abscesses because treatment by antifungals only has limited success. When surgery is not possible, voriconazole or posaconazole can be used, voriconazole is effective in penetrating brain tissue. Combination therapy by triazole, echinocandin, and flucytosine could be a frontline option when surgery is not feasible.

Bone and joint infections typically necessitate both surgical intervention and systemic antifungal therapy. Itraconazole, voriconazole, posaconazole, or liposomal amphotericin B are used in treatment. Catheter removal coupled with systemic antifungal therapy is the recommended for the treatment of peritonitis associated with peritoneal dialysis. Systemic steroids and surgical removal of mucin are used to treat allergic fungal sinusitis. The role of antifungal therapy in allergic fungal sinusitis is still debatable. Refractory allergic sinusitis can be managed with azole therapy. While surgical resection usually suffices for sinus fungus balls. Whereas, in invasive sinusitis-systemic antifungals such as amphotericin B and voriconazole are indicated. Allergic bronchopulmonary mycosis, caused by *Bipolaris* and *Curvularia*, is managed by systemic steroids. The role of azoles in managing this condition is yet to be defined. In essence, phaeohyphomycosis treatment strategies are diverse and tailored to each case's specific characteristics, aiming to achieve optimal patient outcomes.

Prevention and Control

Prevention and control of phaeohyphomycosis presents many challenges due to the fungi's inherent cosmopolitan and ubiquitous nature, making complete eradication not feasible. However, preventive measures are vital for reducing phaeohyphomycosis and other mould infections among people with weakened immune systems. The challenge arises from the ubiquity of the fungi causing phaeohyphomycosis, as they are pervasive inhabitants of soil, making exclusion from hospital environments difficult. Customised prevention strategies for phaeohyphomycoses are yet to be defined. Established measures, such as air-quality precautions encompassing positive-pressure isolation rooms and high-efficiency particulate air filters, are often employed to manage aspergillosis in specific solid-organ and bone marrow transplant units. All these contribute to controlling the hospital surroundings for severely immunocompromised patients.

Case Study

A 57-year-old man underwent a living donor liver transplant to treat decompensated chronic liver disease that was caused by hepatitis B and D co-infection. The post-operative immunosuppressive regimen included prednisolone and tacrolimus and a 3-month course of trimethoprim-sulfamethoxazole and oral fluconazole as prophylaxis. Four months after the transplant, the patient developed a brain abscess, which was observed as a single, hyperintense, ring-enhancing lesion with peripheral vasogenic oedema on contrast-enhanced T2-weighted magnetic resonance imaging (MRI). His blood parameters showed an increased white blood cell count with neutrophilic leukocytosis and an elevated C-reactive protein. He was treated empirically with intravenous meropenem and vancomycin and underwent surgical removal of the lesion along with drainage of thick, creamy yellow pus. The histopathology of the tissue sample showed an inflammatory exudate with numerous fungal hyphae, further identified as *Rhinocladiella. mackenziei* by the slide culture. However, molecular diagnosis and antifungal susceptibility testing were not possible due to a lack of facilities. The patient was initially started on intravenous fluconazole but later switched to intravenous voriconazole due to persistent symptoms. A second surgery was performed to drain the residual abscess, and the patient was subsequently discharged on a one-year course of oral voriconazole. The patient recovered fully without developing any neurological deficits and remained stable 1 year after surgery.

1. What are the significant points of concern in the above clinical vignette?
2. What is the clinical spectrum of *Rhinocladiella. mackenziei* infection?
3. What are the risk factors and modes of transmission?
4. What may be the various methods of identification of the pathogen?

Research Questions

1. How do we develop a standard and suitable animal model to study the pathogenesis of diverse neurotrophic phaeohyphomycetes fungi?
2. Does the genomic and transcriptomic approach help identify common virulence traits and drug resistance markers in phaeoid fungi?
3. Are molecular methods-based environmental studies essential to identify the ecological niche of phaeohyphomycetes fungi?
4. Is multi-gene phylogenetics necessary in clinical settings to identify species as a taxonomy of phaeoid fungi is evolving?
5. Is species identification essential to initiate treatment in phaeohyphomycosis cases due to intrinsic resistance in phaeohyphomycetes agents?
6. Does the determination of the epidemiological cut-off value of different phaeohyphomycetes fungi improve the judicial use of antifungal treatment and decrease mortality?

Further Reading

Arcobello JT, Revankar SG (2020) Phaeohyphomycosis. Semin Respir Crit Care Med 41(1):131–140

Ben-Ami R, Lewis RE, Raad II, Kontoyiannis DP (2009) Phaeohyphomycosis in a tertiary care cancer center. Clin Infect Dis 48(8):1033–1041

Brandt ME, Warnock DW (2003) Epidemiology, clinical manifestations, and therapy of infections caused by dematiaceous fungi. J Chemother 15(Suppl 2):36–47

Chen SC, Blyth CC, Sorrell TC, Slavin MA (2011) Pneumonia and lung infections due to emerging and unusual fungal pathogens. Semin Respir Crit Care Med 32(6):703–716

Chowdhary A, Meis JF, Guarro J, de Hoog GS, Kathuria S, Arendrup MC, et al. (2014) ESCMID and ECMM joint clinical guidelines for the diagnosis and management of systemic phaeohyphomycosis: diseases caused by black fungi. Clin Microbiol Infect 20 (Suppl 3):47–75

He Y, Zheng HL, Mei H, Lv GX, Liu WD, Li XF (2022) Phaeohyphomycosis in China. Front Cell Infect Microbiol 12:895329

de Hoog GS, Guarro J, Gene J, Figueras MJ. (2000) Atlas of Clinical Fungi, 2nd edn. Utrecht, Reus: Centraalbureau voor Schimmelcultures, Universitat Rovira i Virgili.

Moreno LF, Vicente VA, De Hoog S (2018) Black yeasts in the omics era: achievements and challenges. Med Mycol 56(Suppl_1):S32–S41

Nosanchuk JD, Casadevall A (2003) The contribution of melanin to microbial pathogenesis. Cell Microbiol 5(4):203-223

Revankar SG, Patterson JE, Sutton DA, Pullen R, Rinaldi MG (2002) Disseminated phaeohyphomycosis: review of an emerging mycosis. Clin Infect Dis 34(4):467–476

Seyedmousavi S, Guillot J, De Hoog GS (2013) Phaeohyphomycoses, emerging opportunistic diseases in animals. Clin Microbiol Rev 26(1):19–35

Seyedmousavi S, Netea MG, Mouton JW, Melchers WJG, Verweij PE, de Hoog GS (2014) Black yeasts and their filamentous relatives: principles of pathogenesis and host defense. Clin Microbiol Rev 27(3):527–542

Teixeira MM, Moreno LF, Stielow BJ, Muszewska A, Hainaut M, Gonzaga L, et al. (2017) Exploring the genomic diversity of black yeasts and relatives (Chaetothyriales, Ascomycota). Stud Mycol 86:1–28

Feline Sporotrichosis

14

Leila Lopes-Bezerra, Fabiana Monti, Camila Stefanie Fonseca de Oliveira, Maria Griselda Perona, Brenda Kischkel, Júlia Marcondes Figueiredo, and Marconi Rodrigues de Farias

Learning Objectives
By the end of the chapter, the readers should be able to:

1. Learn the main advances about zoonotic sporotrichosis published in the literature.
2. Understand new *Sporothrix* pathogenic species and their genomic aspects.
3. Describe the morphology, pathology, immune response, clinical–epidemiological information, diagnostic and treatment protocols.

Introduction

Sporotrichosis was known as a benign subcutaneous disease affecting humans since the first clinical description by Benjamin Schenck in 1898. It is the "old" rose-gardener disease, in other words, *Sporothrix schenckii* infection as it was known. Currently, due to the scientific advances made in the twenty-first century and the discovery of new pathogenic species, the paradigm of this disease has dramatically changed and will be described here.

Feline sporotrichosis caused by *Sporothrix brasiliensis* has been associated with a larger and

L. Lopes-Bezerra (✉)
BIDiagnostics® at the Technology-Based Incubator of São Paulo at São Paulo University (USP), São Paulo, Brazil
e-mail: leila.lopes@bidiagnostics.com.br

F. Monti
Postgraduate Program in Animal Science at the Pontifical Catholic University of Paraná, Paraná, Brazil

C. S. F. de Oliveira
Department of Preventive Veterinary Medicine, Veterinary School, Universidade Federal de Minas Gerais, Belo Horizonte, Minas Gerais, Brazil

M. G. Perona
Molecular Biology, Chemistry Institute, São Paulo University, São Paulo, Brazil

B. Kischkel
Department of Internal Medicine and Radboud Institute for Molecular Life Sciences, Radboud University Medical Center, Nijmegen, The Netherlands

J. M. Figueiredo
Institute of Biomedical Sciences, São Paulo University, São Paulo, Brazil

M. R. de Farias
Veterinary Dermatology and Postgraduate Program in Animal Science at the Pontifical Catholic University of Paraná, Paraná, Brazil

S. C. Parija, S. M. Rudramurthy (eds.), *Textbook of Fungal Zoonoses and Sapronoses*, Microbial Zoonoses, https://doi.org/10.1007/978-981-97-3214-2_14

more complex clinical–epidemiological case series of sporotrichosis with high cat–human transmissibility. Infection by this species poses a threat of global spread in the absence of policies for epidemiological containment and/or control of zoonotic transmission, including the lack of modern tools for diagnosis and epidemiological screening of new cases. Dogs are rarely affected by *S. brasiliensis*, however, diagnosis in dogs is a major challenge and new cases are being reported. The number of human cases has also increased progressively, usually associated with feline transmission.

Infections by *S. schenckii* in rats, cats, dogs, rabbits, and even horses were described in the literature in the early nineteenth century, and at that time, the possibility of human infection by contact with animals was suggested. Infection in cats has been reported since the early 1950s in distinct continents with no epidemiological impact.

History

For more than a century since its discovery by Benjamin Schenck in 1898, sporotrichosis was attributed to a single etiological agent: *S. schenckii* sensu stricto. A natural infection by this species in rats (*Rattus norvegicus*) was reported in 1907 by Lutz and Splendore who also described the parasitic form of *S. schenckii*, the unicellular yeast form (Fig. 14.1a). However, in 2007, a radical change occured by the discovery of other pathogenic cryptic species, showing morphological similarities but distinct genotypes. Taxonomic studies were refined, and four pathogenic species, *S. schenckii*, *S. brasiliensis*, *S. globosa*, and *S. luriei*, were grouped within a pathogenic clade of the genus *Sporothrix*. Among the pathogenic species, *S. schenckii* and *S. brasiliensis* are the only two important zoonotic species.

Taxonomy

Medically relevant *Sporothrix* (Ascomycota: Ophiostomatales) were classified as a "species complex," previously called the *S. schenckii* complex. However, with progress in knowledge of the biology of *Sporothrix* species, the denomination of a species complex is no longer appropriate. Currently, it is known that instead of being a monotypic taxon formed by a single species, it is a diverse genus with marked genetic, phenotypical, and ecological diversity (Table 14.1). In 2016, de Beer and collaborators redefined the *Sporothrix* genus. The newly defined *Sporothrix*

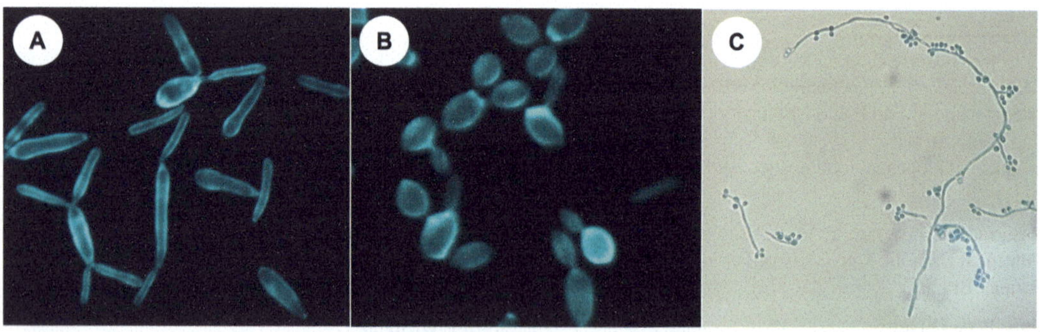

Fig. 14.1 Yeast and mycelium morphological phases of *Sporothrix* spp. (**a**) Immunofluorescence image of *Sporothrix schenckii* yeast cells showing a cigar-shaped morphology. The cell wall was labeled with calcofluor white (CFW); (**b**) immunofluorescence image of *Sporothrix brasiliensis* yeast cells labeled with CFW showing a round or oval morphology; and (**c**) light microscopy of the mycelium phase obtained by a cotton blue stained *Sporothrix* spp. microculture. The image shows the typical specialized conidiophores of *Sporothrix* spp. with ovoid-shaped conidia in a rosette-like arrangement, as well as the presence of sessile conidia (a conidium arising directly from the surface of a hypha)

includes 51 species, and among these, four species were classified within the Pathogenic Clade, as described earlier.

Developments in basic and clinical research of *Sporothrix* reveal significant differences among *Sporothrix* species exhibiting different associations between organisms and their hosts or niches. These include morphological, physiological, genetic, epidemiological, and virulence traits. These also include varying susceptibility to antifungals, among other aspects that enable the recognition of distinct species rather than a complex of cryptic species.

Molecular studies based on calmodulin, β-tubulin genes, and ribosomal DNA provided evidence of at least six putative phylogenetic species prevalent in different geographical regions. The proposed species are distributed in five main clades (named I to V). In addition, those clades present physiological differences, such as differences in sucrose, raffinose, and ribitol metabolism and in thermotolerance.

The clades were separated as follows. (a) Clade I included *S. brasiliensis* isolates from the Brazilian territory. (b) Clade II included practically all the American isolates of *S. schenckii* and was divided into two subclades (IIa and IIb). Subclade IIa groups most of the isolates from the United States and South America (Argentina, Bolivia, Colombia, Peru, and Venezuela), while subclade IIb refers to the rest of the isolates from South America (Peru and Argentina) and one isolate from South Africa. (c) Clade III comprises *S. globosa* identified in isolates from China, India, Italy, Japan, and Spain. (d) Environmental isolates form the two basal clades, clades IV and V. Clade IV groups isolate from Mexico (*Sporothrix mexicana*), and clade V groups two isolates identified as *S. pallida* (also known as *S. albicans*). (e) Clade VI is represented by *S. luriei*. This is a very rare pathogen reported in human infections on four occasions, including one reported Africa.

Table 14.1 Classification of *Sporothrix*

Kingdom	Fungi
Division	Ascomycota
Class	Sordariomycetes
Order	Ophiostomatales
Family	Ophiostomataceae
Genus	*Sporothrix*

All clinically relevant species of *Sporothrix* spp. were included in clades I, II, and III. Hence, it is more appropriate to use the expression "clinical clade" or "pathogenic clade" to refer to *S. brasiliensis*, *S. schenckii*, *S. globosa*, and *S. luriei*, which were isolated from human and/or animal cases. The remaining *Sporothrix* species are nested in an "environmental clade." They are often associated with substrates that vary from soil and decomposing organic matter to insects and plants. Indeed, these environmental species are rare agents of infections in mammals.

Genomics and Proteomics

Genetic diversity among *Sporothrix* spp. also translates into different virulence mechanisms as well as different antifungal susceptibility and thermotolerance. Comparative proteomic analysis revealed a particular protein profile in pathogenic and non-pathogenic isolates of *Sporothrix* spp. Those unique patterns may explain the differences in virulence among them.

Genomics and proteomics have provided insights to conclude that the higher virulence of *S. brasiliensis* and its capacity to infect a broad range of mammalian species may be due to metabolic differences from other species from the *Sporothrix* genus and/or to the cell surface conformation and antigen expression. Significant differences were found comparing the protein expression levels among *S. brasiliensis*, *S. schenckii*, and *S. globosa*, with more than 100 identified proteins being differentially expressed, mostly related to carbohydrate and amino acid metabolism as well as to stress response.

Like many other fungi, *Sporothrix* species present complex genomes and small chromosomes that are difficult to visualize with traditional karyotyping techniques. Because of this limitation, the number of chromosomes for each species wasn't precisely determined yet.

The Fungus Morphology

The Sporothrix genus encompasses thermal dimorphic and zoonotic species which shows differences in the parasitic yeast form (Fig. 14.1a, b). The saprophytic mycelial phase is the environmental form of *Sporothrix* spp., which can be cultivated in vitro at temperatures of 25–28 °C. The mycelium consists of septate hyaline hyphae. The conidiophores rise from the hyphae with a characteristic structure in the form of a "flower" or "daisy" appearance, specialized structures that produce oval or pyriform conidia (Fig. 14.1c). *Sporothrix* spp. also have sessile conidia that arise directly from the hyphae (Fig. 14.1c). Conversely, the single-celled yeast form has an elongated morphology ("cigar-shaped") as observed in *S. schenckii* cultures (Fig. 14.1a), whereas *S. brasiliensis* often presents yeasts with oval or round morphology (Fig. 14.1b). The yeast form can be cultured in vitro at 36–37 °C.

Life Cycle of Fungus

Sporothrix has an asexual life cycle. The fungus is present in the environment as a mycelium that produces conidia. These conidia can infect humans and animals by traumatic inoculation into the skin, and then morphogenesis occurs to the yeast parasitic phase (dimorphism). The yeast's asexual life cycle is based on budding growth, like other yeasts. The *S. brasiliensis* cycles of zoonotic transmission are illustrated in Fig. 14.2.

Pathogenesis

The severity of the disease in cats is attributed to the virulence of the emerging zoonotic species, *S. brasiliensis*. However, the immune response of cats is unknown.

Sporothrix Virulence Factors

Melanin is an important factor is associated with virulence profiles and resistance against conventional antifungal drugs in different species of *Sporothrix* spp.

Sporothrix species can produce three types of melanin as secondary metabolites. These include DHN–melanin, L-DOPA—melanin, and pyomelanin. In the infected host, *Sporothrix* spp. synthesizes DHN and L-DOPA. These two melanins protect the fungi against antifungal agents or from macrophage killing by oxygen radicals. The production of this pigment is higher and faster in *S. brasiliensis* than in *S. schenckii*. These characteristics explain the differences in virulence and drug-resistance profiles among these two species. *S. brasiliensis* is the most resistant species.

Thermotolerance is attributed as an important virulence factor of *Sporothrix* species. The thermal resistance exhibited by *S. brasiliensis* but not by *S. schenckii* and *S. globosa* is possibly an important adaptation mechanism as cats have an average body temperature between 38 and 39.2 °C. These factors that may contribute to the high pathogenicity of *S. brasiliensis*, with the potential to cause serious infections in animals.

The lower thermotolerance of *S. schenckii* impairs its virulence and, in consequence, to cutaneous clinical manifestations. Cutaneous lesions in cats are localized at colder body sites, such as ear tips and nose. Moreover, in the case of *S. schenckii* infection, there is also a reduced probability of developing lymphocutaneous or disseminated forms due to the host's body temperature.

Immunology

S. brasiliensis and *S. schenckii* induce varying immune responses in infected hosts. The immunological response of domestic animals (cats and dogs) in response to *Sporothrix* infection is still poorly understood.

Peptido-rhamnomannan (PRM), a major cell wall component, is present on the outer cell wall layer of *S. brasiliensis* and *S. schenckii*. Human PBMCs stimulated with PRM from *S. brasilien-*

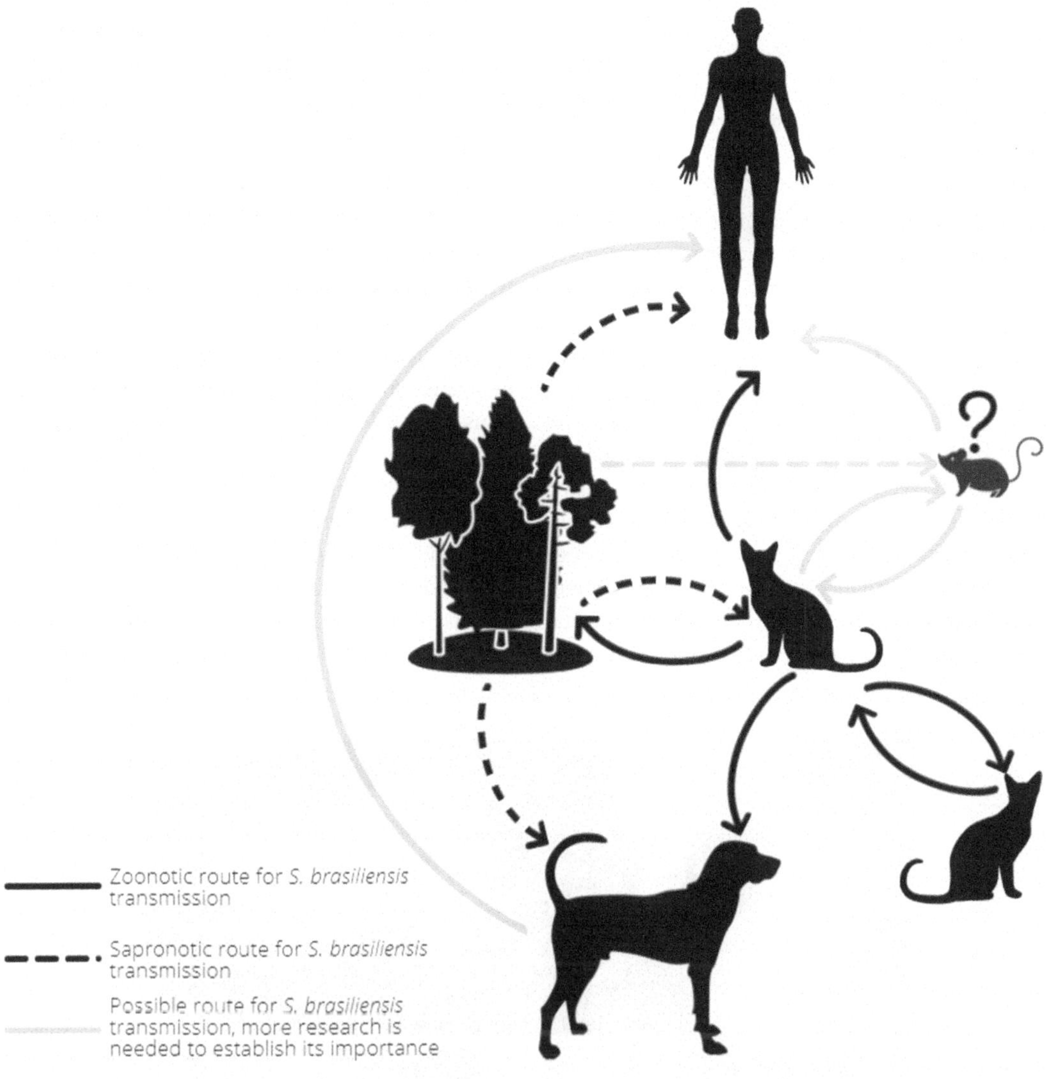

Fig. 14.2 Zoonotic cycle of feline sporotrichosis transmission. The pathogen *S. brasiliensis* has high cat-to-human transmissibility, and feline infection can be transmitted to dogs. This peri- and intra-domiciliary cycle defines a new model of transmission and the spreading of sporotrichosis within the concept of One Health

sis were able to induce a robust inflammatory response, with high production of pro-inflammatory innate immune cytokines: TNF-α, interleukin (IL-6), and IL-1β and induction of T-helper cytokines IFN-γ, IL-17, and IL-22. In contrast, *S. schenckii* induces higher concentrations of IL-1 antagonist receptor (IL-1Ra).

The IL-1 family is a key cytokine responsible for the inflammatory response induced by *S. brasiliensis* infection. They trigger and/or are key initiators of necrosis-initiated inflammation, amplifying the inflammation in response to infection-associated tissue injury. The IL-1 cytokine plays a significant role in innate and acquired immunity. It acts by promoting the resolution of infection.

PRM is responsible for the recognition of *S. schenckii* and *S. brasiliensis* by human macrophages via CR3. Complement activation, most likely initiated by the opsonization of yeasts by heat-labile complement factors present in the human serum, is crucial for developing the immune response against sporotrichosis. This is regulated by toll-like receptors (TLRs). Importantly, differences in the structure of the PRM isolated from these two species have been observed. These findings suggest that the PRM from *S. brasiliensis* has unique rhamnose residues that may help to explain the differences in the induced immune response between *S. brasiliensis* and *S. schenckii*.

Infection in Humans

Clinical presentations of sporotrichosis in humans are either cutaneous or extra-cutaneous. The cutaneous manifestations based on lesion topography are subdivided into fixed cutaneous, disseminated cutaneous, and lymphocutaneous varieties. Extracutaneous forms include ophthalmic, bone, pulmonary, and systemic manifestations. Humans acquire infections either by contact with contaminated soil and vegetation debris or from scratches and contact with secretions of infected felines (Fig. 14.2). *S. brasiliensis* is mainly related to extracutaneous forms, and *S. globosa* is rarely reported to cause extracutaneous and disseminated infections.

Infection in Animals

S. brasiliensis and *S. schenckii* are the species that cause sporotrichosis with varying manifestations in cats. It is important to highlight here that *S. brasiliensis* usually causes a severe disease in animals with a high mortality rate.

The fixed cutaneuous manifestations are illustrated in Fig. 14.3a. This condition occurs as isolated lesions, typically restricted to the site of the fungus inoculation. It manifests as plaques, nodules, or tumors, which develop liquefactive

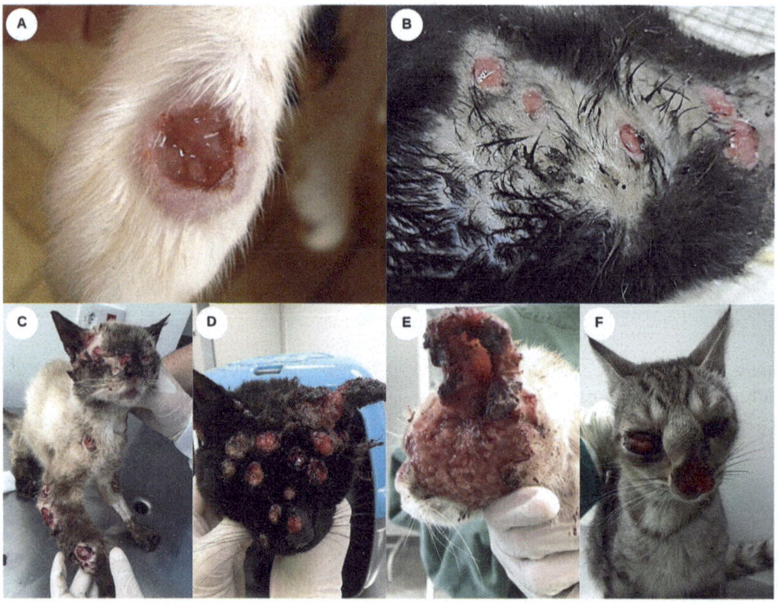

Fig. 14.3 Clinical manifestations of feline sporotrichosis caused by *S. brasiliensis*. (**a**) Nodular lesion with raised edges and an ulcerative–necrotic center in a 3-year-old male cat with fixed cutaneous sporotrichosis. (**b**) Ulcerative and exudative lesions associated with ascending lymphangitis and regional lymphadenitis in an adult mixed-breed male cat with lymphocutaneous sporotrichosis. (**c**) Lymphocutaneous form associated with the disseminated cutaneous form in an adult Siamese male cat with sporotrichosis. (**d**) Multiple ulcero-gummous lesions on the face of an adult mixed-breed male cat with sporotrichosis. (**e**) Ulcerative, crateriform, intensely exudative, crusted, and facial deforming lesions on an adult mixed-breed male cat with disseminated cutaneous sporotrichosis. (**f**) Two-year-old mixed-breed male cat with tumoral, ulcero-gummous, facial deforming lesion associated with granulomatous blepharoconjunctivitis secondary to sporotrichosis

necrosis ulcerative lesions with raised edges (*sporotrichotic chancre*). The lesions are usually present on the limbs, especially the thoracic limbs, and on the nasal plane and can cause the loss of facial architecture.

The lymphocutaneous form of sporotrichosis (Fig. 14.3b) is characterized by the presence of multiple, ascending tegumentary lesions. These ulcerative lesions are parallel to the lymphatic path and associated with lymphangitis and regional lymphadenitis. This manifestation is typically associated with disseminated cutaneous lesions.

The disseminated cutaneous forms of sporotrichosis (Fig. 14.3c, d) may result from self-inoculation or lymphatic or hematogeneous dissemination of yeasts. This is the most frequent clinical manifestation in cats. It is characterized by extensive ulcerative and ulcero-gummatous lesions, as well as multiple plaques or by isolated or coalescent nodules. The lesions are crateriform, with inflamed edges, intensely exudative, and topped by thick crusts with a meliaceous appearance. These lesions affect the skin, mucocutaneous junctions, and respiratory mucous membranes and are associated with a high tissue burden (Fig. 14.3e). Sporotrichosis in the respiratory system is characterized by lesions in the upper airways that manifest as stridor, sneezing, naso-ocular secretions, intranasal polyps, stenosis, nodulo-tumoral formations in the nasal plane and facial deformity (Fig. 14.3f). In clinical studies reported by the authors, 23% of *S. brasiliensis*-infected cats presented the fixed cutaneous form of sporotrichosis while 71% presented the disseminated cutaneous form. In the latter group, 79% had concomitant respiratory involvement, and 77% developed lymphadenomegaly, emaciation, hyporexia/anorexia, lethargy, and digestive disorders.

Feline sporotrichosis mostly involves the face (nasal bridge and periorbital, frontal, temporal, and auricular pavilions), thoracic limbs, nail bed, and cervical region, places described as more vulnerable to injury from fights. In addition, involvement of the pelvic limbs, tail, and external genitalia is also common. In studies conducted by the authors, the frequency observed in 294 cases of feline sporotrichosis was as follows: 240 (82%) presented lesions on the face; 167 (57%) on the limbs; 111 (38%) on the back; and 34 (12%) on the tail.

Epidemiology and Public Health

S. brasiliensis and *S. schenckii* are the zoonotic species that are primarily reported in Brazil and Malaysia, respectively. The first evidence of cat-to-human zoonotic transmission was described in the United States in 1986, with a report of five infected cats that led to seven cases of human infection. Subsequently, more feline cases have been described in Brazil, Spain, and Japan.

Cats are victims but a reservoir for *S. brasiliensis*, transmiting the fugus to humans through bites and scratches. However, in some human cases, there was no history of accidents with cats. Dogs still play a minor role in the transmission of infection. Retrospective phylogenetic studies have demonstrated that the true etiological agent of the anthropozoonotic sporotrichosis outbreak described in Brazil in 2000 is *S. brasiliensis*, not *S. schenckii*, as suggested earlier.

In Brazil, in the southeast and south regions, there was a rapid spread of the disease caused by *S. brasiliensis*, with reports of hundreds to thousands of feline sporotrichosis cases. Based on data from the Zoonosis Control Centers in Brazil from 2019 to 2023, zoonotic sporotrichosis has spread to almost all 27 Brazilian Federal States. Currently, zoonotic sporotrichosis caused by *S. brasiliensis* has been reported in Chile, Argentina, and other South American countries.

Cases of feline sporotrichosis caused by *S. schenckii* have been described more frequently in Asia but have low animal-to-human transmissibility. A single case of feline sporotrichosis caused by *S. globosa* was recently described in Japan. Feline sporotrichosis is mostly caused by *S. brasiliensis* in South America. Previous phylogenetic studies using the ITS region of rDNA and calmodulin gene sequencing showed that isolates from cats with skin lesions in Malaysia belong to *S. schenckii* clinical clade D, diverging from Brazil where the cases are associated with *S. brasiliensis*.

Due to the difficulty in obtaining environmental isolates of *S. brasiliensis*, the low number of individuals contaminated with plant or soil propagules and the high occurrence of infections caused by cat transmission suggest that the Brazilian outbreak of zoonotic sporotrichosis is possibly due to the adaptation of *S. brasiliensis* to higher temperatures, i.e., body's high temperature of cats, rather than an environmental expansion of *S. brasiliensis* in nature. This fungus has been isolated from cats' nasal and oral cavities as well as claws and feces. The environmental fungal population had spread in all Brazilian cities and municipalities due to the excretion of *S. brasiliensis* in cat feces, increasing soil and vegetation contamination. Despite the treatment of feline cases, euthanasia has been followed for infected animals. These are buried and not cremated, thereby increasing the environmental contamination cycle. Euthanasia may be only considered if all treatment alternatives are exhausted.

The geographical expansion of sporotrichosis caused by *S. brasiliensis* is possibly due to several factors, including a low level of responsible pet ownership, the low socioeconomic and sanitary conditions of peripheral urban areas with greater risk of exposure to *S. brasiliensis*, the absence of an epidemiological control program, the uncontrolled growth of street cats and rodent population, and the lack of knowledge in the management of sporotrichosis caused by *S. brasiliensis*. One of the main goals of this chapter is to update and spread the information about this new and emerging infection to health professionals.

In conclusion, the main etiological agent of zoonotic sporotrichosis with epidemiological importance is *S. brasiliensis* and not *S. schenckii*. *S. brasiliensis* may spread to all continents and to other vertebrate animal hosts. There is already an increasing number of cases being reported in domestic dogs.

Diagnosis

In Animals

The clinical manifestations of sporotrichosis in cats are nonspecific and may mimic nocardiosis (Fig. 14.4a), mycobacteriosis (Fig. 14.4b), cryptococcosis (Fig. 14.4c), histoplasmosis (Fig. 14.4d), pemphigus foliaceous, phaeohyphomycosis (Fig. 14.4e), and squamous cell carcinoma (Fig. 14.4f). Even though the clinical signs, history and epidemiological data of the patient may suggest sporotrichosis, a definitive diagnosis is based on laboratory tests, such as the isolation of the fungus by culture.

Cytopathology

Cytopathology is a presumptive test. The specimens for the test include (1) slide imprint on the surface of an ulcerated lesion, (2) swabs collected from lesional exudate, (3) aspirates of abscesses or nodules, and (4) biopsy samples. Slides can be stained using the fast panoptic method, a Romanowsky stain. Yeasts are detected within leucocytes and macrophages cytoplams or the extracellular milieu. They are pleomorphic and vary in shape from round to oval, measuring 3–5 μm in diameter, 5–9 μm in length, and surrounded by a clear halo (Fig. 14.5a).

The advantage of cytopathology is its low cost. It usually reveals a large number of yeasts in cat lesions. Cytopathology in animals with single lesions or with little disseminated lesions or respiratory and chronic conditions shows low sensitivity due to the presence of a low number of yeasts (Fig. 14.5b). Furthermore, cytopathological findings can be misdiagnosed with *Cryptococcus neoformans* (Fig. 14.5c) or *Histoplasma capsulatum* (Fig. 14.5d). An expert pathologist is necessary.

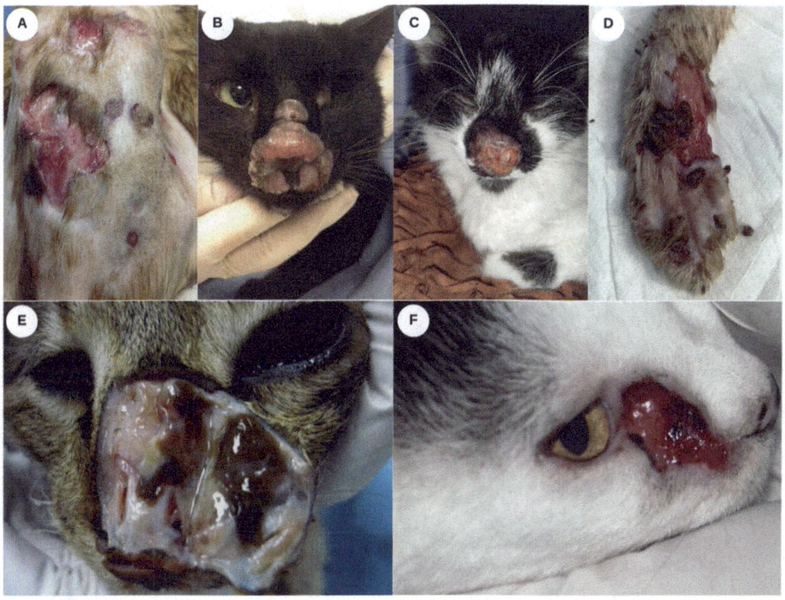

Fig. 14.4 Cases with clinical presentation similar to feline sporotrichosis and that should be considered for differential diagnosis based on clinical–epidemiological evidences. (**a**) Cellulitis associated with ulcerative lesions, draining sinuses and exudation in an adult mixed-breed female cat with cutaneous nocardiosis. (**b**) Nodulo-tumoral, coalescent, alopecic, and erythematous lesions on the nasal and nasolabial plane in a 2-year-old mixed-breed male cat with mycobacteriosis. (**c**) Tumor, circumscribed, erythematous, and floating lesion on the nasal plane of an adult mixed-breed female cat with cryptococcosis. (**d**) Ulcero-gummous lesion on the skin under the extensor surface of the right carpus in an adult mixed-breed male cat with histoplasmosis. (**e**) Ulcerative lesion with gelatinous, blackish, and intense exudate on the nasal bridge of a 5-year-old mixed-breed male cat with phaeohyphomycosis. (**f**) Ulcerative, crateriform, facial deforming lesion on the nasal bridge and malar region of a 9-year-old mixed-breed male cat with squamous cell carcinoma

Histopathology

Hematoxylin and eosin are used routinely to observe the inflammatory infiltrate associated with the infection; however, they are not able to reveal the *Sporothrix* spp. yeasts. Grocott methenamine silver (GMS) or periodic acid-Schiff (PAS) are stainning methods that can be used to identify *Sporothrix* spp. The specimens are collected from the edges of active lesions with 3- or 4-mm punches or from tissue fragments obtained during necropsy and should be fixed in 10% buffered formalin.

In vitro Culture

The gold-standard diagnosis of sporotrichosis is the isolation and identification of *Sporothrix* spp. in culture. A biosafety level 2 laboratory is needed for *Sporothrix* spp. biosafety manipulation and cultivation. Culture takes nearly 15 to 30 days.

Clinical specimens include exudates from the lesion, tissue samples of nonulcerated abscesses, nasal swabs, or biopsy fragments. The specimens cannot be fixed with 10% formaldehyde, and they should be transported in saline with antibacterial agents to the laboratory. The collected swabs are stored in Stuart's medium for better preservation. The specimens can be kept refrigerated at 4 °C for a maximum of 8–10 h.

The specimens are inoculated on Sabouraud dextrose agar and Mycosel agar and then incubated at low temperatures, between 25 and 30 °C. The mycelial growth of *Sporothrix* spp. can be evidenced in 1 or 2 weeks, but the correct diagnostic depends on the identification of the typical conidiophores (Fig. 14.1c) and the mycelial conversion to the yeast form in blood agar at

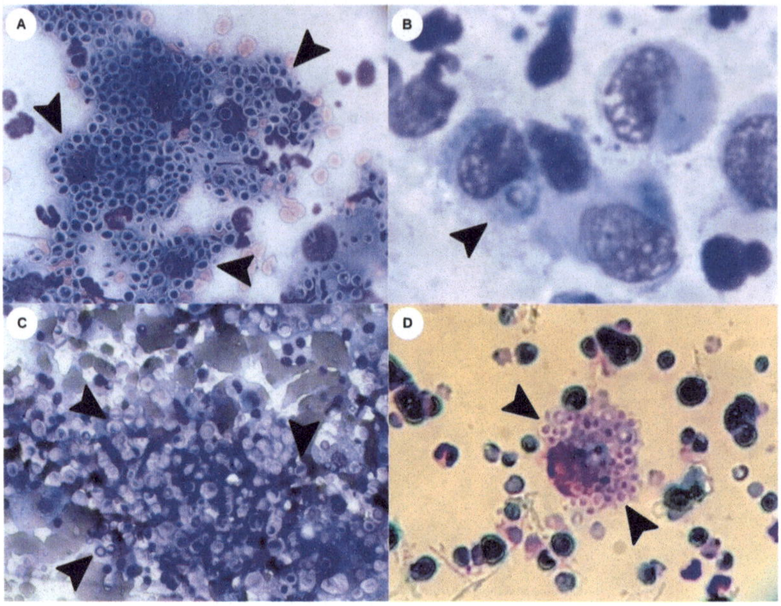

Fig. 14.5 Cytopathological diagnosis revealing the presence of intracellular *Sporothrix* spp. yeast and/or to the causative agents of cryptococcosis and histoplasmosis. (**a**) Multiple pleomorphic intra- and extracellular yeasts surrounded by a clear halo associated with a pyogranulomatous infiltrate from a cat with disseminated cutaneous sporotrichosis (Panoptic 40×). (**b**) Presence of a circumscribed yeast surrounded by a clear halo in the cytoplasm of a macrophage from an adult mixed-breed male cat with nasal sporotrichosis (Panoptic 100×). (**c**) Multiple oval yeasts of varying diameters, some surrounded by a clear capsule of a mucopolysaccharide nature, associated with a pyogranulomatous infiltrate from a cat with nasal cryptococcosis (Panoptic 10×). (**d**) Multiple oval yeasts of varying diameter phagocytized by a giant cell from a cat with cutaneous histoplasmosis (Panoptic 100×)

37 °C. A study with 196 cats revealed that mycological culture was more sensitive (95.2%) than cytopathological examination (52.6%). In a study conducted by the authors, fungal cultures from 245 cats were used to diagnose sporotrichosis and showed a sensitivity of 85%.

Serodiagnosis

Serodiagnosis is useful for the determination of anti-*Sporothrix* IgG levels in the cat's serum and to follow up treatment and mycological cure. The Cat *Sporothrix* IgG ELISA test (BIDiagnostics®, São Paulo, Brazil) has a high efficacy determined by the ROC curve (receiver operating characteristic curve) with an AUC (area under the curve (AUC) of 87,3%, and is currently available for diagnosis in private laboratories of veterinary diagnosis, in Brazil.

In Humans

Conventional methods of diagnosing sporotrichosis in humans are like those described above for felines with the exception of cytopathology. Histopathology has a low sensitivity as the number of yeasts in human specimens is very low and, in consequence, PCR methods will have a low sensitivity. Intradermal skin tests using mycelial or yeast phase culture filtrate-derived antigens (sporotrichin) can only be used to determine the endemicity of sporotrichosis. By the new regulations these intradermal tests are now forbidden in several countries, including Brazil.

A commercial latex agglutination test (IMMY, USA) to detect anti-*S. schenckii* antibodies was recommended for invasive and extracutaneous sporotrichosis, not the cutaneous form. A new commercial serological test is the unique test validated for *S. brasiliensis* infection, Human

Sporothrix IgG ELISA (BIDiagnostics®, São Paulo, Brazil), and is already available for diagnosis in a private reference Brazilian chain of human laboratories.

Treatment

Itraconazole (ITZ) and potassium iodide (KI) are used for the treatment of sporotrichosis in cats. Ketoconazole is less effective and is associated with more adverse effects than ITZ.

Chemotherapy is carried out for at least one month after clinical cure, as established by the absence of all signs and symptoms and the resolution of cutaneous lesions.

Itraconazole

ITZ is the most widely used drug in the treatment of feline and human sporotrichosis. For the treatment of human sporotrichosis caused by *S. brasiliensis* infection, the initial dose is 200 mg/day and in refractory cases can reach up to 400 mg/day.

It is a triazole antifungal and acts through the inhibition of the synthesis of ergosterol, which is an important component of the fungal plasma membrane. It has good oral absorption, a long half-life, good tissue distribution, and is considered a relatively safe drug. Its absorption depends on food intake and is reduced with hypochlorhydria. The recommended dose for the treatment of cats with sporotrichosis is 25 mg/day for animals weighing < 1 kg, 50 mg/day for animals weighing between 1 to 3 kg, and 100 mg/day for animals weighing over 3 kg.

ITZ shows good in vitro activity against *S. brasiliensis* strains isolated from cats and dogs. However, in vivo, the therapeutic response to ITZ is quite variable. Up to date, resistance to ITZ monotherapy is frequently reported in cats. The cure rate observed by a group of feline cases followed up by the authors was only 32% when using ITZ monotherapy, a cure rate similar to that achieved by other clinical groups (38%) that used the same ITZ protocol.

Potassium Iodide

KI is administered at a dose of 2.5–5.0 mg/kg orally every 24 h. The combination of ITZ and KI was successfully used in the treatment of feline sporotrichosis and is recommended in cases of resistance to ITZ monotherapy, revealing a cure rate of 96% with this therapeutic combination. In the authors' opinion, the combination of KI and ITZ is necessary in most cases to achieve a clinical cure, especially in cases of disseminated cutaneous lesions associated with upper respiratory tract manifestations caused by *S. brasiliensis* in cats (Fig. 14.6).

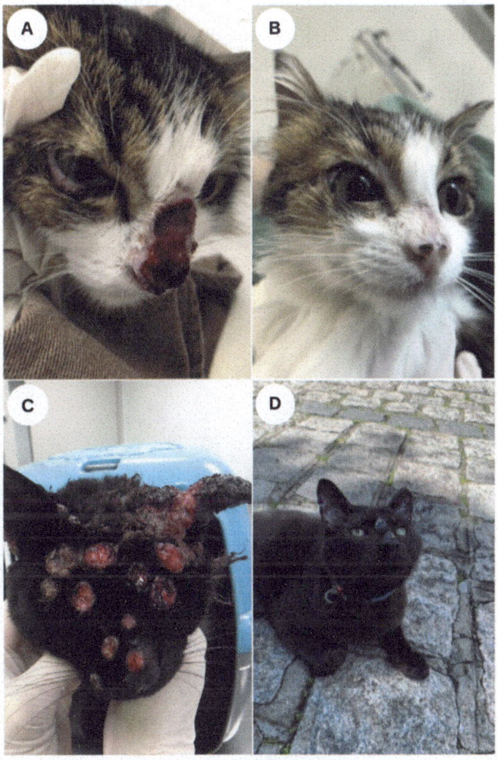

Fig. 14.6 Cases of feline sporotrichosis successively treated with a combination of ITZ and KI. (**a**) Two-year-old mixed-breed female cat diagnosed with sporotrichosis, with an ulcerative lesion on the nasal plane and bridge. (**b**) Photo of the same animal (**a**) after 3 months of treatment with ITZ combined with KI. (**c**) Adult mixed-breed male cat with diagnosis of disseminated cutaneous sporotrichosis, presenting nodular lesions with an ulcero-necrotic center and hematic crusts on the cephalic region. (**d**) Photo of the same animal (**c**) after 12 months of treatment with ITZ combined with KI

Amphotericin B

Amphotericin B deoxycholate (D-AMB) is derived from *Streptomyces nodosus*. This drug has low oral bioavailability, and thus, intravenous, intrathecal, intraarticular administration or administration at surgical sites is required. The drug exerts its antifungal effects by oxidative damage and by binding to ergosterol in the fungal cytoplasmic membrane. This results in the formation of pores and increased plasma membrane permeability, leading to cell death. D-AMB may be nephrotoxic when administered intravenously. Thus, intralesional administration has been proposed and successfully used in refractory cases of feline sporotrichosis. This form of administration promotes high tissue concentrations of the drug and increases its effectiveness, minimizing side effects associated with nephrotoxicosis, hyperkalemia, and cardiac arrhythmias.

In a study conducted by the authors for 47 animals treated with ITZ and KI, 16 (34%) animals required complementary therapy with weekly intralesional application of D-AMB due to the severity of the lesions, mainly located on the nasal plane and bridge (Fig. 14.7a, b). In these animals, the combination of ITZ with KI was not effective for the total involution of clinical signs and symptoms.

AMB is also the drug of choice for severe cases of human sporotrichosis and refractory cases due to *S. brasiliensis* infection.

Cryosurgery

Treatment with a combination of cryosurgery and ITZ was shown to be effective in the treatment of sporotrichosis and reduced the treatment time when compared to the drug therapy alone. In our study group, we observed that cryosurgery was useful as a complementary therapy to ITZ and KI, especially in chronic or relapsed lesions located on the nasal plane, reducing the time of drug therapy.

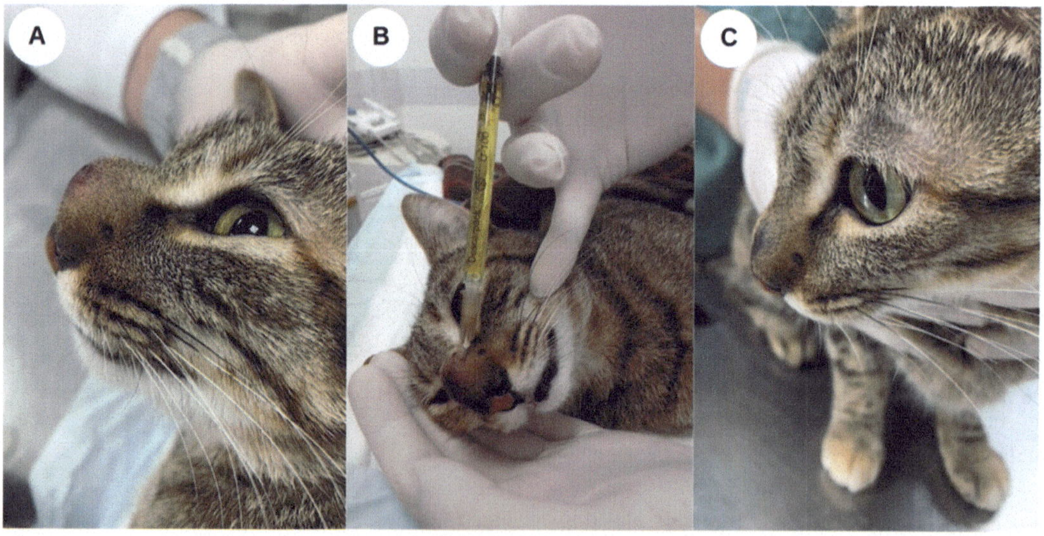

Fig. 14.7 Intralesional application of D-AMB as a complementary treatment to the combination of ITZ and KI. (**a**) Form of weekly intralesional application of D-AMB. (**b**) Nodulo-tumoral lesion on the nasal bridge of an adult mixed-breed female cat diagnosed with sporotrichosis. (**c**) Same animal after treatment with a combination of ITX and KI, plus weekly intralesional applications of amphotericin B. Resolution of the nasal lesion after 3 months of therapy

Case Study

Cat, male, adult, 3.3 kg, mixed breed, rescued from the street, was taken to a veterinary consultation with disseminated skin lesions, mainly on the face, after 5 weeks of evolution. Clinical examination revealed ulcerous, exudative, and crusted lesions in the nasal planum, left eyelid, and frontal face (Fig. a), associated with sneezing, stridulum, and nasal obstruction. The cytopathological examination showed several pleomorphic yeasts associated with a pyogranulomatous infiltrate (Fig. b), and mycological culture was positive for *Sporothrix* spp. Treatment with itraconazole was instituted at a dose of 100 mg/cat, associated with potassium iodide, at a dose of 2.5 mg/kg. Treatment was maintained for 2 months after the clinical cure. The treatment required 8 months (Fig. c), and so far, the patient has not relapsed.

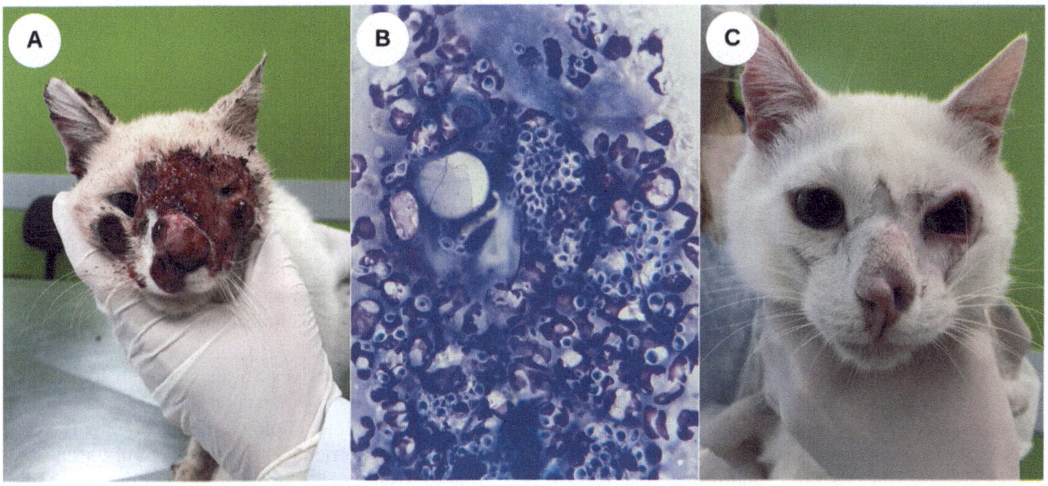

1. How could this cat have acquired the disease?
2. What clinical presentations of feline sporotrichosis were developed by the described patient?
3. In addition to cytology and fungal culture, what other diagnostic tests could be performed on the patient described?
4. What is the reason for the association of itraconazole with potassium iodide in the above case?
5. Are recurrences of sporotrichosis in cats common? What factors favor its appearance?
6. What general guidelines and measures should be established to prevent the spread of sporotrichosis from this cat to other animals and its dermatozoonotic transmission?

Research Questions

1. What is the main species of *Sporothrix* spp. responsible for the major urban outbreak of feline sporotrichosis?
2. What are the main epidemiological and clinical aspects of feline sporotrichosis caused by the newly emerging species of *Sporothrix* spp.?
3. What is the best-recommended treatment for feline sporotrichosis?
4. What are the disease prevention and control protocols?

Further Reading

Alvarez CM, Oliveira MME, Pires RH (2022) Sporotrichosis: a review of a neglected disease in the last 50 years in Brazil. Microorganisms 10(11):2152. https://doi.org/10.3390/microorganisms10112152

Boechat JS, Pereira SA, de Sá Machado AC, Viana PG, Almeida-Paes R, Zancopé-Oliveira RM, Gremião IDF, de Oliveira MME (2021) Canine sporotrichosis: polyphasic taxonomy and antifungal susceptibility profiles of *Sporothrix* species in an endemic area in Brazil. Braz J Microbiol 52(1):135–143. https://doi.org/10.1007/s42770-020-00328-8

Bustamante B, Hidalgo JA, Campos PE (2017) Antifungal drugs. In: Mora-Montes HM, Lopes-Bezerra LM (eds) Current progress in medical mycology, 1st edn. Springer, pp 29–89

de Beer ZW, Duong TA, Wingfield MJ (2016) The divorce of *Sporothrix* and *Ophiostoma*: solution to a problematic relationship. Stud Mycol 83:165–191. https://doi.org/10.1016/j.simyco.2016.07.001

de Souza CP, Lucas R, Ramadinha RH, Pires TB (2016) Cryosurgery in association with itraconazole for the treatment of feline sporotrichosis. J Feline Med Surg 18(2):137–143. https://doi.org/10.1177/1098612X15575777

de Souza EW, Borba CM, Pereira SA, Gremião IDF, Langohr IM, Oliveira MME, de Oliveira RVC, da Cunha CR, Zancopé-Oliveira RM, de Miranda LHM, Menezes RC (2018) Clinical features, fungal load, coinfections, histological skin changes, and itraconazole treatment response of cats with sporotrichosis caused by *Sporothrix brasiliensis*. Sci Rep 8(1):9074. https://doi.org/10.1038/s41598-018-27447-5

Gremião ID, Miranda LH, Reis EG, Rodrigues AM, Pereira SA (2017) Zoonotic epidemic of sporotrichosis: cat to human transmission. PLoS Pathog 13(1):e1006077. https://doi.org/10.1371/journal.ppat.1006077

Gremião IDF, da Silva M, da Rocha E, Montenegro H, Carneiro AJB, Xavier MO, de Farias MR, Monti F, Mansho W, de Macedo Assunção Pereira RH, Pereira SA, Lopes-Bezerra LM (2021) Guideline for the management of feline sporotrichosis caused by *Sporothrix brasiliensis* and literature review. Braz J Microbiol 52(1):107–124. https://doi.org/10.1007/s42770-020-00365-3

Gremião IDF, Oliveira MME, Monteiro de Miranda LH, Saraiva Freitas DF, Pereira SA (2020) Geographic expansion of sporotrichosis, Brazil. Emerg Infect Dis 26(3):621–624. https://doi.org/10.3201/eid2603.190803

Han HS, Kano R (2021) Feline sporotrichosis in Asia. Braz J Microbiol 52(1):125–134. https://doi.org/10.1007/s42770-020-00274-5

Han HS, Toh PY, Yoong HB, Loh HM, Tan LL, Ng YY (2018) Canine and feline cutaneous screw-worm myiasis in Malaysia: clinical aspects in 76 cases. Vet Dermatol 29(5):442–e148. https://doi.org/10.1111/vde.12672

Kischkel B, Lopes-Bezerra L, Taborda CP, Joosten LAB, Dos Santos JC, Netea MG (2022) Differential recognition and cytokine induction by the peptidorhamnomannan from *Sporothrix brasiliensis* and *Sporothrix. schenckii*. Cell Immunol 378:104555. https://doi.org/10.1016/j.cellimm.2022.104555

Lopes-Bezerra LM, Mora-Montes HM, Zhang Y, Nino-Vega G, Rodrigues AM, de Camargo ZP, de Hoog S (2018a) Sporotrichosis between 1898 and 2017: the evolution of knowledge on a changeable disease and on emerging etiological agents. Med Mycol 56(Suppl_1):126–143. https://doi.org/10.1093/mmy/myx103

Lopes-Bezerra LM, Walker LA, Niño-Vega G, Mora-Montes HM, Neves GWP, Villalobos-Duno H, Barreto L, Garcia K, Franco B, Martínez-Álvarez JA, Munro CA, Gow NAR (2018b) Cell walls of the dimorphic fungal pathogens *Sporothrix schenckii* and *Sporothrix brasiliensis* exhibit bilaminate structures and sloughing of extensive and intact layers. PLoS Negl Trop Dis 12(3):e0006169. https://doi.org/10.1371/journal.pntd.0006169

Macêdo-Sales PA, Souto SRLS, Destefani CA, Lucena RP, Machado RLD, Pinto MR, Rodrigues AM, Lopes-Bezerra LM, Rocha EMS, Baptista ARS (2018) Domestic feline contribution in the transmission of *Sporothrix* in Rio de Janeiro State, Brazil: a comparison between infected and non-infected populations. BMC Vet Res 14(1):19. https://doi.org/10.1186/s12917-018-1340-4

Marimon R, Cano J, Gene J, Sutton DA, Kawasaki M, Guarro J (2007) *Sporothrix brasiliensis*, *Sporothrix. globosa*, and *Sporothrix. mexicana*, three new *Sporothrix* species of clinical interest. J Clin Microbiol 45:3198–3206

Rodrigues AM, Gonçalves SS, de Carvalho JA, Borba-Santos LP, Rozental S, Camargo ZP (2022) Current progress on epidemiology, diagnosis, and treatment of sporotrichosis and their future trends. J Fungi (Basel) 8(8):776. https://doi.org/10.3390/jof8080776

Santos AF, Azevedo MI, Amaral CI, Grom NA, Marinho F, de Oliveira CSF, de M Soares DF, MHF M, Brandão ST, Menezes RC, Ecco R (2023) Feline sporotrichosis: characterization of cutaneous and extracutaneous lesions using different diagnostic methods. Vet Pathol 29:3009858231189448. https://doi.org/10.1177/03009858231189448

Travassos LR, Lloyd KO (1980) *Sporothrix schenckii* and related species of *Ceratocystis*. Microbiol Rev 44(4):683–721. https://doi.org/10.1128/mr.44.4.683-721

Zhang Y, Hagen F, Stielow B, Rodrigues AM, Samerpitak K, Zhou X, Feng P, Yang L, Chen M, Deng S, Li S, Liao W, Li R, Li F, Meis JF, Guarro J, Teixeira M, Al-Zahrani HS, Pires de Camargo Z, Zhang L, de Hoog GS (2015) Phylogeography and evolutionary patterns in *Sporothrix* spanning more than 14 000 human and animal case reports. Persoonia 35:1–20. https://doi.org/10.3767/003158515X687416

Part IV

Systemic Fungal Zoonotic Infections?

Coccidioidomycosis 15

Pooja Gurram and Palash Samanta

Learning Objectives

By the end of the chapter, the readers should be able to:

1. Understand the current epidemiology and risk factors of coccidioidomycosis.
2. Learn the clinical manifestations and diagnostic modalities of coccidioidomycosis.
3. Identify the populations at high risk of dissemination.
4. Gain knowledge of management and preventive strategies of coccidioidomycosis.

Introduction

Coccidioidomycosis, also known as "Valley fever," named after California's San Joaquin Valley or "Desert rheumatism," is an endemic mycosis of the Western hemisphere caused by thermally dimorphic fungus, *Coccidioides* spp. It has been a significant problem in endemic areas. Over the years, the number of reported cases increased significantly secondary to rising awareness, improved diagnostics, an increase in the number of susceptible populations with an overall increase in the number of organ transplantations, and the use of newer biologics in the management of autoimmune disorders and increased reporting to the state. Coccidioidomycosis puts a significant economic burden on healthcare systems. This chapter describes the epidemiology, mycology, pathogenesis, risk factors, clinical features, diagnosis, management, and prevention strategies for coccidioidomycosis.

History

Coccidioides infection was first described by a medical student, Alejandro Posadas, in a skin lesion of a patient in Argentina in 1892 and shortly after in the United States (US) in a patient from San Joaquin Valley in California. The first laboratory exposure was recorded in 1929 when a medical student accidentally inhaled a positive culture plate, leading to pneumonia. This incident resulted in a series of significant research findings, including an epidemiological link between dust exposure and the development of coccidioidin skin testing. The first outbreak was reported in 1942 by a group of research students digging holes in search of rattlesnakes. Since then, the relationship between environmental disturbances and the causation of this disease has been established, and several outbreaks have been reported in the US and Latin American countries.

P. Gurram
Division of Infectious Diseases, University of Florida, Gainsville, FL, USA

P. Samanta (✉)
Transplant Infectious Diseases, Division of Infectious Diseases, University of Pittsburgh Medical Center, Pittsburgh, PA, USA
e-mail: samantap@upmc.edu

S. C. Parija, S. M. Rudramurthy (eds.), *Textbook of Fungal Zoonoses and Sapronoses*, Microbial Zoonoses, https://doi.org/10.1007/978-981-97-3214-2_15

Taxonomy

Coccidioides belong to the phylum Ascomycota, order Onygenales, and family Onygenaceae. The order Onygenales contains other pathogenic dimorphic fungi, *Paracoccidioides brasiliensis*, *Histoplasma capsulatum*, and *Blastomyces dermatitidis*. However, *Coccidioides* is the only pathogenic dimorphic fungi under the Onygenaceae family. Initially, it was wrongly identified as protozoa. Animal research in the 1900s led to its correct taxonomic classification. Genomic analysis suggests that *Coccidioides* possibly existed for 40–50 million years and split into two species: *Coccidioides immitis* and *Coccidioides posadasii*, approximately 5 million years ago. *C. immitis* is prevalent in California and eastern Washington State, whereas *C. posadasii* predominates in the rest of the southwestern states in the United States and most Latin American countries, including Mexico (Table 15.1). While these two species are genetically different, they are phenotypically similar, and there are no identifiable clinical differences in infections caused by them. Molecular methods are the only methods available to differentiate between these species.

Genomics and Proteomics

Given the broad range of disease severity, it is essential to understand the genotypes within the pathogen population and human host variations. This helps to understand the dynamics of outbreaks, identify the virulence factor, and predict the disease's progression and severity. The strain, originally isolated in 1951 from San Joaquin Valley, CA was submitted to American Type Culture Collection (ATCC) as a type strain for the genus (ATCC 28868). In 2001, based on genomic analyses, *C. immitis* and *C. posadasii* were officially designated as two distinct species, and the type strain at ATCC became *C. posadasii* strain Silveira. The whole genome of strain Silveira was sequenced in 2010. However, detailed genomic and proteomic studies are sparse for *Coccidioides* spp. due to the requirement of rigorous category-three laboratory biosafety conditions. A recent study provided insight into transcriptome sequencing (RNA sequencing) on this strain. This study revealed evidence for 837 genes expressed at the protein level, of which 169 proteins were putative proteins and 103 were not annotated in the Silveira genome. There is a recent discovery of hybrid strains of this organism. Other studies also showed that hybridization occurred between two species, resulting in differentially expressed transcripts between two species and morphological states. The ongoing studies would provide some insights into the molecular pathogenesis of coccidioidomycosis in the near future.

Table 15.1 Prevalence of distribution of *Coccidioides* spp.[a]

Coccidioides spp.	Global distribution
Coccidioides immitis	Northern and southern California, USA
Coccidioides posadasii	Arizona, Texas, California, USA Mexico South America

[a] This is based on phylogenetic analyses and population genomics data. Both species appear to be sympatric in Southern California and Mexico. The division is more blurred due to immigration of patients, who may have been infected in other areas and remain asymptomatic

Fungus Morphology

Coccidioides is a thermally dimorphic fungus that grows as hyphae in the soil in semiarid regions. The septate hyphae mature into barrel-shaped, single-celled arthroconidia or spores of the size 3–5 μm. Within the host, the arthroconidia initiate thermal parasitic cycles and convert them into spherules unique to the *Coccidioides* genus. When the spherule maturates, it reaches up to 100 μm and may contain 100–300 single-celled endospores. Endospores are released when the spherules rupture and each endospore can produce a spherule, thus continuing the parasitic cycle. Both hyphae and arthroconidia are asexual forms. Molecular data suggest the existence of a sexual phase.

Mating type alleles (MAT) identified in *Coccidioides* are typical of heterothallic ascomycetes. Studies do indicate that *Coccidioides* undergo sexual recombination. However, a sexual form has not been described for either species of *Coccidioides*.

Cultivation of Fungi

Coccidioides is listed as a "select agent" for potential bioterrorism. Given high-risk, laboratory-acquired infection, biosafety level 3 (BSL-3) containment is recommended for the manipulation of spores and soils containing or suspected to contain *Coccidioides* spores. The clinical tissue samples should be minced, not ground and the specimens should be cultured in tubes rather than plates. *Coccidioides* spp. is not a fastidious organism; thus, it readily grows in common media such as Sabouraud dextrose agar, brain heart infusion agar, or potato dextrose agar. The addition of cycloheximide suppresses common saprophytic fungi, thus making it a selective agent. Cultures are incubated at 25–35 °C. Cultures should be examined daily for 7 days, weekly thereafter for 3 weeks. Cultures usually become positive within 2–7 days as white, creamy colonies with concentric rings with filamentous areas (Fig. 15.1). The tease mount of colonies with lactophenol cotton blue (LPCB) stain shows thin, hyaline, septate hyphae, with side branches producing barrel-shaped arthroconidia. Isolates morphologically resembling *Coccidioides* spp. need to be confirmed using molecular methods.

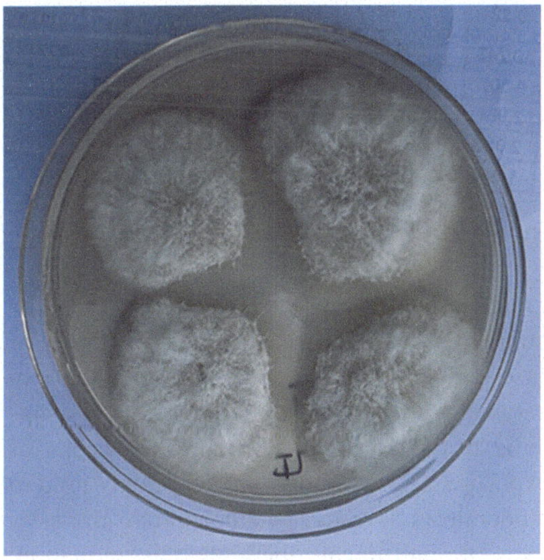

Fig. 15.1 *Coccidioides posadasii* forming creamy colonies with concentric rings on Sabouraud dextrose agar plate

Animal Models

Animal BSL-3 is recommended for animal studies for *Coccidioides* spp. Mice remain well-established animal models for various preclinical studies including antifungal efficacy. Mice are inoculated either by pulmonary, intravenous, intraperitoneal, intrathecal, or intracerebral route to study pathogenesis and host responses in coccidioidal infections. Other rodent species, such as rats, are also used, but much less frequently. Inoculation of rabbits via intracisternal injection is an efficient and reliable model for the study of coccidioidal meningitis/arteritis. The rabbit model mimics well for human coccidioidal meningitis to study host response as well as antifungal response. Cost and increased need for animal care limit the use of other larger animals such as dogs and nonhuman primates.

Life Cycle of the Fungus

Both *C. immitis* and *C. posadasii* have asexual life cycles. Sexual forms have not been identified. They exist in the environment in saprobic form and as a parasitic form in human hosts. Asexual forms include the mycelial stage, typically in rainy weather, which then develops arthroconidia during dry weather. Arthroconidia become airborne when the soil is disturbed and, when inhaled by a susceptible host, initiates the parasitic phase. Over the next 24–96 h, arthroconidia enlarge, divide by fission, and mature into endospores and spherules. Mature spherules rupture, releasing more endospores, which form more spherules in the host.

Pathogenesis and Pathology

The disease, coccidioidomycosis, is primarily transmitted by inhalation of arthroconidia, which are present in the soil. The loosely attached arthroconidia aerosolize very easily with disturbances in soil and are inhaled by humans or other animals such as rodents, dogs, cats, livestock, etc. Studies showed as few as 10–50 arthroconidia could cause disease in nonhuman primates in 4–6 weeks. The median incubation period is 4–31 days. Owing to their small size, the fungi can reach even up to the terminal bronchiole. Once inside the warm and moist environment of the host, the arthroconidia initiate thermal parasitic cycles and are converted into spherules, unique to the *Coccidioides* genus (Fig. 15.2). Each spherule contains 100–300 single-celled endospores. Spherules eventually rupture, releasing the endospores, which are then engulfed by alveolar macrophages resulting in local host response and acute inflammation. Each endospore can produce spherules, thus continuing the parasitic cycle in the host until controlled by the host immune system or therapeutic intervention. Occasionally, in a susceptible host, the spherules or endospores are carried by macrophages via hematogenous or lymphatic routes to extrapulmonary sites leading to disseminated disease. The primary route of infection is almost always pulmonary. However, primary cutaneous coccidioidomycosis rarely occurs if fungal spores are directly deposited through exposed skin. There are some reports of transmission from inanimate objects, such as cotton, hay, grain, fruit, and animal products originating from areas where *Coccidioides* spp. are endemic. There was also one report of intrauterine transmission of coccidioidomycosis during pregnancy.

Although several possible virulence factors are described in the literature, the exact virulence factor is largely unknown. Among the many virulence factors reported, the outer wall glycoprotein of the spherules, urease activity with ammonium production, and melanin production are most important in the pathogenesis of coccidioidomycosis.

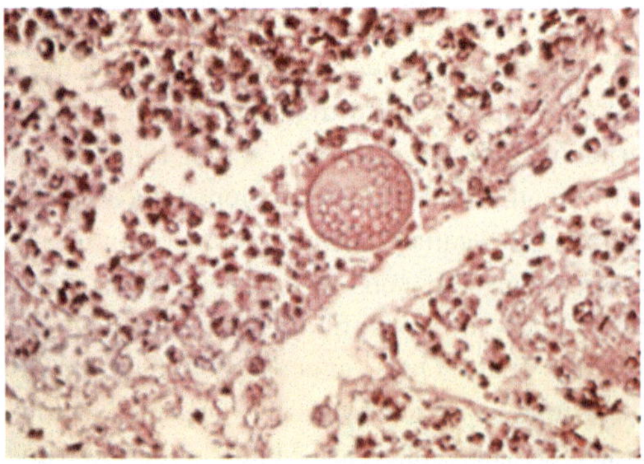

Fig. 15.2 Lung tissue with *Coccidioides* spherule. (Courtesy CDC)

Immunology

T-lymphocytes, particularly T helper 1 cells (Th1), are the mainstay of defense and confer long-term protective immunity. The effectors cells in innate immunity, such as natural killer cells, neutrophils, and monocytes, are also helpful in the early stages of infection when arthroconidia reach the lung tissue, when spherules are small, or when the spherules release the endospores. Eosinophilia may be present in this earlier stage of infection. In the later stages, lymphocytes and plasma cells predominate. Chronic granulomatous inflammation around the unruptured spherules is typically found in the infected host tissue without any progression of the disease.

Infection in Humans

Coccidioidal infections are asymptomatic or minimally symptomatic in up to two-thirds of cases. Pulmonary infections predominate in patients with symptomatic disease, with symptoms often developing within 1–4 weeks of exposure. The symptoms are commonly mild to moderate in severity, which improve irrespective of antifungal therapy. Dissemination, spread beyond lungs and pleura, occurs in <1% of cases and could involve any organ, but cutaneous, musculoskeletal, and central nervous system (CNS) involvement are more commonly seen. The risk of progression and dissemination increases several folds in those with immunosuppressing conditions, including but not limited to solid organ transplantation (SOT), hematopoietic transplantation, usage of immunosuppressants, human immunodeficiency virus/acquired immunodeficiency syndrome (HIV/AIDS) (with CD4 <250 cells/μL), and third trimester of pregnancy. The risk is also higher in those with African or Filipino ancestry, irrespective of immune status. Nevertheless, complications and disseminated diseases may occur in the absence of pulmonary signs and symptoms.

Pulmonary Infections

Coccidioidal Pneumonia

It is one of the common causes of community-acquired pneumonia in endemic areas and is often a challenge to differentiate from other etiologies based on clinical symptoms alone.

Fever and cough are the most common symptoms, followed by chest pain, shortness of breath, fatigue, weight loss, arthralgias, and headache. The presence of headache may indicate dissemination of the fungus to the meninges, but it typically improves with time and is not as progressive or persistent as in coccidioidal meningitis. In case of doubt, cerebrospinal fluid (CSF) examination helps to differentiate between the two entities. Cutaneous manifestations such as rash, erythema nodosum, or erythema multiforme may be part of early symptomatology and are immunological in nature. The triad of fevers, arthralgias, and erythema nodosum constitutes the characteristic desert valley fever/desert rheumatism. Symptoms eventually resolve in most cases, irrespective of antifungal treatment. Fatigue, however, is the last to improve, even being more disabling and debilitating at times. Radiographic findings may reveal unilateral or bilateral infiltrates, consolidations, cavities, parapneumonic effusions, and hilar adenopathy. Laboratory findings may be nonspecific, with leukocytosis and elevated inflammatory markers. The presence of eosinophilia is not universal, but, when present, is often suggestive of coccidioidomycosis and helps to differentiate it from pneumonia of other etiologies.

Other Pulmonary Manifestations

Most symptomatic infections in immunocompetent hosts are mild to moderate. However, in patients with underlying immunosuppression associated with defects in cell-mediated immunity, the presentation may be severe. The condition presents as diffuse pneumonia and manifests as acute respiratory distress syndrome (ARDS) and sepsis associated with fungemia. Morbidity and mortality are often high in such cases.

Pulmonary infections resolve in most cases but may develop nodules or cavities in up to 5% of cases. Nodules often are asymptomatic, do not require antifungal therapy, and are hard to differentiate from malignant nodules, even among patients from endemic areas. Positive coccidioidal serology is inadequate to rule out malignancy effectively. In such situations, a biopsy of the nodule for histopathology and cultures or follow-up with serial imaging is essential. The nodules usually resolve over a period of 2 years; other nodules may persist or evolve into a thin-walled cavity, often close to the periphery. Many of these cavities are asymptomatic, and their evolution is independent of antifungal therapy. They may manifest as pleuritic chest pain, hemoptysis, or develop superimposed bacterial infections or mycetoma (either with coccidioidal mycelia or other molds such as *Aspergillus* species). The thin-walled cavities are benign, and hence do not require any surgical intervention, but may require surgical intervention in the patients, in which the cavities rupture into the pleural cavity resulting in pneumothorax or empyema.

Some patients may develop chronic, fibrocavitary disease associated with thick-walled cavities, surrounded by pulmonary infiltrates. Patients are often symptomatic with fevers, cough, and weight loss. Diabetes mellitus, chronic tobacco use, and underlying pulmonary disorders are the risk factors. Prolonged antifungal therapy may be needed, often requiring surgical management.

Extra-thoracic Infections

Dissemination to extra-thoracic sites often occurs primarily by hematogenous spread, less frequently by lymphatics. Dissemination occurs within 2 years of infections, mostly within the first few weeks to months. The dissemination may occur beyond the 2 years to skin, soft tissue, or even the central nervous system (CNS), in patients with underlying immunosuppression or those who have been previously treated with antifungal therapy.

Cutaneous Infections

Cutaneous manifestations may be immunological or fungal invasion of the skin and soft tissue. The reactive form in primary pulmonary coccidioidomycosis manifests as erythema nodosum, erythema multiforme, acute exanthem, and rarely with Sweet's syndrome or granulomatous dermatitis. Organism-specific manifestations may occur following dissemination or direct fungal inoculation resulting in primary cutaneous coccidioidomycosis.

Skin and soft-tissue infection may be seen in up to two-thirds of the disseminated cases. The condition presents as maculopapular lesions, nodules, abscesses, ulcers, and verrucous lesions. The face, neck, scalp, chest wall, and extremities are the common sites. Biopsy and histopathological studies suggest granulomatous inflammation laden with spherules with endospores.

Musculoskeletal Infections

Bone and joints are other common sites of dissemination. Dissemination to the joints is secondary to fungal invasion, mostly asymmetric, and is associated with synovitis and effusion. Knees and vertebrae, predominantly lumbar, are the most common sites, although any bones or joints may be involved. Synovitis may be associated with adjacent bone involvement. Vertebral osteomyelitis and discitis may result in paravertebral, psoas, and epidural abscesses.

Central Nervous System Infections

Dissemination to the meninges is the most serious complication, with high morbidity and mortality, if left untreated. Basilar meningitis mimicking tuberculosis is the most common manifestation. Significant exudative inflammation of the meninges and subsequent fibrosis may result in cranial neuropathies and obstruction to cerebrospinal fluid (CSF) outflow. Persistent headaches, nausea, and vomiting are frequent. Asymptomatic involvement of CNS is uncommon; hence, CSF analysis is not recommended, even with dissemination of infections to other organs.

Communicating hydrocephalus, often associated with high mortality, is an important complication of coccidioidal meningitis. Fibrosis, after the resolution of active inflammation, may also result in hydrocephalus. Thus, it may present at any time during the illness. Patients may manifest altered sensorium, gait abnormalities, nausea, vomiting, and other symptoms that frequently overlap with meningitis. Intracerebral abscesses and cerebral vasculitis with resultant stroke syndrome are rare manifestations. Other rare manifestations include arachnoiditis and syringomyelia, which may present with neuropathies and radiculopathies.

Endovascular Infections

Endovascular coccidioidomycosis is exceedingly rare. A high degree of suspicion is required for any early diagnosis, especially in patients from endemic areas presenting with culture-negative endocarditis. Only a very few cases of endovascular infections have been reported so far.

Infection in Animals

Coccidioides spp. can infect a variety of rodents, mammals (dogs, cats, sheep, cattle, pigs, horses, dolphins, sea lions), and reptiles. Dogs and cats are most commonly affected. The clinical manifestations are variable. The condition varies from asymptomatic to symptomatic infections including disseminated infections, which are often fatal. There have been reports of placental infections leading to abortions.

Epidemiology and Public Health

Coccidioidomycosis is endemic in the southwestern United States, southcentral Washington, northern Mexico, and some parts of Latin America, including Brazil, Argentina, Venezuela, Colombia, Paraguay, Bolivia, Honduras, Guatemala, and Panama (Table 15.1) (Fig. 15.3). Historically, coccidioidin skin testing has been helpful in determining the endemicity of coccidioidomycosis. In the United States, the Central Valley of California and Arizona are the most endemic areas. However, endemicity has been established sporadically in other areas such as Utah, Nevada, Texas, and New Mexico. These endemic regions have similar environmental conditions characterized by arid/semiarid conditions, alkaline soil, low precipitation, and gentle winter.

Apart from the endemic areas, the fungus was also isolated in the soil of Washington and Oregon States. Thus, the possibility of endemicity in the surrounding areas to endemic states cannot be ruled out completely, especially while managing high-risk individuals. Indeed, spherules were identified histopathologically in Bison excavated from Nebraska, far away from the current area of endemicity. According to genetic analysis, *Coccidioides* spp. were probably introduced by mammals from North America to South America around 9000–14,000 years ago. In the background of globalization and tourism, cases of coccidioidomycosis have also been reported in Canada, Europe, Australia, and Asia, known to be non-endemic for the disease. These cases are often misdiagnosed or diagnosed late in the disease course as this diagnosis is often not in the differential in the non-endemic areas. Thus, travel history is extremely important to suspect coccidioidomycosis as a possible diagnosis, especially in pneumonia or other invasive diseases not responding to standard antimicrobial therapy.

The risk of coccidioidomycosis steadily increases with advancing age, with the highest incidence of 209 cases per 100,000 population over the age of 70 years to 7.7 cases per 100,000 population in the age group of 1–4 years. Men

Fig. 15.3 World map showing the global distribution of coccidioidomycosis

appear to have a higher incidence and risk than women. The incidence and risk of dissemination are also higher amongst patients with African, Filipino, or Hispanic ancestry. Patients with different immunosuppressive conditions and on immunosuppressive medications are also at higher risk of *de-novo* infection or reactivation and dissemination. In the study mentioned earlier, the two largest outbreaks among the 47 outbreaks from 1940 to 2015 occurred after an earthquake (Northridge earthquake in 1994) and a large dust storm (Bakersfield dust storm in 1977), respectively, in California, USA. Thus, environmental factors appear to play a significant role in the incidences of coccidioidomycosis. The studies also showed that more than one-third of outbreaks of the infection occurred in areas not known to be endemic and half of the outbreaks happened after occupational exposures. Different risk factors for coccidioidomycosis have been summarized in Table 15.2.

The incidence of coccidioidomycosis and the reports of outbreaks are rising, especially in the last two decades. One recent study described 47 coccidioidomycosis outbreaks resulting in 1451 symptomatic and 13 asymptomatic cases worldwide from 1940 to 2015. In this study, the outbreak was defined as ≥2 human cases linked to a common source, activity, or event. A total of 85% of outbreaks were associated with environmental exposure. Clinical attack rates ranged from 0.03% to 100% (median 43%). Although there is substantial year-to-year variation in incidences, nearly 10,000 cases are reported annually in the United States and >95% of cases are reported from Arizona and California only. The true incidences of coccidioidomycosis are most likely underreported as most of the cases remain undiagnosed due to the asymptomatic nature of the infection. One study estimated that the actual number of cases is probably 6–14 times higher than the reported cases. In the US, as of 2017, coccidioidomycosis is reportable in 22 states. The Center for Disease Control and Prevention (CDC) receives cumulative case counts and data

Table 15.2 Risk factors of coccidioidomycosis

Host factors	Male gender
	Age ≥40 years
	Ethnicity
	  African American
	  Filipino
	  Hispanic
	Immunocompromised hosts
	HIV
	Diagnosis of AIDS
	CD4 count ≤250/μL
	Organ transplantation
	History of prior coccidioidal infection
	Solid organ transplantation
	Bone marrow transplantation
	Immunosuppression augmentation for rejection treatment
	De novo infections
	Donor-derived infections
	Patients with hematological malignancy on chemotherapy
	Other immunosuppressive treatments
	  Steroids
	  Anti-cytokine therapy
	  TNF-α inhibitors
	  Biologics
	Other medical conditions
	  Pregnancy
	  Diabetes mellitus
	  Cardiopulmonary diseases
	Occupational hazards
	  Construction workers
	  Outdoor activities (Hunting/sports)
	  Archeological/field study workers
	  Laboratory exposure
	Temporary move from non-endemic areas to endemic areas
	Incarceration
	Tourists
	Military posting
	Specific gene mutations
	818del4 mutation in gene encoding IFN-γ receptor
	Mutation encoding IL-12 receptor B1

Environmental factors	Earthquake Dust storms Seasonal precipitation

HIV human immunodeficiency virus, *AIDS* acquired immunodeficiency syndrome, *TNF* tumor necrosis factor, *IFN* interferon, *IL* Interleukin from different states through the National Notifiable Diseases Surveillance System (NNDSS).

The potential role of animals such as rodents in the disease transmission process is debated. Early studies noted the correlation between infections among people and animals digging in and around rodent burrows leading to a concept that desert rodents could be a reservoir of *Coccidioides* spp. in the environment. However, there is no evidence that *Coccidioides* spp. will pass through the guts of predators of desert rodents. It is also doubtful if transmission of coccidioidomycosis happens from animals to humans. This has seldom been reported in the literature. There was one report of possible transmission to humans via inhalation due to manipulation and aerosolization of endospores during the autopsy of a horse with disseminated diseases. There was another report of questionable transmission by the bite of a cat with a documented, florid, disseminated coccidioidomycosis. In this case, it was postulated that spherules and/or endospores might be present in the oral/respiratory secretion of the cat, leading to local inoculation in the tissue.

Coccidioidomycosis also has a significant impact on healthcare costs. One study estimated that the average lifetime healthcare cost was ~$57,000 per patient for all disease manifestations, ~22,000 USD per patient for uncomplicated pneumonia, ~132,000 USD per patient for chronic pneumonia, and ~$1,000,000 per patient for meningitis/disseminated diseases. There was also a significant loss of school and work hours.

Diagnosis

Coccidioidomycosis has protean clinical manifestations mimicking many other diseases. Diagnostic confirmation is essential for decisions regarding the management of acute presentation and future implications, such as the risk of dissemination and follow-up.

A definitive diagnosis could be made by identifying the spherules with endospores or by culturing the organism from tissue samples or other clinical specimens. However, serological diagnosis is more commonly used, including in ambulatory settings.

Microscopy

Dimorphic *Coccidioides* spp., appears as spherules in tissue specimens and as mycelia in the culture. The spherules containing endospores (Fig. 15.2), the pathognomonic of coccidioidal infection, are possible on bronchoalveolar lavage (BAL) fluid or histopathological analysis of tissue specimens. Calcofluor white stain of potassium hydroxide (KOH) preparation, Grocott's methenamine silver (GMS) stain, and periodic acid Schiff (PAS) stains are helpful in the identification of spherules. Mycelial forms may rarely be seen in patients with lung cavities.

In Vitro Culture

The culture of BAL or tissue samples is often necessary for severe infection to make a definitive diagnosis, particularly if serology is nondiagnostic. *Coccidioides* spp. typically grow well on bacteriologic or fungal media within 7 days of incubation. It is hard to differentiate the mycelial forms of *Coccidioides* spp. by morphology alone. Ribosomal RNA probes may be used to identify it and, more recently, by matrix-assisted laser desorption/ionization time-of-flight mass spectrometry (MALDI-TOF). Currently, identification of *C. immitis* from *C. posadasii* is not routinely done unless needed for epidemiological purposes, and this is available in reference laboratories only. Identification of different species does not have any clinical implications in terms of management or prognosis.

Skin Testing

Coccidioidal skin testing is based on delayed-type hypersensitivity reaction, which requires active cellular immunity. A positive test does not differentiate between an active from a past infection. It might be a useful preventive strategy in determining who might be at risk for coccidioidomycosis in residential facilities in endemic areas. Skin testing is often used in assessing the risk of developing the active disease in cases of laboratory exposures.

Serodiagnosis

Serodiagnosis is the most common method of diagnosing coccidioidomycosis in patients with clinical symptoms and exposures suggestive of it. The most common methods currently in use are immunodiffusion (ID) and the complement fixation test (CFT). Other serological tests include tube precipitin (TP) and enzyme immunoassay (EIA).

Antibodies against *Coccidioides* spp. typically begin to appear 1–3 weeks after symptom onset for IgM and up to 7 weeks after for IgG antibodies. Antibodies, including IgG, eventually disappear with time, unlike other infections where IgG antibodies may last longer even after clinical improvement. Importantly, the absence of antibodies does not rule out coccidioidomycosis. They may be negative early in the illness, signifying the need to repeat the tests in a few weeks. Serology testing can also be negative in up to 20% of severely immunosuppressed patients.

EIA detects both IgM and IgG antibodies. This test becomes positive earlier in the illness and has very high sensitivity but may have false positives, especially if IgM is the only positive test. It is often used as an initial screening test. Still, it is necessary to have a confirmatory test with either ID or CF. Patients with discordant results, with positive EIA, but negative ID or CF could either have a false positive test or an early infection. Clinical suspicion for coccidioidomycosis helps in interpreting it in such scenarios. If the concern for coccidioidomycosis is more serious, it is prudent to consider it a true positive. One may repeat ID or CFT in a few weeks or opt for direct identification or culture.

The TP test evaluates for the ring of precipitation when the patient's serum is incubated with coccidioidal test antigen. It is often quoted as an IgM test as an IgM antibody with its high avidity to antigens results in the formation of a precipitation ring. ID has now replaced this in many laboratories, which use a similar mechanism.

CFT predominantly detects IgG antibodies. The patient's serum, test antigen, and complements are incubated, followed by the addition of sensitized red blood cells. Lysis of red blood cells is considered a negative test, while the absence of lysis is positive. The presence of IgG antibodies against *Coccidioides* spp. results in the formation of immune complexes and subsequent consumption of complements. Thus, when sensitized red cells are added, there are no complements left to lyse the cells. Once the qualitative test is positive, it can be followed by a quantitative test for a titer that uses doubling dilutions of the serum. The last dilution with the absence of lysis of red cells gives the titer of IgG antibodies. This test is of diagnostic and prognostic significance, as higher the titer higher the risk of dissemination or severe disease. A titer of >1:16 is commonly used as the cutoff associated with a higher risk of dissemination, but this may not always be the case as dissemination can occur with lower titers, and higher titers may be seen in diseases limited to the thorax. CF titers normally peak within 2–3 months of exposure and begin to decline thereafter. Monitoring titers are useful for assessing the progress of the infection, with worsening titers after initial improvement signifying persistence or relapse of the disease. CF tests can be used in CSF and synovial samples as well. This is more useful in diagnosing coccidioidal meningitis than CSF cultures, as the latter has the limitation of low sensitivity.

ID tests can detect both IgG and IgM and are one of the most commonly used serological tests in addition to CFT methods. Testing for coccidioidal antigen could be done in serum, urine, and CSF. It tends to be positive in disseminated and severe infections rather than in mild-to-moderate pulmonary disease.

Molecular Diagnosis

Molecular diagnostics include real-time polymerase chain reaction (RT-PCR) and antigen testing. Culture remains the gold standard for the diagnosis of coccidioidomycosis. However, the turnaround time is up to 3 weeks and may delay the initiation of appropriate antifungal therapy. Moreover, laboratory personnel may be at risk of infection if adequate biosafety level 3 precautions are not taken. Identification of *Coccidioides* spp. by PCR is sensitive, specific, and rapid. It may also be performed on formalin-fixed tissue. Molecular diagnostics may, thus, help overcome some of the difficulties faced with other diagnostic modalities.

PCR and coccidioidal antigen testing are not widely available and are limited to reference laboratories.

Supportive Diagnostics

Other modalities include lumbar puncture with CSF analysis in patients with symptoms suggestive of CNS or spinal cord involvement, magnetic resonance imaging (MRI), computed tomography (CT), and bone scans depending on the symptomatology. In patients with CNS manifestations, CSF analysis generally reveals pleocytosis with lymphocyte predominance but may have neutrophilic predominance in early stages, elevated protein, and low glucose. Eosinophils may be elevated, and their presence increases the likelihood of it being coccidioidal meningitis. Testing for antibodies via CF of CSF is more sensitive than CSF cultures.

Diagnostic modalities for coccidioidomycosis are summarized in Table 15.3.

Table 15.3 Diagnostic approaches in coccidioidomycosis

Diagnostic modality	Characteristics
Direct examination of sputum/BAL/tissue specimens using calcofluor white, GMS, PAS stains	Spherules (50–100 μm) with up to 100–300 endospores each
Culture of sputum/BAL/tissue specimens	Mycelia (identification using ribosomal RNA probes or MALDI-TOF)
Serology	Screening tests EIA (both IgM and IgG) – Sensitivity > specificity Confirmatory tests Immunodiffusion assay – IDTP (IgM) – IDCF (IgG, like CF) CFT – Qualitative and quantitative – Diagnostic and prognostic significance – Can be done in serum, CSF, and synovial fluid – CSF titers more sensitive than CSF cultures in meningitis
Molecular diagnostics	Real-time PCR (sputum, BAL, tissue specimens) Coccidioidal antigen (serum, urine, CSF, synovial fluid)
Skin testing	Delayed-type hypersensitivity Cannot differentiate past versus present infection May help determine the risk of future infection in endemic areas
Supportive diagnostics	Imaging methods (CT, MRI) Lumbar puncture with CSF analysis (pleocytosis with lymphocytic predominance but could be neutrophilic in early stages, elevated protein, lower glucose, possible eosinophils) Eosinophilia (supportive but not mandatory)

BAL bronchoalveolar lavage, *RNA* ribonucleic acid, *MALDI-TOF* matrix-assisted laser desorption/ionization time of-flight mass spectrometry, *EIA* enzyme immunoassay, *IDTP* immunodiffusion tube precipitin, *IDCF* immunodiffusion complement fixation, *CF* complement fixation, *PCR* polymerase chain reaction, *CSF* cerebrospinal fluid, *CT* computed tomography, *MRI* magnetic resonance imaging, *GMS* Grocott's methenamine silver, *PAS* periodic acid Schiff

Treatment

Antifungals Active Against Coccidioidomycosis

Azoles and polyenes are active against coccidioidomycosis. Current Food and Drug Administration (FDA)-approved antifungals against coccidioidal infections are ketoconazole and amphotericin-B, which are associated with significant adverse effects. Ketoconazole has been replaced predominantly by fluconazole, while other azoles such as itraconazole, voriconazole, and posaconazole are also being used. Amphotericin-B deoxycholate has largely been replaced by liposomal amphotericin-B in areas where the supply is not limited. However, amphotericin-B deoxycholate is still used intrathecally in severe or refractory coccidioidal meningitis cases or coccidioidal meningitis in pregnant women in earlier trimesters.

Fluconazole is the most common azole used for coccidioidomycosis, including meningitis. It is readily available, affordable, well-tolerated, and has excellent bioavailability and tissue penetration with good CSF levels. It is often the drug of choice for coccidioidomycosis in mild-to-moderate cases, except for skeletal infections, where itraconazole may be preferable. There was no statistically significant superiority of itraconazole over fluconazole in the randomized trial. Still, a trend towards higher efficacy and lower relapse rates was noted with itraconazole, especially with skeletal infections. One caveat is that the fluconazole dose used in the trial was 400 mg. Thus, higher doses may have better efficacy and outcomes, but this has not been studied.

Itraconazole has poor CSF penetration but appears to be efficacious in treating meningitis. The drawback is its poor absorption and dependency on food and gastric pH for the capsule form, and more gastric side effects and higher cost with the solution form. These issues may be overcome with SUBA (super bioavailable) itraconazole, but this is not widely available.

Voriconazole and posaconazole have been used in refractory cases of coccidioidomycosis and appear to be efficacious. Voriconazole has good CNS penetration and has been used for meningitis. Experience with posaconazole in meningitis is limited, but success has been reported. Isavuconazole, the newest azole on the market, has been used for other fungal infections with CNS involvement. Although its experience in coccidioidomycosis is limited to case series, it may be used in refractory cases.

All azoles have the common side effects of hepatotoxicity and QTc prolongation except for isavuconazole, which shortens QTc. Fluconazole has cutaneous side effects such as xerosis and alopecia. In contrast, voriconazole has the potential for visual disturbances, neurotoxicity, and other long-term consequences such as the increased risk of cutaneous malignancies and fluorosis/periostitis. Itraconazole may be associated with negative inotropic effects, and both posaconazole and itraconazole may lead to pseudo-hyperaldosteronism. All azoles are associated with drug–drug interactions of varying degrees depending on their potential to inhibit various cytochrome (CYP) enzymes. Assessing these interactions is of utmost importance, especially in immunosuppressed patients on calcineurin inhibitors. Therapeutic drug monitoring is often recommended for itraconazole, voriconazole, and posaconazole, while it is not typically recommended for fluconazole, given its predictable pharmacokinetics. There isn't enough data on isavuconazole monitoring for its widespread use.

Polyenes are often used in severe or refractory cases of coccidioidomycosis. Amphotericin-B deoxycholate use has largely been replaced by liposomal formulations, given their better side effect profile. However, for severe and refractory CNS infections, intrathecal amphotericin-B deoxycholate is used as intravenous administration has poor CNS penetration. Polyenes are also used for coccidioidomycosis in all types in pregnant women in earlier trimesters, given the teratogenic potential of azoles.

Antifungal susceptibility testing for coccidioidomycosis is not routinely performed and is often reserved for refractory cases. An in vitro study was done to assess the minimal inhibitory concentrations (MICs) of the available azoles, and amphotericin-B was tested on various isolates referred to the mycology laboratory at the University of Texas, San Antonio, USA. Several isolates had higher MICs against fluconazole when compared to other mold-active azoles (voriconazole, posaconazole, and itraconazole) and amphotericin B. However, the clinical significance of this finding is yet to be determined. This study supports the recommendation of higher doses of fluconazole for treating coccidioidomycosis and a switch to alternative azoles in cases refractory to fluconazole.

Indications for Antifungal Therapy in Coccidioidomycosis

Many patients with uncomplicated pulmonary coccidioidomycosis recovered without complications or dissemination in the pre-azole era. Amphotericin-B deoxycholate-associated toxicity was high enough not to warrant therapy for uncomplicated cases. Since fluconazole was introduced, many patients in the endemic area have been treated for uncomplicated coccidioidomycosis as well, given the safety profile of the drug. However, there are no randomized trials comparing azole therapy versus clinical monitoring in this population. An observational study comparing patients with pulmonary coccidioidomycosis treated with antifungals versus clinical observation alone showed that the groups did not differ in terms of time to symptom resolution. Moreover, some of the patients in the antifungal group had relapsed, and one had dissemination with meningitis after stopping antifungals. Another study found that patients treated with antifungals may not develop an IgG response. This raises the concern that antifungals may blunt the immune response to *Coccidioides* spp. and may result in delayed dissemination or relapse once the drug is discontinued. Thus, the decision to treat uncomplicated cases must be individualized.

Infectious Diseases Society of America (IDSA) guidelines recommend treating patients with pulmonary coccidioidomycosis based on the severity of symptomatology and underlying risk factors for severe disease and risk of dissemination. Patients with severe symptoms requiring hospitalization, weight loss of >10%, protracted night sweats, bilateral infiltrate or greater than 50% infiltrates in one lung, and persistent hilar adenopathy are some of the symptoms that warrant treatment. Treatment is also recommended in patients with diabetes mellitus and other immunosuppressing conditions such as SOT, hematopoietic transplantation, HIV/AIDS, third-trimester pregnancy, immunosuppressive medication use, African or Filipino ancestry, and those with high CF titers (>1:16). All cases of extrapulmonary coccidioidomycosis are to be treated as well.

The decision regarding azole versus polyene use depends on the severity or refractoriness of the disease. Duration of therapy varies based on pulmonary versus extrapulmonary disease and is described in Table 15.4. Surgical management is often needed in patients with chronic cavitary disease, empyema, skeletal disease, and hydrocephalus. Patients with coccidioidal meningitis require life-long fluconazole, given that the risk of relapse is up to 80% after stopping therapy. Long-term follow-up of up to 2 years or longer is essential to monitor for signs of dissemination and recurrence.

Extra-thoracic dissemination may occur within the first 2 years of infection, often within the first few months. However, in those treated with antifungals for primary pulmonary coccidioidomycosis, this may occur years after stopping antifungals. Moreover, the risk of relapse after stopping antifungals is higher in patients with extra-thoracic disease, reaching up to 80% in CNS disease. Hence, close clinical, radiological, and serological monitoring is essential in all cases of coccidioidomycosis. The exact duration of follow-up is uncertain and variable, with a minimum of 2 years for those who are untreated to up to a year or longer after stopping therapy in those treated with antifungals.

Table 15.4 Management of coccidioidomycosis

Coccidioidomycosis clinical spectrum	Management/anti-fungal therapy	Duration of treatment
• Pulmonary coccidioidomycosis • Uncomplicated with mild to moderate symptoms in immunocompetent hosts • Uncomplicated pulmonary coccidioidomycosis in diabetics or other immunosuppressing conditions[a,] African or Filipino ancestry • Uncomplicated pulmonary coccidioidomycosis with the following symptomatology: – Requiring hospitalization – >10% weight loss – Protracted night sweats – >50% infiltrates in one lung or bilateral infiltrates – CF titers >1:16 • Severe complicated pneumonia or diffuse pneumonia/ARDS • Pulmonary nodule • Pulmonary cavity: thin walled and asymptomatic • Pulmonary cavity : chronic symptomatic cavitary pneumonia	Clinical monitoring and close follow up for up to 2 years for early recognition of progression or dissemination Fluconazole (no consensus on the dose but not lower than 400 mg per day) or itraconazole (200 mg twice daily) Fluconazole (no consensus on the dose but not lower than 400 mg per day) or itraconazole (200 mg twice daily) Liposomal amphotericin B (3-5 mg/kg per day) intravenously Essential to rule out malignancy. No antifungal therapy indicated No antifungal therapy indicated. Fluconazole or itraconazole Intravenous liposomal amphotericin B in severe cases Surgical management may be needed in cases unresponsive to antifungal therapy	3–6 months 3–6 months Up to 12 months May need prolonged course of therapy
Skin/soft tissue coccidioidomycosis	Fluconazole or itraconazole Alternative azoles in refractory cases[b] Severe cases may require intravenous liposomal amphotericin B	6–12 months
Musculo-skeletal coccidioidomycosis	Itraconazole or fluconazole (higher doses preferred with fluconazole dose at least 800 mg daily) Alternative azoles in refractory cases[b] Severe cases with concern for limb or cord compromise or synovial involvement require intravenous liposomal amphotericin B Early surgical consultation	3 years to lifelong (based on severity)
Meningeal coccidioidomycosis Hydrocephalus	Fluconazole (higher doses of 800–1200 per day, adjusted per renal function) or itraconazole Alternative azoles in refractory cases[b] Intrathecal amphotericin-B deoxycholate in refractory cases (only in experienced centers) Liposomal amphotericin-B intravenously may be useful (limited experience) Early neurosurgical interventions for shunt placement in addition to antifungal therapy	Duration of treatment depends on clinical improvement and is often prolonged. Lifelong secondary prophylaxis is essential and treatment doses of azoles are often continued.

[a] Solid organ transplant and hematopoietic transplant recipients, HIV/AIDS, pregnant women in the third trimester, and patients on immunosuppressive medications

[b] Alternative azoles in refractory cases include voriconazole, posaconazole and isavuconazole. Limited experience with the latter two in meningeal disease

Prevention and Control

Immunity developed after a natural infection usually persists lifelong, preventing reinfection. This indicates that the vaccine could be effective. Although there are studies ongoing in animal models, there is no vaccine available so far. Prevention efforts should be targeted to the population at risk for coccidioidomycosis or those at higher risk for reactivation, severe or disseminated disease. Studies have suggested that dust control measures, including planting grass, paving, and applying oil to the soil in athletic areas, might help in reducing the risk of infection. Serology testing could be an essential tool to screen out persons at risk of acquiring the infection. As a measure to prevent infection, inmates with immunosuppression of African or Filipino descent are not housed in prisons located in endemic areas for coccidioidomycosis. California Department of Public Health and the United States Geological Survey recommend strategies for construction workers and supervisors such as education to increase awareness, use of respiratory protection, and dust control measures, e.g., wetting soil before disrupting and proper cleaning of transported materials. In addition to the challenges in implementing these measures, their efficacy is yet to be determined. Laboratory exposures have been reported as positive *Coccidioides* cultures are highly infectious. Laboratories must report if there is clinical suspicion of coccidioidomycosis so that proper precautions can be taken while handling clinical specimens and culture plates in biosafety cabinets by trained staff. *Coccidioides* are also listed as biological select agents and toxins. Thus, their handling, storage, culture, and transport need registration with CDC.

Prevention of Coccidioidomycosis in Patients at Risk of Dissemination

Solid Organ Transplantation Recipients

The incidence of coccidioidomycosis after SOT varies from 1.4% to 6.9% in endemic regions. SOT recipients with coccidioidomycosis also have a heightened risk of extrapulmonary dissemination of up to 75% compared to fewer than 1% risk in immunocompetent persons. As per the American Society of Transplantation guidelines, all transplant candidates should be interviewed for a history of residence in or travel to an endemic region. If history is suggestive, the candidate should undergo a thorough evaluation for active symptoms of coccidioidomycosis and serology testing. EIA is the preferred screening test given its higher sensitivity. If there is any evidence of prior infection or positive serology, patients should be evaluated by infectious disease specialists to determine candidacy and effective prophylactic measures. Given the degree of immunosuppression for grafts survival is at its peak during the first year of transplant, the incidence of coccidioidomycosis appears to be higher during this period. Coccidioidomycosis in this population could be due to reactivation, de novo disease, and rarely donor-derived. Given the high rates of dissemination and mortality associated with coccidioidomycosis in organ recipients, fluconazole prophylaxis is recommended in all organ-transplant recipients in endemic areas. The duration is variable and could be from 6 to 12 months. Those who develop active coccidioidomycosis after transplant or recipients with recent active disease prior to transplant and positive serology may need life-long secondary prophylaxis. All types of coccidioidomycosis in transplant recipients must be treated with antifungals, including uncomplicated pulmonary coccidioidomycosis. A reduction in immunosuppression may be essential to control the disease effectively.

Patients with active coccidioidomycosis pretransplant must have a symptomatic resolution, imaging resolution or stability, and serological response before being considered for transplantation.

Hematopoietic Transplant Recipients

The risk of disseminated coccidioidomycosis in patients undergoing hematopoietic transplantation is like that of SOT recipients. Thus, the current recommendations are to give azole prophylaxis for up to 2 years or longer, depending on the degree of immunosuppression.

People Living with HIV

Before the advent of potent combined antiretroviral therapy (cART), morbidity and mortality associated with coccidioidomycosis in people living with HIV/AIDS were extremely high, given the depressed cellular immunity, which is paramount to controlling the disease. In addition to the risk of dissemination in patients with CD4 <250 cells/μL, coccidioidomycosis may present as a severe diffuse pulmonary disease indistinguishable from pneumocystis pneumonia and is associated with high mortality despite antifungal treatment. In the era of cART, control of HIV viral load and increase in CD4 count have improved outcomes to the point that patients with HIV and a CD4 count higher than 250 cells/μL have similar outcomes as immunocompetent persons.

If patients with newly diagnosed HIV are started on appropriate cART, there is no indication for antifungal prophylaxis based on available literature. It is recommended that they be monitored by serology and chest radiographs on an annual basis, with the initiation of azole therapy in patients with positive serology and CD4 <250 cells/μL.

In patients with active coccidioidomycosis with a CD4 of <250 cells/μL, antifungals are continued for up to 6–12 months or longer based on the severity of disease, dissemination and until the HIV viral load is suppressed on potent cART and CD4 count is ≥250 cells/μL, while those with CD4 ≥250 cells/μL at the time of active coccidioidomycosis are treated the same as immunocompetent hosts.

Pregnant Women

The risk of disseminated disease is higher in the later trimesters of pregnancy, while azoles may be teratogenic when used in the first trimester. Thus, treatment of coccidioidomycosis in pregnancy is dependent on the stage of pregnancy in addition to the severity of the disease. Patients in the first trimester with uncomplicated mild coccidioidomycosis may be monitored with close follow-up. However, any severity of coccidioidomycosis in later trimesters warrants therapy, given the risk of dissemination. Intravenous liposomal amphotericin-B is the preferred drug during the first trimester for non-meningeal diseases. Azoles may be used in later trimesters for mild-to-moderate disease with liposomal amphotericin-B in severe cases. Intrathecal amphotericin-B is the preferred drug in coccidioidal meningitis in the first trimester.

In conclusion, coccidioidomycosis continues to be a significant threat to public health. The fungus could be present in the soil of many arid areas throughout the US and Latin America outside the defined endemic region. Thus, monitoring outbreaks is crucial to identify emerging epidemiology and assess if new areas of endemicity could be identified. A better understanding of epidemiology would enable a better assessment of public health risks and their mitigation in the future. Awareness among employers in high-risk occupations and providers is also critical to reducing the incidence of coccidioidomycosis and future outbreaks. Clinical manifestations, severity, and dissemination depend on the host and the virulence factors of the fungus. Therapeutic management is based on the severity of the infection and the host's susceptibility. Effective prevention measures, including the development of vaccines, are the need of the hour to reduce the overall burden of the disease.

Case Study

A 45-year-old Caucasian man with diabetes and HIV infection presented to the emergency department with fevers, progressively worsening headaches, nausea, vomiting, and confusion for 3 weeks. He was diagnosed with HIV infection 3 years ago and was started on bictegravir/emtricitabine/tenofovir alafenamide. However, he was lost to follow-up and has been noncompliant with his medication. The current CD4 count was 80 cells/μL with a viral load of 80,000 copies/mL. He lived in Arizona, USA, until a year ago, after which he moved to Pennsylvania, USA. Information regarding prior serologies or anti-infective prophylaxis was unavailable at the time of HIV diagnosis. A complete blood picture showed eosinophilia (1500 cells/μL). Erythrocyte sedimentation rate and c-reactive protein were elevated. A CT scan of the head showed no evidence of mass lesions. Lumbar puncture with CSF analysis showed elevated opening pressure, lymphocytic pleocytosis, high eosinophil count, protein, and depressed glucose. CSF Gram stain, bacterial culture, cryptococcal antigen, and Venereal Disease Research Laboratory (VDRL) were negative. Given prior residence in an area endemic to coccidioidomycosis, serum EIA, ID/CF studies, and CSF complement fixation were done. Serum EIA was positive with a CF titer of 1:64. CSF CF antibodies were also positive, confirming the diagnosis of coccidioidal meningitis. CSF fungal cultures remained negative. MRI brain had signs of basilar meningitis. He was initiated on high-dose fluconazole of 800 mg/day along with placement of a VP shunt for hydrocephalus. Although delaying cART to prevent immune reconstitution, the inflammatory syndrome is controversial in coccidioidomycosis, given the CNS involvement and a CD4 of <100 cells/μL, he was resumed on cART 4 weeks after initiating antifungal therapy. CF titers 12 weeks after starting treatment decreased by twofold. His symptoms continued to improve, and he has been compliant with cART. His CF titers were undetectable after 1 year, along with suppressed HIV viral load with a CD4 of 150 cells/μL on cART. He was continued on high-dose fluconazole indefinitely, given a higher than 80% risk of relapse with discontinuation.

Research Questions

1. What are the robust surveillance methods that are needed to identify outbreaks, especially after natural disasters and outside of historically endemic regions?
2. How can we identify if other species of *Coccidioides* or any hybrid species exist?
3. What are the investigation pathways to investigate virulence factors, molecular pathogenesis, and identification of genotypic and phenotypic markers to predict severe or disseminated diseases?
4. How can specificity and sensitivity in diagnostics be improved in the early diagnosis of coccidioidomycosis?
5. Can we conduct randomized clinical trials to help guide treatment plans?
6. How can a clinical vaccine to prevent coccidioidomycosis be developed?

Further Reading

Akram SM, Koirala J (2022) Coccidioidomycosis. In: StatPearls. StatPearls Publishing, Treasure Island, FL

Ampel NM (2007) The complex immunology of human coccidioidomycosis. Ann N Y Acad Sci 1111:245–258

Bays DJ, Thompson GR 3rd (2021) Coccidioidomycosis. Infect Dis Clin N Am 35(2):453–469

Blair JE et al (2006) Serologic testing for symptomatic coccidioidomycosis in immunocompetent and immunosuppressed hosts. Mycopathologia 162(5):317–324

Crum NF et al (2004) Coccidioidomycosis: a descriptive survey of a reemerging disease. Clinical characteristics and current controversies. Medicine (Baltimore) 83(3):149–175

Fisher MC et al (2002) Molecular and phenotypic description of *Coccidioides posadasii* sp. *nov.*, previously recognized as the non-California population of *Coccidioides immitis*. Mycologia 94(1):73–84

Freedman M et al (2018) Coccidioidomycosis outbreaks, United States and worldwide, 1940-2015. Emerg Infect Dis 24(3):417–423

Galgiani JN et al (2000) Comparison of oral fluconazole and itraconazole for progressive, nonmeningeal coccidioidomycosis. A randomized, double-blind trial. Mycoses Study Group. Ann Intern Med 133(9):676–686

Galgiani JN et al (2016) 2016 Infectious Diseases Society of America (IDSA) clinical practice guideline for the treatment of coccidioidomycosis. Clin Infect Dis 63(6):e112–e146

Graupmann-Kuzma A et al (2008) Coccidioidomycosis in dogs and cats: a review. J Am Anim Hosp Assoc 44(5):226–235

Hernandez H, Erives VH, Martinez LR (2019) Coccidioidomycosis: epidemiology, fungal pathogenesis, and therapeutic development. Curr Trop Med Rep 6(3):132–144

Kauffman CA et al (2014) Endemic fungal infections in solid organ and hematopoietic cell transplant recipients enrolled in the Transplant-Associated Infection Surveillance Network (TRANSNET). Transpl Infect Dis 16(2):213–224

McCotter OZ et al (2019) Update on the epidemiology of coccidioidomycosis in the United States. Med Mycol 57(Supplement_1):S30–S40

Miller R, Assi M, American Society of Transplantation Infectious Diseases Community of Practice (2019)

Endemic fungal infections in solid organ transplant recipients-guidelines from the American Society of Transplantation Infectious Diseases Community of Practice. Clin Transpl 33(9):e13553

Nguyen C et al (2013) Recent advances in our understanding of the environmental, epidemiological, immunological, and clinical dimensions of coccidioidomycosis. Clin Microbiol Rev 26(3):505–525

Stevens DA (1995) Coccidioidomycosis. N Engl J Med 332(16):1077–1082

Thompson GR 3rd, Barker BM, Wiederhold NP (2017) Large-scale evaluation of in vitro amphotericin B, triazole, and echinocandin activity against coccidioides species from US Institutions. Antimicrob Agents Chemother 61(4):e02634-16

Wheeler C et al (2018) Risk stratification with coccidioidal skin test to prevent valley fever among inmates, California, 2015. J Correct Health Care 24(4):342–351

Wilson L et al (2019) The rise of valley fever: prevalence and cost burden of coccidioidomycosis infection in California. Int J Environ Res Public Health 16(7):1113

Wright PW et al (2003) Donor-related coccidioidomycosis in organ transplant recipients. Clin Infect Dis 37(9):1265–1269

Paracoccidioidomycosis 16

Beatriz L. Gómez, Ángela M. Tobón, and Ángel González

Learning Objectives

By the end of the chapter, the readers should be able to:

1. Enhance awareness about the importance of recognizing signs and symptoms of the causal agent of paracoccidioidomycosis, *Paracoccidioides* spp., and distinguishing them from those of other mycoses and tuberculosis.
2. Obtain an overview of the epidemiology, current taxonomy or phylogenetic positioning, morphology, life cycle, pathogenesis and thermal dimorphism of the genus *Paracoccidioides* spp.
3. Acquire fundamental knowledge of aetiology, clinical manifestations, at-risk groups and current recommendations for treating paracoccidioidomycosis.
4. Gain insight into the successes and continuing challenges of diagnosing the pathogen *Paracoccidioides* and distinguishing it from other genera of the Onygenales order.

B. L. Gómez (✉)
Studies in Translational Microbiology and Emerging Diseases (MICROS) Research Group, School of Medicine and Health Sciences, Universidad del Rosario, Bogotá, Colombia
e-mail: beatriz.gomez@urosario.edu.co

Á. M. Tobón
Internal Medicine, Universidad CES and Colombian Tropical Medicine Institute, Medellín, Colombia

Á. González
Basic and Applied Microbiology Research Group (MICROBA), School of Microbiology, Universidad de Antioquia, Medellín, Colombia

Introduction

Paracoccidioidomycosis (PCM) is one of the most important systemic and endemic mycoses affecting Latin America, covering Mexico to Argentina. Brazil has had more than 80% of all reported cases, followed by other countries, including Colombia, Venezuela, Ecuador, Argentina and Bolivia. So far, this mycosis has not been reported in Chile, Nicaragua, Surinam, Belize, Trinidad, Granada or Guadeloupe.

At present, five phylogenetically diverse *Paracoccidioides* species (*Paracoccidioides brasiliensis*, *Paracoccidioides restrepiensis*, *Paracoccidioides venezuelensis*, *Paracoccidioides americana* and *Paracoccidioides lutzii*) have been identified as etiologic agents of this mycosis. The later *P. lutzii* diverted from the other species around 30 million years ago (MYA), whereas *P. venezuelensis* and *P. restrepiensis* appeared closely associated with the latest divergence of less than 0.2 MYA. *Paracoccidioides* spp., a thermally dimorphic fungus, exhibits two morphotypes, mould at temperatures under 25 °C and yeast in cultures at 35–37 °C or in host tissue.

S. C. Parija, S. M. Rudramurthy (eds.), *Textbook of Fungal Zoonoses and Sapronoses*, Microbial Zoonoses, https://doi.org/10.1007/978-981-97-3214-2_16

Yeast cells reproduce by multiple budding resembles a pilot's wheel.

Paracoccidioidomycosis is characterized by the presence of two clinical presentations as follows: (1) a subclinical infection which generally courses as asymptomatic and (2) the disease which could be acute or chronic with the involvement of several organs and systems, including the lungs, mucosa, skin and adrenal glands, among others.

History

Adolfo Lutz first described this mycosis and its etiologic agent in 1908, and later, it was done by Splendore, who described clinical observations and presented detailed comparisons of several fungal isolates from Latin American patients and foreign cases. Initially, the fungus was erroneously classified as a strain of *Coccidioides immitis*. Later, de Almeida, in 1930, properly differentiated the causal agent as *Paracoccidioides brasiliensis*. Originally, this mycosis was described as the "Brazilian disease", but when other cases were described in other countries of Latin America, it was re-named South American blastomycosis, thus accentuating its peculiar geographical limits. The term PCM was later made official by Latin American scientists during the First Pan American Symposium in Uruguay.

Taxonomy

The genus *Paracoccidioides* is only known in its asexual (anamorph) stage. However, molecular techniques have revealed the presence and expression of the mating type locus, thus allowing its classification in the phylum Ascomycota, order Onygenales, family Onygenaceae, close in the phylogenetic tree to the dimorphic fungi *Blastomyces dermatitidis*, *Histoplasma capsulatum* and *Emmonsia parva*, all of which present a teleomorphic or sexual stage in the genus *Ajellomyces*. Phylogenetic analysis of this fungal pathogen showed that the *Paracoccidioides* genus could be classified into at least three species bound to endemic regions: S1, PS2 and PS3. The highly divergent "Pb01-like" group was named "*Paracoccidioides lutzii*", and, recently, names have also been proposed for PS2 (*P. americana*), PS3 (*P. restrepiensis*) and a fifth phylogenetic species restricted to Venezuela, PS4 (*P. venezuelensis*) (Table 16.1).

Genomics and Proteomics

The first database for whole genomes of the genus *Paracoccidioides* was completed in 2009 by the BROAD Institute. Later, the genome sequences were updated with additional Illumina short-read sequencing and a new algorithm to correct base-level errors in the original sequences and revise annotations. The assembled sequences and their annotations are available at https://www.ncbi.nlm.nih.gov/genome/browse/#!/eukaryotes/334/.

In addition to the four reference species/strains that were initially chosen, additional strains from different regions of South America have been sequenced for population genomic studies, and proteomics studies have also been performed.

Fungus Morphology

Paracoccidioides spp. grows as a mould at room temperature and needs a prolonged incubation for up to 30 days. Isolates incubated on rich media produce thin septate hyphae (1–3 μm in width), and, occasionally, they produce chlamydospores. Interestingly, some isolates can produce conidia if cultured in poor nutritional conditions, which vary in shape and size (<5 μm) and can be observed as intercalary, septate or pedunculated (originating from a narrower base). Conidia respond to temperature changes; thus, at 20–24 °C, conidia can germinate into hyphae or transform into yeasts at 36 °C when appropriate media are used. The yeast morphotype can exhibit an oval or rounded form, and characteristically, it can be observed as a mother cell surrounded by multiple buds, resembling a ship's "pilot wheel".

Table 16.1 Distribution of *Paracoccidioides* infection in the South American population

Paracoccidioides spp.	Distribution
Paracoccidioides brasiliensis	Brazil, Argentina, Paraguay, Peru and Venezuela
Paracoccidioides lutzii	Brazil and Ecuador
Paracoccidioides americana	Brazil, Uruguay and Peru
Paracoccidioides venezuelensis	Venezuela
Paracoccidioides restrepiensis	Colombia

Cultivation of Fungi

The fungus could be cultured in several culture media, including Sabouraud dextrose agar, brain–heart infusion (BHI) plus glucose and a selective medium such as Mycosel™. *Paracoccidioides* spp. grows as a mould at room temperature (19–24 °C) and requires a long incubation period, between 2 and 6 weeks; the yeast phase could be recovered by incubating at 37 °C for 3 or 7 days.

Life Cycle of the Fungus

Although the specific niche of *Paracoccidioides* is unknown, it is believed that the fungus remains in environments with the factors that allow it to grow. In nature, the fungus is found in the form of mould, and once it accidentally comes into contact with the host, the conidia or fragments of hyphae can be inhaled by the host, reaching the respiratory tract where it is transformed into yeast, considered the pathogenic form. Likewise, the secretions produced in the host and containing the yeast morphotypes can be deposited in the environment, where these pathogenic forms of the fungus germinate, forming hyphae, remain in the environment and start their life cycle again when they find the host. So far, humans and armadillos are recognized as the "natural" hosts prone to developing the disease. It has also been suggested that free-living amoebae could play an important role in the life cycle of this and other fungi found in nature.

Animal Models

Animal models have contributed largely to the knowledge of the pathogenesis of PCM and the development of therapeutic treatments. These animal models have allowed the study of *Paracoccidioides* infections from the beginning to the development of chronic stages that mimic what is happening in humans. Various animal models for this fungal infection have been reported, which include studies in mice, hamsters, guinea pigs and rats. Nonetheless, the mouse has been the most studied model, and several strains with different susceptibility patterns have been reported, with the A/SN and A/J strains being the most resistant, whereas B10D2/nSn, B10.A and IB10D2/oSn are the most susceptible strains to *Paracoccidioides* infection. Interestingly, BALB/c and C57BL/6 mice strains showed intermediate susceptibility to fungal infection. Of note, as described in humans, gender is an important factor in developing the disease, and indeed male BALB/c and B10D2/nSn mice are more susceptible to infection than females of these strains.

Moreover, genetically modified mice have also been used to study the host–fungus interactions during *Paracoccidioides* infection, mainly focusing on cells or molecules involved in the immune response. Such modified mice include athymic or nude (nu/nu) mice lacking T cells, knockout mice for certain cytokines, nitric oxide, caspase-1, TLRs (2, 4 and 9), CCR5 and transgenic mice.

Additionally, non-mammalian models have been implemented to investigate the virulence factors of *Paracoccidioides* spp. Thus, invertebrate models have been developed using *Galleria mellonella* and *Caenorhabditis elegans*. More recently, employing *in vitro* studies, it was reported that different amoebas, including *Acanthamoeba castellanii*, *Acanthamoeba* spp., *Allovahlkampfia spelaea* and *Vermamoeba vermiformis*, can ingest and destroy *Paracoccidioides*. All these suggest that this fungus can inhabit a complex environment with several predators that could exert a selective pressure to guide the evolution and virulence trait of this fungal pathogen. These invertebrate models have the advantages of having low cost and fewer ethical issues, allowing easier experimentation, being time-efficient and offering the possibility of using many animals per experiment compared to mammalian models developed. However, care should be taken when interpreting and extrapolating results obtained using such models to findings observed in humans.

Pathogenesis and Pathology

Once the infectious fungal particles of *Paracoccidioides* spp. enter the host, they interact with different cell types, including lung epithelial cells and alveolar macrophages. This facultative intracellular pathogen can initiate its morphological transition from conidia to yeast, a process that is more effective and faster at the intracellular level. The phase transition responsible for the fungal dimorphism is a specialized and regulated process that is triggered once the fungus recognizes and responds to the host environment. This is essential for the pathogenicity of the fungus. The ability to undergo this morphological change is considered a virulence factor and is obligatory for establishing infection and developing the disease.

To survive within the host, the fungus has developed different mechanisms that prevent its elimination by the immunological response developed by the host. Among them, the expression of adhesins on the fungal surface not only allows the fungus to adhere to the epithelium through the recognition of extracellular matrix (ECM) proteins and receptors present on the surface of epithelial and alveolar cells, including macrophages and dendritic cells but also confers on the fungus the ability to invade or spread to other organs and tissues. In addition to morphological change and expression of adhesins, other virulence factors are also involved in the pathogenesis of PCM. These include enzymes that confer to the fungus the ability to hydrolyse ECM proteins and melanin production and biofilm formation. All these confer resistance to the host's fungicidal mechanisms and antifungal therapy.

PCM is generally characterized by developing a chronic inflammatory response with the formation of granulomas. These structures consist of giant, epithelioid and mononuclear cells that surround the yeast and are considered one of the distinctive histological marks of PCM. Finally, this chronic inflammatory process ends with the development of fibrosis, considered one of the most important sequelae of this mycosis, because of the activation and uncontrolled production of ECM proteins (mainly collagen) by fibrocytes and macrophages.

The initial interaction of the fungus with the host cells is through receptors, including the complement receptor (CR), mannose receptor (MR), dectin-1, toll-like receptors (TLRs)-2 and TLR-4, among others. These, in turn, recognize different molecules or components, such as polysaccharides, proteins and lipids, present in the wall of the fungus. The host cells become activated and capable of producing pro-inflammatory chemokines and cytokines, thus initiating the inflammatory process, in which phagocytic cells such as neutrophils, macrophages and other immune system cells reach the site where the infection is located, mainly the lungs.

Subclinical infection may develop if the host cannot control the fungal growth, or the host may develop a clinically manifested disease. Such a fungal infection can become dormant but may be reactivated afterwards under the influence of ill-defined conditions, including chronic alcoholism, malnutrition and smoking. The subclinical infection involves mainly the lungs and often courses as asymptomatic.

The symptomatic infection is characterized by the involvement of several organs and systems and is subclassified into two clinical forms: (1) the acute/subacute or juvenile and (2) the chronic or adult form. Additionally, a residual form is recognized, characterized by fibrotic sequelae in almost 50% of patients, who suffer from the chronic form. Patients with the disease present a gamut of constitutional symptoms, which include adynamia, asthenia, fever, weight loss, general malaise and other symptoms associated with the affected organ. In the increased number of patients with a marked immunocompromise, mainly in those with human immunodeficiency virus/acquired immunodeficiency syndrome (HIV/AIDS), this mycosis is frequently diagnosed. In these patients, the clinical manifestation could not be classified as chronic or acute form. More recently, this presentation has been described as a mixed form. Brazilian studies indicate a mortality rate of 0.9 per 1 million inhabitants; in other countries, this mortality is unknown.

Immunology

Macrophages, neutrophils, eosinophils and natural killer (NK) cells are the main effector cells in the innate immune response. Macrophages represent the major cellular defence against *P. brasiliensis*. They are extremely important for the host, not only for infection control but also for induction and regulation of the inflammatory response. Although neutrophils are considered important players in the innate immune response against *P. brasiliensis*, recent reports using a PCM experimental model suggested that neutrophils play a dual role in this mycosis. This is essential during the early infection and detrimental during the chronic course of the disease. Eosinophils seem to play a detrimental role; thus, in afflicted patients, the severity of the disease is associated with eosinophilia. Dendritic cells and keratinocytes should play an important role in innate and adaptive immunity due to their function as antigen-presenting cells. However, internalization of the fungus into keratinocytes appears to represent a possible escape mechanism. Additionally, it has been described that *Paracoccidioides* spp. induces apoptosis when they invade epithelial or phagocytic cells, a process that benefits their survival at the intracellular level.

During the development of an adaptive or a specific immune response, there is a dichotomy between the humoral and cellular immune responses, in which an increase in the production of antibodies is associated with a hyperproliferative response of T cells, which suggests the development of a Th2-like immunological pattern. In the case of patients with the severe form of the disease, an increase in the production of cytokines of the Th2 profile has been observed, mainly interleukins (IL)-10, IL-4 and IL-5, and a decrease in the profile of Th1 (mainly interferon-γ (IFN-γ) and IL-2). All these suggests that the Th1 profile exerts a protective function. Genetic studies in which the disruption of genes that code for cytokines or their neutralization using specific monoclonal antibodies have shown the cytokines' protective role, such as IL-12, tumour necrosis factor-α (TNF-α) and IFN-γ. At the same time, IL-4 and IL-18 appear deleterious or play a dual role. Recently, it has been reported that the Th17 profile appears detrimental and is related to the development of an exacerbated inflammatory response.

Among the different subpopulations of T lymphocytes, to control the fungal load through cytotoxic mechanisms, CD8+ subsets participate. In contrast, CD4+ participates in developing a delayed hypersensitivity response and production of specific antibodies. In the polyclonal activation of B lymphocytes, gamma delta T lymphocytes seem to participate and T-regulatory cells (CD4+, CD25+, FoxP3+) participate in the development of a systemic and local immune response.

On the contrary, B lymphocytes are activated and produce specific IgA, IgG and IgE antibodies against the fungus. IgA and IgG2 antibodies are increased in patients with the chronic form, while IgG4 and IgE antibodies predominate in patients with the acute form of this mycosis.

Infection in Humans

Paracoccidioidomycosis is classified as a subclinical or symptomatic infection.

Subclinical Infection

Individuals with this presentation are mostly asymptomatic and are detected mainly by reactive skin or serological testing or, in some cases, by abnormal chest radiographs. Moreover, the fungus may remain in a latent stage in the infected host, who may develop the disease years after the initial contact. This disease development could result from a debilitating immunological process in those individuals. As proof of this, the disease has been diagnosed in individuals who live outside of areas that are endemic for the mycosis but who visited such areas and were infected several years earlier.

Symptomatic Infection

The disease manifestations can vary among individuals, and this variation depends on various factors that include age, sex and immune response among others.

Acute/subacute Disease

The acute/subacute disease or juvenile form affects mainly children, young people (<30 years old) and immunocompromised individuals of both sexes and accounts for at least 15% of all cases. This clinical form is characterized by the involvement of the reticuloendothelial system, and affected patients present lymphadenopathy, hepato- and/or splenomegaly. Multiple skin lesions have been described in approximately 31% of the cases. Constitutional abnormalities include general malaise, fever, weight loss, anaemia, hypoalbuminaemia, eosinophilia and hypogammaglobulinaemia. Moreover, this acute form has common digestive tract and abdominal manifestations, including diarrhoea, vomiting, abdominal masses or distention, abdominal pain, ascites and/or lymph node enlargement. Although the fungus may be seen or isolated in the respiratory secretions, respiratory symptoms and abnormalities in Transverse aortic constriction (TAC) studies are infrequent (5–11%). Without a prompt diagnosis, the mortality rates in patients with this form could reach up to 10%.

Chronic Disease

The chronic or adult form of the disease is the most common presentation (80–90%). This clinical presentation occurs mainly in male adult individuals of ages above 30 years (ratio 13 male:1 female), especially in people engaged in agriculture-related activities. The chronic form of the fungal disease is characterized by pulmonary affectation (80%) and extra-pulmonary dissemination, mainly to the skin and mucous membrane (90%). To note, the lesions are mostly ulcerative, infiltrated and painful, showing an appearance of "mulberry stomatitis" (Figs. 16.1 and 16.2).

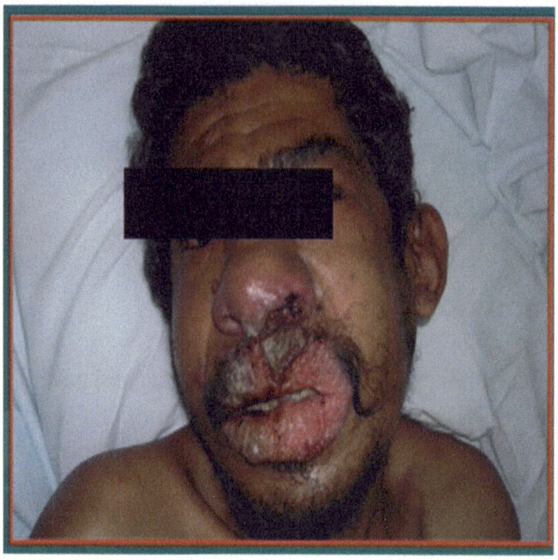

Fig. 16.1 Mucosa–skin involvement in paracoccidioidomycosis. Ulcerative-crusted lesions involving the lips, oral and nasal mucosa with accompanying edema as a characteristic of involvement by *Paracoccidioides brasiliensis*

Oropharyngeal affectations are common, presenting dysphagia and dysphonia, as well as sialorrhea and gingivitis with the lost tooth in some cases. Due to the hypertrophy of the lymph nodes, self-draining fistulae formation is common. The adrenal glands are also involved in at least 48% of the cases, with decreased functional reserve. The central nervous system (CNS) could be compromised in 25% of the cases with a pseudotumoral lesion and meningeal or spinal compromise.

Lung abnormalities have been described in 98% of the patients, as revealed by high-resolution computed tomography (CT) studies showing alteration of the pulmonary architecture, as well as reticulate and septal thickening and emphysema. Chest radiographs also show mixed infiltrates, which are predominantly bilateral and located in the central and lower fields of the lungs, mostly without affectation of the apices, in one-third of the patients (in contrast to tuberculosis) (Fig. 16.3). Pulmonary fibrosis is the most frequent sequela and is observed in almost 30% of the patients, who suffer from the chronic form, significantly affecting pulmonary function. Interestingly, in this clinical form, a clinical–radiological dissociation is frequently observed; thus, it has been described as an important lesion seen in imaging studies but

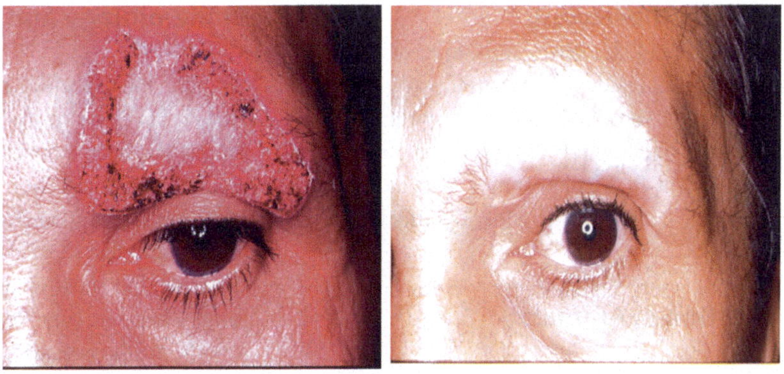

Fig. 16.2 Skin involvement in paracoccidioidomycosis. (**a**) Ulcerative lesion with well-defined border, granulomatous and with hemorrhagic dot. (**b**) Hypochromic scar after treatment

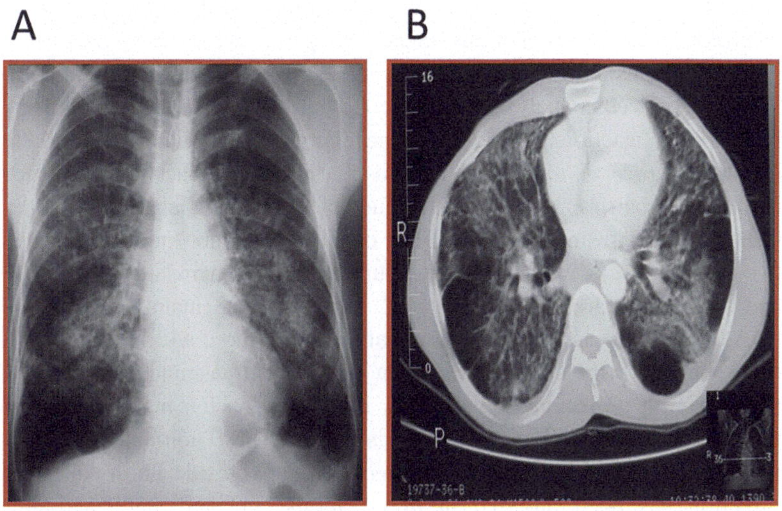

Fig. 16.3 Pulmonary involvement in paracoccidioidomycosis. (**a**) Chest X-ray. (**b**) CT scan of the chest: mixed (alveolar and interstitial) bilateral infiltrates in middle and lower fields. Bullae and fibrosis

without severe symptoms. Without timely treatment, mortality in patients with this clinical form could reach up to 45%.

Paracoccidioidomycosis (PCM) in Immunocompromised Patients

Even though PCM is uncommon in immunocompromised patients, HIV-PCM coinfection has been reported in endemic areas.

In the last 30 years, 136 cases of PCM reported in patients with HIV/AIDS infection have been retrieved from the literature. Nearly, 80% with CD4 counts <200 cells/mm^3 and a mixed clinical form of presentation with fever, widespread lymphadenopathy, splenomegaly, skin lesions and pulmonary involvement are demonstrated in 70.3% of the cases. This is due to inefficient immune control of the initial subclinical pulmonary foci, followed by lympho-hematic dissemination. In highly endemic areas, this affection represents 5% of all PCM cases in patients with AIDS, and mortality is present in 35% versus 7.9% in patients without HIV infection. Prophylaxis with trimethoprim/sulfamethoxazole does not rule out PCM infection.

Among the neoplastic hemopathies, PCM is more often associated with lymphoma. Also, PCM has been confused with respiratory tract carcinoma, with its progression in the same anatomic site of the fungal infection in 58% of the cases.

In transplant recipients, PCM is mostly related to the late phase of kidney transplantation, along with the involvement of skin, mucosae and lungs with infiltrates accompanied or not by lymph node enlargement.

In patients receiving immunotherapy, the development of the chronic form of PCM has been informed after 1 year of biological therapy.

PCM and Tuberculosis

In 15–20% of the cases, this mycotic infection can be associated with tuberculosis (TB). This disorder also produces similar respiratory abnormalities but minor (0.4%) mucocutaneous involvement in TB and minor affectation in the lung apices, sites more frequently attacked in TB.

Infection in Animals

Paracoccidioides have been isolated from the armadillos *Dasypus novemcinctus* and *Cabassous centralis*. So far, isolation of this fungal pathogen in those animals' burrows has been unsuccessful. However, molecular analysis using a nested polymerase chain reaction (PCR) allowed the identification of *Paracoccidioides* spp. in aerosol at the entry of the burrows. Moreover, the fungus has been isolated, in unique cases, from dogs and two-toed sloths, as well as from penguin faeces and the intestinal tract of bats.

Additional evidence of infection of various domestic and wild animals has been contributed through molecular, serological, intradermal and histopathological methods. Thus, serological detection of *Paracoccidioides* infection has been reported in pigs, rabbits, dairy goats, dogs, cats, sheep, chickens, horses and primates. In contrast, molecular detection has allowed the identification of *Paracoccidioides* DNA in wild animals, including raccoons (*Procyon cancrivurus*), Brazilian guinea pigs (*Cravia aperea*), spiny tree porcupine (*Sphiggurus spinosos*), ferrets (*Galictis vittate*) and tayras (*Eira barbara*).

The above reports not only indicate a broad distribution and adaptation of *Paracoccidioides* spp. to different mammalian hosts but also suggest that these animals could be used as sentinels or epidemiological markers for the pathogens in the environment. Additionally, the reports reinforce the importance of the contact of the host with soil as a risk factor for infection.

Epidemiology and Public Health

So far, the specific environmental habitat of *Paracoccidioides* spp. is unknown, even though the fungus has been isolated from the soil in endemic areas. Fungal infection has been reported from Mexico to Argentina. Nonetheless, in some countries, including Chile, Nicaragua, Belize, the Guyanas and most of the Caribbean islands, this fungus has not been reported, apart from scarce reports. In the countries where the entity is diagnosed, the cases are distributed irregularly throughout the territory. Of note, the ecologic characteristics of these areas have been relatively well-defined: (1) humidity is high and forests abound, (2) tropical and subtropical forest, (3) large watercourses, (4) high rainfall, (5) mild temperatures and (6) coffee or tobacco crops. Infection is usually acquired by the inhalation of conidia, but the incubation period is not known. This fungal pathogen is known to remain latent for very long periods in its host's tissue (mainly the lymph nodes) after a primary infection, which is usually asymptomatic. PCM is not transmitted from person to person.

It is important to mention that approximately a 100 non-autochthonous cases have been reported outside Latin American countries, including the United States and some countries from Europe and Japan. These patients previously visited or resided in the endemic areas in Latin American countries, mainly Brazil, for a mean period of 14 years before they presented the characteristic clinical manifestations. These reports confirmed the capability of this fungal pathogen to remain latent in the host.

Diagnosis

The classical PCM diagnosis combines clinical evaluation and laboratory studies that include routine direct examination, culture, histopathology, immunological-based techniques (antigen and antibody detection) and nucleic acid-based methods such as polymerase chain reaction (PCR) assays (Table 16.2). More recently, commercial platforms have also been developed, including matrix-assisted laser desorption ionization time-of-flight mass spectroscopy (MALDI-TOF MS). The latter is based on analyses of the spectrum of organic biomolecules/proteins in the fungal isolates and does not depend on genetic information. Nonetheless, the final diagnosis of PCM depends on microscopic examination and positive culture, the culture-based methods being considered the "gold standard". However, it may take 2–6 weeks to isolate and identify *Paracoccidioides* spp. from clinical specimens. Thus, using a combination of methods to diagnose this mycosis is the best approach.

Microscopy

Characteristic fungal cells can often be found in respiratory samples or biopsy material from other tissues. Yeast cells could be observed in up to 85% of the assessed clinical specimens using fresh preparations or treated with potassium hydroxide (KOH). Other assays, including calcofluor or immunofluorescence preparations, are often useful. *P. brasiliensis* can be observed as an oval to round, translucent-walled yeast cell, and characteristically it presents multiple buds surrounding the mother cell, with prominent intracytoplasmic lipidic vacuoles that also can be found.

Table 16.2 Laboratory diagnosis of paracoccidioidomycosis

<table>
<tr><th>Assay</th><th>Characteristics</th><th>Sensitivity (%)</th></tr>
<tr><td>Direct examination:
Wet mount, KOH test
Calcofluor white
fluorescence</td><td>Under the microscope, "pilot's wheel" multiple budding yeast (diameter: 4–60 μM)</td><td>80–90</td></tr>
<tr><td>Histopathology:
GMS
PAS
H&E</td><td>Under the microscope, "pilot's wheel" multiple budding yeast (diameter: 4–60 μM)</td><td>90–95</td></tr>
<tr><td>Culture:
Sabouraud agar
Mycosel agar
Yeast extract agar
BHI agar</td><td>18–24 °C:
Macroscopic: white cottony colonies, adhering to agar and covered with short fluffy mycelia
Microscopic: thin hyaline hyphae, septate with or without chlamydoconidium
36 °C:
Macroscopic: cream-coloured cerebriform colonies
Microscopic: multiple budding yeast</td><td>60–80</td></tr>
<tr><td>Immunodiagnostics:
ID
CF
CIE
ELISA
WB
Dot blot
Latex agglutination
Antigenemia
Antigenuria</td><td>Antibody detection
Antigen detection</td><td>65–100
84–94</td></tr>
<tr><td>Molecular tests:
Conventional PCR
Nested PCR
Real-time PCR</td><td>Detection of nucleic acids: rDNA, ITS, 27 kDa protein, gp43</td><td>91–100</td></tr>
</table>

GMS Gomori's methenamine silver stain, *PAS* periodic acid-Schiff stain, *H&E* haematoxylin-eosin stain, *BHI* brain–heart infusion, *ID* immunodiffusion, *CF* complement fixation, *CIE* counterimmunoelectrophoresis, *ELISA* enzyme-linked immunosorbent assay, *WB* Western blotting, *rDNA* ribosomal DNA, *ITS* internal transcribed spacer, *gp43* 43 kDa glycoprotein

Additionally, other yeast cell shapes could be observed, forming chains with single buds or, in some cases, large yeasts. Staining of wet preparations with lactophenol cotton blue, calcofluor white methylene blue or Gram stain is often useful for observing fungal morphotypes. Moreover, staining of tissue sections can be done with haematoxylin and eosin (H&E), Gomori methenamine-silver stain (GMS), or periodic acid-Schiff (PAS); these stain techniques are very helpful, as they reveal fungal structures (Fig. 16.4).

In Vitro Culture

As described before, *Paracoccidioides* spp. exhibits dimorphism. Thus, it grows as a mould at room temperature (19–24 °C) and requires a long incubation period, between 2 and 6 weeks; the yeast phase of *Paracoccidioides* species could be recovered by incubating at 37 °C. It is highly recommended to isolate *P. brasiliensis* from different clinical specimens or tissue fragments as isolation success is quite high, between 86% and 100%. It requires several kinds of culture media, including Sabouraud dextrose agar, brain–heart infusion (BHI) plus glucose and selective medium such as Mycosel™. When clinical materials such as sputum are analysed, they may be contaminated with bacteria; thus, media containing antibiotics, including chloramphenicol or gentamicin, should be used.

Moreover, media containing cycloheximide help avoid the growth of saprophytic fungi; the latter is useful because it can also be used as a differential tool for isolation and identification. In general, colonies develop within 3–7 days, but some strains might need to get incubated for a longer period, which could be as long as 4–6 weeks. Isolates display slow growth when incubated at 25–30 °C and produce a variety of forms ranging from glabrous, leathery, brownish, flat colonies with a few tufts of aerial mycelium to folded, wrinkled, floccose colonies and velvety, white-to-beige forms. Without conidial production, most fungal strains grow for long periods. Under the microscope, mould morphotype shows thin septate hyphae and occasionally, chlamydospores (15–30 μm) may be observed. When cultured in simple media (poor nutritional conditions), some isolates can produce conidia, which vary in shape and size (<5 μm). The mycelial morphotype is not distinguishable from those of other fungi, and dimorphism must be confirmed. The resulting yeast colonies exhibit a cerebriform appearance when cultures are performed at 37 °C on rich media such as BHI, Pine's or Kelley's agar. Yeasts are between 2 and 30 μm in diameter and are oval, round or irregular, displaying the characteristic "pilot's wheel". Conversion to the yeast phase is also very slow,

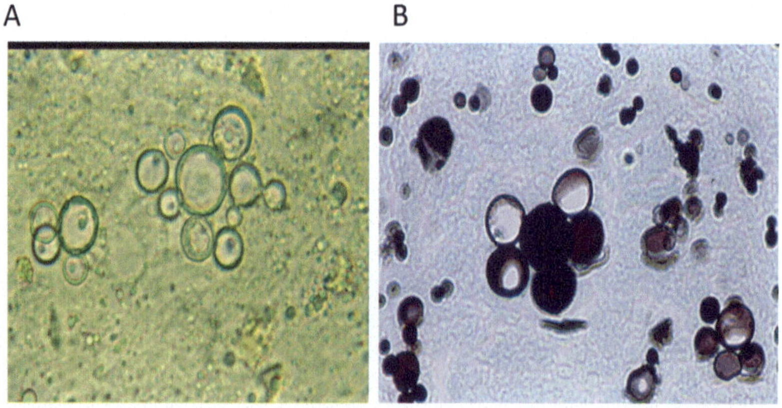

Fig. 16.4 Wet mount of (**a**) Sputum in KOH wet mount showing blastoconidia of *Paracoccidioides brasiliensis*. Magnification ×200. (**b**) Gomori methenamine-silver stain of lung

and culture plates should be held for at least 6 weeks.

To handle this fungal pathogen, a biosafety level 2 facility is recommended for processing clinical specimens. Nonetheless, procedures that involve manipulating sporulating cultures should be taken care of in a biosafety level 3 facility.

Serodiagnosis

Immune-based methodologies, including antibody and antigen detection, are helpful for the diagnosis and monitoring of the patient's treatment course. The traditional methods used to diagnose paracoccidioidomycosis are immunodiffusion (ID) and complement fixation (CF). However, other tests, such as enzyme-linked immunosorbent assays (ELISA), counterimmunoelectrophoresis (CIE), dot blot and immunoblotting, have also been developed. Antibody detection has been accomplished using antigen preparations from mycelial or yeast cell morphotypes, available in several reference laboratories. A series of immunoreactive antigens (27, 43, 70, 87 kDa) are present in the lysates or derived from culture filtrates of mycelial-phase or yeast-phase broth cultures. Nevertheless, the predominant antigen is the 43 kDa glycoprotein, gp43. Compared to culture filtrate antigens, cell wall antigens are less useful, as wall antigens exhibit cross-reactivity with galactomannan. In endemic countries, the gel immunodiffusion test is the most popular method because of its simplicity. This test allows the detection of circulating antibodies in 65–95% of the cases and is highly specific. The CF method is more precise for evaluating the patient's response to treatment; However, compared to the ID method, it is less specific, and cross-reactions have been observed in patients with histoplasmosis. Nonetheless, CF titres of ≥1:8 could be considered presumptive evidence of PCM, with falling titres being predictive of successful treatment. For ID or CF, no commercial kits are available.

In diagnosing PCM, antigen detection methodologies exhibit certain important advantages over antibody detection, especially in immunocompromised patients and those previously exposed to *Paracoccidioides* spp.; in the latter, they may have pre-existing antibody titres. As a routine diagnostic test, paracoccidioidomycosis antigen testing is not available. Two main proteins, the gp43 glycoprotein and the 87-kDa heat-shock protein, have been reported as helpful targets to be detected in patients' sera. Notably, gp43 has also been detected in urine, BAL samples and cerebrospinal fluid. Several reports have indicated that serum antigen levels diminished or could disappear during successful treatment.

Improvements in serodiagnosis involve the detection of antibodies against chemically characterized and/or recombinant *P. brasiliensis* antigens, notably gp43, pb27 and the 87-kDa heat shock protein. There is an increased sensitivity (92%) and specificity (88%) due to a combination of two recombinant products.

It is important to mention that interlaboratory reproducibility remains an important issue despite efforts to standardize techniques and characterize antigen targets. With the advances in *Paracoccidioides* genus characterization, a higher complexity has been added to the standardization of laboratory techniques for the diagnosis of PCM. In this scenario, classical laboratory techniques do not have enough discriminatory power over cryptic agents. For example, studies have demonstrated that *P. brasiliensis* and *P. lutzii* may elicit different host antibody responses as they have distinct antigenic profiles. Several studies are underway to find new markers; new epitopes, techniques and approaches have recently been described.

Molecular Diagnosis

Various molecular diagnostic methods involving PCR, nested PCR, real-time (RT) PCR, PCR-enzyme immunoassay and loop-mediated isothermal amplification to detect specific *P. brasiliensis* DNA sequences have been successfully used. However, no commercially available systems exist to detect *P. brasiliensis* DNA in human clinical samples. So far, there are a few molecular tests, all "in-house", and they have not been validated. The main target genes used for these molecular tests are gp43 and the rDNA regions, especially 5.8 and 28S subunits and ITS regions, the latter allowing one to discriminate between *Paracoccidioides* spp. and other human pathogenic fungi. A real-time PCR employing the ITS 1 region as a target has been reported with 100% sensitivity and specificity. More recently, a real time-polymerase chain reaction amplifying the gene for recombinant protein PB27 was designed. Primer and probe sequences were deposited in the Brazilian Coordination of Technological Innovation and Transfer (CTIT) under the patent reference number BR1020160078830. Preliminary results using 78 clinical samples from confirmed cases showed a sensitivity of 94% and a specificity of 100%. More studies are needed to validate this test further.

As mentioned before, a new challenge where extra care must be taken regards *Paracoccidioides* and its recently described cryptic species, which is not always acknowledged in experimental design. It should be noted that there is still a need to validate these techniques and molecular platforms, as well as the extraction protocols and the targets to be used.

MALDI-TOF Mass Spectroscopy

Matrix-assisted laser desorption ionization-time of flight mass spectrometry (MALDI-TOF MS) is a new tool that has been applied for *Paracoccidioides* identification by Nobrega de Almeida Jr. et al. This is aimed to standardize an assay that will differentiate the spectra obtained from proteins expressed by the *P. brasiliensis* complex and *P. lutzii*. In their study, *Paracoccidioides* spp. isolates, previously characterized by molecular techniques (ITS, MLST and PCR), were identified by MALDI-TOF, and no misidentifications were found among all the strains tested. This technology is not currently available for the identification of *P. brasiliensis* directly from clinical samples.

Treatment

Treatment includes sulphonamides, polyenes (amphotericin B) and some azoles (Table 16.1). Usually, the treatment of this mycosis is long, mainly for two reasons: (1) the patient's immune status and (2) the ability of the fungus to survive in the host's tissues even after antifungal treatment. The implementation of the therapy should consider the severity of clinical presentation, and standard therapy protocols are needed. Moreover, additional measures must include (1) adequate nutrition and (2) suspension of both smoking and alcohol intake. Additionally, corticosteroids are recommended when a significant tissue inflammatory response is present, which can result in severe clinical complications (e.g. laryngeal narrowing, compressive syndromes).

Sulphonamides have been widely used due to their effectiveness and low cost. This class of drug is often employed in Brazil, typically as trimethoprim-sulfamethoxazole. The dosage is different for adults (80 mg of trimethoprim and 400 mg of sulfamethoxazole per tablet) than for children (8–10 mg/kg/day of trimethoprim component). In less severe forms of the disease, this drug should be administered for 2 years, and as a result, there are high dropout rates during treatment (Table 16.3). Among the advantages is that this combination could be used via parenteral administration when necessary, and its tolerability is good. One of the main side effects is myelotoxicity, but it can be controlled by folinic acid administration without affecting treatment. This treatment can also be used initially in association with amphotericin B (AmB).

Amphotericin B (AmB) deoxycholate is recommended for treating severe PCM forms (Table 16.3). However, to administer this polyene, patients must be hospitalized because of the side effects caused by this drug. The duration of the treatment depends on when clinical improvement is observed in the patient and is always continued with oral therapy. A liposomal AmB formulation (AmBisome) may sometimes be used to avoid the nephrotoxicity associated with AmB deoxycholate. Moreover, two other AmB lipid formulations [AmB colloidal dispersion

Table 16.3 Antifungals recommended for the treatment of paracoccidioidomycosis

Antifungals	Dose (daily)	Route	Duration
Itraconazole (tablets or solution)	Adults: 200–400 mg Child (<30 kg or >5 years old): 5–10 mg/kg	PO	6–12 months
Trimethoprim/sulfamethoxazole tablet 80 mg/400 mg	2–3 tablets bid	PO	18–24 months
Amphotericin B	Deoxycholate formulation 0.7–1 mg/kg/day 1–2 g (cumulative dose)[a] Lipid formulation: 3–5 mg/kg[a]	IV	2–4 weeks (until clinical improvement)

PO oral, *IV* intravenous
[a] To be followed by an oral medication

(ABCD) and AmB lipid complex (ABLC)] have been developed and are now available in several countries.

Most azoles are effective for treating PCM; among these drugs, itraconazole is considered the best therapy option for this mycosis. Several studies have described that this azole induces a more rapid clinical improvement with the serological cure than other antifungal drugs. This azole is regularly administered at a dose of 200–400 mg/day for up to 6 months, and the duration of the treatment depends on the cure parameters. Moreover, higher doses of this drug can be used in severe PCM cases in adult patients. Itraconazole is effective in up to 95% of the cases with minimal side effects, and relapses have been documented in up to 5% of the cases. With the administration of itraconazole capsules, other drugs that reduce gastric acidity should be avoided. For better bioavailability and absorption, the use of itraconazole in the oral solution presentation is recommended. Although used in a few patients, isavuconazole and voriconazole have been shown to have inhibitory action in vitro and in vivo. Fluconazole has been less used due to its higher in vitro minimum inhibitory concentration (MIC), and echinocandins have poor antifungal activity against *P. brasiliensis*.

Despite most antifungal drugs' high effectiveness in treating PCM, none reduced the pulmonary fibrosis sequelae.

Cure, if ever attained considering *P. brasiliensis* long-lasting latency periods, should be evaluated according to clinical, mycological, radiological and serological parameters: (1) regression or absence of signs and symptoms, the disappearance of the fungus, especially from respiratory samples, (2) presence of scarring and fibrosis images in radiologic studies and (3) detection of antibody titres negatives or stabilized at low levels.

Prevention and Control

Preventive measures are difficult to implement because the source of *P. brasiliensis* infection is unknown. Avoiding aerosols and being cautious when removing soil or hunting armadillos in known endemic areas is recommended. This mycosis is not transmissible from person to person.

Antifungals combined with immunomodulatory compounds, biological therapies or prophylactic vaccination employing a *P. brasiliensis* recombinant antigen may successfully attenuate pulmonary inflammation and fibrosis and is a promising therapeutic strategy. It will be necessary to move these experimental studies to clinical trials.

Case Study

The following case study exemplifies a relatively rare involvement of the central nervous system (CNS; neuroparacoccidioidomycosis) and illustrates the severity of problems that can ensue if diagnosis and treatment of PCM are delayed.

A 55-year-old male car driver from Medellín, Colombia, who had once been a farmer in a rural region, and a heavy smoker with regular alcohol intake, was admitted. He was admitted after progressive problems during the previous month, including left hemiplegia, paresthesias, right palpebral ptosis and, in the last days, also tonic–clonic seizures of the left extremities with secondary generalization, global headache and persistent vomiting. In the periumbilical zone of the abdomen, cutaneous ulceration (3 × 2 cm) with raised borders and granulomatous appearance was observed. A year earlier, the patient had already noticed but then disregarded the lesion, as it caused no major discomfort. The patient was alert and cooperative. His vital signs were normal, as were his memory and cortical function, and he presented no signs of meningeal irritation. Blood parameters and tests were also normal except for leukocytosis and neutrophilia. However, an ocular funduscopic examination revealed venous congestion with papilledema. Pupillae were isochoric and reactive; common ocular motor nerve palsy was observed. There was central right facial paresis and left hypoesthesia and hemiparesis. CT scans showed a ring-enhancing lesion with a large area of perilesional oedema. After a biopsy was taken, the new CT revealed that the lesion (18 × 16 mm with a necrotic centre) involved not only the right thalamus but also the mesencephalon. Thoracic X-rays showed bilateral interstitial infiltrates in both basal fields.

The first diagnostic impressions did not include suspicion of PCM. The patient was initially treated with fluconazole because the findings were compatible with cryptococcosis. A lack of response after 3 weeks prompted further studies, which finally identified *P. brasiliensis* as the causal agent. PCM diagnosis was firmly established, and treatment with itraconazole (ITZ; 400 mg/day) was initiated.

The patient had been receiving sodium valproate (total of 600 mg/day) to avoid seizures. A month later, the clinical response was still minimal (with undetectable ITZ levels in plasma), the sodium valproate was suspended and replaced by clonazepam (1 mg/day). Fifteen days later, ITZ had reached acceptable levels, and the ITZ dose was lowered to 300 mg/day. The abdominal ulceration and the common ocular nerve palsy healed within the next 2 months. After prolonged therapy (with ITZ dose lowered to 200 mg after 5 months), the headache disappeared, hemiparesis became less noticeable and intracranial hypertension was resolved. After the end of the therapy at 1 year, the patient was followed for another year with no relapses and presented substantial CT scan improvement.

A corollary is that possible CNS involvement should be considered in patients with PCM to allow diagnosis and pertinent treatment and to prevent possible permanent brain damage or death of the patient.

Questions

1. What could be a reasonable explanation for why ITZ was not detectable in plasma?
2. What clinical and microbiological characteristics or similarities do paracoccidioidomycosis and cryptococcosis share so that they can be confused?
3. What laboratory test could be useful to perform the diagnosis?
4. From the medical records, what risk factors could be associated with mycosis?

Research Questions

We mention only three of the many interesting open questions in *Paracoccidioides* research.

1. Despite much progress in understanding the parasitic (pathogenic) yeast phase of *Paracoccidioides* and other related thermally dimorphic fungi, little is known about its modes of existence and dispersal and its preferred habitat as a saprophyte found in soil, in its mycelial phase. It is also unclear how and for what purposes the fungus perfected its ability to infect and survive in mammals. One theory that has been proposed is that these abilities evolved as a defence mechanism

against non-mammalian (e.g. protozoan) species that posed a threat to the fungus; another view might be that mammals and other species offer efficient dispersal.

2. The mating of *Paracoccidioides* spp. is still largely an open question. Until now, no mating has been observed in this genus, even though mating-type genes are present in *Paracoccidioides* genomes.
3. There has been a recent emergence in South Africa and other countries of a fungal pathogen affecting immunocompromised individuals, *Emergomyces* spp. and a rise in the prevalence of the disease it causes, Emergo mycosis. Since this genus is closely related to the pathogens *Paracoccidioides* and *Histoplasma*, and especially to *Blastomyces* and *Emmonsia*, cross-reactions (i.e. ambiguities in identification) have recently been observed. When assays for those related species are used, confounding diagnoses and lowering specificity. There is, therefore, an interest in finding new assays that can reliably distinguish genera or species within the Onygenales order.

Further Reading

Albano APN, Klafke GB, Brandol TM, Da Hora VP, Minrllo LF, Jorge S et al (2014) Wild animals as sentinels of *Paracoccidioides brasiliensis* in the state of Rio Grande do Sul, Brazil. Mycopathologia 177:207–215

Benard G (2021) Pathogenesis and classification of paracocidioidomycosis: new insights from old good stuff. Open Forum Infect Dis 8(3):ofaa624

Bocca AL, Amaral AC, Teixeira MM, Sato PK, Shikanai-Yasuda MA et al (2013) Paracoccidioidomycosis: eco-epidemiology, taxonomy and clinical and therapeutic issues. Future Microbiol 8:1177–1191

Boyce KJ, Andrianopoulos A (2015) Fungal dimorphism: the switch from hyphae to yeast is a specialized morphogenetic adaptation allowing colonization of a host. FEMS Microbiol Rev 39:797–811

Cacere CR, Mendes-Giannini MJ, Fontes CJ, Kono A, Duarte AJ, Benard G (2008) Altered expression of the costimulatory molecules CD80, CD86, CD152, PD-1 and ICOS on T-cells from paracoccidioidomycosis patients: lack of correlation with T-cell hyporesponsiveness. Clin Immunol 129:341–349

Calich VL, Vaz CA, Burger E (1998) Immunity to *Paracoccidioides brasiliensis* infection. Res Immunol 149(4–5):407–417; discussion 499–500. https://doi.org/10.1016/s0923-2494(98)80764-5

Cano LE, Singer-Vermes LM, Vaz CA et al (1995) Pulmonary paracoccidioidomycosis in resistant and susceptible mice: relationship among progression of infection, bronchoalveolar cell activation, cellular immune response, and specific isotype patterns. Infect Immun 63:1777–1783

Chaves EGA, Parente-Rocha JA, Baeza LC, Araújo DS, Borges CL, de Oliveira MAP, Soares CMA (2019) Proteomic analysis of *Paracoccidioides brasiliensis* during infection of alveolar macrophages primed or not by interferon-gamma. Front Microbiol 10:96

da Silva JF, de Oliveira HC, Marcos CM, Assato PA, Fusco-Almeida AM, Mendes-Giannini MJS (2016) Advances and challenges in paracoccidioidomycosis serology caused by *Paracoccidioides* species complex: an update. Diagn Microbiol Infect Dis 84(1):87–94

De Almeida JN Jr, Peçanha-Pietrobom PM, Colombo AL (2019) Paracoccidioidomycosis in immunocompromised patients: a literature review. J Fungi 5(1):2

de Pina DR, Alvarez M, Giacomini G, Pavan ALM, Guedes CIA, de Souza Cavalcante R et al (2017) Paracoccidioidomycosis: level of pulmonary sequelae in high resolution computed tomography images from patients of two endemic regions of Brazil. Quant Imaging Med Surg 7(3):318–325

Gonzalez A (2020) The therapy of pulmonary fibrosis in paracoccidioidomycosis: what are the new experimental approaches? J Fungi (Basel) 6(4):217

Losnak DO, Rocha FR, Almeida BS, Batista KZS, Althoff SL, Haupt J et al (2018) Molecular detection of fungi of public health importance in wild animals from southern Brazil. Mycoses 61(7):455–463

Martinez R (2017) New trends in paracoccidioidomycosis epidemiology. J Fungi (Basel). 3(1):1

Niño-Vega GA, Camacho E, Restrepo Moreno A, Tobón AM, Gómez BL, Teixeira MM, Barker BM (2017) Chapter 9: *Paracoccidioides* spp. and paracoccidioidomycosis. In: Montes HM, Lopes-Becerra L (eds) Current progress on medical mycology. Springer International Publishing, pp 281–308

Pinheiro BG, Hahn RC, Camargo ZP, Rodrigues AM (2020) Molecular tools for detection and identification of *Paracoccidioides* species: current status and future perspectives. J Fungi (Basel). 6(4):293

Restrepo A, Tobón AM, Gonzalez A (2019) Chapter 267: Paracoccidioidomycosis. In: Bennett JE, Dolin R, Blaser MJ (eds) Mandell, Douglas, and Bennett's principles and practice of infectious diseases, 9th edn. Elsevier, Philadelphia, pp 3211–3221

Romaneli M, Tardelli NR, Tresoldi AT, Morcillo AM, Pereira RM (2019) Acute-subacute paracoccidioidomycosis: a paediatric cohort of 141 patients, exploring clinical characteristics, laboratorial analysis and developing a non-survival predictor. Mycoses 62(11):999–1005

Thompson G, Gómez B (2019) *Histoplasma*, blastomyces, coccidioides, and other dimorphic fungi causing systemic mycoses. In: Carroll KC, Pfaller MA, Landry ML, McAdam AJ, Patel R, Richter SS et al (eds) Manual of Clinical Microbiology, 12th edn. ASM Press, Washington, DC, pp 569–605, 2109–2127

Villa LA, Tobón A, Restrepo A, Calle D, Rosero DS, Gómey BL, Restrepo A (2000) Central nervous system paraccocidioidomycosis. Report of a case successfully treated with itraconazol. Rev Inst Med Trop S Paulo 42(4):231–234

Zoonotic Paracoccidioidomycosis Ceti

17

Mendoza Leonel and Vilela Raquel

Learning Objectives

By the end of the chapter, the readers should be able to:

1. Learn new nomenclatural issues related to paracoccidioidomycosis ceti (PCMC) in dolphins.
2. Understand the history of the disease.
3. Highlight the epidemiological, clinical, and pathological features of PCMC in dolphins.
4. Learn laboratory diagnosis of PCMC and review dolphin cases treated with antifungal drugs.

Introduction

Currently, the etiologic agent of skin granulomas in dolphins is known as *Paracoccidioides ceti*, and the disease is named paracoccidioidomycosis ceti (PCMC). Based on earlier phenotypic features, it was assumed that the aetiology of dolphin granulomas was *Lacazia loboi*. Since then, numerous cases of cutaneous granulomas with identical yeast-like cells in the infected tissues have been found in different dolphin species around the coasts of Asia, Europe, and North and South America.

History

The infections caused by *Paracoccidioides ceti* (by mistake, the species in the original proposal was spelled with two "i" by Vilela et al. 2021) were first reported by Migaki et al. in 1971 in a bottlenose dolphin (*Tursiops truncatus*) dwelling in US coastal waters. The lesions in the infected dolphin were restricted to the subcutaneous tissues, with clinical and histopathological features similar to the human skin disease described by Jorge Lôbo in 1931. Therefore, Migaki et al. assumed that they were dealing with the same etiologic agent as that reported by Lôbo. Supporting this tenet was the finding that the dolphin aetiology resisted culture as well, a feature in common with the aetiology of human skin disease.

Soon after, several cases of the disease, developing skin granulomas, were also reported in other dolphin species: *Lagenorhynchus obliquidens*, *Tursiops aduncus* and *Tursiops truncatus* dwelling in many oceans, and *Sotalia guianensis*

M. Leonel (✉)
Microbiology, Immunology and Genetics, Biomedical Laboratory Diagnostics, Michigan State University, East Lansing, MI, USA
e-mail: mendoza9@msu.edu

V. Raquel
Microbiology, Immunology and Genetics, Biomedical Laboratory Diagnostics, Michigan State University, East Lansing, MI, USA

Faculty of Pharmacy, Federal University of Minas Gerais, Belo Horizonte, MG, Brazil

S. C. Parija, S. M. Rudramurthy (eds.), *Textbook of Fungal Zoonoses and Sapronoses*, Microbial Zoonoses, https://doi.org/10.1007/978-981-97-3214-2_17

captured in the Surinam River estuary. These cases were diagnosed as "lobomycosis or lacaziosis" and were believed to be caused by the same etiologic agent as in humans, *Lacazia loboi*.

In our understanding, we found no statistical support to separate the pathogen of dolphins from that of humans using a phenotypic approach (unpublished data). In addition, Taborda et al. reported another difference between the human and dolphin pathogens. They stated that, except for *Lacazia loboi* (the name used at that time), *Paracoccidioides* species lack melanin in their cell walls. Probably, the authors were not aware that melanin is an important virulence factor in all *Paracoccidioides* species, including *Paracoccidioides americana*, *Paracoccidioides brasiliensis* (sensu lato), and *Paracoccidioides lutzii*. Since the etiologic agent causing skin granulomas in humans and dolphins was believed to be the same fungus, the disease and its aetiology in dolphins were known by the same names as that in humans. Lôbo's disease, lobomycosis, lacaziosis, and others are examples of the diseases caused by *Loboa loboi*, *Lacazia loboi* and others. Recently, phylogenetic analyses of DNA sequences showed that the etiologic agent in dolphins was a *Paracoccidioides* species, suggesting the binomial *P. ceti*.

Taxonomy

Ever since Jorge Lôbo reported the first human case of the disease in 1931, its uncultivated aetiology has been at the centre of endless controversies. For instance, Fonseca and Area Leão in 1940 claimed that they had isolated the etiologic agent of the disease from the same patient originally studied by Jorge Lôbo. Based on the phenotypic features of the isolate, they introduced a new genus and species, *Glenosporella loboi*. The authors followed nomenclature requirements of that time, including a Latin description of the proposed name. The recovered isolate, however, was subsequently identified as a pure culture of *P. brasiliensis*, thus invalidating the proposal. Nine years later, Fonseca studied a new human case and observed phenotypic features in the infected tissues different from that proposed by Fonseca and Leão, suggesting a different aetiology. Based on this concept, he introduced a new binomial *Glenosporopsis amazonica* for a filamentous fungus isolated from the patient, however the same isolate was later identified as *Aspergillus penicillioides*. Consequently, the uncultivated nature of the pathogen led to the publication of numerous proposals of disease names and binomials for its aetiology, culminating in the proposal of *Lacazia loboi* in 1999.

Genomics and Proteomics

Earlier molecular analyses of several DNA coding proteins and rDNA sequences supported the placement of the uncultivated etiologic agent of skin disease in humans and dolphins in its own genus. However, Rotstein et al. using genomic DNA from a US dolphin with the disease, first challenged this notion. They found that the PCR-amplified 28S LSU rDNA from the yeast-like cells in the infected tissues had more in common with the DNA sequences of *P. brasiliensis* than with that of the uncultivated human pathogen known as *L. loboi*. Subsequently, several investigators using various DNA sequences from different infected dolphin species confirmed their findings. More recently, Vilela et al. in population genetic analyses with several DNA sequences from protein-coding and rDNA, concluded that the dolphin pathogens were indeed a *Paracoccidioides* species they named *P. ceti*, different from the one causing skin granulomas in humans, labelled by them as *P. loboi*. However, the taxonomic validity of the *P. loboi* proposal was questioned. This was due to the many genera and species introduced in the last 92 years for the uncultivated human pathogen, and therefore it is unclear which of the proposed binomials is the true basionym of the human etiologic agent. Recently, Vilela et al. (2023) proposed the binomial *Paracoccidioides lobogeorgii* for the human pathogen to end 92 years of taxonomic uncertainties for this uncultivable species.

Fungus Morphology

In the infected tissues of both the humans and dolphin, *L. loboi* displays the presence of numerous, uniform in size, lemon-shaped yeast-like cells (10–12 μm), forming short or long yeast chains connected by small bridge-like structures (Fig. 17.1). It has been postulated that the size of yeast-like cells in humans and dolphins differs. However, we have had the opportunity to measure numerous specimens of humans and dolphins and found that the etiologic agent in both species cannot be discriminated by their yeast-like cell size alone.

Cultivation of Fungi

P. ceti and *P. lobogeorgii* have yet to be cultured in vitro (Table 17.1). The first report on culturing the fungus from dolphin clinical specimens was that of Caldwell et al. They inoculated several media with macerated tissue sections containing numerous *P. ceti* yeast-like cells. Some of the inoculated cultures were negative, while others grew several well-known saprophytic fungi. The same results were obtained later by De Vries and Laarman and by various other investigators with similar cases of PCMC in dolphins. More recently, Schaefer et al. utilising novel media, cell culture and artificial skin matrices, were unsuccessful in recovering the etiologic agent of PCMC. In essence, *P. lobogeorgii* and *P. ceti* resist culture.

Animal Models

Successful experimental inoculation of the uncultivated aetiology of Jorge Lôbo disease, causing skin infection in humans, in different animals, including humans, has been successfully published. However, currently no literature is available on the successful experimental inoculation of *P. ceti* in animals.

Experimental Infection from Infected Dolphin to Mice: Caldwell et al. in 1975, using fresh yeast-like cells from an infected dolphin, were the first to inject monkeys (*Macaca mulatta*) and mice. The monkeys were subcutaneously inoculated in different anatomical skin areas with a heavy suspension of yeast-like cells. The animals were observed daily, and after 1 week, small non-

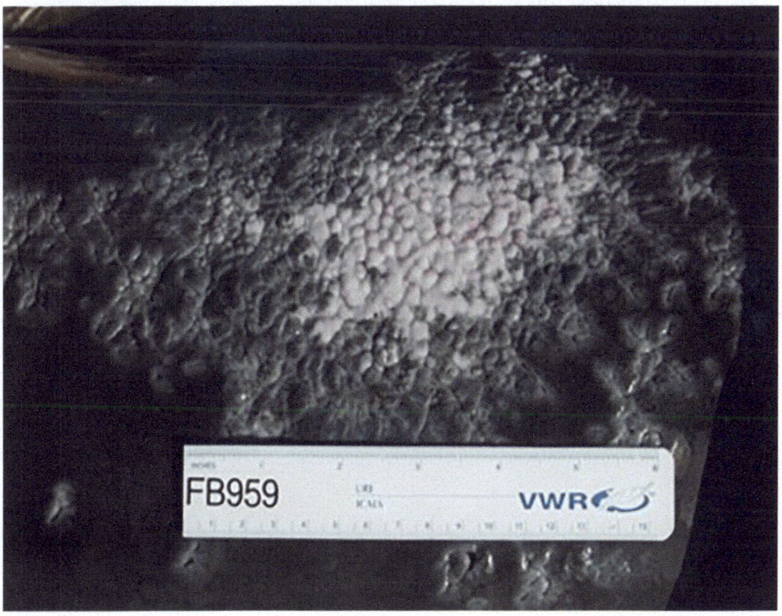

Fig. 17.1 The figure shows extensive multifocal granulomatous lesions on the body of an infected dolphin. Note 1.0–2.0 cm in diameter new lesions (left lower section), as well as the extensive grey verrucous lesions surrounding and old white granulomatous centre. The skin on the verrucous areas bleed easily as can be observed around the white verrucous centre (Courtesy of Dr. Gregory D. Bossart)

Table 17.1 Laboratory diagnosis of paracoccidioidomycosis ceti

Diagnostic approaches	Methods	Targets	Remarks
Direct microscopy	10% Potassium hydroxide (KOH)	Yeast-like cells	Economical and fast The yeast appears as single 5–20 µm in diameter cells forming chains connected by small bridges
Histopathology	Haematoxylin and Eosin (H&E)	Infected tissue	Fibrotic tissue, granulomatous inflammatory reaction, neutrophils and giant cells with numerous ghost yeast cells
	Gomori methenamine silver (GMS)	Yeast cells	Yeast cells appears as dark oval to lemon-shape structures forming small and large chains connected by small bridges
	Periodic acid-Schiff (PAS)	Yeast cells	Yeast cells appear as purple oval to lemon-shape structures forming small and large chains connected by small bridges
In vitro cultures	*Paracoccidioides ceti* resists culture	NA	*Paracoccidioides ceti* does not growth on culture
Immunodiagnostics	Western blot (WB)	Gp43 protein	PCMC dolphins recognise a Gp43-like antigen in WB
Molecular assays	Polymerase chain reaction (PCR)	18S rRNA and ITS1–5.8 s rRNA-ITS2, Ribosylation factor, Gp43, CHS4 and others	High sensitivity and specificity Used to confirm histopathological findings

erythematous 5 mm in diameter nodules were detected at each injection sites. Subsequently, the nodules gradually decreased in size and finally disappeared. After 18 months of daily observations, no lesions developed in the challenged animals. The dolphin yeast-like cells were also inoculated into four mice. The mice were challenged with three-hundredths of 1.0 mL of the inoculum injected into the right rear footpad. The animals were evaluated daily for the development of cutaneous granulomas. After 12 months of daily observation, no lesions developed during that period. The experimental infection was successfully transmitted to three generations of mice. This outcome indicates that the etiologic agent of PCMC could infect other species (Dr Libero Ajello, personal communication) (Fig. 17.2).

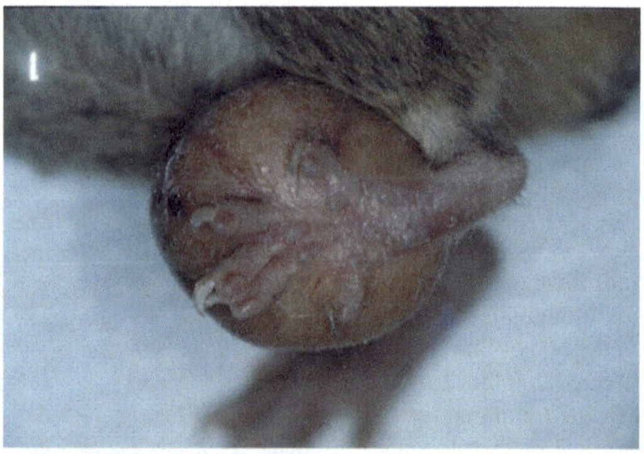

Fig. 17.2 Dr. Libero Aiello's original photograph of a mouse experimentally inoculated with *Paracoccidioides ceti* yeast like cells from an infected dolphin. *Paracoccidioides ceti* yeast-like cells were injected in the footpad and after 12 months, the mouse developed a 1.5 cm in diameter granuloma. In histopathology, numerous *Paracoccidioides ceti* yeast-like cells detected in the infected areas (data not shown)

Life Cycle of Fungi

The life cycle of *P. ceti* is yet to be described. It is assumed that this fungus is closely associated with aquatic environments. Difficulty in the isolation of this fungus limits the understanding of the life cycle of this fungus.

Pathogenesis and Pathology

The disease is acquired by the inoculation of *P. ceti* through small skin cuts. Once inside the tissue, *P. ceti* develops small yeast cells in branching chains that reproduce very slowly by budding. Bossart et al. proposed that *P. ceti* could be transmitted dolphin-to-dolphin, since most infected dolphins have been inhabiting the coastal areas of Florida their entire lives. Perhaps infected dolphins migrating from Latin America to the coastal areas of Florida carry the pathogen, which is then transmitted to healthy dolphins by direct contact through bleeding wounds containing the pathogen.

In the infected tissue, *P. ceti* triggers a granulomatous response with the production of abundant fibrotic tissue (responsible for the parakeloidal aspect of the lesions), numerous macrophages, giant cells, mast cells, natural killer cells, and lymphocytes. The pathogen is present between the superficial layers of the skin and the subcutaneous tissues. The infected areas contain numerous 5 to ~18 µm in diameter yeast-like cells, forming short and long branching chains. The yeast-like cells at the base are larger than the ones at the end of the chains. These cells could be round or lemon-shaped, with their cell wall refractile and prominent. Each yeast cell is connected to the next by small bridges, sometimes forming multiple branching chains. In wet mounts (10% KOH preparation), the cytoplasm of the yeast-like cells shows one or more spherical inclusions, displaying Brownian movement. In haematoxylin and eosin, *P. ceti* yeast-like cells are poorly stained (ghost cells) (Table 17.1). In this stain, the cytoplasm appears slightly eosinophilic, sometimes with small basophil inclusions. In Gomori methenamine silver (GMS) and Periodic acid-Schiff (PAS) stains, *P. ceti* yeast cells stain very well with strong dark and purple colours, respectively.

Immunology

Several authors have linked the susceptibility of dolphins to *P. ceti* propagules, to immunological abnormalities and water pollution. For instance, Reif et al. recently postulated that dolphin with PCMC had multiple abnormalities in immunologic parameters when compared to healthy dolphins. They found that the absolute number of circulating lymphocytes and the serum albumin concentration in these animals was low, whereas neutrophils, alpha 1, total beta, total gamma and total globulins had increased. The authors also found that innate immunity was relatively intact. However, plasma lysozyme concentrations, a pro-inflammatory marker, were elevated in dolphins with PCMC. This could be attributed to secondary contamination of the skin lesions with bacteria.

Reif et al. found that cell immunity parameters were significantly low in the infected dolphins. They reported low counts in the absolute numbers of CD4+ helper T cells and CD19+ and CD21+ B cells. They also found that the ratios of CD2+ T cells to CD4+ cells and CD2+ to CD21+ cells increased, and the numbers of lymphocytes expressing MHC class II molecules decreased in infected dolphins. When lymphocyte proliferation was tested, it was found to be reduced in response to stimulation with lipopolysaccharide and concanavalin A. These results indicate that dolphins with PCMC exhibit significant impairment in adaptive immunity. Host immunology, together with external parameters, such as water pollution, would have played a role in the susceptibility of the dolphin species to *P. ceti* infection.

Infection in Humans

Naturally acquired infections due to *P. lobogeorgii* in humans inhabiting endemic areas have been encountered with increased frequency. Tourists from non-endemic areas visiting endemic Latin American countries were diagnosed with skin infections caused by *P. lobogeorgii*. In addition, an accidental infection of a physician dealing with experimentally infected mice and successful inoculation in a human volunteer have also been recorded. These reports suggest that humans are susceptible to *P. lobo-georgii)* infections from naturally infected sources, including material from infected hosts.

Zoonotic Transmission in Humans

Since 1971, numerous cases of PCMC in dolphins have been reported; however, transmission from infected dolphins to humans is a rare event. The first documented case was reported in 1983. Symmers reported the presence of skin granulomatous lesions in an aquarium attendant who was in touch with a PCMC-infected dolphin. The dolphin was captured in the Gulf of Gascony, France, and was found below the range of weight normal for its size. After careful examination, the dolphin showed opaque plaque-like lesions on the skin, some showing ulcerated areas. The lesions were distributed on the dorsal and ventral areas, around the eyes, and tail flukes. The dolphin was later sacrificed after the histopathological evaluation of the dolphin's skin diagnosed an advanced case of skin cancer.

During the time in the aquarium, the dolphin, which was sacrificed, was under the supervision of a 24-year-old man. Three months later, he noticed an itchy papule on the skin of the medial part of the dorsum of one hand. In the following weeks, the lesion grew, becoming rough and scaly. He did not seek any medical treatment because he thought it was a common wart. Five months later, the lesion measured 1.0 cm in diameter, was well-defined and raised 0.5 cm above the smooth skin. It was also noted that the supratrochlear lymph node on the same side of the granulomatous lesion was enlarged, firm and tender. The man had never been outside Europe. A histological preparation in H&E from the skin of the infected man evaluated by Symmers (1983) showed giant cells containing numerous yeast-like cells. The diagnosis of the condition as Jorge lobo's disease was established, and thus, the review of the histological section of the infected dolphins was requested. Upon re-examination, it was found that the earlier diagnosis of cancer in dolphin was a case of Jorge Lobo's disease. Since France is not endemic for Jorge Lobo's disease, the aquarium attendant likely acquired the infection after the traumatic implantation of *P. ceti* yeast-like cells. This unusual case report suggests *P. ceti* could, in fact, infect human hosts. Support for Symmers' case report was provided by Dr. Libero Ajello (personal communication), who replicated the infection in experimentally inoculated mice with yeast-like cells from an infected dolphin (Fig. 17.2). In addition, *P. loboi* can also be experimentally reproduced in several animals, including humans, suggesting that *P. ceti* could naturally infect humans.

Norton, in 2006, reported an interesting experience. He was assigned to Hawaii, where he had the opportunity to examine a dolphin with PCMC lesions. During the collection of biopsied tissue, the dolphin bucked and the surgical blade cut his thumb. His first thought was that some yeast elements had entered his thumb, and because the disease does not have treatment, he considered amputating his finger. Instead, he decided to wait. Subsequently, in 2 weeks, the wound healed normally, and until now he did not develop lesions at the site of injury. Reif et al. are of the opinion that if *P. ceti* can cause zoonotic infection in humans, but the number of cases is extremely low. They argue that in the coastal areas of Florida, particularly in the Indian River Lagoon, 10% of the dolphins had been diagnosed with PCMC, but no cases of human had ever been reported in that area, even in individuals who entered these waters. Therefore, the case reported by Symmers of zoonotic transmission may suggest humans are resistant to the infections, but there is a small population that may be susceptible to this uncultivated pathogen. The fact that the infection can be experimentally replicated in mice (Dr. Libero Ajello, personal communication) suggests that *P. ceti* can be transmitted to other species.

Infection in Animals

Most dolphins diagnosed with PCMC have been reported in stranded animals or in severe cases showing multiple multicentric lesions and thus captured by veterinarians for further analysis. Lesions on the studied dolphins with PCMC have been usually observed around their faces, right and left flippers, dorsal fin, right and left flukes, and other anatomical areas. In early cases, the lesions appear as small (1.0 cm in diameter) and verrucous without ulceration. In chronic cases, extensive lesions are found with the formation of granulomatous multifocal plaques (Fig. 17.1). Although the crusty verrucous plaques are characteristically hard to the touch, they bleed easily. This could release the yeast-like cells present in the infected verrucous areas and could contaminate other dolphins and humans. Macroscopically, a thickened ulcerated epidermis is observed, followed by larger circumscribed nodular areas deep into the reticular dermis, with the enlargement of regional lymph nodes.

Epidemiology and Public Health

Due to the intractability of *Paracoccidioides* to culture, the ecological niche of *P. ceti* is unknown. *Paracoccidioides* infections occur only in South America; hence, it is assumed that dolphins are first infected when swimming near large rivers in the coastal areas of Mexico, Central and South America. Susceptible dolphins swimming in the estuaries of these rivers may be exposed, through small skin wounds, to the propagules of *P. ceti*. Migrating dolphins could carry the pathogen to other geographical locations to non-endemic areas of coastal North America, Europe, South Africa, Australia, and Japan, where cases of PCMC in dolphins have been reported.

Diagnosis

Microscopy

Infected dolphins microscopically showed thickened epidermis due to severe acanthosis, large discrete histiocytic granulomas in the papillary area of the dermis and fibrotic tissue. The granulomas consist of histiocytes, mononuclear macrophages, neutrophils and giant cells. Macrophages are usually found adjacent to the granulomatous lesions and displayed foamy cytoplasm. Within the granuloma, numerous lemon-shaped yeast-like cells connected by slender bridges can be observed (Table 17.1). The yeast cells in silver stains measure between 5 and 12 μm in diameter and are characterised by the formation of long branching chains consisting of two or more yeast-like cells (Fig. 17.3). Some cells could measure more than 15 μm. The mother cell multiplies by budding, and usually, the daughter cell is smaller and remains attached to the mother cell by small tubular bridges. Haubold et al. using phenotypic measurements, recorded differences in cell size between the dolphin and the human aetiology, as mentioned earlier. However, Minakawa et al. reported that the diameter of the mother yeast-like cells infecting *Lagenorhynchus obliquidens* in Japan could reach 7–25 μm. H&E yeast-like cells appear as round ghost structures that do not retain the H&E colours.

In Vitro Culture

Although the granulomatous lesions consist of a large number of yeast-like cells, it has not been possible to recover the etiologic agents of PCMC in pure culture. Recently, Vilela et al. proposed a theory to explain why the pathogens affecting the skin of dolphins and humans resist culture. They postulated that possibly the genes required for hyphae-to-yeast conversion and that involved in yeast-to-yeast maintenance in the infected tissues are successfully expressed during infection. However, the genes required for yeast to hyphae conversion might hold abnormalities (mutations?), resulting in a failure to grow in culture media.

Serodiagnosis

In a study, Western blot (Table 17.1) was used for the detection of serum antibodies in 21 sera collected from PCMC dolphins captured on the coast of Florida, USA. They found that all the

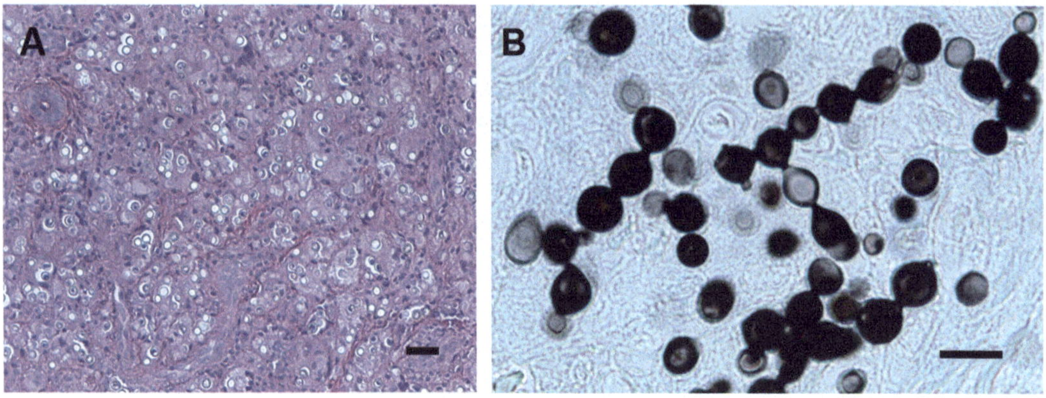

Fig. 17.3 Panel A shows a Haematoxylin and Eosin (H&E) low magnification tissue section from the skin of a dolphin infected with *Paracoccidioides ceti*. Small empty unstained areas containing numerous *Paracoccidioides ceti* yeast-like cells is a common finding. Panel B depicts a silver staining magnification of *Paracoccidioides ceti* from an infected dolphin. Note the long-branched chains of yeast like cells of uniformed size connected by small bridges

tested sera of dolphins possess anti-IgG antibodies against *P. brasiliensis* Gp43 antigen.

Molecular Diagnosis

Molecular methodologies have also been used to unveil the phylogenetic traits of *P. ceti* and its taxonomic connection with cultivable *Paracoccidioides* species (Table 17.1).

Treatment

Dolphins and humans infected with *P. ceti* and *P. lobogeorgii*, respectively, responded poorly to antifungal treatment. It appears that both uncultivable pathogens are intrinsically resistant to most antifungal drugs.

Cases of Jorge Lobo's disease in humans have been treated by surgically removing the skin-affected areas. However, in most cases, the lesions recurred. This treatment is not recommended for dolphin cases of PCMC for obvious reasons. Although most antifungal drugs tested in humans with Jorge Lobo's disease failed, there is one successful report of the use of miconazole in a *Tursiops truncates* with active PCMC. The majority of captured dolphins with PCMC lesions receive only palliative treatment.

Prevention and Control

Until now, a protocol to prevent the pathogen from spreading from PCMC-positive dolphins with skin lesions inhabiting coastal areas is yet to be available. Hence, control methods for PCMC-positive cases in the infected areas are lacking. Even though an increasing number of dolphin PCMC cases in the Florida Indian River Lagoon, USA have been reported so far, no cases of dolphin-human transmission have been diagnosed in this USA-infected coastal area. As a matter of fact, there has been only one dolphin-to-human PCMC transmission case reported by Symmers in 1983.

Case Study

An Atlantic bottlenose adult male dolphin (*Tursiops truncates*) showing numerous cutaneous lesions was captured after being tangled in a fishing line on the coast of Florida, USA. The animal was transported for medical treatment. The dolphin showed firm, whitest verrucous lesions over the rostral melon that laterally extended to the blowhole affecting the right eye. Multicentric lesions were also observed on the edges of both pectoral and dorsal fins. Some lesions were ulcerated and bled easily upon contact.

Blood examination showed leukocytosis with neutrophilia and lymphopenia. Other laboratory reports were within normal reference ranges. Biopsies were collected and submitted pathology. In H&E, the skin tissue showed acanthosis with chronic granulomatous inflammation, including histiocytes, multinucleated giant cells, and numerous neutrophils. Within the infected tissues, numerous poorly stained ghost yeast-like cells were observed. In GMS and PAS stains, abundant dark 5–20 μm in diameter yeast-like cells, some of them forming branching chains of yeast cells connected by small fridges, were the main feature. The pathology report included the diagnosis of PCMC. To validate the report, a fresh biopsy was submitted to the molecular laboratory for DNA analysis. The DNA sequencing showed 100% identity with *P. ceti* DNA sequences available at the National Center for Biotechnology Information (NCBI), confirming the pathology report. Despite palliative treatment with antibiotics and cimetidine, the dolphin died 10 days after its capture. This is a typical example of a common outcome of dolphins suffering from PCMC in infected areas. A comprehensive strategy is necessary to develop a systematic approach to treat and take care of dolphins infected with *P. ceti*.

1. What are the different methods for the diagnosis of the PCMC?
2. What are the key points in the above case study?
3. What are the current treatment plans available for PCMC?
4. Is the pathogen geographically restricted or not?

Research Questions

1. Do you think the clinical diagnosis of *P. ceti* infection in dolphins must be confirmed with laboratory tests?
2. Which tests are available for the diagnosis of this infection?
3. How common is PCMC among dolphins?
4. Are there any geographical niche where this disease could be present?

Further Reading

Almeida FP, Lacaz CS (1949) Blastomicose "Tipo Jorge Lôbo". An Fac Med S Paulo 24:5–37

Belone AFF, Madeira S, Rosa PS, Opromolla DVA (2001) Experimental reproduction of the Jorge Lobo's disease in BALB/c mice inoculated with *Lacazia loboi* obtained from previously infected mouse. Mycopathologia 55:191–194

Bossart GD (1984) Suspected acquired immunodeficiency in an Atlantic bottlenose dolphin with chronic-active hepatitis and lobomycosis. J Am Vet Med Assoc 185:141–1415

Burns RA, Roy JS, Woods C, Padhye AA, Warnock DW (2000) Report of the first human case of lobomycosis in the United States. J Clin Microbiol 38:1283–1285

de Macedo PM et al (2019) Clinical features, and background of the sympatric species *Paracoccidioides brasiliensis* and *Paracoccidioides Americana*. PLoS Negl Trop Dis 13:e0007309. https://doi.org/10.1371/journal.pntd.0007309

Haubold EM et al (2000) Comparative morphology of *Lacazia loboi* (syn. *Loboa loboi*) in dolphins and humans. Med Mycol 38:9–14

Mendoza L et al (2008) Use of sera from humans and dolphins with lacaziosis and sera from experimentally infected mice for western blot analyses of *Lacazia loboi* antigens. Clin Vaccine Immunol 15:164–167

Muñoz JF et al (2016) Genome diversity, recombination, and virulence across the major linages of *Paracoccidioides*. mSphere 1:e00213–e00216. https://doi.org/10.1128/mSphere.00213-16

Norton SA (2006) Dolphin-to-human transmission of lobomycosis? J Am Acad Dermatol 55:723–724

Reif JS et al (2013) Lobomycosis: risk of zoonotic transmission from dolphins to humans. Vector-Borne Zoonotic Dis 13:689–693

Schaefer AM et al (2016) Toward the identification, characterization, and experimental culture of *Lacazia loboi* from Atlantic bottle-nosed dolphin (*Tursops truncatus*). Med Mycol 54:659–665

Teixeira MM et al (2009) Phylogenetic analysis reveals a high level of speciation in the *Paracoccidioides* genus. Mol Phylogenet Evol 52:273–283

Vilani-Moreno FR et al (2005) Immunohistochemical characterization of the cellular infiltrate in Jorge Lobo's disease. Rev Iberoam Micol 22:44–49

Vilela R et al (2005) Molecular model for studying the uncultivated fungal pathogen *Lacazia loboi*. J Clin Microbiol 43:3657–3661

Vilela R, Martins JE, Pereira CN, Melo N, Mendoza L (2007) Molecular study of archival fungal strains isolated from cases of lacaziosis (Jorge Lobo's disease). Mycoses 50:470–474

Vilela R et al (2009) Molecular phylogeny of animal pathogen *Lacazia loboi* inferred from rDNA and DNA coding sequences. Mycol Res 113:851–857

Vilela R et al (2016) Cutaneous granulomas in dolphins caused by a novel uncultivated *Paracoccidioides brasiliensis*. Emerg Infect Dis 22:2063–2069

Vilela R, Mendoza L (2018) Paracoccidioidomycosis ceti (Lacaziosis/lobomycosis) in dolphins. In: Seyedmousavi S et al (eds) Emerging and epizootic fungal infections in animals. Spring International Publishing AG, pp 177–196

Vilela R et al (2021) The taxonomy of two uncultivated fungal mammalian pathogens is revealed through phylogeny and population genetic analyses. Nat Sci Rep 11:18119. https://doi.org/10.1038/s41598-021-97429-7

Vilela R et al (2023) A taxonomic review of the genus *Paracoccidioides*, with focus on the uncultivable species. PLOS Trop Negl Dis 17(4):e0011220. https://doi.org/10.1371/journal.pntd.0011220

Blastomycosis

18

Malini R. Capoor and Sheetal Sharma

Learning Objectives
By the end of the chapter, the readers should be able to:

1. Describe the etiological agents of blastomycoses.
2. Understand the pathogenesis and clinical manifestations of the disease.
3. Learn different diagnostic modalities of blastomycoses.
4. Understand management strategies for blastomycoses.

Introduction

Blastomyces dermatitidis, a thermo-dimorphic fungus that causes blastomycosis, is endemic in North America. *B. dermatitidis* is the asexual form of *Ajellomyces dermatitidis* (teleomorphic stage) that exists in nature and in culture as a mould at room temperature. Upon transformation to the conidia, which become airborne and are inhaled, they cause infection in humans. The conidia reach up to the alveoli, where they are transformed into yeast. The yeast is found in infected mammalian tissue and in culture at 37 °C. This transition from mould to yeast is associated with oxidative phosphorylation, leading to partial uncoupling. *B. dermatitidis* infections are diverse, mimicking many other illnesses, and although potentially fatal, they are treated effectively with antifungal therapy.

History

Blastomycosis was previously known as Gilchrist disease in honour of Gilchrist and Stokes, who first identified the causative fungus. Initially, the fungus *Blastomyces*, was misinterpreted by Gilchrist to be a protozoan. The condition was earlier called North American blastomycosis to differentiate it from South American blastomycosis (paracoccidioidomycosis). Subsequently, this terminology was dropped because it was observed that blastomycosis is not confined to North America but is also found elsewhere. The infection was also referred to as Chicago disease earlier since the earliest cases were found in the Chicago region. However, the name is currently obsolete. In India, *B. dermatitidis* has been isolated from the liver and lungs of the same bat, *Rhinopoma hardwickii*. It is yet to be known whether this fungus can infect humans from bats as a reservoir.

M. R. Capoor (✉) · S. Sharma
Department of Microbiology, VMMC and Safdarjung Hospital, New Delhi, India
e-mail: malinircapoor@gmail.com

S. C. Parija, S. M. Rudramurthy (eds.), *Textbook of Fungal Zoonoses and Sapronoses*, Microbial Zoonoses, https://doi.org/10.1007/978-981-97-3214-2_18

Taxonomy

The family Onygenales (Ajellomycetaceae) consists of mammal-associated fungi belonging to the genera *Histoplasma*, *Blastomyces*, *Paracoccidioides,* and *Emmonsia.* A proposed revision of the taxonomy resulted from the combination of morphological and phylogenetic analyses to identify boundaries between species within the Ajellomycetaceae that have clinical significance. This includes the incorporation of the new genus *Emergomyces,* encompassing two novel species, *Emergomyces orientalis* and *Emergomyces africanus*, the identification of the new species *Blastomyces percursus*, and ongoing efforts to define further species that are like *Blastomyces* and *Emmonsia.* Within the Ajellomycetaceae, dimorphic species are usually limited to certain ecological niches and geographic areas. For instance, *Paracoccidioides brasiliensis, B. dermatitidis,* and *E. africanus* are limited to Latin America, North America, and South Africa respectively. However, *Emergomyces pasteurianus* and *Histoplasma capsulatum* are believed to have a global range. *Ajellomyces crescens* is the teleomorph stage of *Emmonsia crescens* that share the same genus as *H. capsulatum* and *B. dermatitidis*. Moreover, the genomic research verified *Emmonsia's* relatedness to *Blastomyces,* more so for *Emmonsia parva* compared to *E. crescens*. Nonetheless, the designation of *Emmonsia* as a distinct genus was maintained, designating *E. parvum* and *E. crescens* as two different species. Due to the sequencing process, the genus *Emmonsia* underwent significant changes, and the new genus "*Emergomyces*" was created. This new genus includes five species: *Emergomyces pasteurianus*, *Emergomyces europaeus*, *E. orientalis*, *Emergomyces canadensis* and *E. africanus* . Additionally, *E. Crescens,* and *Emmonsia sola* remained in the genus *Emmonsia,* while *E. parva* was moved from the *Emmonsia* to the *Blastomyces*. The presence of tiny budding yeast cells and thermotolerance were found to be the primary physiological characteristics that distinguished *Emmonsia* from *Emergomyces*.

Genomics and Proteomics

The genomic sequence of *Blastomyces* is significantly expanded, revealing long, and often clearly delineated stretches of truncated GC sequences. The isochoric regions, such as low GC gypsy elements, are rich and differ in overall size amongst virulent pathogens such as *B. gilchristii* strain SLH14081 compared to non-pathogenic *B. dermatitidis* strain ER-3, where they are unlikely to be prominent. These isochoric-like regions may have recently evolved in the ancestor of the *Blastomyces* lineage, as evidenced by the absence of indistinguishable regions in closely related species. Genetic content is preserved in abundance among *Blastomyces* and fungi related to *Blastomyces*. Nevertheless, copy number changes are observed in the gene associated with host interactions, such as characterised antigens and proteases.

Subsequent changes in *B. dermatitidis* gene expression have been demonstrated with the interaction of the infective yeast phase with macrophages and in mouse models. The highest expressed gene in vivo, BAD1, enciphers a yeast stage-specific virulence factor that supports attachment to host cells and evasion from host immune protection. The abundance of BAD1 was observed more in the yeast cultured with macrophages at 37 °C than in the yeast-less macrophages. These studies illustrate a potent antioxidant defensive reaction in *Blastomyces* and the in vivo increase of dioxygenase. Hence, it implies that antioxidant-generated dioxides are available for amino acid metabolism. An increase in many functional categories was found in vivo studies. For example, *Blastomyces* may contain important virulence factors such as secretory zinc acquisition proteins, tryptophan, and cysteine metabolism that have been overlooked earlier in vitro studies. The loss of genes for acquiring zinc and variations in the amino acid metabolism indicate *Blastomyces* having host environment adaptations. These findings provide insight into the dynamics of genome evolution and the elements that possibly influence *Blastomyces* virulence.

The Fungus Morphology

The morphology of *B. dermatitidis* varies depending on the stage during its life cycle and on growth conditions. It is a dimorphic fungus, existing as a mould at 25 °C and yeast at 37 °C. The yeast is adapted for growth and survival within the host, while the mould form is better suited for survival in the environment. The morphology of *B. dermatitidis* may vary slightly between different strains and geographic locations.

Mould: In the environment, *B. dermatitidis* exists as a mould, primarily in soil enriched with organic matter. The mould is characterised by its ability to produce conidiophores, the specialised structures that bear conidia (asexual spores) at their tips. The mould is filamentous and produces hyphae, which are hyaline, branching, and septate. These hyphae measure about 2–4 μm in diameter.

Yeast: The yeast form of *B. dermatitidis* is found in the infected tissues and in the culture at 37 °C. The yeast cells are typically spherical to ovoid in shape and measure approximately 8–15 μm in diameter. They have a thick and double-contoured cell wall that appears refractile by microscopy. The yeast cells reproduce asexually by budding, forming daughter cells attached to the parent cell. Each daughter cell can develop into a new yeast cell, contributing to the spread and dissemination of the infection.

Conidia: Conidia are the asexual spores produced by *B. dermatitidis*. They are formed on the conidiophores in the mould form and are responsible for the airborne transmission of the fungus. The conidia are typically small, single-celled, and round to oval. They measure approximately 2–5 μm in diameter. The conidia are released into the air when the mould is disturbed, allowing them to be inhaled by humans or animals, leading to infection.

The dimorphic conversion or switch between the yeast and mould forms depends on environmental conditions. This transition occurs in the host tissues, particularly in response to the body temperature (37 °C) of the infected host and plays a crucial role in its pathogenesis. The yeast is adapted for growth and survival within the host, while the mould form is better suited for survival in the environment. The morphology of *B. dermatitidis* may vary slightly between different strains and geographic locations.

Cultivation of Fungus

Blastomyces thrives in areas with damp, acidic terrain, particularly in locations with a lot of vegetation close to rivers or other water bodies. Most of the early attempts to identify *Blastomyces* from soil samples relied on culture or recovery after intravenous injection into mice, both of which were challenging processes with a few encouraging results.

Potato dextrose agar, Sabouraud dextrose agar or inhibitory mould agar are used for the isolation of *B. dermatitidis* in specimens from sterile sites such as tissue and aspirates. Both basal and inhibitory/selective media contain antibiotics like cycloheximide to prevent the growth of saprophytic fungi. Selective media is used when a sample is likely to be contaminated with bacteria and saprophytic fungi. Brain heart infusion agar with antibacterial agents is recommended for tissue specimens.

Animal Models

Animal models have made significant contributions to our understanding of fungal infections and the therapies that are available to treat them. When it comes to studying experimental blastomycosis, mice and rabbits have several benefits as a model system. The onset and course of the disease may now be quantitatively evaluated in the murine model. By using yeasts for intranasal inoculation, microconidia growth is not delayed, and standard measures may be taken without the need for specialised biohazard equipment and facilities. The goal of growth under in vitro conditions is to consistently titre inocula using visual techniques. The challenge inocula can be used to titrate median survival. Replication gradually advances, at a pace indicated for different challenge sizes, to a limit of 107–108 colony-forming units per gram of cells not compatible with life, even though primary post-challenge lung cultures are not invariably positive. Above a threshold, lung weight and colony-forming units per gram have a strong correlation. Lethality is correlated with both age and starting body weight. Weight accurately predicts death during illness. In cases of prolonged infections, extrapulmonary spread happens before death. Documentation reveals a broad range of virulence in fungal strains. The model's quantification of these characteristics makes it valuable for researching host response or treatment interventions, since the model's efficacy may be linked to pathogen load and illness stage.

Life Cycle of *Blastomyces*

Blastomyces exists as a free-living saprophytic mould with septate aerial hyphae in the environment. These hyphae produce spores (conidia), which, when disturbed, are released into the environment. Aerosolised spores are either inoculated into the skin or inhaled by the susceptible host. The favourable conditions (temperature, pH, humidity) in the host facilitate the transformation of conidia to pathogenic broad-based budding yeast. The yeast may continue to colonise the lungs and disseminate in the bloodstream and other parts of the body, such as the skin, bones, joints, and central nervous system.

Pathogenesis and Pathology

Blastomyces are moulds that produce fungal spores in the environment. Animals and humans who breathe in the spores run the risk of contracting blastomycosis. After entering the lungs, the spores become yeast and subsequently, from the lungs, can disseminate to other organs of the body. When pulmonary host defences fail to contain the infection, it leads to reticuloendothelial and hematogenous dissemination, with the centre of infection reaching the skin, genitourinary system of the male, and skeletal systems.

The ability to transform mould into yeast is important in the pathogenesis of *Blastomyces* spp. infection. Phagocytosis of *B. dermatitidis* conidia by neutrophils, alveolar macrophages, and monocytes provides natural resistance to infection in the lungs. Alveolar macrophages prevent conidia from transforming into pathogenic yeast. It is difficult for scavenger cells to phagocytose and subsequently kill the yeast that possesses thick capsules. In an experimental study, processing of inhaled conidia was found better in polymorphonuclear leucocytes than in macrophages but ineffective for yeast forms. BAD1 (blastomyces adhesion-1, previously known as WI-1) and DRK1 (dimorphism-regulated kinase-1) are two genes required for phase transition and virulence in yeast.

To convert mould to yeast in *B. dermatitidis* and *B. gilchristii* responding to the temperature change from 22 °C to 37 °C requires a hybrid variant of histidine kinase, which is encoded by DRK1. When the DRK1 gene is absent, *Blastomyces* spores develop as hyphae at 37 °C and do not change to yeast form. Mutants with DRK1 null gene defect do not express BAD1. This serves as an important virulence factor and has a varying amount of discrepancy in their cell wall carbohydrate distribution, for example, chitin and α-(1,3)-glucan. Mutants with downregulation of DRK-1 processing are avirulent in lung infection of mouse models.

BAD1 expression in *B. dermatitidis*, a 120 kDa protein, promotes attachment and immune evasion during the yeast stage. BAD1 is secreted into the extracellular matrix and adheres to the chitin present on the cell wall surface. It acts as an adhesin that binds heparin sulphate to bind yeast cells to host tissues and suppress TNF production, thereby enabling immune evasion. TNF-α is a key cytokine that promotes the host's defensive response against infection by *Blastomyces*. When the mould gets transitioned into yeast, it results in an increase of α-(1,3)-glucan in the cell wall. It has been observed that β(1,3)-glucan ranges between 40% and 50% in mould form and 5% in yeast form. Reduced concentration of β(1,3)-glucan in the cell wall of yeast shows important therapeutic and diagnostic attributes. These limit the use of assays of β-(1,3)-glucan for the diagnosis and the use of echinocandins in management.

Immunology

B. dermatitidis infection is characterised by the development of a suppurative reaction, followed by non-caseating granulomas in the lung as well as in the extrapulmonary sites. Reactivation of infection at pulmonary or extrapulmonary sites can occur following the initial resolution of pulmonary foci and even after treatment. Cell-mediated immunity is considered important in progressive blastomycosis prevention. It is described by the demonstration of yeast inside monocytes, giant cells, and macrophages in the form of pyogranulomatous response to blastomycosis. Most information regarding cell-mediated immunity is available from experimental studies in animal models, which correlates with DTH in vitro due to the lack of reliable antigens for skin tests. Humoral immunity does not appear to play an important role in blastomycosis.

The characteristic immune response to *B. dermatitidis* is the formation of non-caseating granulomas involving giant cells (predominantly foreign body type) and epithelioid histiocytes with a cluster of neutrophils and mixed inflammatory reaction. Immunocompetent people who suffer from rapidly progressing and severe diseases may only have neutrophils as part of the inflammatory response. In contrast, immunocompromised people have many organisms in the infected tissue with less inflammatory cells. The presence of pseudo-epitheliomatous hyperplasia and microabscesses in the skin and mucosal surface involvement is characteristic of *B. dermatitidis* infection.

Infection in Humans

Clinical manifestations of blastomycoses vary from asymptomatic infections to symptomatic manifestations even leading to metastasised debilitating illness.

Pulmonary Blastomycosis

Acute pneumonia, chronic pneumonia, or asymptomatic radiographic abnormalities are all signs of pulmonary blastomycosis.

Acute blastomycosis presents flu-like symptoms with cough, fever, arthralgia, myalgia, pleurisy, and acute pneumonia. Chills, fever, and productive cough in the presence or absence of haemoptysis characterise acute pneumonia.

Chronic pneumonia is characterised by the presence of these symptoms, along with night sweats, weight loss, and chest pain persisting for 2–6 months. Upper lobes of the lungs are involved, with the presence of nodular lesions simulating tumours or infiltrates. The radiographic patterns of solitary, miliary, or reticular nodules are less often noticed than the alveolar infiltrate patterns. The cavity formation, nevertheless, is rare. Chronic pneumonia may be misdiagnosed clinically as chronic pulmonary histoplasmosis, pulmonary tuberculosis, or cancer. The intrabronchial spread of infection or miliary blastomycosis can progress to non-cardiogenic pulmonary oedema or diffuse pulmonary infiltrates, presenting as acute respiratory distress syndrome.

Extrapulmonary Blastomycosis

The condition can develop in the skin, bone, genitourinary system, and rarely other tissues.

Cutaneous blastomycosis is characterised by the skin lesions found in nearly 40%–80% of cases. The skin lesions can be verrucous and ulcerative. The subcutaneous abscess has warty lesions with crusted, raised, sharp edges, and irregular shapes with the development of microabscesses. These may resemble giant keratoacanthoma, squamous cell, or basal cell carcinoma in their gross appearance. When an abscess in subcutaneous tissue ruptures and drains through the skin, it may lead to a skin ulcer with raised edges, with or without an exudate.

Bone, next to the skin, is the most common site of infection, approximately in 6%–48% of cases. Ribs, vertebrae, skulls, and long bones are commonly affected. Paravertebral or psoas muscle abscesses are common following infection from the spine. These subcutaneous lesions discharge pus, progressing to prolonged discharge-producing ulcers or sinuses.

The genitourinary system is infected in 10%–30% of cases. The condition manifests as tubo-ovarian abscess in females or infection of prostate, testis, and epididymis in males.

Rare sites involved in blastomycosis include infection of the mouth, nose, or larynx; the eye and middle ear; breast tissue; myocardium, sinuses, pericardium, and gastrointestinal tract; and thyroid and adrenal glands.

Infection in Animals

Animals acquire the infection by inhaling the infectious spores present in the environment. Following inhalation of spores, *B. dermatitidis* converts from mould stage to yeast in the lungs, the pathogenic phase responsible for establishing infection. In the lungs, the yeast form undergoes replication, eliciting an inflammatory response characterised by the infiltration of neutrophils, macrophages, and lymphocytes. The yeast can subsequently disseminate to other organs via the bloodstream, causing systemic infection.

The clinical presentations of blastomycosis in animals vary depending on the animal species, duration of infection, and organs affected. The respiratory system is predominantly affected in dogs, the most affected animals. Clinical presentations include cough, difficulty breathing, and fever. Skin, bones, and eyes are the other organs affected. Skin infection presents as draining suppurative abscesses with ulcers. Blastomycosis is less common in cats but, if present, can manifest as in dogs.

Epidemiology and Public Health

Blastomycosis, primarily a North American disease, is found in the Mississippi River and Ohio River valleys, in the Midwestern states, in the Saint Lawrence Riverway in Canada, and in the Great Lakes. Mississippi, Arkansas, Tennessee, Kentucky, North Carolina, Illinois, Louisiana, and Wisconsin historically had the most cases reported. The disease is hyperendemic in parts of northern Ontario and northcentral Wisconsin, Canada. Only five states in the United States currently mandate reporting of blastomycosis infections: Arkansas, Louisiana, Michigan, Minnesota, and Wisconsin. In Vilas County, northcentral Wisconsin, the annual incidence is 40 cases of blastomycosis per 100,000 people found to be living close to the waterway. Every third patient reported that a minimum of one of their dogs owned had been diagnosed with blastomycosis 6 months prior to the onset of infection. This observation suggests that blastomycosis in canines can be the main source for acquiring the disease in humans.

Cold weather seasonality has also been described in Wisconsin, Manitoba (Canada), and Ontario. Southeastern Mississippi County has reported 12 cases of blastomycosis per 100,000 people, where it was commonly found, with higher rates among African Americans. Genuine autochthonous cases of blastomycosis have been documented in the African continent, specifically Zimbabwe and South Africa. A few cases of autochthonous infections have also been documented in India.

Reports of blastomycosis occurring in patients living in areas not suited for *B. dermatitidis* may be multifactorial. First, physicians unfamiliar with the disease in non-endemic areas may misdiagnose the infection; second, transmission by fomites from naturally occurring areas to non-endemic areas; and third, endogenous reactivation in individuals that move from endemic to non-endemic areas.

Blastomyces survive in moist soil and decomposing organic matter, especially in areas near water bodies or in wooded areas. People who work outside, go hunting or fishing, or who have recently been exposed to places where soil and plants have been disturbed (due to building, excavation, etc.) are frequently reported to be suffering from blastomycosis. Dogs, especially sports or hunting dogs living in these endemic areas, are at higher risk of suffering from the infection due to their high exposure to the soil by virtue of their digging/sniffing behaviours.

Isolation of *B. dermatitidis* from soil was first done by Klein and colleagues during an outbreak involving 48 cases of blastomycosis among adults and school children who travelled to Beaver Dam on the Eagle River of Wisconsin. The fungus was isolated from environmental sources by the same group during different epidemics. The subsequent study, which collected geographically diverse *B. dermatitidis* isolates and were tested by PCR, discovered that the Eagle River soil isolates were different from the variants that caused most of the cases. Nevertheless, cases suggest that the ecological niche of *B. dermatitidis* includes rotting wood, moist earth enriched with animal droppings and decaying organic matter. Infectious conidia get released into the air and inhaled by a susceptible host upon the disturbance of organic matter or moist soil containing *B. dermatitidis*.

The most common mode of transmission for *B. dermatitidis* is air-borne conidia inhalation. Direct skin inoculation occurs less frequently because of outdoor lacerations, bites from affected dogs, or laboratory accidents. The occurrence of disease in canines and humans is only because of their shared environmental surroundings. *B. dermatitidis* transmission among humans and animals is unlikely. Transmission among humans is not reported, and blastomycosis is generally non-contagious. The vicinity of open water bodies in endemic areas is as important a risk as occupational exposure for the residents. The disease is predominantly found among adult males because of their involvement in activities like hunting, forestry work, or fishing, which are associated with exposure to *B dermatitidis*. Sex, race, and age do not have a direct effect on blastomycosis susceptibility.

Diagnosis

The diagnosis of blastomycosis in humans is based on patient history, clinical diagnosis, and laboratory tests.

Specimens include bronchoalveolar lavage fluid (BAL), sputum, and biopsy from skin lesions. In disseminated infection, specimens such as bone, prostate, brain, etc., may be used. Isolation of *B. dermatitidis* by culture remains the gold standard for blastomycosis diagnosis.

It is important to stain biopsy specimens with specialised stains like Grocott-Gomori Methenamine Silver Nitrate (GMS) to search for yeast morphology. This is because acanthosis and pseudo-epitheliomatous hyperplasia presentations can mislead the diagnosis.

Microscopy

Microscopy allows rapid diagnosis by observing the characteristic broad-based budding yeast cells with the *figure-of-eight appearances* of *B. dermatitidis*. Wet mount preparation is the simplest procedure for direct microscopy. Potassium hydroxide (KOH) wet mount is useful for skin scrapping containing cells, tissue debris, and sputum. KOH is used as it dissolves tissue material. Calcofluor white, a fluorochrome compound that stains chitin present in the cell wall of fungi, is used in fluorescence microscopy. *B. dermatitidis* on exposure to UV light fluoresces when observed by a fluorescence microscope.

Histopathology

Based on the European Organisation for Research and Treatment of Cancer and the Mycoses Study Group Education and Research Consortium (EORTC/MSGERC) consensus guidelines, diagnosis is made by the presence of characteristic yeast form of *B. dermatitidis* in the tissues section (Fig 18.1). Distinct yeast forms are relatively large (usually 8–15 μm), thick-walled, and have a single budding daughter cell adhered to the parent cell by a broad base. Periodic acid-Schiff (PAS) stain or methenamine silver stain is used for tissue section to delineate the organism. Haematoxylin and eosin stains are used for body fluids and aspirates, which shows a refractile thick yeast cell wall. Calcofluor white or Papanicolaou staining of cytological specimens is best to appreciate *B. dermatitidis*. Patients with

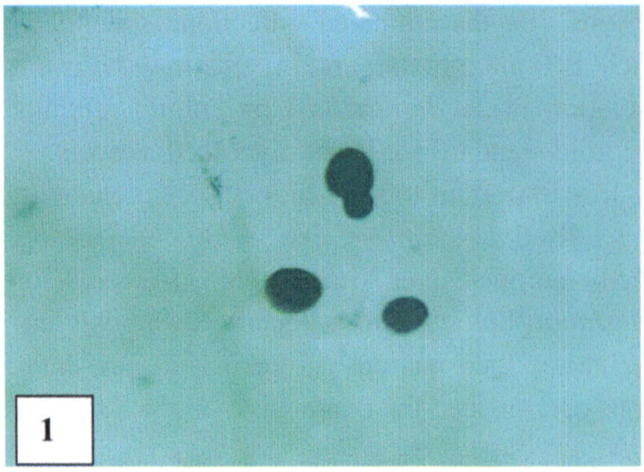

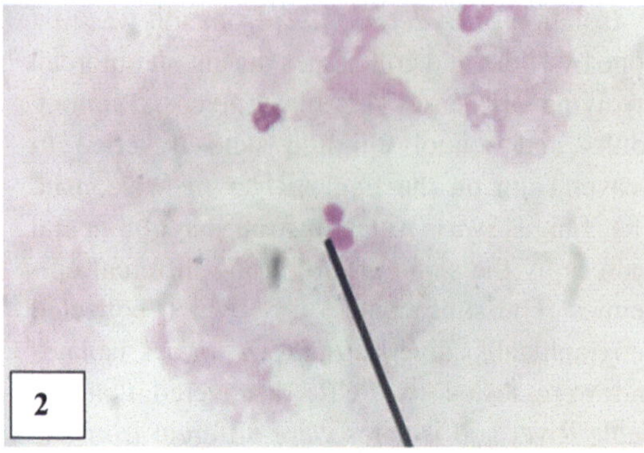

Fig. 18.1 Gomori Methenamin stain (GMS) stain (20×) of biopsy of an exophytic lesion, shows thick-walled yeasts approximately 8–10 μm in diameter, (1) one of which shows broad-based budding typical of *Blastomyces dermatitidis* are seen. (2) An Haematoxyline and Eosin (H&E) stain showing figure of eight appearances

blastomycosis exhibit distinctive pseudo-epitheliomatous hyperplasia or pyogranulomas on the tissue reaction. These methods take less time than the culture. Differentiating features of *B. dermatitidis* from other fungi are listed in Table 18.1.

In vitro Culture

The culture remains the gold standard for the diagnosis of blastomycosis and is more sensitive than direct examination of specimens by microscopy. Potato dextrose agar, Sabouraud dextrose agar, and inhibitory mould agar are used for the isolation of *B. dermatitidis* in specimens from sterile sites such as tissue and aspirates. Brain heart infusion agar with antibacterial agents is recommended for tissue specimens. Inoculated plates are incubated for 4–6 weeks at 25–30 °C. *B. dermatitidis* growth is generally apparent in 7–28 days on fungal isolation media. Colonies are white to off-white initially and glabrous or waxy in appearance, and as they develop with age, aerial hyphae become grey to brown. The mould form of *B. dermatitidis* can be seen under a microscope and is distinguished by delicate, septate hyphae that are 1–2 μm in diameter. Demonstration of oval or pyriform single-celled conidia that are 2–4 μm in diameter and are placed solitary at short or long conidiophores like a lollipop is diagnostic.

Converting the mould to the yeast form, even though it is a reliable method of identification, takes a long time and, hence, is rarely performed in laboratories.

Serodiagnosis

Complement fixation (CF) and immunodiffusion (ID) assays, which are useful for the diagnosis of histoplasmosis, are not used for the diagnosis of blastomycosis because they are not sensitive or specific.

BAD-1 enzyme immunoassay (EIA) is the most frequently used test for the diagnosis of blastomycosis. Serum, urine, CSF, and BAL samples can be tested by using this assay. The test using serum is highly sensitive and specific. The sensitivity of the test in serum ranges from 56% to 82%, heating and EDTA pre-treatment may boost sensitivity. Antigen-based enzyme immunoassay that detects serum galactomannans in the cell wall of *B. dermatitidis* is available. The test shows occasional, cross-reactivity with the cell wall antigens of *H. capsulatum*. The sensitivity and specificity of the enzyme immunoassay for urine samples range from 76% to 90%. The test can be used for antigen detection in BAL fluid and CSF. Antigen detection has been used as a prognostic and diagnostic marker.

Table 18.1 Differentiating features of *Blastomyces dermatitidis*

Fungus	Microscopy	Culture	Serology	Molecular
Blastomyces dermatitidis	Yeast is large (8–15 µm) broad-based budding, with thick refractile cell wall "Figure of eight morphologies" Mucicarmine stain and Alcian blue stain- mildly positive Masson Fontana-negative	Dimorphic At 25 °C downy or white fluffy within 14 days later may become tan coloured At 37 °C—White and fluffy at first later moist, pasty and wrinkled	BAD-1 (EIA) Sensitivity (88%) & specificity (99%) Cross reactivity, with *Histoplasma capsulatum*	Nor standardised, neither widely accessible. *Blastomyces dermatitidis* Polymerase chain reacion (PCR) testing is provided by a few reference laboratories
Cryptococcus neoformans (capsulated)	India ink wet mount shows round yeast cell (5–20 µm) with narrow base budding. With surrounding thick gelatinous refractile capsule Mucicarmine stain and Alcian blue—Shows rose coloured and light blue colour capsule prominently	Monomorphic At 37 °C—Mucoid, cream to buff appearing growth on SDA within 48–72 h Produce brown colour effect when grown on media containing Niger/ sunflower seed Urease test rapid positive Actidione in SDA is –inhibitory	Enzyme immunoassay (EIA) (meridian biosciences), latex agglutination test (meridian biosciences), lateral flow assay (IMMY) for detection of antigen in serum and CSF Cross reactivity with *Trichosporon* spp., *Capnocytophaga* spp., *Stomatococcus* spp.	
Cryptococcus neoformans (non-capsulated)	Masson Fontana stain for melanin production highly useful for cell deficient in capsule			
Immature spherules of *Coccidioides immitis*/*Coccidioides posadasii* particularly which are poorly pigmented	Potassium hydroxide (KOH)/ calcofluor white (CFW)/ Papanicolaou stain stains spherules Branching septate hyphae and chains of barrel-shaped alternate arthroconidia present Endospores (2–5 µm) can be confused with yeast cell of *Blastomyces dermatitidis* or *Cyptococcus. neoformans* Mucicarmine negative Alcian blue negative	Dimorphic but grow as mould at both 25 °C and 37 °C At 25 °C—Moist, smooth colony, 3–5 days later become downy/ suede greyish white with brownish-tan underside arthroconidia take up to 10 days to develop	Antibody testing Precipitin test—Primary diagnosis or exacerbation of infection Latex particle agglutination and complement fixation are most important for disseminated infection	

(continued)

Table 18.1 (continued)

Fungus	Microscopy	Culture	Serology	Molecular
Microforms *Histoplasma capsulatum*	2–4 µm narrow based budding, thin cell wall Tuberculate (spiny) macroconidia (8–15 µm) with septate hyphae Mucicarmine stain barely visible (negative)	White fluffy growth within 2–4 weeks sometimes it might delay to 6 weeks	EIA (IMMY ALPHA *Histoplasma* EIA) in vitro diagnosing testing/MiraVista *Histoplasma capsulatum* galactomannan third-generation EIA Can be used for diagnosis or monitoring the treatment outcome Cross reactivity with other dimorphic fungi Antibody testing by enzyme immunoassay, immunodiffusion, and complement fixation	Currently there is no known PCR-based tests approved by the FDA

Molecular Diagnosis

B. dermatitidis PCR is a highly specific and sensitive test employed to confirm the detection of *B. dermatitidis* in clinical specimens. However, the test is available as an in-house test in only a few reference laboratories. Repetitive-sequence-based PCR and post-culture DNA sequencing are additional methods to confirm the identification of *B. dermatitidis* isolates.

The chemiluminescent DNA probe is another confirmatory test available commercially. With sufficient growth in the culture, this test can detect fungus in a matter of hours. The limitation of the test is that all the isolates of *Paracoccidioides brasiliensis*, *Gymnascella hyaline-spora*, *Emergomyces canadensis*, and *Emmonsia parva* give positive results, even though they differ morphologically from *Blastomyces*.

An assay that targets the BAD1 adhesin gene has been evaluated. This test has been frequently used to detect *B. dermatitidis* in environmental samples, paraffin-embedded tissues from dogs with suspected blastomycosis, and human tissue samples.

Diagnosis of Blastomycosis in Animals

Diagnosis depends on a combination of clinical manifestations, history, and laboratory tests. Fungal culture remains the gold standard for definitive diagnosis. Tissue samples, such as respiratory secretions, skin lesions, or lymph nodes, are cultured for the isolation of *B. dermatitidis*. Microscopy may reveal characteristic yeast cells with broad-based budding. In addition, serological tests, such as antibody detection or antigen detection, facilitate diagnosis.

Treatment

Treatment of blastomycosis is carried out by induction therapy with amphotericin B (deoxycholate or liposomal) for a period of 1–2 weeks. This is followed by maintenance therapy with itraconazole for a duration of at least 12 months. In immunosuppressed individuals, the duration of therapy is usually prolonged. Therapeutic drug monitoring is recommended for patients on azoles for more than 2 weeks.

Discontinuation of azole therapy for AIDS patients is recommended if the CD 4 cell count is above 150 cells/l for at least 6 months while on highly active antiretroviral therapy and have completed 12 months of antifungal therapy. For pregnant women with blastomycosis, lipid amphotericin B at a dose of 3–5 mg/kg/day is recommended by IDSA guidelines. Children tolerate conventional amphotericin better, and maintenance drug includes itraconazole at 10 mg/

kg body weight up to a maximum of 400 mg. Drug interactions, toxicity, and pregnancy influence the therapy of blastomycosis. There is no role of antifungal drug susceptibility as the susceptibility breakpoint is yet not defined for dimorphic fungi.

Pulmonary Blastomycosis

For moderately severe to severe disease in adults, a lipid formulation of amphotericin B (AmB) at a dosage of 3–5 mg/kg per day or AmB deoxycholate at a dosage of 0.7–1 mg/kg per day is used for the initial treatment of 1–2 weeks or until improvement. This is followed by oral itraconazole, 200 mg thrice a day for 3 days and then 200 mg twice a day for 6–12 months.

Extra-Pulmonary Blastomycosis

For moderately severe to severe disease—lipid formulation AmB, 3–5 mg/kg per day, or AmB deoxycholate, 0.7–1 mg/kg per day, for 1–2 weeks or until improvement is recommended, followed by oral itraconazole, 200 mg thrice a day for 3 days, followed by a dosage of 200 mg twice a day for at least 12 months is recommended. For mild to moderate disease, oral itraconazole in a dose of 200 mg is given thrice a day for 3 days and then once or twice daily for 6–12 months.

Treatment of Blastomycosis in Animals

Treatment of blastomycosis in animals is by systemic antifungal therapy. The most used drugs include itraconazole and fluconazole. Treatment duration is generally long-term, often lasting several months or more. In severe cases or cases following complications, combined therapy with amphotericin B may be necessary. Supportive treatment is symptomatic treatment.

Prevention and Control

No vaccine against blastomycosis is currently available. In the areas where blastomycosis is prevalent in the environment, it might not be possible to avoid being exposed to the fungus that causes the disease. Immunocompromised people, however, may avoid activities that involve disrupting soil in these areas (Tables 18.2 and 18.3).

Case Study

A 30-year-old man was admitted to the emergency department with weakness in his left leg, which was diagnosed as left knee arthritis, and he was sent home. On this readmission, he presented with a 2-month history of worsening bilateral leg weakness. He could not walk or get out of bed. He denied fever, chills, or cough but reported

Table 18.2 *Blastomyces dermatitidis* infection in immunocompetent and immunocompromised hosts

Blastomycosis	Immunocompetent host	Immunosuppressed host (solid organ transplant (SOT), malignancy, HIV, chemotherapy, etc.)
Infection rate	Subacute to chronic presentation	Often acute presentation
Cutaneous manifestation	More commonly seen	
Pulmonary manifestation	Pneumonia is most common presentation	Acute pulmonary infection and respiratory failure occurred more frequently.
Mortality		More frequent
Disseminated diseases	Rate of dissemination is similar in both conditions. It shows that pathogen virulence plays greater role in its dissemination potential, rather than host immune response	

Table 18.3 Global distribution of *Blastomyces* species

Species	Global distribution	Comments
Blastomyces dermatitidis	United States, North America, including southern, midwestern, south eastern the St. Lawrence seaway, and midwestern Canadian provinces	Most common agent of blastomycosis
Blastomyces gilchristii	Primarily in Ontario and certain areas of Wisconsin	Can be distinguished from *Blastomyces dermatitidis* only by phylogenetic analysis
Blastomyces helicus	United States and western Canada	Rare, especially in hosts with impaired immune systems
Blastomyces parvus	America (south and north), Australia. Eastern Europe	Causes adiaspiromycosis in rodents and rarely in humans
Blastomyces percursus	Africa, Middle East	Compared with typical blastomycosis, there are more cutaneous and bone involvement
Blastomyces emzantsi	Africa (south)	Similar geographic distribution as *Blastomyces percursus*
Blastomyces silverae	Western Canada	Unrelated to any human illness

night sweats. He lived in Milwaukee and worked as a clerk in the forest department. On his head, torso, and legs, red, annular, raised, and crusted skin lesions appeared for several months, which did not heal with topical treatments. He is afebrile, with a blood pressure of 109/70 and a heart rate of 75. Motor strength was absent in the left leg and barely detectable in the right leg. In the laboratory tests and radiology, the WBC count was 8500/μL with a normal differential count. An MRI of the spine showed partial marrow replacement of the T8 vertebral body and near complete marrow replacement of the T7, L2, and L3 vertebral bodies. An adjacent epidural mass was 6.5 cm long and 11 mm thick. No abscess was noted. Chest radiographs showed an infiltrate in the upper left lobe. The presumptive diagnosis was neoplasm. A granulomatous disease was detected using a lung biopsy. In lung and skin biopsies, fungal elements were observed stained with PAS. A broad-based budding yeast with a thick refractile cell wall was seen in skin scrapings incubated in Mycosel medium at 37 °C.

1. What are the tell-tale pointers in this case?
2. How will you confirm the diagnosis of blastomycosis?
3. How can you differentiate different species implicated in blastomycosis?
4. What are the other diagnostic modalities available?
5. How will you manage the case of pulmonary and disseminated blastomycosis?

Research Questions

1. What is the gold standard for diagnosis of blastomycosis?
2. *Blastomyces dermatitidis* occurs in both mould and yeast form; which one is pathogenic?
3. Why is the selection of antifungals difficult for blastomycosis?
4. What preventive measures could be taken to target the disease in its early stages?

Further Reading

Amini M, Shams WE, Barklow T, Sarubbi FA (2007) A 45-year-old man with fever and adult respiratory distress syndrome. Clin Infect Dis 44(566–67):615–616

Axelson GK, Giorgadze T, Youngberg GA (2008) Evaluation of the use of Congo red staining in the differential diagnosis of *Candida* vs. various other yeast-form fungal organisms. J Cutan Pathol 35:27–30

Babady NE, Buckwalter SP, Hall L, Le Febre KM, Binnicker MJ, Wengenack NL (2011) Detection of *Blastomyces dermatitidis* and *Histoplasma capsulatum* from culture isolates and clinical specimens by use of real-time PCR. J Clin Microboil 49:3204–3208

Bariola JR, Hage CA, Durkin M, Bensadoun E, Gubbins PO, Wheat LJ et al (2011) Detection of *Blastomyces dermatitidis* antigen in patients with newly diagnosed blastomycosis. Diagn Microbiol Infect Dis 69:187–191

Baumgardner DJ, Buggy BP, Mattson BJ, Burdick JS, Ludwig D (1992) Epidemiology of blastomycosis in a region of high endemicity in north Central Wisconsin. Clin Infect Dis 15:629–635

Bradsher RW (2014) Blastomycosis. In: Bennett JE, Dolin R, Blaser MJ (eds) Mandell, Douglas, and Bennett's principles and practice of infectious diseases, 8th edn. Elsevier, Amsterdam, pp 2963–2973

Chapman SW, Dismukes WE, Proia LA, Bradsher RW, Pappas PG, Threlkeld MG, Kauffman A (2008) Clinical practice guidelines for the management of blastomycosis: 2008 update by the Infectious Diseases Society of America. Clin Infect Dis 46(12):1801-12:15

Chu JH, Feudtner C, Heydon K, Walsh TJ, Zaoutis TE (2006) Hospitalizations for endemic mycoses: a population-based national study. Clin Infect Dis 42:822–825

Crampton TL, Light RB, Berg GM, Meyers MP, Schroeder GC, Hershfield ES, JMEmbil. (2002) Epidemiology and clinical spectrum of blastomycosis diagnosed at Manitoba hospitals. Clin Infect Dis 34:1310–1316

Donnelly JP, Chen SC, Kauffman CA, Steinbach WJ, Baddley JW, Verweij PE et al (2020) Revision and update of the consensus definitions of invasive fungal disease from the European Organization for Research and Treatment of cancer and the mycoses study group education and research consortium. Clin Infect Dis 71:1367–1376

Di Salvo F, Klein BS (2009) Blatomycosis. In: Merz WG, Hay RJ (eds) Topley wilson's microbiology and microbial infections, 10th edn. Hodder Arnold Publication, London, pp 481–501

Jiang Y, Dukik K, Muñoz JF, Sigler L, Schwartz IS, Govender NP, Kenyon C, Feng P, van den Ende BG, Stielow JB, Stchigel AM (2018) Phylogeny, ecology and taxonomy of systemic pathogens and their relatives in Ajellomycetaceae (Onygenales): *Blastomyces, Emergomyces, Emmonsia, Emmonsiellopsis*. Fungal Divers 90(1):245–291

Linder KA, Kauffman CA (2020) Current, and new perspectives in the diagnosis of blastomycosis and histoplasmosis. J Fungus 7(1):12

Maini R, Ranjha S, Tandan N, Regmi MR, Ullah S, Stone T, Nayeemuddin F, Pervin N (2020) Pulmonary blastomycosis: A case series and review of unique radiological findings. Med Mycol Case Rep 1(28):49–45

McBride JA, Gauthier GM, Klein BS (2017) Clinical manifestations and treatment of blastomycosis. Clin Chest Med 38(3):435–449

Muñoz JF, Gauthier GM, Desjardins CA, Gallo JE, Holder J, Sullivan TD, Marty AJ, Carmen JC, Chen Z, Ding L, Gujja S (2015) The dynamic genome and transcriptome of the human fungal pathogen *Blastomyces* and close relative *Emmonsia*. PLoS Genet 11(10):e1005493

Reiss E, Shadomy HJ, Lyon GM (2012) Blastomycosis. In: Blackwell W (ed) Fundamental Medical Mycology. John Wiley and Sons Publication, New Jersey, pp 125–139

Saccente M, Woods GL (2010) Clinical and laboratory update on blastomycosis. Clin Microbiol Rev 23(2):367–381

Schwartz IS, Kauffman CA. Blastomycosis. In: Seminars in respiratory and critical care medicine. Thieme Medical Publishers; 2020:41(1):31–41

Smith JA, Gauthier G. New developments in blastomycosis. In: Seminars in respiratory and critical care medicine. Thieme Medical Publishers. 2015:36(5);715–728

Histoplasmosis

19

Shiv Sekhar Chatterjee, Sayantan Banerjee, and Nishith Kumar Pal

Learning Objectives

By the end of the chapter, the readers should be able to:

1. Enumerate and identify the agent(s) of histoplasmosis.
2. Describe the pathogenesis and the pathology of histoplasmosis.
3. Understand the epidemiology of histoplasmosis.
4. Learn the clinical presentation of histoplasmosis.
5. List and describe various diagnostic modalities in the diagnosis of histoplasmosis.
6. Describe various strategies for the management of histoplasmosis.

Introduction

Histoplasmosis, also known as Darling's Disease and named after its discoverer Samuel Taylor Darling, is an endemic mycosis. The etiologic agent, *Histoplasma capsulatum* sensu lato, or simply the erstwhile *Histoplasma capsulatum*, is a dimorphic fungus. It is present in human and animal tissue as intracellular unipolar budding yeast cells and in the environment as filamentous fungi producing typical echinulate (tuberculate) macroconidia. Yeast forms in tissues range from the small (2–4 μm diameter) to the large forms (8–15 μm diameter). Histoplasmosis occurs mainly in restricted geographical foci distributed around the world, especially the Mississippi and Ohio River Valley, USA, Mexico, Panama, Bolivia, Ecuador, Brazil, Cuba, and the Ganges Brahmaputra plains of India. Current scientific evidence indicates that *H. capsulatum*, a single species with at most three varieties: *capsulatum*, *duboisii*, and *farciminosum*, is probably a conglomerate of at least five cryptic species and nine or more clades. Wild as well as domestic animals are commonly infected by the fungus.

History

The disease entity of histoplasmosis was first identified during the seminal work on autopsy patients by Samuel Taylor Darling in 1905. Darling noted the superficial similarity of the etiological agent to *Leishmania donovani* amastigotes and *Plasmodium* species. Autopsy findings of all three patients of Darling were like visceral leishmaniasis with hepatosplenomegaly and mucocutaneous

S. S. Chatterjee (✉) · S. Banerjee
Department of Microbiology, All India Institute of Medical Sciences (AIIMS), Kalyani, West Bengal, India
e-mail: shivsekharchatterjee@gmail.com

N. K. Pal
Department of Microbiology, Jagannath Gupta Institute of Medical Sciences and Hospital, Kolkatta, West Bengal, India

S. C. Parija, S. M. Rudramurthy (eds.), *Textbook of Fungal Zoonoses and Sapronoses*, Microbial Zoonoses, https://doi.org/10.1007/978-981-97-3214-2_19

ulcers. He, however, also noted the characteristic absence of kinetoplasts and the artifactual presence of a hyaline halo surrounding the microbe. In 1913, Henrique da Rocha Lima suspected *Histoplasma* to be yeast due to its similarity to the etiological agent of epizootic lymphangitis in horses. Darling, during his autopsy work, ascribed *H. capsulatum* as a distinct protozoal infection occurring in Central America. Katherine Dodd (1929) established the first antemortem diagnosis in a child. In 1934, de Manbrum identified the property of dimorphism in *Histoplasma*. The untiring works of Amos Christie (1934–35) provided evidence of histoplasmosis being a common respiratory illness and the suspicion of the existence of this pathogen in natural environments. Chester Emmons established the environmental habitat for the fungus. Kwong Chung, in 1973, discovered the perfect state of this pathogen.

Taxonomy

McGinnis and Katz established that the teleomorph of *Histoplasma* belongs to the genus *Ajellomyces*. The scientific name for the teleomorphic state of *H. capsulatum* is *Ajellomyces capsulatus*. *H. capsulatum* is included in the Kingdom Fungi, Subkingdom Dikarya, Phylum Ascomycota, Class Eurotiomycetes, Order Onyngenales, Family Onyngenaceae, and genus *Histoplasma*.

As per the disease pattern, morphology, and restricted host pattern, *H. capsulatum* was previously classified into three varieties with morphologic and pathologic peculiarities: *H. capsulatum* var. *capsulatum*, a human pathogen with small-sized yeast morphology; *H. capsulatum* var. *duboisii*, a human pathogen restricted to Africa with a large yeast phase; and *H. capsulatum* var. *farciminosum*, a pathogen of horses and mules. This classification, however, is no longer preferred.

Genetic analysis of this fungus at multiple loci [18S, 5.8 S, 28S rRNA genes, EF1 (Elongation Factor 1), RPB1 and RPB 2 (RNA polymerase II subunits 1 and 2), ITS-2] has shown the genus to consist of at least eight clades, of which seven have been ascribed phylogenetic species status. The clades include Panama, North American 1 and 2 (Nam 1, Nam 2), Latin American A and B (LAmA and LAmB), Australia, Netherlands (Indonesia), Eurasia, and Africa (Table 1). Further studies, including whole genome sequencing studies on more numerous isolates, have described more clades (Table 1), with 11 current phylogenetic species. The newer phylogenetic species include RJ (previous LAmA), LAmA1, LAmA2, LAmB1, LAmB2, and BAC1 (now designated Nam3). Names and descriptions have been provided for four species from America [*H. capsulatum* sensu stricto (Panama), *H. mississippiense* (Nam 1), *H. ohiense* (Nam 2), and *H. suramericanum* (LAm A)] (Fig 19.1).

Genomics and Proteomics

The diversity of *H. capsulatum* strains has been documented by employing multiple typing approaches like ribosomal DNA polymorphisms, Restriction Fragment Length Polymorphism (RFLP) analysis of the *yps-3* gene, and arbitrary primer PCR. Major differences in the chromosome number of individual strains (ranging from three to seven) have been documented by different techniques like field-inversion gel electrophoresis (FIGE) and contourclamped homogeneous electric field electrophoresis (CCHEFE). Reassociation kinetic experiments show a haploid genome size of 22 Mb with less than 0.5% repetitive DNA. The absolute difference between the varieties *capsulatum* and *duboisii* does not hold true in genetic studies, including those of Mitochondrial DNA restriction patterns. ONT sequencing, followed by additional polishing using short reads from Illumina sequencing, has been used to fully assemble *Histoplasma* genomes (genome size ~31–40 Mb).

The causation of the *H. capsulatum* dimorphism is yet to be fully understood. Important factors involved in the dimorphism of this pathogen include changes in incubation temperature, redox, and CO_2 tension. Sulfhydryl compounds appear to play an important role in dimorphism. *H. capsulatum* specified by multicellular or unicellular developmental states can be attributed to unique transcriptional programs, niche-specific highly transcribed genes that are unique to opposite morphologic states, hyphal or yeast. Many such genes, in addition, have a role in infectivity

Table 19.1 Distribution of *Histoplasma* phylogenetic species in the global population

Histoplasma capsulatum Clade	*Histoplasma* phylogenetic spp. if applicable, Holotype (type strain)	Host	Disease association in humans and special remarks	Distribution
Panama lineage H81	*Histoplasma capsulatum* sensu stricto, ST Darling, 1906, G186A	Humans	Disseminated and pulmonary disease	Central America (Panama), northern South America (North Columbia), and Brazil
NAm1	*Histoplasma mississippiense*, Down's strain	Humans, animals, and environment	Disseminated and pulmonary disease	North America, primarily Mississippi river valley
NAm2	*Histoplasma ohiense*, strain G217B	Humans, animals, and environment	Disseminated and pulmonary disease	North America, Ohio river valley
LAmA1	*Histoplasma suramericanum*, MZ5 strain	Humans, animals, and environment	Disseminated and pulmonary disease. Previous LAmA clade now divided into LAmA1 and LAmA2	Central America including Mexico, Guatemala, Honduras, and Panama
LAmA2	*Histoplasma suramericanum*,	Humans, animals, and environment	Disseminated and pulmonary disease. Previous LAmA clade now divided into LAmA1 and LAmA2	South America including Columbia
RJ genotype in LAmA clade	*Histoplasma suramericanum* RJ type	Humans, animals, and environment	Disseminated and pulmonary disease	South eastern Brazil (Brazilian Atlantic)
African clade including *Histoplasma capsulatum* var. *duboisii*	*Histoplasma capsulatum* Africa	Humans, animals, and environment	Associated with pulmonary and disseminated disease other than skin lesions	Tropical Africa (western and central areas of sub-Sahara), and Madagascar
Australian clade	No species name yet	Humans, animals, and environment	Disseminated and pulmonary disease	Australia
Netherlands (Indonesia) clade	No species name yet	Humans, animals, and environment	Not known	Europe and South East Asia
LAmB1 clade	No species name yet	Humans	LAmB clade now divided into LAmB1 and LAmB2	South America (Columbia, Mexico, Argentina, Brazil)
LAmB2 clade	No species name yet	Humans	LAmB clade now divided into LAmB1 and LAmB2	South America (Columbia, Argentina, Brazil)
NAm3 clade (previous BAC1 clade)	No species name yet	Bats	No human isolate yet	Mexico, Brazil
Lam clade	Not elevated to phylogenetic species level	Humans, animals, and environment	Disseminated and pulmonary disease	Brazil, especially Northeast Brazil
Eurasian clade	Not elevated to phylogenetic species level	Humans, animals, and environment	Disseminated and pulmonary disease. Almost all equine strains belong to this clade	Europe, Thailand, China, India, and North Africa (Egypt)
Indian clade	Not elevated to phylogenetic species level	Human isolates only tested	Disseminated and pulmonary disease	India

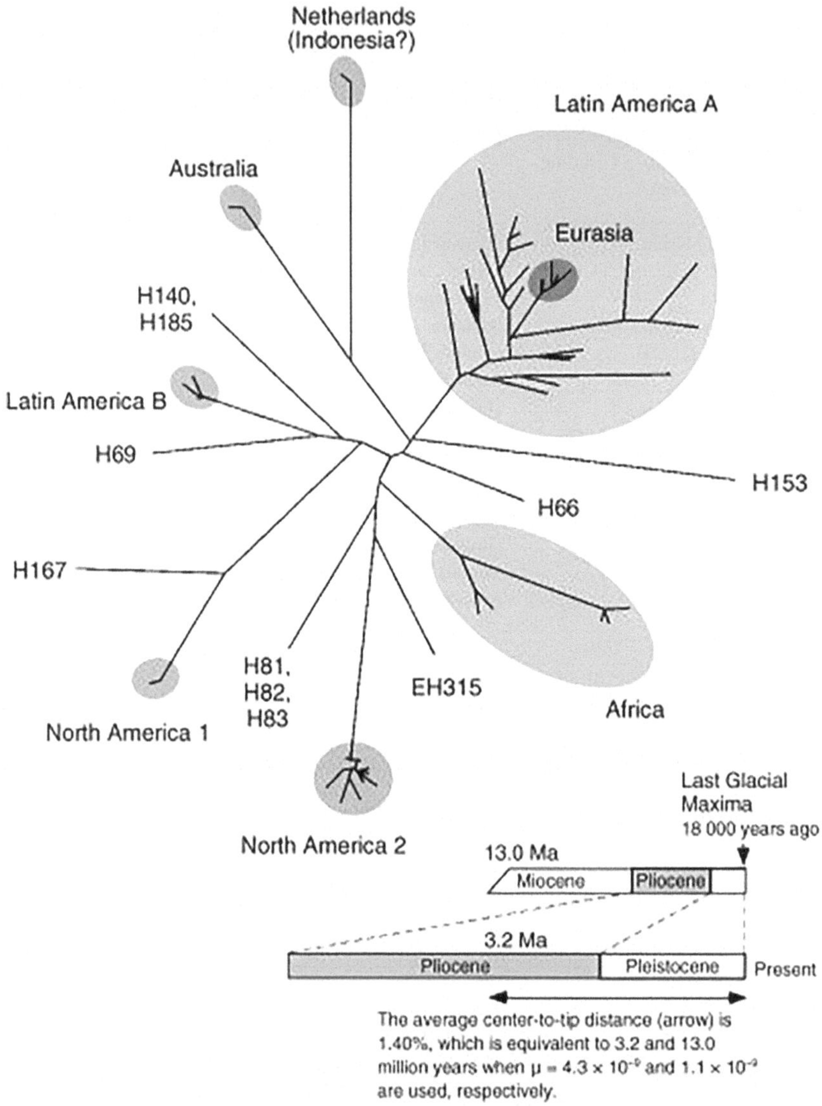

Fig. 19.1 Cladogram of *Histoplasma capsulatum* isolates (A neighbor-joining tree diagram showing the star phylogeny relationship between *Histoplasma* species and clades. The average genetic distance among distinct lineages [NAm 1, NAm 2, LAm A, LAm B, Australia + Netherlands (Indonesia) group, Africa, H81 and H153] was calculated by Kasuga T et al. as 2.8%). (Reproduced with permission from Kasuga T, White TJ, Koenig G, McEwen J, Restrepo A, Castañeda E, Da Silva Lacaz C, Heins-Vaccari EM, De Freitas RS, Zancopé-Oliveira RM, Qin Z, Negroni R, Carter DA, Mikami Y, Tamura M, Taylor ML, Miller GF, Poonwan N, Taylor JW. Phylogeography of the fungal pathogen *Histoplasma capsulatum*. Mol Ecol. 2003; 12(12):3383–401)

or pathogenicity. Yeast-specific genes (*yps3* and calcium-binding protein gene *CBP1*) are preferentially expressed in the yeast phase, while mold-specific genes (*MS8*, *M46,* and the stress response *DDR48* genes) are expressed at higher levels in the environmental mold state. Nearly 2% of *H. capsulatum* transcripts exhibit 5′ leader sequences that differ markedly in length between filamentous and yeast morphologic states.

Due to intracellular compartmentalization within the phagosome of macrophages upon mammalian infection, nutritional challenges exist for yeast growth and replication. *Histoplasma* isolates can *de novo* synthesize all vitamins except thiamine. *RIB2* genes (riboflavin biosynthesis) and *PAN6* genes (pantothenate synthesis) have been found to be closely related to *Histoplasma* virulence.

Among cell wall glycosylated proteins, peptidogalactomannan (pGM) from the yeast phase cell wall is important in phagocytosis by macrophages. The pGM exposure further leads to the release and synthesis of Interleukin 10 (IL-10) and Interferon-gamma (IFN-γ) secretion by the infected macrophages and can induce delayed-type hypersensitivity (DTH) when injected in guinea pigs. It is the carbohydrate moiety on this pGM molecule that is responsible for antigenicity rather than the protein portion.

The cell wall β-1,3-glucan can bind Dectin-1 and DC-SIGN (CD209) receptors on Dendritic cells, neutrophils, cytotoxic T cells, and natural killer (NK) cells and trigger innate immune responses against *Histoplasma*. Another cell wall component, α-1,3-glucan, is expressed as an outermost layer and masks the innate immune recognition (from Dectin-1) of the β-1,3-glucan molecule and regulates yeast-form multiplication inside phagocytes. *Histoplasma* yeasts secrete two β-1,3-glucanases; one named Eng1 is an endoglucanase, while Eng8 is an exoglucanase. Eng1 is mostly responsible for the removal of exposed cell wall β-1,3-glucans. Secreted *Eng1* along with the cell wall α-1,3-glucan allows the *Histoplasma* yeasts to avoid immune surveillance, especially when present within the macrophages.

Heat shock proteins (hsp) found in *H. capsulatum* isolates include *hsp60*, *hsp70*, and *hsp82*, playing a chaperoning role for intracellular proteins and supervising adequate protein folding. *H. capsulatum hsp60* protein effectively binds to the CD11b/CD18 receptor (CR3 molecule) present in macrophages, neutrophils, monocytes, and eosinophils. This binding plays an important role in the immune recognition of these pathogenic yeasts and their phagocytosis by macrophages and neutrophils. Chitins, *yps-3* protein, cell surfaces Histone-2B like 17 kDa protein, M antigen, H antigen, lipids like ceramide monohexosides (CMHs), and cell wall melanin are other important components of the cell wall, many of which are believed to have a role in the virulence of the organism.

The Fungus Morphology

H. capsulatum sensu lato is a thermally dimorphic fungus. A filamentous mold form exists in the environment at temperatures ~25°C. Two types of conidia may be present in the mold phase. Macroconidia are thick-walled with a diameter of 8–15 μm and possess characteristic tubercles/projections on the surface. Smooth-walled microconidia have a diameter of 2–4 μm. Hyphae of the mycelial phase are relatively thin (1.25–2 μm in width).

The yeast phase exists in tissue and at temperatures above 35°C. The yeast phase develops as small (2–5 μm), ovoid, and unipolar budding yeasts. The yeast cells found in African histoplasmosis are thick-walled and are considerably larger (8–15 μm). All yeast forms have typical narrow-based budding starkly dissimilar to the broad-based budding yeast cells of *Blastomyces dermatitidis*.

The cell wall of *Histoplasma* has an inner skeletal layer of chitin, covered by layers of β-1,3-glucan and α-1,3-glucan. The cell wall has smaller components of peptidogalactomannan, heat shock protein molecules, cell surfaces Histone-2B like 17 kDa protein, M antigen (B catalase), H antigen (probably a β-glucosidase), *yps-3* protein, lipids like ceramide monohexosides (CMHs), lectin-like compounds and cell wall melanin. Extracellular vesicles composed of lipid bilayers are secreted through the cell wall. Carbohydrates and lipids constitute around 80% and 3%–10% of cell wall dry weight, respectively.

Cultivation of Fungi

The most useful media to grow *Histoplasma* species include brain heart infusion agar containing sheep blood, sheep blood agar (SBA), Sabouraud's dextrose agar (SDA), potato dextrose agar (PDA), and Smith and Goodman's yeast extract agar with ammonium hydroxide.

Most primary clinical isolations require 7–21 days for visible growth of *H. capsulatum* to appear at 25–30°C. Classically, colonies on SDA or PDA at ambient temperatures are initially off-white to white fluffy mold. These colonies turn buff color with age. These two common types of colonies are termed morphologic types A (albino) and B (brown). In subculture, the brown type shows sectoring and eventually converts to the albino type. The albino colonies reveal white and fine aerial hyphae. The brown type has flat colonies that slowly turn light tan or dark brown on the obverse in 7 days. Albino type may turn buff with the passage of time. At 37°C, colonies grow as granular to rough, mucoid, and cream-colored, and turn brown in a couple of weeks.

Animal Models

Animal models used to study pathogenesis and preclinical therapeutic studies of histoplasmosis include mice, rabbits, and guinea pigs. Routes of infection in these models are usually inhalation by either intranasal inoculation or through aerosol chambers and intratracheal instillation. The intravenous route in mice, rats, rabbits, and guinea pigs has been used to study systemic histoplasmosis. Intracerebral inoculation in mice has been useful during therapeutic evaluations of antifungal drugs for central nervous system (CNS) histoplasmosis.

Life Cycle of *Histoplasma*

Histoplasma exists as a mold with aerial hyphae producing macroconidia and microconidia in the environment. These microscopic conidia circulate in the air once the contaminated soil or guano is disturbed. On inhalation by humans or animals, the conidia enter alveolar macrophages and change into yeast form at a higher temperature (Dimorphism). Within the macrophages, the yeasts survive and even multiply in the phagosome primarily by inhibiting phagosome reactive oxygen and nitrogen species, thereby preventing phagosome acidification and possibly phagolysosomal fusion. *Histoplasma* yeasts aggregate around the cell nucleus around 5 h after infection and induce programmed cell death of the host macrophages. Cytokines like Tumour necrosis factor-alpha (TNF-α) and IL-10 and activation of intracellular caspases 1 and 3 enable the macrophage to release *Histoplasma* yeasts, thereby infecting other macrophages. The intra-macrophage yeasts can then travel first through lymphatics to lymph nodes and then through either lymphatics or hematogenous routes to different organs.

Pathogenesis and Pathology

In the environment, *Histoplasma* exists mostly as a mycelial form. On exposure during various activities such as construction, demolition, clearing of wooded areas, uprooting of trees, shrubs, and cleaning attics contaminated with an infected bat/bird guano, it is the microconidia that gain first access to the respiratory system through inhalation. Neither human-to-human nor direct animal/bird-to-man transmission has ever been documented.

By virtue of their small size and the negative pressure in the respiratory tree during inhalation, the microconidia overcome the nasal and pharyngeal mucus and mucociliary clearance, attributes of the innate protective measures of the infected host. Microconidia subsequently reach the terminal bronchioles and alveoli, encounter higher mammalian body temperature, become dormant, and finally transform into yeasts. Shunt pathways mediated by cysteine and sulfhydryl compounds become active in hours to days and contribute to this dimorphism. The pathogen must wade through the fungicidal surfactant proteins, such as calcium-dependent surfactant protein, which are prime cellular factors in innate immunity. These mediate an increase in the permeability of the transforming yeasts. Dendritic cells (DCs), neutrophils, and natural killer (NK) cells restrict *Histoplasma* growth through phagocytosis and degradation. The β-1,3-glucan, a major cell wall component, is a putative pathogen-associated molecular pattern (PAMP) of the yeast. Nevertheless, the transforming yeasts mostly localize rapidly within the alveolar macrophages via phagocytosis through complement receptors.

The binding of *Histoplasma* heat shock protein 62 kDa (*hsp60*) to the CR3 receptor is one among many unique interactions that enable the binding of the yeast phase to macrophages without triggering additional inflammatory responses. The α-1,3-glucan moiety in the cell wall masks the immunodominant β-1,3-glucan moiety from recognition by the Dectin-1 molecules of the macrophages. Further, the yeast-secreted endoglucanase *Eng1* destroys the β-1,3-glucan moiety, thus allowing the pathogen to escape immune response within the macrophages.

In the lungs of immunocompetent healthy adults, *Histoplasma* infections are characterized by epithelioid cell granulomas, which usually progress to caseation necrosis. These individual caseating granulomas coalesce to produce large areas of consolidation or even liquefy, typically in chronic obstructive airway disease (COAD) patients to form cavities. Spontaneous resolution of infection is the rule in immunocompetent hosts. Both spontaneous resolution and treatment of histoplasmosis with antifungal therapy lead to fibrosis and concentric calcification of lesions, but few yeasts may persist in sequestered niches.

Immunosuppressed patients with fulminant presistent disseminated histoplasmosis (PDH) show focal accumulations of mononuclear phagocytes filled with typical yeast cells throughout tissues and organs of the body. Epithelioid granulomas are absent or rare in the tissues of such infected patients. Persons with underlying illness, especially those with depressed cell-mediated immune systems, are at particular risk of disseminated histoplasmosis. Patients with advanced HIV infection, systemic corticosteroid therapy, allograft transplants requiring immunosuppressants like corticosteroids, mycophenolate or calcineurin inhibitors, very young neonates, smoking-induced COAD, and even COVID-19 infection are also at high risk to the infection (Fig. 19.2).

Infection in Humans

The incubation period is variable (1–3 weeks) and shorter in case of reinfection. Individuals with defective T lymphocyte functioning/numbers are predisposed to a prolonged infection with *Histoplasma*. In cases of reinfection, the disease presents earlier, after 4–7 days of exposure. Rarely, the disease may occur much later, after 18 months or even 10–12 years of migration from an endemic area (Fig. 19.3).

Most *Histoplasma* infections are asymptomatic and self-limiting. The clinical manifestations of disease in symptomatic cases primarily depend on (a) the intensity of exposure, (b) the immune competence of the infected person, and (c) the pulmonary status and anatomy of the infected person. Clinical presentations of classical histoplasmosis are varied and include many conditions.

Asymptomatic and Mild/Self-Limiting Illnesses

Low-level exposure in otherwise healthy immunocompetent individuals results in asymptomatic or mild and self-limited disease. Depending on the level of endemicity, a large proportion of adults residing in endemic areas have a skin test and/or radiographic evidence of the previous infection without symptoms. When smokers from endemic areas are screened for lung cancers, asymptomatic lung nodules representing controlled pulmonary histoplasmosis are often found.

Pulmonary Histoplasmosis

Symptoms of acute histoplasmosis in immunocompetent hosts manifest usually after 1–4 weeks of exposure. Fever, chills, sweats, headache, myalgia, anorexia, cough, dyspnea, and chest pain are the important symptoms after heavy exposure, and may mimic a flu-like illness. Radiological findings reveal minimal focal consolidation (on exposure to low numbers of the fungus), diffuse consolidation (heavy exposure to the fungus), pleural effusion, and hilar or mediastinal adenopathy. Histoplasmomas show up as solitary lung nodules on radiology. Some patients may even present with multiple pulmonary nodules. Such nodules may persist indefinitely as

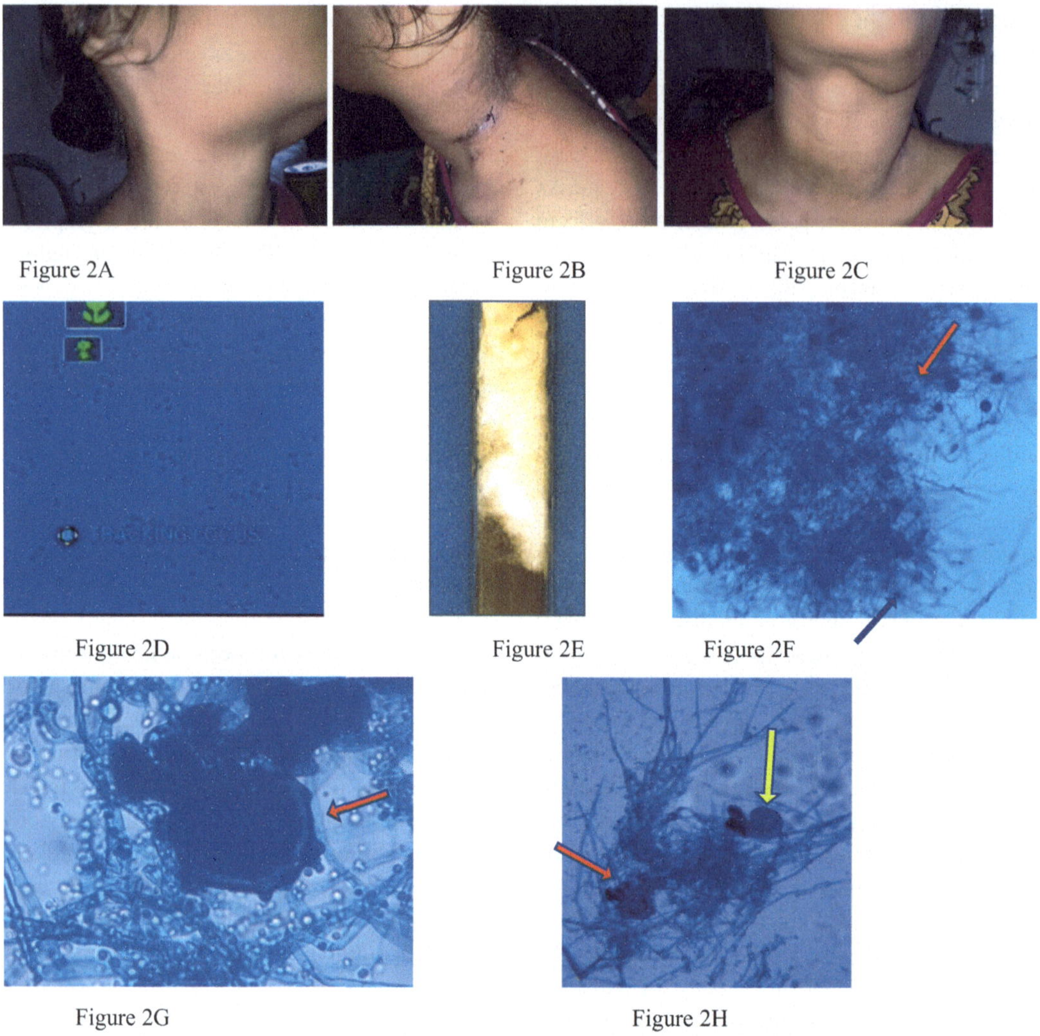

Fig. 19.2 A 32-year-old HIV patient presented with on and off high-grade fever with chills, swelling in the neck glands (**a**–**c**), remarkable weight loss, and fatigue. Previous lymph node biopsy had failed to reach a diagnosis. On fine needle aspiration (FNA), few narrow-based budding yeasts were demonstrated (**d**). On FNA culture in SDA, *Histoplasma capsulatum* could be isolated (**e**) showing multiple echinulate Macroconidia (**f**–**h**, red arrows), very few microconidia (**f**, mauve arrow), and non-tuberculate Macroconidia (**h**, yellow arrow)

foci of infection. Calcification at the center of the suspected histoplasmoma or in concentric rings is characteristic but may take a long time to develop and is not found in all the cases.

In a subset of patients with acute histoplasmosis, there is progressive granulomatosis and/or fibrosis around a peripheral parenchymal lesion or hilar and mediastinal lymph nodes due to aberrant hypersensitivity to the fungus. Mediastinal granulomatosis represents coalescence and excessive fibrosis of lesions in one or several mediastinal lymph nodes—several nodes that become matted may break down and be encapsulated to form a mediastinal mass up to 10 cm in size. The wall of this lesion is 4–5 mm thick or even greater in rare cases. Such a large mass exerts a pressure effect on vital adjacent structures like the superior vena cava, pulmonary artery, pulmonary veins, and bronchi, usually in the right peritracheal areas. Bronchocentric granulomatosis due to *Histoplasma* is a series of granulomata developing in the bronchi in response to

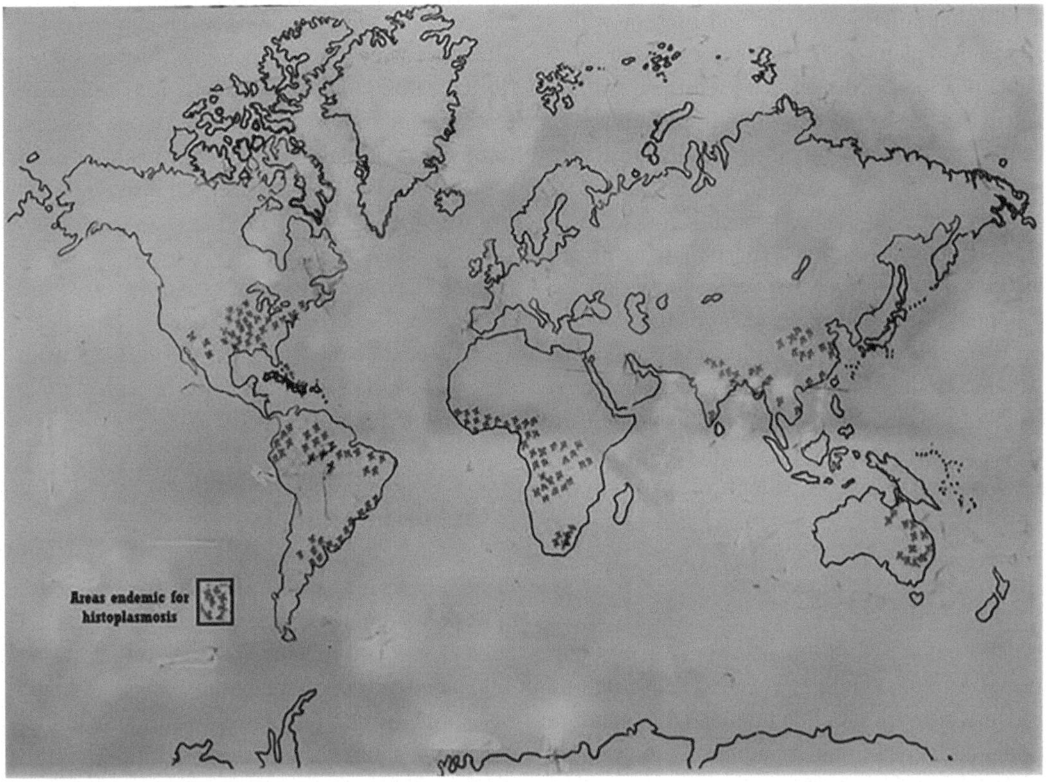

Fig. 19.3 World map showing the areas endemic for histoplasmosis

the fungus resulting in stenosis and mucoid impaction of the bronchus. Histoplasmomas may form in the peripheral lung parenchyma as an enlarging fibrous mass that develops around a healed primary focus.

Extrapulmonary features such as arthralgia, arthritis, uveitis, etc., may occur but usually disappear fast. Hilar and mediastinal lymphadenopathy present in many patients enlarges and may undergo necrosis. Coalescence of such necrotic nodes results in large mediastinal masses leading to external compression of great vessels, proximal airways, and the esophagus or, rarely, bronchopleural fistulas. An indolent form of disseminated histoplasmosis may occur in immunocompetent patients with a special predilection toward oral ulceration and adrenal involvement.

Chronic cavitary pulmonary histoplasmosis (COAD): Smokers who have structural lung pathology (COAD) or pure bullous emphysema often develop a chronic cavitary variety of *Histoplasma* infections. The condition is characterized by chronic productive cough, dyspnea, low-grade fever, night sweats and weight loss, upper lobe infiltrates, cavitation, and pleural thickening on radiographs, and strongly mimics secondary pulmonary tuberculosis.

Progressive disseminated histoplasmosis (PDH): Histoplasmosis in patients with low immunity, especially cell-mediated immunity, typically presents with persistent disseminated histoplasmosis (PDH), the most common presentation. Risk factors for the onset of PDH include acquired immunodeficiency syndrome (AIDS with CD4 <200 cells/μL), extremes of age, administration of immunosuppressive drugs (corticosteroids, mycophenolate mofetil, calcineurin inhibitors like tacrolimus) to prevent or treat rejection following transplantation, and the use of methotrexate, anti-TNFa agents, and other biological response modifying agents for autoimmune disorders. Laryngeal involvement, meningitis, and endocarditis can also occur. The most fulminant PDH infections are acute, have a rapidly deteriorating course, and present with a radiological picture of diffuse interstitial or reticulonodular lung infiltrates or even diffuse military mottling. These patients rapidly progress to respiratory failure, shock, coagulopathy, and multi-organ failure. Some PDH patients, however, present with a more subacute course and a focal organ distribution. These patients often report prolonged fever and weight loss and, on examination, show features of hepatosplenomegaly, lymphadenopathy, and/or thrombocytopenia. Chronic meningitis, focal brain lesions, ulcerations of the oral mucosa, gastrointestinal ulcerations, bleeding, and multiple cutaneous lesions are also present in some patients.

Cutaneous Histoplasmosis

Primary cutaneous histoplasmosis is a rare presentation. The condition occurs following an inadvertent injection of contaminated material in a laboratory or during an autopsy, and a thorn prick while cleaning bat guano. A subset of patients with healed histoplasmosis presents with broncholithiasis, where calcified mediastinal lymph nodes or pulmonary nodules can erode the wall of the airways resulting in hemoptysis and expectoration of calcified material. *Histoplasma* infections can occur as obscure causes of a single organ disease, mostly as a mildly chronic or even recurrent manifestation—bone disease as single or multiple punched-out lytic lesions of the humerus, or ribs or even as an isolated laryngeal affection with thick plaque-like lesions.

African Histoplasmosis

African histoplasmosis caused by *H. capsulatum* var. *duboisii* is clinically distinct and characterized by skin and bone involvement. The condition presents as an indolent disease at onset, with the predominant site of predilection being skin, bone, and lymph nodes. Multiple cutaneous lesions develop in many patients, especially on the face and trunk. These enlarge and ulcerate. Punched-out bony lesions or even chronic osteomyelitis occurs in some patients, further spreading to nearby joints or into contiguous soft tissue, resulting in a purulent subcutaneous abscess. Rarely do patients with African histoplasmosis present with widely disseminated disease.

Other Forms of Histoplasmosis

These include ocular histoplasmosis, CNS histoplasmosis, etc., seen frequently in patients with disseminated histoplasmosis.

Infection in Animals

Histoplasma infection has been documented in domestic animals (dogs, cats, cattle, sheep, and horses) and wild rodents, marsupials, bats, armadillos, badgers, and chinchillas. Dogs are highly susceptible to the fungus, showing dermatological manifestations, oral/gingival lesions, multifocal lymphadenopathy, and mycotic abortions. *Histoplasma* infection in cats often presents as weakness, lethargy, emaciation, fever, anorexia, focal ocular signs, skeletal involvement, and respiratory problems. Primary gastrointestinal histoplasmosis presenting as hematochezia and melena has also been seen in both dogs and cats. *H. capsulatum* var. *farciminosum* is an agent of disease of horses and mules (epizootic lymphangitis). Animal infections of epizootic lymphangitis are seen worldwide, although the maximum prevalence is seen in Egypt and India.

The natural infection has been documented in bats, baboons, and badgers. Baboons have been documented to have a natural infection with *H. duboisii*. *Histoplasma* is a normal intestinal content of bats and birds. Bats when suffering from the disease have many intestinal ulcerative lesions discharging the fungi through their guano. Chicken, reptiles, and amphibians are resistant to the disease. Most birds are usually resistant to infection, but rare reports of avian histoplasmosis causing oral lesions and osteomyelitis have been documented.

Epidemiology and Public Health

Histoplasmosis is worldwide in distribution. The condition is reported mostly in areas with fertile alluvial plains surrounding large rivers. Histoplasmosis is highly endemic in the Mississippi, Missouri, Ohio, and St Lawrence River valleys in North America; and Mexico and Panama in Central America. Other endemic areas include South America (Bolivia, Ecuador, Brazilian Amazon, and Cuba), Malaysia, the Philippines, China, Africa, and the Ganges Brahmaputra plains and southern coastal plains of India. Skin test positivity in parts of America is as high as 90%. In Asia, skin test positivity ranges from 0% to 12.3% in India, 3% in Malaysia, 26% in the Philippines, and 19% in China. The geographic distribution of histoplasmosis is related to the acidic and humid nature of the soil. Soils enriched with bat or bird guano promote the growth and sporulation of *H. capsulatum*.

Many cases of histoplasmosis are also documented in countries such as Europe, outside the countries, endemic for the disease. These cases represent (a) imported diseases from endemic areas of the Americas, Africa, or Asia by travel to these countries (b) an increasing number of patients receiving immunosuppressive therapies for autoimmune diseases, cancers, and organ transplantations. A study of 213 cases reviewed from the Indian subcontinent showed that 56% of affected people were apparently immunocompetent and 31% were immunocompromised. The majority of Indian patients (55%) with disseminated histoplasmosis had their adrenal gland involved. Strong male predisposition is seen in disseminated histoplasmosis, including in Indian patients (6:1 male-to-female ratio).

Human infections with *H. capsulatum* var. *duboisii* are restricted to Africa. Most cases are reported from Nigeria, while many are also reported from Niger, Senegal, Congo, Zaire, and Uganda. However, it is postulated that the disease can be transmitted to newer areas, as documented through the infection of baboon colonies imported from Africa to North America.

Histoplasmosis is transmitted by inhalation, ingestion, or inoculation in the skin through breaks in the epidermis. Occasional cases of transmission during organ transplantations have also been documented. The most important factor responsible for infection is environmental exposure. Activities involving the clearing of wooded areas and shrubs, excavation of soil, construction, demolition, cleaning of chicken coops, attics, and ledges, and exploration of caves are known to transmit the infection, if the soil has been contaminated with bat or bird guano.

H. capsulatum is found in soils of endemic regions spread across the globe. Soils enriched with bird or bat guano allow optimal growth of the pathogenic fungus. This growth is attributed to the soil's high nitrogen content. Bird and bat droppings are an important nutrient source for this fungus. Soil samples from bird roosting sites are found to remain contaminated for at least a decade after clearing the roost.

The organism has not been isolated from bird feces, indicating that birds are not reservoir hosts or carriers of *Histoplasma*. It is believed that bird and bat excreta added to the soil already contaminated with *Histoplasma* create an ideal growth medium for fungi, allowing the organism to grow at the environmentally hardened mycelial stage. Bats and armadillos are considered important vectors for the dissemination of the fungus. An Indian study found that among patients with proven histoplasmosis, the majority of 87% were either farmers or had significant occupational or household exposure to birds. Infectious microconidia are released during activities causing disruption of the soil surface, like excavation, spelunking (cave exploration), construction, bird handling, and recreational activities. Released microconidia are then inhaled, causing pulmonary or disseminated histoplasmosis. Common environmental and wilderness-related risk factors for histoplasmosis are bird and bat watching, cave and cave entrance exploration, and bamboo removal and burning. Road construction, roofing, bridge and water tower work, demolition, and masonry are those occupations in which people employed may be more susceptible to *Histoplasma* infection.

Diagnosis

Laboratory diagnosis of histoplasmosis is based on the isolation and identification of *H. capsulatum* (Table 19.1). Serologic techniques may be useful but are not considered sacrosanct. Similarly, the detection of *Histoplasma* DNA by molecular methods and *Histoplasma* antigen detection have been considered definitively diagnostic. Delayed hypersensitivity to histoplasmin antigen using the skin test is not useful for diagnosis.

Microscopy

Demonstration of the fungi in stained sputum/aspirated material/biopsy/bone marrow confirmed by immunofluorescence is a reliable affirmation of the diagnosis in the absence of culture.

Direct demonstration in sputum samples by KOH mount often fails and is not recommended for diagnosis of pulmonary histoplasmosis. However, the direct demonstration by microscopy of Giemsa-stained smears of methyl alcohol fixed sputum smeared slides is highly useful. The organism, 2–5 μm in size, ovoid, unipolar narrow-based budding yeasts, with budding at the smaller end and a halo of unstained material around the cell, representing the cell wall is demonstrated within macrophages. A large vacuole that is crescent-shaped and a red-staining mass of protoplasm is present within the yeast, at the larger terminus of the cell. The buds are delicate and usually detach during the staining procedure.

Bone marrow, buffy coat, a sediment of blood (collected in EDTA or heparin), biopsy material, and aspirate from lesions are the specimens of choice for the diagnosis of suspected cases of progressive disseminated histoplasmosis or organ-specific histoplasmosis. Special fungal stains such as Giemsa/May-Grunwald Giemsa, Gomori's Methenamine Silver (GMS), and Periodic Acid Schiff (PAS) stain are used to demonstrate *Histoplasma* from the abovementioned specimen. *Histoplasma* yeasts are stained uniformly black by GMS and red on PAS stains. Fine Needle ASspiration Cytology (FNAC) preparations stained by Papanicolaou (PAP) stain, *Histoplasma* yeasts are difficult to demonstrate and may appear as punched-out "holes" in the cytoplasm of the macrophage.

Histopathology

The histopathology of the lesion depends on the duration of the infection whether recent or old, severity, and nature of the disease. In the acute, rapidly fatal varieties of PDH, numerous yeast cells are found only within histiocytes. This condition is known as histiocytomycetic disease. Neutrophils, plasma cells, and lymphocytes are not abundant and do not contain yeast cells. Yeast cells within the histiocytes are budding and of a uniform small (3 μm) size and are demonstrated by Gram, Giemsa, and H&E stains. Epithelioid granulomas that contain plasma cells, lymphocytes, macrophages, neutrophils, and giant cells are demonstrated in less fulminating forms of the disease. These are found in smaller numbers than in fulminant disease. In chronic cavitary disease, yeasts are even fewer in number, especially in the necrotic tissues rather than in deeper tissue, and are very rarely seen as bud forms. In some areas of the old chronic cavities, yeasts may be abundant but are very irregular in size and morphology. In old lesions and coin lesions, histoplasmomas, and calcified nodules, the yeasts are moribund but viable and very few. In some such lesions, yeasts are so few that cultures are negative and diagnosis is dependent on histopathology only. In these cases, specific fluorescent antibody staining and molecular identification may be attempted directly on tissue sections (Table 19.2).

In Giemsa-stained smears, *Histoplasma* yeasts resemble amastigotes of *Leishmania donovani* but lack the rodlike kinetoplast. The yeasts also need to be differentiated from that of *Toxoplasma gondii* zoites, which must also be ruled out. In special fungal stains, GMS, and PAS, neither *L. donovani* nor *Toxoplasma* zoites take up the stain. It is suggested to differentiate the yeast cells of

Table 19.2 Laboratory diagnosis of histoplasmosis

Diagnostic approaches	Methods	Targets	Remarks
Direct microscopy	Direct examination of aspirates, sputum, and excised tissue	Intracellular unipolar narrow-based budding yeasts, usually 2–4 μm in diameter	Cheap and fast technique, material needs to be fresh, without chemical preservatives. GMS and PAS stain are most useful
	Histopathology	Intracellular unipolar narrow-based budding yeasts, usually 2–4 μm in diameter	The damage caused by the fungus during its pathogenesis in the tissue is observed. GMS and PAS stain are most useful
Culture of sputum, bone marrow, lymph node aspirate, and lesion aspirate	Smith and Goodman's yeast extract agar with ammonium hydroxide, Sabouraud dextrose agar, and brain heart infusion agar	Mycelial phase colonies with "tuberculate" macroconidia and microconidia at 25°C	Probe based or molecular detection and confirmation of isolate or yeast conversion must be to confirm diagnosis
Blood and bone marrow culture	Lysis centrifugation system. Culture on Sabouraud sheep blood agar and Sabouraud dextrose agar	Growth of fungus	Probe based or molecular detection and confirmation of isolate or yeast conversion. Most useful for diagnosis of acute histoplasmosis and progressive disseminated histoplasmosis in AIDS
Immunodiagnostics	Antigen detection	Antigen in urine or serum	Most useful for diagnosis of acute histoplasmosis and progressive disseminated histoplasmosis in AIDS
	Antibody detection by complement fixation test.	Complement fixing antibodies. Titer evaluation	1:32 titers significant. Positivity starts by the end of 2 weeks. Some patients have lower titers of 1:8 and 1:16. A rising titer suggests dissemination. A third of progressive disseminated histoplasmosis patients may be negative
	Antibody detection by immunodiffusion test	Anti-H and anti-M antibodies	Anti-H antibodies suggest active infection
Molecular methods	Polymerase chain reaction	ITS1/ITS2/ITS 5.8S (pan generic)	Highly sensitive and specific for generic diagnosis and management, not useful for phylogenetic identification

H. capsulatum from newly released spores and young spherules of *Coccidioidis* species, *Cryptococcus neoformans* (Mayer's mucicarmine may be useful), the small cell form of *B. dermatitidis* (broad-based budding and multinucleate state) and intracellular *Candida glabrata* yeasts (culture or immunofluorescence on tissue sections).

In Vitro Culture

Specimens of choice for culture isolation include sputum, gastric aspirate, bone marrow, blood, lymph node, and lesion aspirates. In the case of primary progressive histoplasmosis, sputum is the specimen of choice followed by gastric lavage done early in the morning before feeding the patient. Bone marrow, aspirated material, and biopsy specimens are collected from the patients with PDH. Blood culture in manual lysis centrifugation systems is a diagnostic method with a high yield in disseminated histoplasmosis. SBA without antibiotics, SDA, and Smith and Goodman's media are used for contaminated samples like sputa. Yeast extract-phosphate medium is also used with a drop of ammonium hydroxide placed on one side of the inoculated plate. Inoculated media are incubated for 6–12 weeks. Demonstration of characteristic tuberculate macroconidia, microconidia, and conversion to the yeast state (on blood agar slants at 37°C) is diagnostic. Exoantigen testing, specific DNA probes, and specific gene amplification and sequencing are the methods currently used to detect mycelial to yeast conversion.

Serodiagnosis

Complement fixation test (CFT) using *Histoplasma* antigen was of great value in diagnosis as well as for the prognosis of the disease. Complement-fixing antibodies appear as early as 2 weeks following infection, and most immunocompetent patients have demonstrable titers by 4 weeks. A titer of at least 1: 32 is considered diagnostic, but some patients with active disease have titers of only 1:8 or 1:16. A rise in titer indicates dissemination of infection. Following the resolution of the disease, titers fall gradually and usually disappear in the ninth month. Complement-fixing antibodies, however, are not demonstrated in nearly a third of PDH patients.

The immunodiffusion test is generally positive in disseminated disease. Titers start appearing by the third or fourth week following infection. An H line closer to the well-containing serum corresponds to the presence of active infection. An M line is present in those in whom the infection has resolved or very early in infection.

Histoplasma antigen detection in urine and serum is useful for diagnosis of acute pulmonary and PDH forms of the disease. Antigen detection is less useful in chronic cavitary lung histoplasmosis. Depending on the platform (ELISA or LFA) and reagents (polyclonal or monoclonal antibodies) used, cross-reactions occur in patients with penicilliosis, blastomycosis, and paracoccidioidomycosis and rarely in coccidiodomycosis and aspergillosis. Antigen tests have the advantage of being used as prognostic tests to detect the level of serum antigen to monitor the therapy of histoplasmosis.

Western blot assays are the most frequently used test, as an in-house test for the detection of histoplasmosis infections during the COVID-19 pandemic.

Molecular Diagnosis

Nucleic acid amplification of *H. capsulatum* directly from clinical samples is available in reference laboratories and research settings. A nested PCR assay for a gene encoding the 100 kDa protein has been used for the detection of the fungus directly in clinical samples. Identification of *H. capsulatum*-specific 18S rRNA gene sequences, the PPK gene, or the CFP4 gene is also being evaluated.

Treatment

Induction, Consolidation, and Maintenance Antifungal Regimens for Disseminated Histoplasmosis in People Living with HIV or Other Immunocompromised Patients

In patients with severe or moderately severe histoplasmosis, treatment with Liposomal Amphotericin B (L-AmB) (3.0 mg/kg) for 2 weeks is the recommended regimen. Induction therapy is given for 2 weeks. The duration may be reduced depending on clinical presentation and the toxicity of amphotericin B therapy. In settings where L-AmB is unavailable, d-AmB (0.7–1.0 mg/kg) is recommended for 2 weeks. In patients with renal failure or at risk of renal injury, the prevention and monitoring of toxicities are important. Induction therapy with itraconazole 200 mg twice daily is recommended as an alternative for the management of mild-to-moderate histoplasmosis in PLHIV.

Itraconazole (400 mg/daily) for 12 months is the recommended regimen for maintenance therapy of histoplasmosis.

Antiretroviral therapy need not be delayed for patients diagnosed with disseminated histoplasmosis. The most common first-line ART in resource-limited settings is TDF + FTC/3TC + EFV or TDF + FTC/3TC + DTV (TDF = tenofovir disoproxil fumarate, FTC = emtricitabine, 3TC = lamivudine, EFV = efavirenz, DTV = dolutegravir). Both TDF and EFV have significant interactions with itraconazole and warrant alternative therapy. Efavirenz may reduce itraconazole serum concentration significantly. Itraconazole is not used concurrently with or within 2 weeks of efavirenz therapy (Table 19.3).

Therapy for Tuberculosis for Patients Coinfected with Tuberculosis, HIV, and Histoplasmosis

Patients should receive TB therapy in PLHIV with TB-histoplasmosis coinfection according to the WHO treatment guidelines. One significant challenge for physicians in resource-limited settings is the coadministration of rifampicin and Itraconazole. The use of itraconazole concurrently with or within 2 weeks of any CYP3A4 inducer [in order of potency/risk: rifampicin > rifapentine > rifabutin] is not recommended. If such a combination cannot be avoided, monitor patients closely for evidence of diminished clinical response to itraconazole (reduced antifungal efficacy).

Fluoroquinolone (levofloxacin or moxifloxacin), isoniazid, Pyrazinamide, and ethambutol can be the recommended first-line antitubercular therapy in the presence of itraconazole maintenance therapy. Another alternative can be rifabutin. However, itraconazole might lead to diminished metabolism of rifabutin and lead to rifabutin toxicity, e.g., Uveitis—which needs monitoring. This potential interaction is much less in severity than severely compromised or undetectable itraconazole levels with rifampicin.

Case Study

A 43-year-old farm worker was admitted to the hospital with high-grade fever of 3 weeks' duration, cough, anorexia, and loss of weight. On examination, pallor, hepatosplenomegaly, and bilateral cervical lymphadenopathy were found. The cervical lymph nodes were palpable, small to large, soft to firm, and in some places matted. Both horizontal as well as vertical groups of cervical lymph nodes were palpable. The firm lower border of the liver (moving with respiration) was palpable at 7 cm below the costal margin in the midclavicular as well as the anterior axillary and midaxillary planes. The upper border of the liver dullness was established at the fifth intercostal space in the same planes. The spleen was palpable 4 cm below the left costal margin as a firm mass moving with respiration along its axis. Respiratory examination showed mild dullness on percussion and localized inspiratory crackles in the right middle and lower lobes. The cardiovascular system showed no abnormality. Laboratory investigations revealed a hemoglobin of 8.7 gm%, ESR of 65 mm in the first hour, total leucocyte count of 3800 cells/μL with 45% neutrophils, and platelet count of 114,000/μL. The serum was negative for Hepatitis B and Hepatitis C but positive for HIV-1. His CD4 count was 194 cells/μL. Chest X-ray revealed localized interstitial infiltrates in the right middle and lower lobes of the lung. Computerized tomography showed bilateral peribronchial and interstitial reticulonodular infiltrates, more prominent in the right lung. FNAC from lymph nodes revealed necrotic material with plenty of neutrophils and macrophages. Some of the macrophages contained numerous 3 μm size oval capsulated cells, as seen in the Giemsa stain. Grocott-methenamine silver staining of the same revealed these intracellular

Table 19.3 Drug-drug interactions between antiretroviral drugs and itraconazole

Nucleoside reverse transcriptase inhibitors					Integrase strand transfer inhibitors		Non-nucleoside reverse transcriptase inhibitors		Protease inhibitors	
TDF	FTC	3TC	AZT	ABC	DTG	RAL	EFV	NVP	LPV/r	ATV/r
Potential interaction, but clinically significant effect unlikely	No interaction expected	No interaction expected	No interaction expected	No interaction expected	No interaction expected	No interaction expected	Reduces itraconazole levels. No dose recommendation can be made.	Reduces itraconazole levels. Dosing recommendations differ	Increases itraconazole levels	Increases itraconazole and ATV/r levels

TDF tenofovir disoproxil fumarate, *FTC* emtricitabine, *3TC* lamivudine, *AZT* zidovudine, *ABC* Abacavir, *DTG* dolutegravir, *RAL* raltegravir, *EFV* efavirenz, *NVP* Nevirapine, *LPV/r* ritonavir-boosted lopinavir, *ATV/r* ritonavir-boosted atazanavir cells to be yeasts (stained black) with narrow-based budding.

1. How will you confirm the diagnosis of histoplasmosis?
2. What is the differential diagnosis for this patient?
3. How can you differentiate different suspected agents from those that cause histoplasmosis?
4. What are the other diagnostic modalities available?
5. How will you manage a patient with advanced HIV disease and progressive disseminated histoplasmosis?

Research Questions

1. How to evaluate the taxonomic status of human and animal pathogenic isolates of *Histoplasma* and their taxonomic relation with environmental isolates.
2. What are the factors that confine histoplasmosis to certain endemic areas?
3. What strategies are to be followed for designing and innovating media for improved isolation of *H. capsulatum* from specimens by culture?
4. What are the new approaches to develop less toxic, more effective, and cheaper drugs for chemotherapy of histoplasmosis?

Further Reading

Baker J, Setianingrum F, Wahyuningsih R, Denning DW (2019) Mapping histoplasmosis in South East Asia—implications for diagnosis in AIDS. Emerg Microbe Infect 8(1):1139–1145

Banerjee S, Denning DW, Chakrabarti A (2021) One health aspects & priority roadmap for fungal diseases. Indian J Med Res 153:311–319

Blancett LT, Runge KA, Reyes GM, Kennedy LA, Jackson SC, Scheuermann SE, Harmon MB, Williams JC, Shearer G Jr (2021) Deletion of the stress response gene DDR48 from *Histoplasma capsulatum* increases sensitivity to oxidative stress, increases susceptibility to antifungals, and decreases fitness in macrophages. J Fungi (Basel) 7(11):981

Blomley M, Teare EL, De Belder A et al (1990) Itraconazole and Antituberculosis drugs. Lancet 2:1255

Brandt ME, Warnock DW (2011) *Histoplasma, Blastomyces, Coccidioides* and other dimorphic fungi causing systemic mycoses. In: Murray PM, Baron EJ, Joprgensen JH, Landry ML, Pfaller MA (eds) Manual of clinical microbiology, vol 2, 9th edn, pp 1857–1873

Deepe GS Jr (2016) *Histoplasma capsulatum*. In: Bennett JE, Dolin R, Blaser MJ (eds) Mandell, Douglas, and Bennett's infectious disease essentials E-book, 8th edn. Elsevier Health Sciences, pp 2949–2962

Diaz JH (2018) Environmental and wilderness-related risk factors for histoplasmosis: more than bats in caves. Wilderness Environ Med 29(4):531–540

Drayton J, Dickinson G, Rinaldi MG (1994) Coadministration of rifampin and itraconazole leads to undetectable levels of serum itraconazole. Clin Infect Dis 18:266

Gilmore SA, Voorhies M, Gebhart D, Sil A (2015) Genome-wide reprogramming of transcript architecture by temperature specifies the developmental states of the human pathogen *Histoplasma*. PLoS Genet 11(7):e1005395

Gupta A, Ghosh A, Singh G, Xess I (2017 Dec) A twenty-first-century perspective of disseminated histoplasmosis in India: literature review and retrospective analysis of published and unpublished cases at a tertiary care hospital in North India. Mycopathologia 182(11–12):1077–1093

Hay RJ (2014) Fungal infections. In: Farrar J, Hotez PJ, Junghanss T, Kang G, Lalloo WNJ (eds) Manson's tropical diseases, 23rd edn. Elsevier Saunders Publication, pp 441–458

Hensel M, Hoffmann AR, Gonzales M, Owston MA, Dick EJ Jr (2019) Phylogenetic analysis of *Histoplasma capsulatum* var. duboisii in baboons from archived formalin-fixed, paraffin embedded tissues. Med Mycol 57:256–259

Kasuga T, White TJ, Koenig G, McEwen J, Restrepo A, Castañeda E, Da Silva LC, Heins-Vaccari EM, De Freitas RS, Zancopé-Oliveira RM, Qin Z, Negroni R, Carter DA, Mikami Y, Tamura M, Taylor ML, Miller GF, Poonwan N, Taylor JW (2003) Phylogeography of the fungal pathogen *Histoplasma capsulatum*. Mol Ecol 12(12):3383–3401

Maresca B, Kobayashi GS (1989) Dimorphism in *Histoplasma capsulatum*: a model for the study of cell differentiation in pathogenic fungi. Microbiol Rev 53:186–209

Mittal J, Pnce MG, Gendlina I, Nosanchuk JD (2019) *Histoplasma capsulatum*: mechanisms for pathogenesis. Curr Top Microbiol Immunol 422:157–191

Rakislova N, Hurtado JC, Palhares AEM, Ferreira L, Freire M, Lacerda M, Monteiro W, Navarro M, Casas I, Teixeira MM, Castillo P, Rodrigo-Calvo MT, Marimon L, Guerrero J, Varo R, Delgado V, Quintó L, Marco F, Letang E, Vila J, Bassat Q, Menéndez C, Ordi J, Martínez MJ (2021) High prevalence and mortality due to *Histoplasma capsulatum* in the Brazilian Amazon: an autopsy study. PLoS Negl Trop Dis 15(4):e0009286

Rippon JW (1988) Chapters: histoplasmosis (histoplasmosis capsulate) and histoplasmosis Duboisii. In: Medical Mycology: the pathogenic fungi and the pathogenic actinomycetes, 3rd edn, pp 381–432

Rodrigues AM, Beale MA, Hagen F, Fisher MC, Terra PPD, de Hoog S, Brilhante RSN, de Aguiar CR, de Souza Collares Maia Castelo-Branco D, MFG R, JJC S, de Camargo ZP (2020) The global epidemiology of emerging *Histoplasma* species in recent years. Stud Mycol 97:100095

Staffolani S, Buonfrate D, Angheben A, Gobbi F, Giorli G, Guerriero M, Bisoffi Z, Barchiesi F (2018) Acute histoplasmosis in immunocompetent travelers: a systematic review of literature. BMC Infect Dis 18:673–686

Tille PM (ed) (2017) Bailey & Scott's diagnostic microbiology, 14th edn. Elsevier Publications, St. Louis, MO

Part V

Opportunistic Fungal Zoonoses?

Aspergillosis

20

Priyadarshini A. Padaki, Jayanthi Savio,
D. Rathnamma, Shrikrishna Isloor, and R. Sharada

Learning Objectives
By the end of the chapter, the readers should be able to:

1. Understand the pathogenesis and describe clinical manifestations of aspergillosis among animals and birds.
2. Understand the currently available diagnostic modalities for the diagnosis of invasive aspergillosis.
3. Describe the microscopic morphology of clinically relevant species of *Aspergillus*.
4. Understand the need for rational use of antifungal drugs and their implications.
5. To understand the concept of one health.

Introduction

Aspergillus species are widely distributed in nature. Although the genus consists of more than 300 species, only a limited number of these have been implicated in infections in animals and humans. Aspergillosis has been reported in domestic animals, birds, and wild species, causing not only localized infections and fatal disseminated infections but also allergic manifestations. *Aspergillus fumigatus* is responsible for most infections in animals and is most often involved in tissue invasion. The other species of importance include *Aspergillus flavus*, *Aspergillus niger*, *Aspergillus terreus*, and *Aspergillus nidulans*. The most prevalent forms of animal aspergillosis are mycotic abortion and mycotic mastitis in cattle, pulmonary and air sac infection in birds, guttural pouch mycoses in horses, and disseminated and sinonasal infections in dogs and cats. Respiratory infections occur following inhalation of spores. Less commonly, the infection can result from the ingestion of spores or following tissue trauma. Systemic infection is invariably associated with immunosuppression. Species such as *A. flavus*, which elaborate potent toxins when growing in cereals and other foods, cause mycotoxicosis. The data on infections in animals is lacking, and the classical association of immunosuppression and invasive disease, as seen in humans, is not well documented. The additional impact of mortality in animals includes economic loss and emotional trauma, especially in the case of pets. The conventional culture-based diagnosis of veterinary aspergillosis is similar in lines to that of human aspergillosis. However, interpretation of immunological parameters and establishing diagnostic cutoffs is required as the values used for diagno-

P. A. Padaki · J. Savio (✉)
Department of Microbiology, St. John's Medical College and Hospital, Bengaluru, Karnataka, India
e-mail: jayanthi.s@stjohns.in

D. Rathnamma · S. Isloor · R. Sharada
Department of Veterinary Microbiology, Veterinary College, Karnataka Veterinary, Animal and Fisheries Sciences University, Bengaluru, Karnataka, India

S. C. Parija, S. M. Rudramurthy (eds.), *Textbook of Fungal Zoonoses and Sapronoses*, Microbial Zoonoses, https://doi.org/10.1007/978-981-97-3214-2_20

sis in humans cannot be extrapolated to the animal population. Molecular diagnostics tests and antifungal susceptibility testing require standardization.

History

The name *Aspergillus* was introduced by P. A. Micheli, a Florentine priest-microbiologist, in 1729 to describe asexual fungi whose conidiophores resembled an aspergillum, a device used to sprinkle holy water. The genus was validated in 1768 by Von Haller, and Fries sanctioned the generic name in 1832.

Taxonomy

Taxonomy, classification, and nomenclature of *Aspergilli* have evolved over the years and undergone many revisions. The currently proposed phylogenetic approach is supported by phenotypic, physiologic, and/or extrolite data. It is classified under the order Eurotiales, family Aspergillaceae, and subdivided into six subgenera, 27 sections, and 75 series. Currently, around 446 species have been described, with the data being updated very frequently. However, the most important pathogenic species include *A. fumigatus*, *A. flavus*, *A. niger*, *A. terreus*, and *A. nidulans*. Sequencing data helps to classify the different species and has been extensively studied to include newer species under the genus. Taxonomic studies help in understanding evolutionary patterns and epidemiological variations of diseases caused by these fungi.

Genomics and Proteomics

There are several typing methods available such as lipotyping, proteotyping, and metabolotyping, which use different patterns for the classification of fungi. Recommended barcoding genes for phylogenetic analysis include Beta-tubulin (BenA) and Calmodulin (Cal). Whole genome sequencing has been done to study the differences among the various species, and it has been found to have a lot of diversity. Studies have evaluated early pathways activated during the growth and development of *Aspergillus* species which are essential for growth and metabolism. In view of the availability of many methods to classify the *Aspergillus* species, there are several species names suggested by different study groups. However, the recent concept of the 'One Fungus-One Name' policy, which has been proposed, might help to resolve the taxonomical complexities.

The Fungus Morphology

Aspergillus species appear as septate fungal hyphae on direct microscopic examination using potassium hydroxide and calcofluor white stain from various samples. However, the identification of the different species depends on macro-morphology and micro-morphology. Macro-morphology has been described on various media such as Sabouraud dextrose agar, Malt extract agar, Czapek Dox agar, etc. Micro-morphology usually is well appreciated by the slide culture technique. The lactophenol cotton blue (LPCB) mount is used for appreciating the sporulating structures. The microscopic appearance of the important *Aspergillus* species is summarized in Table 20.1).

Mounts are made using LPCB from the growth for microscopic examination. The characteristic sporulating or fruiting heads help in the identification of the genus, but the experience is required for species-level identification. The findings for the individual species are described in Table 20.1. The diagrammatic representation of the microscopic pictures is given in Fig. 20.1.

Cultivation of Fungi

Aspergillus species are generally rapid growers and mature within 3 days. Macroscopic appearance is varied with the color of conidia, colony diameter, mycelial color, reverse color, and solu-

Table 20.1 Morphology of important species of *Aspergillus*

Morphology	*Aspergillus fumigatus*	*Aspergillus niger*	*Aspergillus flavus*	*Aspergillus terreus*
Macroscopic morphology	White fluffy growth initially which rapidly becomes velvety with a greenish color	White when young but soon developing a black pepper-like growth which is due to the color of the conidia. The reverse is buff or cream-colored	Cottony grows when young but soon becomes a yellow-green with a granular obverse	White at first, but develops a cinnamon-buff color with a granular texture as the sporulation develops
Microscopic morphology of conidiophores	Short (<300 μm) smooth The conidiophores are moderate in length and possess a "foot cell" at their bases	Long smooth	Variable length, rough, pitted, spiny	Short (<250 μm)
Microscopic morphology of vesicle and conidia	The vesicles are dome-shaped and the upper half to two-thirds bear phialides from which arise long chains of globose, spiny, green conidia (2–3 μm) are present	It has a large spherical vesicle at the end of the conidiophore. The spherical vesicle bears large metulae at the ends of which phialides are present. The conidia are borne at the ends of phialides which are black and have rough walls	It has round vesicles with phialides covering the entire portion of the vesicle. The conidia are 3–5 μm in diameter, yellowish, elliptical, or spherical, and become spiny with age	The vesicles are small and dome-shaped with phialides covering the upper one-third of the vesicle. The conidia (2–3 μm diameter) are elliptical. Aleuriospores may be present on submerged hyphae

ble pigment. The macroscopic morphology of the important *Aspergillus* species is described in Table 20.1.

Animal Models

Various animal models using mice, guinea pigs, and rabbits have been developed to understand the pathophysiology of aspergillosis, assessment of virulence mechanisms, and use of therapeutic drugs to treat the disease spectrum. These have limitations in terms of reproducing the results and variability in terms of the different animals used due to the difference in the host factors related to the response to the fungus. Some examples include (a) genetic knockout mice models for the study of IL6, IL12, and IFNγ as protective factors against *A. fumigatus,* (b) therapeutic potential of posaconazole against aspergillosis in mice models, (c) endocarditis model in guinea pigs to demonstrate the therapeutic effects of azoles, (d) rat models for studying the course of invasive aspergillosis, (e) rabbit models for fungal keratitis, and (f) turkeys models for demonstrating the effect of metabolites such as gliotoxin by *Aspergillus* species.

Life Cycle of Fungus

The life cycle of *Aspergillus* (Fig. 20.2) includes sexual and asexual reproductive stages. During the sexual cycle, the mycelium forms a cleistothecium, which holds the ascospores, which, when released into the environment, can form a new mycelium. In the asexual stage, the mycelium differentiates into the fruiting body, which bears the asexual spores or conidia. Animals and humans can get the infection by inhalation of conidia.

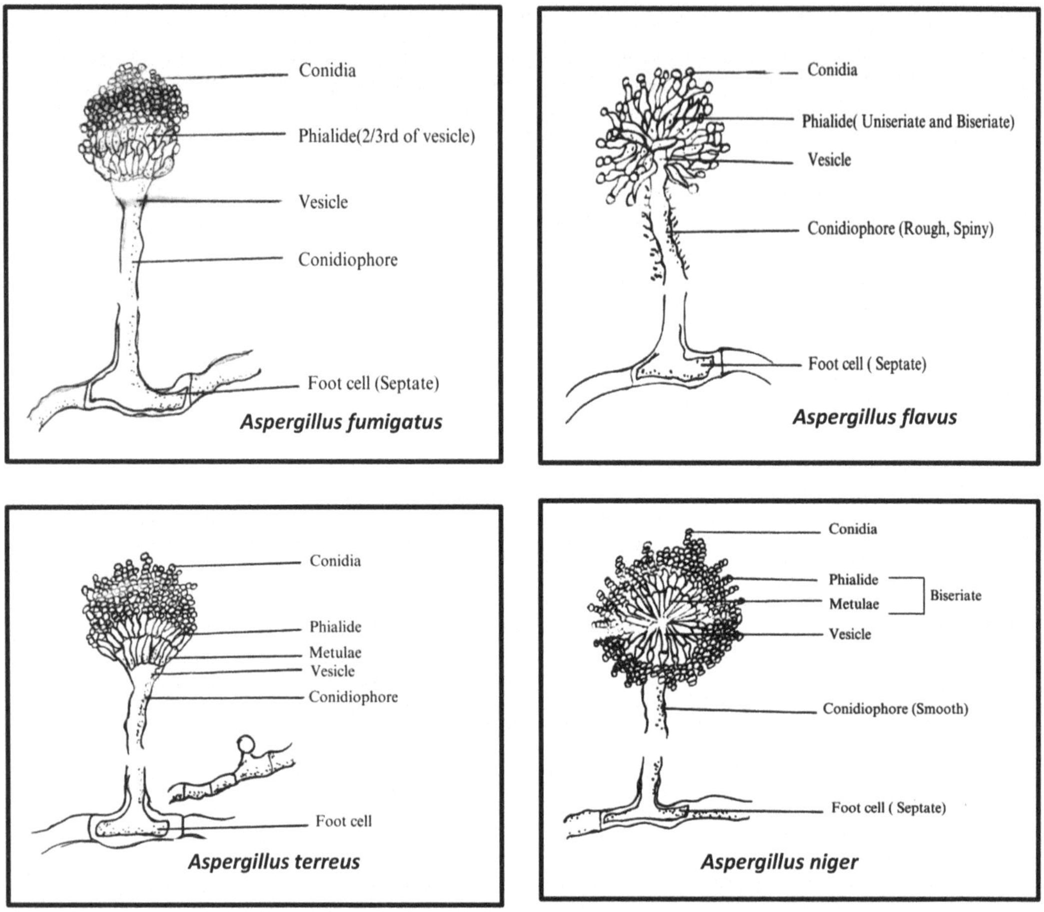

Fig. 20.1 **Microscopic morphology of the different *Aspergillus species*** (Courtesy: Ms. Bineetha, and Dr. Livin J Tessy, Department of Microbiology, St. John's Medical College and Hospital)

Pathogenesis and Pathology

Infection with *Aspergillus* species is seen in many species of animals. The infection is mainly a respiratory infection acquired by inhalation of its spores. The small size of the spores allows them to pass through the upper respiratory tract and reach the bronchioles leading to the accumulation of suppurative exudates in the bronchioles. Germination of these spores and hyphal invasion of deeper tissues depends on a variety of factors. *Aspergillus* spores adhere to collagen, fibrinogen, fibronectin, and laminin. Pathogenic *Aspergillus* species grow well at 37 °C and produce a wide variety of extracellular enzymes. Proteases with elastase, fibrinolytic, and anticoagulative properties are important. The diagrammatic representation of the mode of transmission and development of aspergillosis has been depicted in Fig. 20.3.

A. fumigatus produces many secondary metabolites within the host tissues known as mycotoxins, such as gliotoxin, aflatoxin, and ochratoxin, while others include hemolysins and proteolytic enzymes. Gliotoxin has immunosuppressive properties and can inhibit both the activity of cilia and phagocytosis by macrophages. These potent metabolites are not only important for pathogenicity but also act as a chemical shield to protect the organism by competing with other microorganisms in the environment in which it exists.

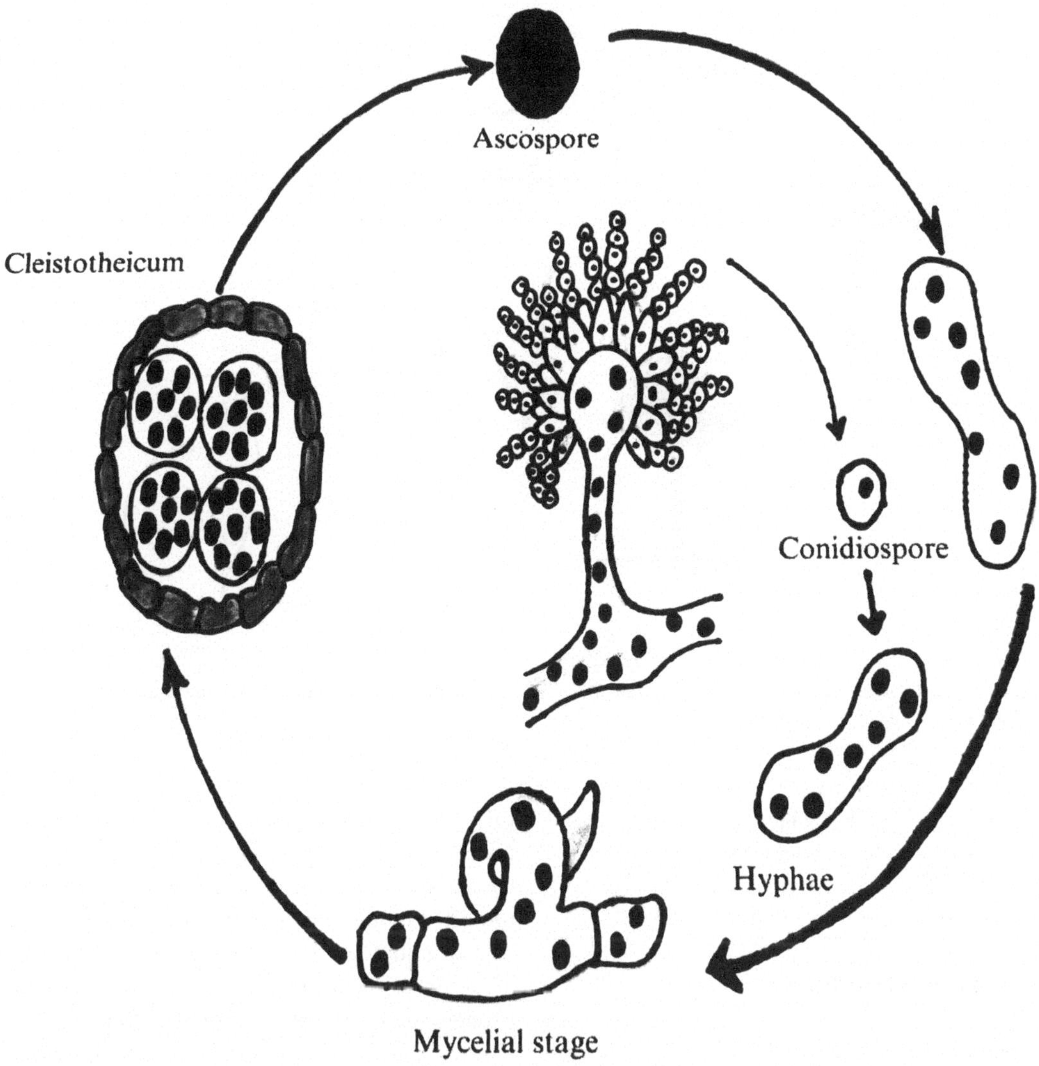

Fig. 20.2 **Life cycle of *Aspergillus*** (Courtesy: Ms. Bineetha, and Dr. Livin J Tessy, Department of Microbiology, St. John's Medical College and Hospital)

Immunology

Immune mechanisms and the immune competence of the host determine the outcome of infection. Factors that modify immune competence include corticosteroid therapy and long-term treatment with antimicrobial drugs. Interference with the function of neutrophils and monocytes predisposes tissue invasion. Fungal metabolites also impair phagocytic functions that would normally cause the destruction of conidial and hyphal forms. Gliotoxin reduces adherence and phagocytosis of fungal elements, while aflatoxin affects phagocytosis, intracellular killing, and production of superoxides. The binding and activation of complement by bound opsonin, which normally enhances phagocytosis, are also affected, making the fungi less susceptible to destruction.

Invasion of the blood vessels by hyphae leads to vasculitis and thrombus formation. Mycotic granulomas may develop in the lungs and other internal organs. Granulomas appear as yellowish-gray nodules. The presence of *A. fumigatus*

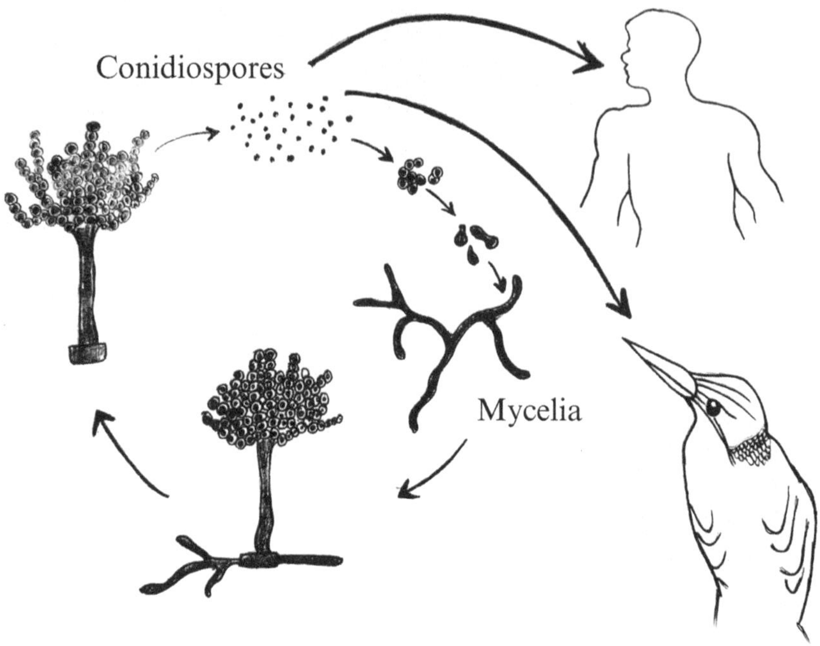

Fig. 20.3 **Pathogenesis of aspergillosis** (Courtesy: Ms. Bineetha, and Dr. Livin J Tessy, Department of Microbiology, St. John's Medical College and Hospital)

spores in the stables where horses are housed may act as an allergen and has been associated with chronic obstructive pulmonary disease.

Infection in Humans

Aspergillosis in humans has a wide spectrum ranging from allergic manifestations to invasive diseases. The most affected organ is the lung in humans. Allergic manifestations in immunocompetent individuals include allergic fungal rhinosinusitis, allergic bronchopulmonary aspergillosis, and chronic pulmonary aspergillosis. Invasive aspergillosis is a more severe manifestation occurring in immunocompromised individuals such as those who have undergone bone marrow transplantation, solid organ transplantation, and patients on cancer chemotherapy and other immunosuppressive agents. During the COVID-19 pandemic, COVID-associated pulmonary aspergillosis became an entity of concern, causing increased mortality among these patients. Several guidelines have been devised for the diagnosis of invasive aspergillosis in humans, which are based on host risk factors, radiological and histologic findings, and microbiological criteria. These guidelines have evolved over a period. They have been worked upon by working groups such as the European Society for Clinical Microbiology and Infectious Diseases (ESCMID) and the European Confederation of Medical Mycology (ECMM).

Infections in Animals and Birds

Clinical Manifestations in Animals and Birds

Aspergillosis among animals and birds is comparatively uncommon and usually sporadic. Birds appear to be more susceptible to aspergillosis than mammals. This susceptibility may be due to the differences in innate and acquired immunity to the fungus in avian and mammalian species. Infections are mainly in the respiratory tract,

Table 20.2 Clinical manifestations of *Aspergillus* species in animals and birds

Hosts	Clinical Condition	Comments
Birds	Brooder pneumonia	Occurs in newly hatched chickens exposed to a large number of spores within incubators
	Pneumonia and air sacculitis	Chickens up to 6 weeks of age are most susceptible. Older birds are sometimes affected
	Generalized or disseminated aspergillosis	Dissemination from the respiratory tract which is the primary source
Horses	Guttural pouch mycosis	Necrotizing infection confined to the guttural pouch. Usually unilateral, may involve carotid blood vessels and the glossopharyngeal nerve.
	Nasal granuloma	Causes nasal discharge and difficulty in breathing. Many fungal species are implicated
	Keratomycosis	Localized corneal involvement following ocular trauma
	Intestinal aspergillosis	Described occasionally in foals
Cattle	Mycotic abortion	Sporadic occurrence in cattle causes thickened placenta and plaques on the skin of the aborted fetus.
	Mycotic pneumonia	Uncommon condition in housed calves
	Mycotic mastitis	Chronic, progressive infection associated with the use of contaminated intramammary antibiotic tubes
	Intestinal aspergillosis	Described occasionally in calves
Dogs	Nasal aspergillosis	Sporadic occurrence, usually in young to middle-aged dolichocephalic breeds. Persistent purulent nasal discharge. May affect turbinate bones and paranasal sinuses
	Otitis externa	Occurs as part of mixed infection
	Systemic aspergillosis	Rare, recorded in the German shepherd breed particularly. Osteomyelitis is often a feature
Cats	Systemic aspergillosis	Rarely encountered; immunosuppressed animals are at risk

although localized infections have been reported in other organs, such as kidneys. Rarely, other fungi such as *Penicillium* species, *Paecilomyces* species, and *Scedosporium apiospermum* may cause opportunistic infections like those caused by *Aspergillus* species. Clinical conditions range from allergic manifestations to invasive diseases like the presentation in humans.

The clinical conditions caused by *Aspergillus* species in domestic animals and birds are summarized below in Table 20.2.

Aspergillosis in Animals and Birds

Guttural Pouch Mycosis (GPM) in Horses

This condition, though sporadic, is life-threatening and is characterized by necrotizing inflammation with the formation of the tough membrane in the guttural pouch. There seems to be no breed disposition. It usually affects adult immunocompetent horses with epistaxis or anatomical defects. Improper ventilation coupled with increased humidity and temperature favors the germination of fungal spores in the environment where horses live. Traumatic implantation of fungi from environmental sources is the predisposing factor. The acts of expiration and deglutination also can favor the entry of spores through the oropharynx. The guttural pouch is a large air-filled extension of the eustachian tube with openings into both the middle ear and oropharynx. It consists of numerous arteries, veins, and nerves. The most common species associated with GPM is *A. fumigatus*. The lesions appear as plaques and develop in the mucosa of the pouch wall. The most common presentation is unilateral nasal discharge and intermittent epistaxis. Accumulation of inflammatory exudates in the pouch probably contributes to the manifestations of postauricular swelling and nasal discharge. Penetration by fungal hyphae into deeper tissues results in erosion of blood vessels, thrombosis, tissue necrosis, and neural damage. Invasion of the internal carotid artery and vein can lead to fatal hemorrhage. Involvement of the glossopharyngeal nerve causes neurological deficits such as dysphagia, laryngeal paralysis, and Horner's syndrome. Most commonly, the lesions are found in the dorsal medial aspect of the guttural pouch. Angioinvasion, erosion of cranial nerves, and tissue necrosis are common pathologic findings.

Keratomycosis in Horses

Keratomycosis in animals commonly occurs in warm climates following corneal injury and implantation of fungal spores from plant or dust particles. It is a common infective corneal condition, but management is difficult. A laterally placed prominent globe with a highly exposed corneal surface makes horses more susceptible than other domestic animals to corneal injury resulting in fungal keratitis. Altered corneal immune protection, inflammatory response to the production of proteases, and immune cell infiltration are postulated mechanisms of pathogenesis. Blepharospasm, photophobia, and increased tear production are the clinical features. Clinical findings include superficial punctate corneal lesions and stromal defects, anterior stromal plaques laden with fungal filaments, intrastromal corneal abscesses, and corneal perforation. The causative fungal agent is determined by geographical location and other environmental factors. Keratomycosis usually follows prolonged topical application of antibiotics and corticosteroids, which can alter the normal conjunctival flora and promote fungal growth.

Sino-Nasal Aspergillosis (SNA) in Dogs

Canine nasal aspergillosis (SNA) is seen in young to middle-aged dogs. This condition is mainly seen in dolichocephalic and mesocephalic dogs, rarely in brachycephalic dogs. German Shepherds and Rottweilers are the most affected breeds. Aspergillosis occurs in 7%–34% of chronic sino-nasal diseases in dogs, as reported by studies. Several species of *Aspergillus* have been implicated, the most common being *A. fumigatus*. The fungus can reside for a long time before the symptom and signs are manifested. A transcriptome-based study has described an interesting view on the pathogenic mechanism of this condition as resulting from a biofilm formation in this warzone area with increased expression IL-8, alarmins, and dysregulation of host TH17 response. The clinical manifestations often include a unilateral, profuse purulent nasal discharge, which is persistent. Sneezing episodes of epistaxis, nasal depigmentation, and turbinate destruction can also be seen. Disseminated aspergillosis affecting multiple organs has been reported, including mottled masses in the kidneys, osteomyelitis, and hematological abnormalities such as anemia, leukocytosis, azotemia, and hypoalbuminemia.

Bovine Mycotic Abortion

This form of abortion occurs sporadically worldwide but affects only a small percentage of a herd. Prevalence may be influenced by poor quality contaminated fodder harvested in wet seasons. Incidence is usually high in the winter season when gravid cows are kept in small sheds and fed with damp hay in poor-quality silage contaminated with *Aspergillus* spores. Infection probably reaches the uterus through hematogenous spread. Placentitis leads to abortion late in the second or third trimester of gestation. Affected cows usually show no signs of systemic illness. Intercotyledonary areas of the placenta are thickened and leathery, and the cotyledons are necrotic. Aborted fetuses may have raised cutaneous plaques resembling ringworm lesions. The cow's fertility following fungal abortion does not seem to be compromised.

Mycotic Pneumonia in Cattle

Mycotic pneumonia is a rapidly progressive fatal disease in ruminants. Clinical signs include increased body temperature, heavy shallow breathing, and wet cough. On autopsy, the lungs are firm and have a mottled appearance. Multiple granulomas like tuberculosis can be observed, and the presence of fungal filaments within the lesions microscopically helps in differentiation. In subacute to chronic mycotic pneumonia, the lungs contain multiple discrete granulomas, and the disease grossly resembles tuberculosis. Coinfection of *Aspergillus* and *Mycobacterium tuberculosis* is also seen in some cases with lymph node involvement.

Aspergillosis in Birds

Aspergillosis in birds is associated with high morbidity and mortality. Birds of all ages are affected irrespective of their immune status. The structural anatomy and physiological functioning of the lung air sac system with poor vascularization and mucociliary function in birds favor the deposition, growth, and penetration of *Aspergillus* conidia. The presence of heterophils instead of neutrophils also helps in the hyphal invasion. Infection by *A. fumigatus* is found in birds held captive or in free-ranging environments. *A. flavus* is the other species commonly associated. Species predilection includes birds of prey, geese, ducks, swans, gulls, penguins, parrots, turkeys, raptors, and waterfowl. Environmental factors include exposure to limited air exchanges, aerosolized toxins resulting in mucosal irritation, and maintenance of improper temperature and humidity. Physical exertion due to migration and immune alteration with the administration of exogenous steroids also plays a role. Aspergillosis is therefore a major cause of morbidity and mortality in birds with far-reaching economic and ecological damages. Further, it can impact avian species in zoological institutions and during rehabilitation.

Brooder Pneumonia in Chickens

This acute aspergillosis affects newly hatched chicks exposed to high numbers of *A. fumigatus* spores resulting in high morbidity and mortality. Chickens develop nonspecific signs such as drowsiness, lack of appetite, respiratory distress, anorexia, lethargy, ruffled feathers, and stunting. Death among many chickens has also been reported. Strict hygiene measures and regular fumigation of incubators are effective control measures.

Aspergillosis in Mature Birds

Infection in older birds occurs due to inhalation of dust from feed or litter contaminated with fungal spores. Poultry, psittacine birds, penguins, and raptors are the species affected. High ambient temperatures and shredded wood bark on aviary floors harbouring the spores, act as predisposing factors for infection in the penguins and raptors. Clinical signs include difficulty in breathing, emaciation, coughing, and sneezing. Encephalitic and meningoencephalitis lesions of the central nervous system led to ataxic gait, muscle spasms of the neck, opisthotonos, paresis, and rarely blindness. Caseous yellowish nodules laden with fungal filaments like that seen in tuberculosis can be observed in the trachea, bronchi, and lungs. Dissemination in other organs can also occur.

Mycotoxicosis

Animals and birds can get affected by ingestion of food contaminated with *Aspergillus* species producing toxins such as aflatoxins. There have been reported cases of mycotoxicosis in turkeys, which created a crisis known as Turkey X disease way back in the 1960s. Ingestion of food with preformed toxin causes mortality among poultry and ruminants. The secondary metabolites of the toxins are highly toxic to the liver and kidney and cause sudden death among ducklings and chickens.

Unusual Manifestations

Localized and disseminated cutaneous and subcutaneous lesions involving unusual sites have been reported. Chronic pulmonary aspergillosis is a clinical entity gaining much attention in recent times among humans and has been described in birds.

Epidemiology and Public Health

Considering the need for study on the epidemiology, diagnosis, and management of zoonotic fungal diseases, The Veterinary Mycology Working Group of the International Society for Human and Animal Mycology (ISHAM-VMWG) was established in 2010.

It is also important to understand the epidemiology from the perspective of it posing a threat as an occupational hazard to staff managing animals and birds, especially for commercial purposes and veterinarians. Public awareness is also important as pet owners must also be aware of the basics of recognizing sick birds and animals.

The concept of One Health in understanding the interaction between the environment, man, and animals/birds has recently been studied in detail. The research question arose while investigating invasive aspergillosis in humans who had infection with azole-resistant strains. Several studies have tried to study the prevalence of azole resistance, the geographical distribution of the pathogenic strains carrying the resistance genes, and the probable source of the fungi in the environment. The findings shed light on the following hypotheses:

- The use of azoles in agriculture to prevent the contamination and destruction of crops has led to the selection of drug-resistant strains of fungi. These chemicals persist in the environment in subinhibitory doses, thus increasing the selection of resistant strains.
- Birds stay near these fields and feed on the crops. These birds might get infected and shed the fungus. Asymptomatic carriage of the resistant fungal strains in the beaks, feathers, and gut of healthy birds can occur. Migration of these birds across the globe can disseminate the strains to wider geographical areas.
- Humans can get infected through the indirect transmission of fungi from environmental sources and through direct contact with poultry.

Hence, the detection of avian aspergillosis can act as an indicator of environmental health. Constant surveillance of disease in animals, the study of migration patterns in birds, use of alternative agents for pest control in agriculture, prevention of dissemination of fungal spores by maintaining hygienic conditions in animal husbandry, prevention of overcrowding, effective control measures for zoonotic aspergillosis, antifungal stewardship in humans and animals are essential to overcome the problem of antifungal resistance.

Diagnosis

Diagnosis of Invasive Aspergillosis in Humans

Conventional diagnosis of invasive mold infections involves microscopy and fungal culture of clinical samples depending on the site of infection. Since invasive pulmonary aspergillosis (IPA) is the most common clinical manifestation, the most common samples received for diagnosis include sputum or bronchoalveolar lavage (BAL) fluid and respiratory biopsy specimens (Table 20.3).

Microscopy

Direct microscopic examination using 10% Potassium hydroxide (KOH) solution: This procedure helps to clear out the debris or tissue material, thus making the fungal filaments clearly visible. Various concentrations (10%–40%) of KOH can be prepared depending on the thickness of the tissue material. The addition of fluorescent stains such as calcofluor white (CFW) increases the sensitivity as the fungal elements are visible as bright apple green/white elements on a dark background. The sample is placed on a clean glass slide. A drop of 10% KOH and a drop of calcofluor white solution are added. A coverslip is applied. The slides are examined with a UV microscope equipped with an excitation filter that transmits wavelengths between 300 and 412 nm (the optimum for CFW is 365 nm). If plain KOH is being used, the slides can be examined using a light microscope. CFW should be covered with dark paper and stored under dark conditions.

Gram stain: Gram stain is routinely performed for most samples received for the culture of bacteria. This stain is useful for the demonstration of fungal hyphae when evidently present. The yeast cells usually show well-stained morphology but filamentous fungi in smears become desiccated and their morphological characteristics may be lost. Gram stain is a less frequently used method for examining a specimen for a fungus. Nevertheless, the stain is useful to detect the host cells, polymorphonuclear neutrophils, and epithelial cells in sputum and to detect the quality of the specimen. Fungal hyphae appear as purple (Gram-positive) filamentous structures.

Lactophenol cotton blue stain for microscopic identification from fungal culture: A drop of lactophenol cotton blue (LPCB) is placed on a labeled clean glass slide. With a bent wire, a small portion of the fungal colony is removed

Table 20.3 Diagnostic approaches in aspergillosis

Diagnostic approaches	Methods	Targets	Remarks
Microscopy	Direct examination using potassium hydroxide and calcofluor white fluorescent stain	Septate fungal filaments can be visualized	Direct examination can be performed from tissues, pus, respiratory secretions
	Histopathological examination using hematoxylin and eosin stain	Fungal filaments and tissue reaction can be observed	Considered as an important diagnostic criterion for tissue invasion in proven cases of invasive aspergillosis
	Special stains such as periodic Schiff	Stains the fungal filaments in tissue sections	
In vitro cultures	The culture media most used for isolation is Sabouraud dextrose agar	Morphological identification	Enables the identification of *Aspergillus* species by macroscopy, microscopy Automated identification using MALDI-TOF and antifungal susceptibility testing can be performed from growth
Serology	Enzyme-linked immunosorbent assay (ELISA)	Galactomannan antigen detection from serum sample	Useful for diagnosis of invasive aspergillosis Needs standardization of assay for use in animals
Molecular diagnosis	Polymerase chain reaction (PCR)	18S rRNA and ITS	High sensitivity and specificity

from the agar surface and placed in the drop of LPCB. With two dissecting needles, the colony is teased apart and a cover slip is placed. The slide is placed under the microscope to appreciate the microscopic morphology of the fungus. This method does not always preserve the original position and structure of the conidia, spores, and other characterizing elements, but it is a very rapid method and a must for identification.

In vitro Culture

In vitro culture from sterile samples such as the biopsy for the fungus is considered the gold standard. However, obtaining such samples from critically ill patients is not feasible. Samples collected aseptically, placed in sterile leakproof containers should be delivered to the laboratory within 2 h. Materials received on swabs are considered unsatisfactory, but specimens on moist swabs from certain sites such as the ear canal, nasopharynx, throat, vagina, and cervix are accepted as they cannot be readily collected by other means. Swabs for collection of material from open wounds or draining lesions are frequently contaminated with environmental microorganisms. Large volumes of fluid are concentrated by centrifuging prior to inoculation for culture. Tissue samples are minced using a mortar/pestle or tissue homogenizer and then inoculated. All specimens of sufficient quantity submitted for fungal culture are examined microscopically. When there is not enough of a specimen to allow both a culture and a direct examination, the culture is done as it is more sensitive than a microscopic examination. Two or four dried sterile Sabouraud dextrose agar (SDA) tubes are used for every sample. All routine fungal cultures are incubated at room temperature (RT) and 37°C for 4 weeks before they are discarded. Cultures are not discarded when a fungus is first isolated but held for the entire period of 4 weeks so that slow-growing fungi are not overlooked. All culture plates and tubes kept both at RT and 37°C are to be examined for growth every day for the first week and thrice weekly thereafter for 3 weeks. Findings are recorded as no growth/growth/yeast/

mold/. Isolates are further identified depending on the type of growth by making LPCB preparations. Fungi that produce sporulating structures are easy to identify, but some fungi might need special media for sporulation such as potato dextrose agar, malt extract agar, Oatmeal agar, V-8 juice gar, etc. The best method for preserving and observing the actual structure of a fungus is slide culture. It is not a rapid technique, but it is extremely useful as a routine means of studying the fine points of the microscopic morphology of the fungi.

Serodiagnosis

Aspergillus GM assay by the ELISA method has been in practice for more than a decade and is an essential microbiological criterion for diagnosis of IA according to the EORTC/MSG guidelines updated as of 2019. The test is validated for use on BAL and serum samples for diagnosis of IA. ELISA (Platelia™ *Aspergillus* antigen, Bio-Rad, Marnes-la-Cocquette, France) is FDA-approved for testing on BAL and serum samples. The test is used for the diagnosis of IA and for monitoring patients on antifungal therapy. The test OD (optical density) cut-off considered positive according to the kit literature is 0.5. However, studies have shown that the diagnostic accuracy of the test is better if done on both BAL and blood using higher cutoffs. One of the major disadvantages of this test is that the time to obtain results varies between centers depending on the number of specimens to be tested, as it requires controls to be run with every batch tested. Therefore, many centers do batch testing to reduce the cost per test. This limitation is overcome using a rapid test, the lateral flow assay (LFA), which can be performed on single samples, and the results are available within an hour. The LFA is based on the principle of sandwich immunochromatographic format, which has specific antibodies conjugated to colloidal nano-gold, which binds to the galactomannan antigen if present in the sample. The test strips must be read using a cube reader, which gives a cutoff value and the result as positive or negative. There are several manufacturers for the assay, and the test can be used for both blood and BAL samples and evaluated for use from human samples.

Molecular Diagnosis

The newer methods of diagnosis include molecular diagnosis with the detection of *Aspergillus* DNA by conventional as well as Real-time PCR, and this is now considered a diagnostic parameter by the EORTC/MSG group. The sensitivity of PCR-based tests is higher and more useful, especially in immunocompromised patients. *Aspergillus* DNA can be detected from site-specific Single plex PCR for individual *Aspergillus* species as well samples as well as whole blood. Pan fungal PCR for all *Aspergillus* species is available. The mere presence of fungal DNA in blood need not imply invasive infection since even dead fungi can be detected. Culture of respiratory samples though technically simple, the yield is poor. Also, most critically ill patients with respiratory involvement are on empiric anti-mold antifungal therapy, making the yield on culture even poorer. In such situations, the only evidence of an invasive respiratory *Aspergillus* infection is the detection of *Aspergillus* GM [*Aspergillus* cell wall component] or DNA in the respiratory samples.

Diagnosis of Invasive Aspergillosis in Animals

Specimens could include pneumonic lung, granulomatous nodules, centrifuged mastitis milk, fetal lesions, fetal stomach contents, cotyledons, ear swabs, skin scrapings, and biopsies from nasal granulomas and fungal plaques in the guttural pouch. The challenge with obtaining the specimens from animals, especially for respiratory samples, is because it might require anesthetizing them to obtain specimens.

Microscopy

Direct microscopic examination of tissue scrapings and other clinical specimens can be done after clearing in 10% KOH. The various microscopic methods explained earlier are used for any specimen with suspected fungal infection. The concentration of KOH required depends on the tissue material. CFW has an increased sensitivity for detection of fungal filaments from samples (Fig. 20.5a).

Histopathology

Due to the ubiquitous presence of the fungus in the environment as well as surfaces, demonstration of the same within deeper tissues by histopathology adds value to diagnosis. For example, the isolation of *Aspergillus* from nasal discharge in a case of nasal granuloma could be attributed to its being a colonizer, whereas septate fungal hyphae, if observed in the tissue section by histopathology, confirm the diagnosis. However, species-level identification of fungi is possible only with culture, and histopathology can be confusing when seen on cut sections of the tissue. Special stains such as Gomori methenamine silver and Periodic acid-Schiff stains are useful for the demonstration of fungi in histopathological sections. Cytology of the fluid from the sino-nasal cavity is useful in dogs. Diagnosis in mycotic abortion of cattle is based on examination of fetal contents and histopathological evidence of mycotic placentitis. Examination of tissue sections in cases of pneumonia in cattle and GPM can reveal granulomatous reactions along with fungal filaments (Fig. 20.4).

In vitro Culture

Isolation of the fungus in culture from the samples is the gold standard for diagnosis of aspergillosis, even in animals. *Aspergillus* species grow on standard laboratory media such as Sabouraud dextrose agar (SDA). Sabouraud dextrose agar, with and without 0.05 g/L chloramphenicol, but not if cycloheximide is used. The inoculated plates are incubated aerobically at room temperature (25 °C) and 37 °C for up to 5 days, although usually the colonies will appear in 2–3 days. Identification of the fungi is based on colony morphology on SDA and microscopic examination using the LPCB wet mount. Newer diagnostic techniques are available, such as MALDI-TOF, for accurate identification of the various species (Fig. 20.5b and c).

Antifungal Susceptibility Testing

The need for susceptibility testing antifungals is well established due to the emerging trends of resistance to antifungal agents. The use of a combination of antifungal agents and therapeutic drug monitoring has become an essential entity. The standard guidelines for in vitro susceptibility assays are given by the Clinical and Laboratory Standards Institute (CLSI) and The European Committee on Antimicrobial Susceptibility Testing (EUCAST). The recommended methods include broth dilution methods where readings are taken visually or by spectrophotometry MIC determination, 50% of MIC, and minimal effective concentration (MEC) measurements are used for the different antifungal agents. Establishing MICs for molds for drugs such as echinocandins can be challenging since no complete macroscopic growth inhibition is observed, but rather partial inhibition is noted, which is associated with abnormal hyphae formation. MEC is used for situations to determine susceptibility. The E-test is performed on an agar plate with a lawn culture of the fungus with a paper strip having a concentration gradient of an antifungal drug placed on it. Results are read by measuring an ellipse zone of inhibition. All these techniques need careful preparation of antifungal drug concentrations, standardization of inoculum, incubation, and interpretation.

Serodiagnosis Aspergillus-Specific IgE and IgG Detection

Serological tests are based on conidial or hyphal-specific galactomannan antigens of *A. fumigatus*. Because of constant exposure, most animals develop antibodies to conidial antigens, and the serological results therefore must be interpreted with caution. Serological assays based on ELISA, agar gel immunodiffusion (AGID), and counter-immunoelectrophoresis (CIEP) are the formats available for serological diagnosis. Due to the wide presence of fungal spores in the environment, the interpretation of positive antibody results may not always suggest invasiveness. In dogs, the most sensitive serological test is ELISA. A commercial human ELISA that detects a cell wall component, β-1,3-D-glucan, in blood

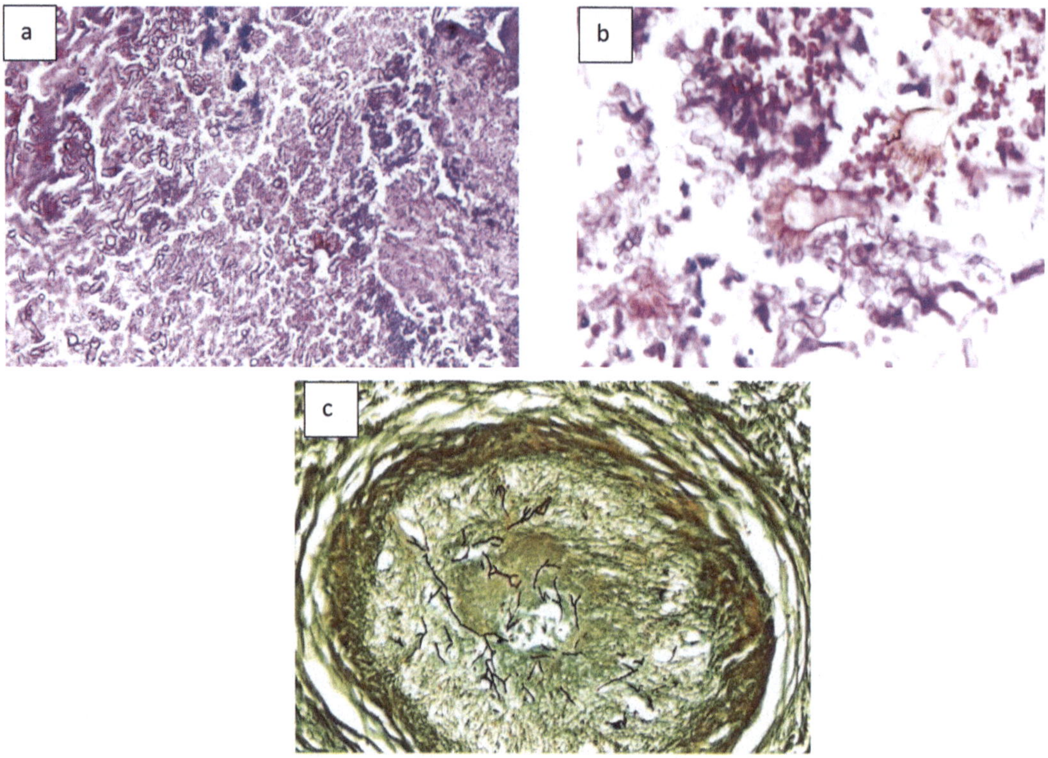

Fig. 20.4 (**a**) Aseptate fungal hyphae of *Aspergillus* in tissue on H&E stain, (**b**) Sporulating *Aspergillus* stained with PAS, and (**c**) Black colored septate fungal hyphae of *Aspergillus* within blood vessel stained with GMS (Department of Histopathology, St. John's Medical College and Hospital)

and bronchoalveolar lavage samples has been applied with some success to avian serum samples. However, the utility of tests such as the galactomannan assay for the diagnosis of invasive aspergillosis in animals needs to be evaluated. These blood-based tests might act as an adjunct to clinical and radiological diagnosis if cutoffs are established for use in animals, especially in situations where invasive sampling may be difficult. There have been attempts to establish the serological diagnosis in cattle with mycotic abortions, but they have not yielded the appropriate cutoffs. Serology may suggest invasive aspergillosis, but culture and histopathological examination are necessary where possible for confirmation.

Detection of antibodies such as *Aspergillus* IgE and IgG for allergic manifestations such as allergic bronchopulmonary aspergillosis and chronic pulmonary aspergillosis has been well established in humans. These tests require standardization for use in zoonotic aspergillosis. Since the exposure levels to the fungus are high in animals and birds, the cutoffs for these tests need to be established in animals. There are several platforms available for performing these tests, such as ELISA, Enzyme-linked fluorescence assay (ELFA), and Chemiluminescence assay (CLIA).

Molecular Diagnosis

Molecular detection techniques can be useful for the detection of fungal DNA from the samples obtained; however, the diagnostic accuracy depends on sterile sampling techniques to avoid detecting fungi from the surface and environment. Sensitive techniques such as PCR are helpful in diagnosis where the other tests are negative.

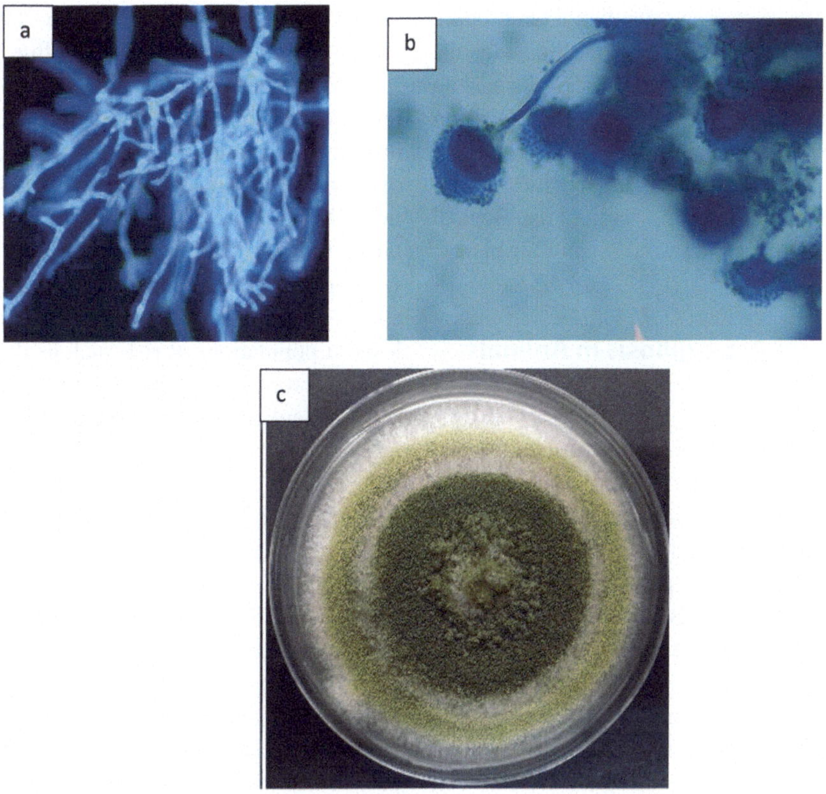

Fig. 20.5 (a) Direct microscopic examination of BAL sample by KOH and Calcoflour showing septate fungal filaments, (b) LPCB mount of tease preparation from culture of *Aspergillus*, and (c) SDA plate with *Aspergillus* growth

Other Methods

Radiological investigations may be carried out based on clinical signs. It is difficult to get these done since general anesthesia is necessary for diagnostic imaging. Specialty referral or veterinary teaching hospitals with advanced equipment and personnel determine the diagnostic modalities.

It is found to be particularly useful for the diagnosis of guttural pouch mycosis, where the presence of fluid accumulated in the pouch alongside an endoscopic demonstration of characteristic lesions is diagnostic. Radiography may reveal an increased radiolucency of turbinate bones in dogs with sino-nasal aspergillosis. Recently, computed tomography imaging is also used widely. Imaging tests can at best be an adjunct to the diagnosis, in addition to identification by biopsy, smear, or culture as conclusive evidence.

Treatment

Treatment of Aspergillosis in Humans

Treatment of aspergillosis in humans is based on the clinical spectrum and presentation. Medical as well as surgical modalities are available. Invasive aspergillosis is managed usually by antifungal agents. The most common azole used is voriconazole. Intravenous alternatives include amphotericin B as lipid formulations, caspofungin, or itraconazole. Oral antifungals include itraconazole and posaconazole. It is recommended to identify the *Aspergillus* species to the species level and perform antifungal susceptibility testing for isolates causing invasive aspergillosis. Itraconazole is useful for allergic bronchopulmonary aspergillosis. Surgical management might be necessary in cases such as aspergilloma of the lung and sinus washes and functional endoscopic sinus surgery in cases of fungal rhinosinusitis. Pulmonary artery embolization is performed to manage certain cases with severe angio-invasion and hemoptysis. Several groups have established guidelines for prophylaxis in specific patient populations such as those with hematological stem cell transplants and solid organ transplants.

Treatment of Aspergillosis in Animals and Birds

Therapeutic options include topical and systemic antifungal administration. Potential toxicity and the high cost of antifungals must be considered before decisions are made. Azoles such as itraconazole, voriconazole, posaconazole, enilconazole, and amphotericin B have been used to treat aspergillosis in animals. Infusion of itraconazole into the pouch is found to be useful for the treatment of guttural pouch mycosis. Surgical management includes procedures such as ligation, balloon catheterization, and trans-arterial coil embolization to manage fatal hemorrhagic events. For sino-nasal aspergillosis, administration of itraconazole through tubes inserted surgically in the sinuses, along with systemic treatment with voriconazole for 6–8 weeks, is useful. The topical application of azoles for infections of the eye, along with systemic therapy, is also useful. Nebulization with amphotericin B and enilconazole is recommended for aspergillosis in birds. Aspergillosis in birds is often difficult to treat and is associated with high mortality. Most of the cases require a combination of these techniques, local as well as systemic therapy. Supportive therapy also plays an important role and might be the only choice in some situations.

Prevention and Control

Prevention Strategies in Humans

As already elucidated, *Aspergillus* causes a variety of allergic manifestations in humans on exposure to spores. The quality of air, both indoor and outdoor, determines the sensitization rates. Damp surroundings within enclosed spaces, improper maintenance of air-conditioning units, and construction activities around hospitals and houses can release large numbers of spores in the air leading to allergic manifestations. Monitoring of air quality plays an important role in prevention.

Prophylactic administration of antifungal agents is helpful in preventing invasive aspergillosis in populations at risk such as those receiving hematopoietic stem cell transplants, acute myeloid leukemia, and febrile neutropenia.

Prevention Strategies in Animals and Birds

Prevention and control strategies are largely applicable to places where animals and birds are kept captive for purposes of husbandry. Fungal contamination of the indoor spaces and cages is facilitated by moisture and animal litter, which aids in the growth and sporulation of the fungi. Appropriate ventilation of the spaces where animals and birds are housed, regular cleaning of litter, and maintaining dry floors help in preventing fungal growth. Control of outbreaks can be done by separating sick animals and birds from healthy ones and ensuring early diagnosis and treatment. Maintaining hygienic conditions and care of poultry plays an important role in the prevention of fungal infections in birds. Good silage facilities and ensuring appropriate storage of fodder are essential to prevent mycotoxicosis. The application of fungistatic agents like thiabendazole, nystatin, or copper sulfate prevents fungal contamination of beds. Enilconazole is available in special formulations for the decontamination of environmental surfaces.

Case Studies

In a hatchery where 5000 birds were housed, many birds started developing signs of respiratory difficulties, dyspnea, anorexia, and drooping feathers resulting in the death of 250 birds. The mortality rate started increasing, and hence a postmortem examination was performed to investigate the possible microorganism responsible. The lungs, heart, liver, and other organs were examined. Microscopic examination of the smear showed severe cellular infiltration with branched septate fungal hyphae. The remaining birds were treated with the antifungal agent griseofulvin (50 mg/20 L) for a period of 1 week. Some of the birds recovered, but there was a huge economic loss due to this scenario.

1. What might be the possible diagnosis in this scenario?
2. How is the diagnosis confirmed? What are the culture methods available for this fungus?
3. What are the treatment options available for this?

Research Questions

The need of the hour is collaborative interdisciplinary research of not only zoonotic fungal infections in general but also aspergillosis. This has wide implications for the concept of one health. The one big deterrent for generating data is the number of genera and species in the zoonotic kingdom. However, well-planned research questions must be formulated and answered for the well-being of the future of not just mankind but the entire ecosystem.

Some of the key questions that need urgent attention include the following:

1. Classification of cases as proven, probable, and possible and their correlation with classical clinical laboratory findings, including radiological findings and molecular diagnosis among the various bird and animal populations.
2. Establishing cutoffs for serological diagnosis.
3. Correlation with culture results and antifungal resistance patterns.
4. Epidemiological studies correlating AFST profiles of various species at the national and international levels.
5. Epidemiological studies on migratory patterns of birds with AFST profiles.

Acknowledgments The contribution of Ms. Rajeswari V and Ms. Amulya A, Department of Microbiology, St. John's Medical College and Hospital, in the preparation of the chapter is acknowledged.

Further Reading

Beernaert LA, Pasmans F, van Waeyenberghe L, Haesebrouck F, Martel A (2010) *Aspergillus* infections in birds: a review. Avian Pathol 39:325–331

Billen F, Peeters D, Peters IR, Helps CR, Huynen P, de Mol P et al (2009) Comparison of the value of measurement of serum galactomannan and aspergillus-specific antibodies in the diagnosis of canine sino-nasal aspergillosis. Vet Microbiol 133(4):358–365

Dagenais TRT, Keller NP (2009) Pathogenesis of *Aspergillus fumigatus* in invasive aspergillosis. Clin Microbiol Rev 22:447–465

Desoubeaux G, Cray C (2018) Animal models of aspergillosis. Comp Med 68(2):109–123

Elad D, Segal E (2018) Diagnostic aspects of veterinary and human aspergillosis. Front Microbiol. 9:1303

Fang W, Latgé JP (2018) Microbe profile: *Aspergillus fumigatus*: a saprotrophic and opportunistic fungal pathogen. Microbiology (United Kingdom) 164(8):1009–1011

Gupta A, Singh NP (eds) (2019) Recent developments in fungal diseases of laboratory animals, fungal biology, pp 49–72. https://doi.org/10.1007/978-3-030-18586-2_4

Leishangthem GD, Singh ND, Brar RS, Banga HS (2015) Aspergillosis in avian species: a review. Journal of poultry. Sci Technol 3:1–14

Lim YY, Mansfield C, Stevenson M, Thompson M, Davies D, Whitney J et al (2022) A retrospective multi-center study of treatment, outcome, and prognostic factors in 34 dogs with disseminated aspergillosis in Australia. J Vet Intern Med 36(2):580–590

Ludwiga A, Gatineaua S, Reynaudb M-C, Cadoréa J-L, Bourdoiseaub G (2005) Fungal isolation and identification in 21 cases of guttural pouch mycosis in horses (1998–2002). Vet J 169:457–461

Markey BK, Leonard FC, Archambault M, Cullinane A, Maguire D (2013) Clinical Veterinary Microbiology, 2nd edn. Mosby Elsevier

Quinn PJ, Markey BK, Leonard FC, FitzPatrick ES, Fanning S, Hartigan PJ (2011) Veterinary Microbiology and Microbial diseases, 2nd edn. Wiley-Blackwell

Schultz RM, Johnson EG, Wisner ER, Brown NA, Byrne BA, Sykes JE (2008) Clinicopathologic and diagnostic imaging characteristics of systemic aspergillosis in 30 dogs. J Vet Intern Med 22(4):851–859

Seyedmousavi S, Guillot J, Arné P, de Hoog GS, Mouton JW, Melchers WJG et al (2018) *Aspergillus* and aspergilloses in wild and domestic animals: a global health concern with parallels to human disease. Med Mycol 53(8):765–797

Sherman AB, Clode AB, Gilger BC (2017) Impact of fungal species cultured on outcome in horses with fungal keratitis. Vet Ophthalmol 20(2):140–146

Tell LA (2005) Aspergillosis in mammals and birds: impact on veterinary medicine. Med Mycol 43:S71–S73

Tsang CC, Tang JYM, Lau SKP, Woo PCY (2018) Taxonomy and evolution of *Aspergillus*, *Penicillium*, and *Talaromyces* in the omics era—past, present and future. Comput Struct. Biotechnol J (16):197–210

Wolf AM (1992) Fungal diseases of the nasal cavity of the dog and cat. Vet Clin N Am Small Anim Pract 22(5):1119–1132

Cryptococcosis

21

Pratibha Kale
and Shivaprakash Mandya Rudramurthy

Learning Objectives

By the end of the chapter, the readers should be able to:

1. Determine the aetiological agents of cryptococcosis.
2. Know the pathogenesis and clinical presentation of cases.
3. Understand various diagnostic modalities in cryptococcosis.
4. Learn various strategies for the management of cryptococcosis.

Introduction

Cryptococcosis is a life-threatening fungal zoonotic disease that affects mammals, birds, amphibians and reptiles. The genus *Cryptococcus* encompasses basidiomycetous yeast with two predominant species, *Cryptococcus neoformans* and *Cryptococcus gattii* species complexes. *C. neoformans* is found in bird excreta, contaminating soil and *C. gattii* is found in plant habitats. Cryptococcosis is caused by the inhalation of yeast cells from the soil with bird excreta or other niches, entering the sino-nasal cavity in animals or lungs in humans. Subsequently, infection spreads haematogenously, leading to infections of the skin or soft tissue, pneumonia and commonly meningoencephalitis.

P. Kale (✉)
Department of Clinical Microbiology, Institute of Liver and Biliary Sciences, New Delhi, India
e-mail: drpratibhapgi@gmail.com

S. M. Rudramurthy
Department of Medical Microbiology, Postgraduate Institute of Medical Education and Research (PGIMER), Chandigarh, India

History

The origin of the fungus dates back to 1894 when *Saccharomyces neoformans* was grown in fruit juice by Sanfelice. The name *C. neoformans* was given by Jean-Paul Vuillemin in 1901, as it lacked ascospores, which were present in *Saccharomyces*. Emmons (1951, 1955) isolated it from pigeon excreta and contaminated soil in that area. *C. neoformans* was subsequently isolated from the avian guano of chickens, sparrows, parrots, skylarks, starlings, etc. during 1960–1970. Later, in 1995–2004, *C. neoformans* was recovered from the nasal cavity and upper respiratory tract of dogs, koalas, ferrets and cats. *C. neoformans* has been recovered from decaying wood and tree hollows; however, *C. gattii* is more strongly associated with such an environmental niche. In 1990, Ellis and Pfeiffer retrieved *C. gattii* (VGI) associated with Eucalyptus camaldulensis trees in Australia. It was reported in India by Chakrabarti et al. in 1997. *C. gattii* has a zoonotic niche in African grey parrots and koalas, bat guano, camel hair and ostrich feathers.

S. C. Parija, S. M. Rudramurthy (eds.), *Textbook of Fungal Zoonoses and Sapronoses*, Microbial Zoonoses, https://doi.org/10.1007/978-981-97-3214-2_21

Taxonomy

Cryptococcus is a basidiomycetous yeast with two predominant pathogenic species, *C. neoformans* and *C. gattii*. *C. neoformans* encompasses varieties such as *C. neoformans* var. *neoformans* and *C. neoformans* var. *grubii*. *C. neoformans* on molecular genotyping has VNI–VNIV and VNB, for var. *grubii*, serotype A or VNI, VNII and VNB strains and for var. *neoformans* serotype D or VNIV strains, and serotype AD strains (VNIII). *C. gattii* includes serotypes B and serotype C, including molecular types VGI, VGII, VGIII and VGIV. Currently, the taxonomy of *Cryptococcus* has been revised to include *Cryptococcus neoformans* (formerly *Cryptococcus neoformans var grubii*) and *Cryptococcus denoeformans* (formerly *Cryptococcus neoformans* var. *neoformans*). *Cryptococcus gattii* is now a species complex containing five species *Cryptococcus gattii* (formerly serotype B), *Cryptococcus baccilosporus* (formerly serotype B), *Cryptococcus deturogattii* (formerly serotype B), *Cryptococcus tertagattii* (formerly serotype C) and *Cryptococcus decagattii* (formerly serotype B).

Genomics and Proteomics

Cryptococcus species that are pathogenic to humans are haploid in nature. However, diploid nature has also been isolated from different geographic regions. The genome size of *Cryptococcus* is 20 Mb, encompassing 14 chromosomes having ~7000 protein-coding genes. *Cryptococcus* is a heterothallic fungus that has defined mating typing locus either denoted as 'a' or 'α', which never switches. The MAT locus measures >100 kb, being encompassed in 20 genes. Next-generation sequencing is a useful tool. It is difficult to differentiate *C. neoformans* from *C. gattii* by restricted fragment length polymorphism (RFLP) fingerprinting with four restriction enzymes. Hybridisation probes help identify *Cryptococcus* at the species level. Intergenic spacer (IGS) and internal transcribed spacer (ITS) combined with topoisomerases and application multilocus sequence typing (MLST) led to the consensus that the MLST scheme has good discriminatory power. MLMT is more reliable and cost-effective than MLST.

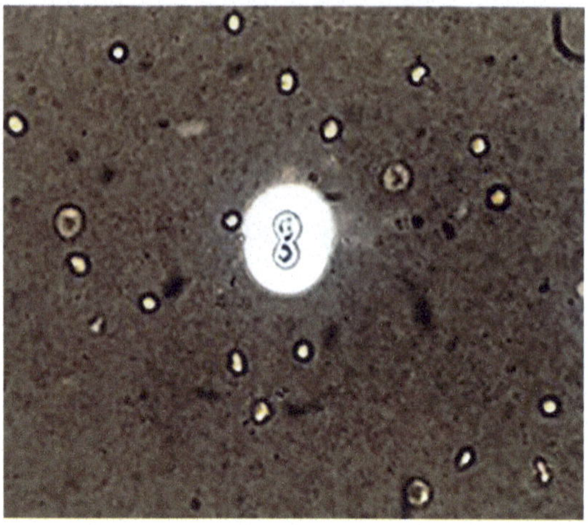

Fig. 21.1 India ink preparation of a CSF sample showing a budding yeast cell with a capsule of *Cryptococcus* (400×)

The Fungus Morphology

C. neoformans yeast is typically round, measuring 4–6 μm in diameter, with a polysaccharide capsule up to 30 μm thick. There is a narrow-based budding. The yeast surrounded by a mucoid polysaccharide capsule is the typical diagnostic feature in the India ink staining (Fig. 21.1).

Cultivation of Fungi

Cryptococcus grows well on Sabouraud dextrose agar, producing rapidly growing white mucoid colonies at 37 °C. *C. neoformans* produces phenol oxidase, which is useful in melanin production, giving the colonies a brown to black colour on caffeic acid agar or Staib's medium. As caffeic acid agar or bird seed agar is selective for *Cryptococcus*, this media is used for its isolation from a colonised specimen such as sputum or BAL. It also helps to differentiate *Cryptococcus* from other yeasts, such as *Candida*, that are present in the sputum/BAL or other colonised specimens. Christensen urea agar and L-canavanine-glycine-bromothymol blue (CGB) agar are used to differentiate between *C. gattii* and *C. neoformans*. Morphologically, *C. neoformans* is globose and *C. gattii* is oval, ovoidal or pyriform. Creatinine deiminase is not suppressed by *C. gattii*. Glycine is utilised as a sole source of carbon by *C. neoformans*, while *C. gattii* utilises it as a nitrogen and carbon source. *C. gattii* grows on media containing L-canavanine with glycine and bromothymol blue (CGB) and is used to differentiate the two species.

Animal Models

Invertebrate Models

Invertebrate models include amoebae, such as *Dictyostelium discoideum* or *Acanthamoeba castellanii*, nematodes like *Caenorhabditis elegans* and *Drosophila melanogaster* and insects such as *Galleria mellonella*. These models are used for signal transduction virulence studies, innate immune interaction, chitin regulation, lipid metabolism and melanin synthesis. The amoeba model is specifically useful to demonstrate the cellular interaction of macrophages with *C. neoformans*. Insect models demonstrate innate immune interaction with *Cryptococcus* and the effect of antimicrobial peptides on yeast survival. However, there are certain limitations in the use of these invertebrate models. These models do not have an adaptive immune response, the route of infection is not an inhalation infection model, and they lack differentiated immune cells, including dendritic cells (DCs) and macrophages.

Vertebrate Models

Vertebrate models include rats, mice, zebrafish and rabbits. Transcriptome analysis has been done in cynomolgus monkeys and mice using RNA sequencing, but these are expensive technologies. The rabbit model is used for studying newer antifungals for meningitis. Rats were used to study the immune interactions with alveolar macrophages, forming granulomas. Rats, being naturally susceptible hosts, are useful for *C. gattii* studies; however, the absence of knockout models and the cost are a major limitation. Mice are frequently used animal models to study *C. neoformans* host-pathogen interactions.

Life Cycle of Fungus

Cryptococcus is predominantly a narrow-based budding yeast cell in patients (in vivo). Haploid yeast cells are of two different mating types, a and α. They fuse during sexual reproduction filamentous forms that are dikaryotic and with the formation of a basidium. There is a fusion of nuclei followed by meiosis and then mitosis. There is budding, and haploid basidiospores are formed. On the contrary, when nuclei of two similar mating types fuse, diploid monokaryotic hyphae are formed. Later, when there is basidium development, meiosis occurs, leading to the formation of haploid basidiospores. These basidiospores mature and germinate into haploid cells after environmental dispersion (Fig. 21.2).

Pathogenesis and Pathology

The main virulence factors of *Cryptococcus* include polysaccharide capsules, melanin production, and other enzymes such as laccase, urease and phospholipases and thermotolerance. The role of virulence factors is similar in human and animal infections. The capsule is composed of polysaccharides and glucuronoxylomannan is antiphagocytic, which helps immune evasion, and the yeast cells may remain latent within the phagolysosome in the thoracic lymph nodes. Melanin protects the yeasts from oxidative damage and also exhibits neurotropism at sites with diphenolic catecholamines. Growth of the yeast at a high temperature, 37 °C, promotes multiple signalling pathways and enzymes for growth *in vivo*. Phospholipase, urease production and multiple enzymes help in survival of *Cryptococcus* within the phagolysosome.

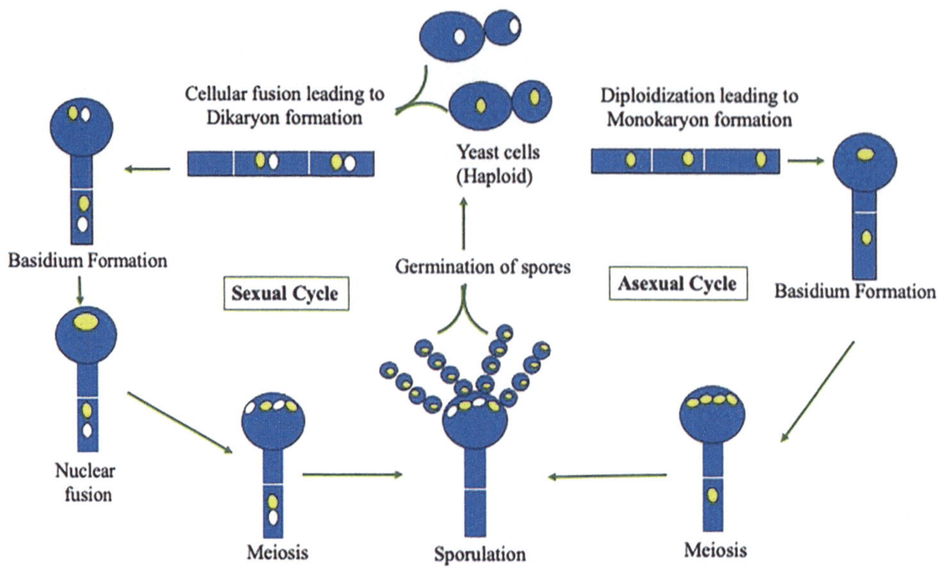

Fig. 21.2 Suggested life cycle of *Cryptococcus* spp.

The cryptococcal infection starts by inhalation of the infectious particles from the environmental niche or guano of infected birds, with deposition in pulmonary alveoli. At times, traumatic inoculation or entry from the gastrointestinal tract also occurs. Primary pulmonary infection usually remains asymptomatic or minimally symptomatic. In some cases, the yeasts are tackled by alveolar macrophages. Host response mainly involves recruitment of helper T cell response and release of cytokines, predominantly tumour necrosis factor (TNF), interferon-γ and IL-2 and IL-12. This immune response leads to the initiation of a granulomatous type of inflammation. Latent infection in the lung or thoracic nodes may be persistent for years, which may reactivate in an immunosuppressed host. Meningeal involvement is by direct invasion of the blood–brain barrier and a trojan horse mechanism via macrophages into the centgral nervous system (CNS).

Immunology

Immune response against *Cryptococcus* depends upon the immune status of the host. In an immunocompetent host, pattern recognition receptors (PRR) in alveolar macrophages interact with fungal spores, activating the macrophages to release chemokines CCL2, leading to the activation and recruitment of DCs and monocytes. These immune cells lyse the yeast forms and antigens presented to CD4 + T-cells. Th1, along with Th17 cells, then release various cytokines, including IL-6, IL-10, interferon γ and granulocyte-macrophage colony-stimulating factor (GM-CSF). This leads to the recruitment of classical (M1) macrophages and the release of reactive oxidative molecules like iNOS, CD80 and TLR4. There is upregulation of TNFα, IL-1β and IL-8. This leads to granuloma formation. In diseased or immunocompromised states, the immune responses are suppressed, leading to excessive fungal growth.

Infection in Humans

In humans, *C. neoformans* infection is mostly associated with HIV-AIDS. In an immunocompromised host, *C. neoformans* var. *neoformans* is an important pathogen, while *C. neoformans* var. *grubii* causes most of the other infections. On the contrary, *C. gattii* predominantly causes infections

Table 21.1 Distribution of *Cryptococcus* infections in the global population

Cryptococcus species	Animal host	Site of infection	Distribution
Cryptococcus neoformans	Dogs, cats, horses, cattle, goats and pigs and wild animals like ferrets, monkeys, cheetahs, koalas, marsupials and marine mammals	Skin, blood and respiratory tract	Worldwide
Cryptococcus gattii	Ferrets and goats	CNS, blood, skin and respiratory tract	Worldwide
Cryptococcus neoformans var. *grubii*	Cats, dogs, koalas and ferrets	CNS, blood, skin and respiratory tract	Worldwide

in immunocompetent individuals. Cryptococcal meningitis is the common clinical manifestation in immunocompromised hosts, followed by pulmonary, disseminated, cutaneous or other localised infections.

Clinical Manifestations

As the first point of entry, pulmonary infection is common, though it may be asymptomatic in most individuals. Pulmonary cryptococcosis may start as asymptomatic colonisation to a pulmonary nodule, pneumonia with Acute respiratory distress syndrome (ARDS) in severe forms. Pulmonary involvement is seen in 10%–55% of patients with HIV-AIDS. The serum cryptococcal antigen is usually negative in case of isolated pulmonary involvement.

CNS cryptococcosis presents with signs and symptoms of meningoencephalitis, including fever, headache, altered sensorium, and signs of meningeal irritation. Those patients with a high load of organisms in Cerebrospinal fluid (CSF) may have acute onset with positive CSF Cryptococcal antigen titres and raised intracranial pressures (ICP). Occasionally, brain parenchymal involvement may occur, forming *cryptococcomas* with signs and symptoms of raised ICP and cranial neuropathies.

Cutaneous infections may present as a part of disseminated disease, more commonly in patients receiving tacrolimus therapy after solid organ transplantation. Infections at other sites, like prostate infection and ocular, bone and joint infections, have been reported in humans.

Infection in Animals

Natural infection is common in birds, amphibians, reptiles and mammals. Cryptococcosis is common in domestic animals like dogs, cats, horses, cattle, goats and pigs; wild animals like ferrets, monkeys, cheetahs, koalas, marsupials and marine mammals. Birds are rarely infected but the fungus can be present in the faeces of canaries, carrier pigeons, budgerigars, parrots and poultry. There are reports of mastitis in cattle, leading to reduced milk production and supra-mammary lymph node swelling. *Cryptococcus* mastitis can occur as an outbreak in cattle and may present sporadically in buffaloes. *C. neoformans* var. *grubii* can transiently colonise nares in cats, dogs, koalas and ferrets. Cryptococcosis caused by *C. neoformans* and *C. gattii* is the most common systemic mycosis in cats. Cats are commonly affected by respiratory tract involvement, subcutaneous granulomata and disseminated infections. Spread and persistence in the environment occur through wild and domestic animals shedding the spores present as nasal colonisation (Table 21.1).

Epidemiology and Public Health

C. neoformans is reported worldwide, predominantly causing infections in the immunocompromised hosts, mainly with HIV-AIDS. The most common presentation is meningitis associated with HIV patients, causing more than six lakh deaths per year. The infections are predominant in areas with a high prevalence of HIV-AIDS, like sub-Saharan Africa, where cryptococcal disease is seen in 20%–30% of AIDS patients. It is also a common AIDS-defining illness in most of these areas. The incidence has drastically reduced with the introduction of retroviral therapies. However, infections are increasing in non-HIV individuals with co-morbidities, like cirrhosis, diabetes, sarcoidosis, corticosteroid therapy, malignancy, chemotherapy, transplant recipients, etc. *C. gattii* is mainly reported from Australia, Central and South America and the southern California coast. Recently, an outbreak was reported from the northwest United States, mainly caused by the hypervirulent *C. gattii* strain.

Cryptococcal species are isolated from domestic and wild animals, as documented in various veterinary studies from Australia, New Zealand, the Hawaiian Islands and Papua New Guinea. In Asia, *C. neoformans* domestic dogs, cats, mice and rats are the host. In Spain, ferrets and goats harbour *C. gattii*, usually associated with eucalyptus trees.

Laboratory Diagnosis

Microscopy

India ink staining of cerebrospinal fluid is the most frequently used microscopic method for the diagnosis of cryptococcosis. The method is based on the negative staining of the capsule, which shows spherical vacuoles of the yeast in Brownian motion (Fig. 21.1) (Table 21.2).

CSF Examination

CSF examination shows low glucose and high protein levels and a normal or slightly elevated white cell count with lymphocyte predominance.

Culture

Cryptococcus grows rapidly at 37 °C on Sabouraud dextrose agar as white mucoid colonies. *C. neoformans* is positive for phenol oxidase production,

Table 21.2 Laboratory diagnosis of *Cryptococcus* infections

Disease	Direct microscopy	Immunodiagnostics	Culture	Remarks
Cryptococcal meningitis	Round narrow budding yeast cells on Gram stain India ink show a capsular halo.	CSF cryptococcal antigen detection by lateral flow assay, immunochromatography or chemiluminescence	SDA with antibiotics, shows cream-coloured yeast colonies.	Rapid growth at 30 ± 2 °C Urease positive. Differentiation from *Cryptococcus gattii* by culture on CGB medium.
Disseminated cryptococcosis	–	Serum/plasma cryptococcal antigen detection	Positive blood culture for budding yeast, subculture on SDA with antibiotics	
Respiratory or Gastrointestinal cryptococcosis	Round narrow budding yeast cells by Gram stain stain India ink show a capsular halo.		Bird seed agar is selective for *Cryptococcus*, can differentiate from *Candida* spp. or other yeasts. Positive cultures from tissues	Histopathology is useful

producing brown to black colonies on phenol oxidase containing caffeic acid agar, tobacco agar and Staib's agar. *C. neoformans* can be differentiated from *C. gattii* by colour differentiation on Christensen urea agar, positive for the former and later positive in L-canavanine-glycine-bromothymol blue (CGB) agar. *C. gattii* assimilates D-proline. Histopathology of tissue sections stained with Gomori's methenamine silver nitrate, Mayer's mucicarmine or periodic acid–Schiff show brown-black staining with a light green background, polysaccharide capsule as bright red and magenta staining with a blue nucleus, respectively.

Serology

Cryptococcal polysaccharide capsular antigen (CrAg) detection is pathognomonic for the diagnosis of cryptococcosis. The CrAg is detected in serum or CSF by latex agglutination test or enzyme immunoassay with 93%–100% sensitivity and 93%–98% specificity, respectively. Recently, antigen detection by immune chromatography test has been developed, with moderate sensitivity and specificity False negatives occur due to the prozone effect with high antigen titres, which can be prevented by testing the sample after dilution. Other causes for false negatives include improper specimen storage, low fungal burden or samples taken at the early stage of infection. Serum in cats demonstrates a high antigen titre (>1:65,536), although a titre of 1:2 is diagnostic of cryptococcal infection.

Molecular Detection

In view of the high sensitivity and specificity of CrAg, detection and diagnosis of *Cryptococcus* by PCR is not given much priority. Evaluation of real-time PCR in diagnosing cryptococcosis using pan-fungal primers has less sensitivity. Recently developed real-time PCR targeting multicopy mitochondrial cytochrome *b* has not only been shown to be highly sensitive (96%) and specific (100%) but also can differentiate *C. neoformans* from *C. gattii.* Fingerprinting techniques such as restriction fragment length polymorphism (RFLP), multilocus sequence typing (MLST), amplified fragment length polymorphism (AFLP) and microsatellite length polymorphism (MLP) using short tandem repeat (STR) markers are the various molecular typing methods used for cryptococcal isolates.

Treatment

Treatment of Human Infections

In humans, the treatment of cryptococcosis depends on the HIV status of the patient. There are increased chances of cryptococcal infection if CD4 count <200 cells/μL, so prevention of acute infection is achieved by giving amphotericin B (0.7–1 mg/kg/d for 2 weeks), with or without flucytosine (100 mg/kg/d in four divided doses 2 weeks), followed by fluconazole (400 mg/d for 8–10 weeks). In patients without AIDS, amphotericin B (0.7–1 mg/kg/d for 6–10 weeks) alone or in combination with flucytosine (100 mg/kg/d in four divided doses for 2 weeks) is the initial therapy. This is followed by fluconazole for at least 10 weeks.

Amphotericin B deoxycholate (AmBd) is the mainstay of treatment. An induction dose of 0.7–1 mg/kg/d is ideal. Liposomal amphotericin B (3–6 mg/kg/d) is preferable as it shows less nephrotoxicity. Flucytosine (5-FC) in a dosage of 100 mg/kg/d in divided doses in combination with AmBd is the recommended first-line therapy in HIV-associated cryptococcal meningitis and for severe disseminated or pulmonary cryptococcosis. The enhanced fungicidal activity of combination therapy has survival benefits compared to AmBd monotherapy. Another useful combination is AmBd plus fluconazole (800 mg/d), which shows fungal clearance and neurologic recovery. The induction phase is followed by consolidation treatment with fluconazole (400–800 mg/d) for 8 weeks. The maintenance with oral fluconazole (200–400 mg/d) is the recommended treatment for cryptococcal meningitis. In non-HIV, non-transplant patients and patients with good prognostic factors, 2-week induction therapy is usually successful.

Treatment of Animal Infections

Antifungal therapy for cryptococcosis in animals includes amphotericin B (Amp B) with 5-fluorocytosine, itraconazole, fluconazole, ketoconazole and terbinafine. The most effective is Amp B. Combination therapy with Amp B plus flucytosine or azole plus flucytosine is recommended for infection in cats. Flucytosine is not preferred in dogs due to toxic reactions like toxic epidermal necrolysis. Monotherapy with flucytosine causes the rapid development of resistance. Fluconazole is recommended in both dogs and cats. In dogs, fluconazole and itraconazole have shown good results for consolidation therapy. *C. gattii* VGII and VGII responds well to other azoles like posaconazole, voriconazole or isavuconazole. Surgery is recommended for large sized lesions to reduce the size and fungal burden. Supportive therapy with steroids in CNS disease is helpful.

Prevention and Control

Animals should be restricted from areas having abundant pigeon droppings, which could be a possible source of *C. neoformans*. Reduced exposure to trees or contaminated soil can prevent *C. gattii* infection. Regular disinfection in a household with hydrogen peroxide, sodium hypochlorite (1.2%), potassium peroxymonosulfate, chlorhexidine (0.5%) or phenolic compounds is helpful. *C. neoformans* gets killed by exposure to heat 121 °C (moist heat) for 20 min or 165–170 °C (dry heat) for 2 h.

Vaccination

Currently, no vaccine is available against cryptococcosis in humans. The few vaccine candidates that have shown promising results in various trials include whole-cell live attenuated or heat-inactivated vaccines such as the chitosan-deficient cda1-3Δ mutant, the hyphal ZNF2oe strain, the ubiquitination E3 ligase fbp1Δ mutant and the sterylglucosidase sgl1Δ mutant.

Case Study

A 45-year-old male patient came to the emergency department with a sudden onset of altered sensorium worsening. The patient was well a week before but developed irrelevant speech with confusion. He had no history of falls, drugs, trauma or intoxication. He had mild fever and headache and no history of breathing difficulty, chest pain, nausea, vomiting or diarrhoea. Detailed examination showed no signs of neck rigidity, photophobia, sinus pain or ear/nasal discharge. No similar episodes were observed in the past. The patient was a chronic alcoholic with chronic hepatitis C and cirrhosis, for which he underwent a living donor liver transplant a year back and was on immunosuppressive therapy. His family members were fine, and his family history was unrevealing. On examination, his BP was 140/100 mmHg, respiratory and oxygen saturation was normal, heart rate was 98 beats per minute and random blood glucose was 168 mg/dL. He was not pale, icteric and had no signs of cyanosis, lymphadenopathy or clubbing. His pupillary reactions were reactive to light with cranial nerves intact, and the sensory system examination was normal. Another systemic examination was unremarkable. Magnetic resonance imaging (MRI) of the brain revealed no abnormalities. Laboratory findings in haematology and biochemistry were within normal range. The CSF was clear on lumbar puncture with raised opening pressure. CSF had 630 white blood cells/mm^3 consisting of 26% polymorphonuclear cells, and 74% were monomorphs, protein 3.29, g/dl CSF sugar 69 mg/dl. Ziehl-Neelsen stain for acid-fast bacilli (AFB) was negative. Gram stain revealed budding yeast cells, with positive India ink showing round yeast cells with a capsular halo. *Mycobacterium tuberculosis* PCR was negative. Cryptococcal antigen was positive in CSF, and India ink staining of the CSF sample revealed yeast cells with a surrounding halo. Fungal culture on SDA showed growth of cream-coloured pasty colonies, smooth to mucoid, that were suggestive of Cryptococcus. The rapid urease test was positive, and there was no colour change in the CGB medium. Vitek identified it as *C. neoformans*. The patient was treated with liposomal amphotericin B and flucytosine for 2 weeks and discharged with clinical improvement on fluconazole 800 mg for 8 weeks.

1. What are the significant points in the above clinical vignette?
2. How do we identify the pathogen by conventional and molecular methods?
3. What are the risk factors for cryptococcal systemic infection?
4. What may be the mode of infection transmission?

Research Questions

1. How to investigate the presence of *Cryptococcus* in various vertebrates, invertebrates and different ecological niches?
2. What are the differences in the pathogenesis of cryptococcal meningitis in HIV and non-HIV patients?
3. What is the immunological reaction to cryptococcal infections in post-transplant patients?
4. Is it prudent to do surveillance of cryptococcal infections in transplant recipients or non-HIV patients?
5. Is it high time to define clinical breakpoints for *Cryptococcus* infections?

Further Reading

Ajello L (1958) Occurrence of *Cryptococcus neoformans* in soils. Am J Hyg 67(1):72–77

Amirrajab N, Haghani I, Rasuli M, Shokohi T (2016) Migratory birds as a potential reservoirs of *Cryptococcus neoformans*. Int J Environ Res 10(3):459–464

Barclay WP, Delahunta A (1979) Cryptococcal meningitis in a horse. J Am Vet Med Assoc 174(11):1236

Beardsley J, Wolbers M, Kibengo FM et al (2016) Adjunctive dexamethasone in HIV-associated cryptococcal meningitis. N Engl J Med 374(6):542–554

Belluco S, Thibaud JL, Guillot J, Krockenberger MB, Wyers M, Blot S, Colle MA (2008) Spinal cryptococcoma in an immunocompetent cat. J Comp Pathol 139:246–251

Berry WL, Jardine JE, Espie IW (1997) Pulmonary cryptococcoma and cryptococcal meningoencephalomyelitis in a king cheetah (Acinonyx jubatus). J Zoo Wildl Med 28:485–490

Buchanan KL, Murphy JW (1998) What makes *Cryptococcus neoformans* a pathogen? Emerg Infect Dis 4(1):71–83

Castella G, Abarca ML, Cabanes FJ (2008) Cryptococcosis and pets. Rev Iberoam Micol 25(1):S19–S24

Chakrabarti A, Jatana M, Kumar P, Chatha L, Kaushal A, Padhye AA (1997) Isolation of *Cryptococcus neoformans* var. *gattii* from *Eucalyptus camaldulensis* in India. J Clin Microbiol 35:3340–3342

Elsegeiny W, Marr KA, Williamson PR (2018) Immunology of cryptococcal infections: developing a rational approach to patient therapy. Front Immunol 9:651

Emmons CW. 1951. Isolation of *Cryptococcus neoformans* from soil. J Bacteriol 62: 685–690

Emmons CW. 1955. Saprophytic sources of *Cryptococcus neoformans* associated with the pigeon (Columba livia). Am J Hyg 62: 227–232.

Frasés S, Ferrer C, Sánchez M, Colom-Valiente MF (2009) Molecular epidemiology of isolates of the *Cryptococcus neoformans* species complex from Spain. Rev Iberoam Micol 26(2):112–117

Klein KR, Hall L, Deml SM, Rysavy JM, Wohlfiel SL, Wengenack NL (2009) Identification of *Cryptococcus gattii* by use of L-canavanine glycine bromothymol blue medium and DNA sequencing. J Clin Microbiol 47(11):3669–3672

Singh K, Ilkit M, Shokohi T, Tolooe A, Malik R, Seyedmousavi S (2018) Cryptococcosis: emergence of *Cryptococcus gattii* in animals and zoonotic potential. In: Seyedmousavi S, de Hoog G, Guillot J, Verweij P (eds) Emerging and epizootic fungal infections in animals. Springer, Cham

Steenbergen JN, Shuman HA, Casadevall A (2001) *Cryptococcus neoformans* interactions with amoebae suggest an explanation for its virulence and intracellular pathogenic strategy in macrophages. Proc Natl Acad Sci USA 98(26):15,245–15,250

Vilgalys R, Hester M (1990) Rapid genetic identification and mapping of enzymatically amplified ribosomal DNA from several *Cryptococcus* species. J Bacteriol 172(8):4238–4246

Yamazumi T, Pfaller MA, Messer SA, Houston A, Hollis RJ, Jones RN (2000) In vitro activities of ravuconazole (BMS-207147) against 541 clinical isolates of *Cryptococcus neoformans*. Antimicrob Agents Chemother 44:2883–2886

22 Adiaspiromycosis

Subhash Chandra Parija and Parakriti Gupta

Learning Objectives

By the end of the chapter, the readers should be able to:

1. Study various aetiological agents of adiaspiromycosis.
2. Understand the pathogenesis and clinical presentation of adiaspiromycosis.
3. Review both clinical and laboratory diagnoses of adiaspiromycosis.
4. Delineate the management strategies for adiaspiromycosis.

Introduction

Adiaspiromycosis is a rare fungal infection of humans caused by *Emmonsia crescens* and *Emmonsia parva*. It is primarily an infection of over 100 mammalian species, including rodents, armadillos and small mammals. Humans are the accidental hosts. Both immunocompetent and immunocompromised individuals acquire the infection via inhalation of conidia. Since *Emmonsia* is a saprobic soil fungus, the infection is more common in individuals with occupational exposure to dust. The patient usually presents with a pulmonary nodule, which might be an incidental finding. Disseminated infection has also been reported in immunocompromised hosts. Diagnosis of adiaspiromycosis is primarily histopathological or molecular-based. On histopathological examination, adiaspores are usually seen as thick-walled double-layered structures. Molecular diagnostics are usually undertaken in patients with culture-negative results or when fungal isolate cannot be identified phenotypically. The infection is mostly self-limiting. Symptomatic cases are treated by antifungal chemotherapy in most of the cases. Since no specific recommended guidelines for the treatment of these cases are available, they have been treated with antifungals such as ketoconazole, thiabendazole, itraconazole, levamisole, pimafucin and amphotericin B.

History

The term "*Adiaspiromycosis*" is used to denote the infections caused by those fungi that form '*adiaspores*' in the tissues. The term '*Adiaspore*' is derived from the Greek letters 'α' that denote 'not' or 'without', 'δια' that means 'by' or 'through', and 'σπορα' that means 'sowing' or 'seed'. This refers to the swollen cells that do not replicate or disseminate beyond the site where they get implanted. Adiaspiromycosis

S. C. Parija (✉)
Sri Balaji Vidyapeeth University, Pondicherry, India
e-mail: subhashparija@gmail.com

P. Gupta
Department of Microbiology, Government Medical College and Hospital, Chandigarh, India

S. C. Parija, S. M. Rudramurthy (eds.), *Textbook of Fungal Zoonoses and Sapronoses*, Microbial Zoonoses, https://doi.org/10.1007/978-981-97-3214-2_22

was initially designated as 'haplomycosis' or 'adiaspirosis' but was substituted by the term 'adiaspiromycosis' as per the recommendation of the International Society of Human and Animal Mycology. Adiaspiromycosis caused by *Emmonsia* species is currently designated as *Emergomyces crescens* and *Blastomyces parva* that produces adiaspores. Although the origin of this genus dates to 1939, the first case of solitary pulmonary nodule caused by *Emmonsia* was reported in France in 1964. Subsequently, many cases of pulmonary and cutaneous adiaspiromycosis have been reported amongst immunocompetent as well as immunocompromised individuals. Brazil has reported the highest number of cases thus far.

Taxonomy

The taxonomy of fungi causing adiaspiromycosis has been evolving extensively since its inception. The fungus was first described from a cystic lesion of a rodent in 1939 in the former Soviet Union. It was initially named *Rhinosporidium pulmonale* owing to its similarity with *Rhinosporidium*. A few years later, in 1942, non-budding cells measuring 14 μm in diameter were observed accidentally in the lungs of a rodent in Arizona by Chester Wilson Emmons and Ashburn during their studies on *Coccidioides*. The researchers believed the fungus to be a member of Zygomycetes, genus '*Haplosporangium*', due to the presence of small asexual conidia within the spherule, analogous to *Haplosporangium bisporale*. Hence, they suggested the name 'haplomycosis' to denote the infection caused by these fungi. Owing to the smaller size of the sporangium, the species was named *Haplosporangium parva*. Subsequently, in 1947, a similar fungus, yet larger in size (300 μm), was reported from the lungs of rodents in Alberta, Canada. Although it resembled *H. parva* morphologically and microscopically, it differed from the former in the production of larger chlamydospores and larger cells in vivo for a second species under the genus.

Change in the taxonomic status of the genus *Haplosporangium* was also made in 1951. Based on conidial characteristics, it was separated from Zygomycetes and allied with *Blastomyces*. In 1959, Ciferri and Montemartini isolated the fungus from the soil in Italy and placed it under the new genus '*Emmonsia*' along with *H. parva*. Later, the isolate from Alberta was identified as *Emmonsia crescens* (*Ea. crescens*) as a new member of this genus since it differed in the size of adiaspores and thermotolerance from that of *Emmonsia parva* (*Ea. parva*). Subsequently, Carmichael suggested a new classification by regrouping all the fungi producing aleurioconidia (conidia released by disintegration blown-out end of hyphal structure or conidiogenous cell and released by rupture below the attachment base) into the genus *Chrysosporium*. Further, *Ea. parvum* and *Ea. crescens* were suggested as the same species with different varieties (*Chrysosporium parvum* var. *parvum, Chrysosporium parvum* var. *crescens*). This new classification, however, has received a mixed response; many workers still retain the older nomenclature with *Emmonsia* as a genus, while few accept the newer one.

Currently, Emergomyces (previously *Emmonsia*) is classified under Ascomycetes along with *Chrysosporium*, Order Onygenales and families Ajellomycetaceae and Onygenaceae, respectively (Fig. 22.1). A few of the other medically significant members of these families include *Histoplasma*, *Blastomyces*, *Paracoccidioides*, *Emergomyces* (*Ajellomycetaceae*) and *Coccidioides* (*Onygenaceae*). *Emmonsia/Emergomyces* is placed with a dimorphic group of fungi, owing to its ability to convert from mycelial form to adiaspores/adiaconidia at body temperature.

The teleomorph stage of *Ea. crescens* was described and named *Ajellomyces crescens*, sharing the same genus as *Blastomyces dermatitidis* and *Histoplasma capsulatum*. Furthermore, the genetic analysis also confirmed the relatedness of *Emmonsia* with *Blastomyces*, more so for *Ea. parva* than for *Ea. crescens*. However, the nomenclature of *Emmonsia*, as a separate genus, was retained with *Ea. parvum* and *Ea. crescens* as two distinct species.

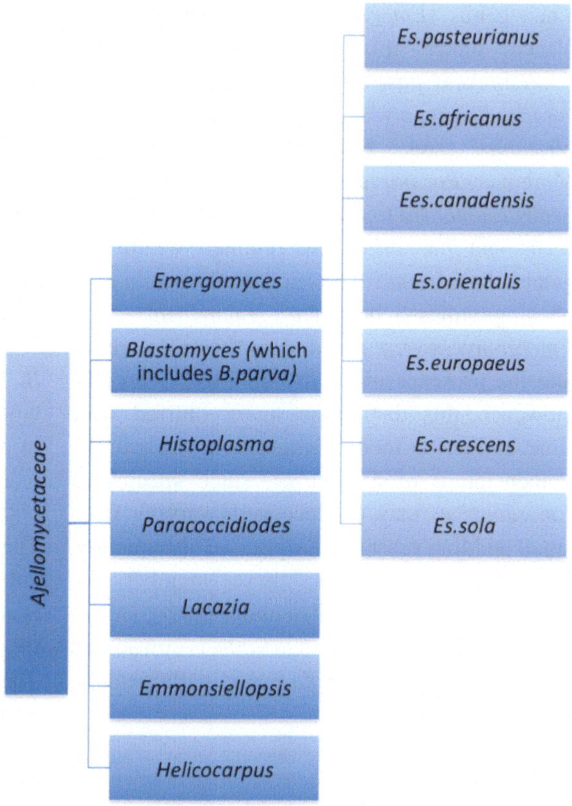

Fig. 22.1 Taxonomic classification of agents causing adiaspiromycosis

Besides *Ea. crescens* and *Ea. parvum*, *Ea. pasteuriana* was also considered a medically relevant species of this genus until recently when it was grouped under the separate genus *Emergomyces* in 2017. *Ea. pasteuriana* was isolated in 1998 from an acquired immune deficiency syndrome (AIDS) patient who presented with cutaneous lesions. Histopathological findings from the biopsy revealed large, thick-walled cells with small budding yeast-like cells and pseudohyphae. Although the cultural characteristics were like *Emmonsia*, these were distinct from the already known species and, thereby, a new species *Ea. pasteuriana* was described. Besides *Ea. pasteuriana*, another new member of the *Emmonsia* genus was described from a human immunodeficiency virus-infected liver transplant patient who presented with cutaneous and disseminated infection. Although the biopsy showed Grocott-Gomori's methenamine silver (GMS)-stained organisms, these were not identified as adiaspores. Despite growth in culture, the isolate could not be identified precisely even after molecular analysis, which exhibited 95% homology with *Ea. crescens*, 91% with *Ea. pasteuriana* and 94% with *Ea. parva*. Many *Emmonsia*-like organisms that are not placed under the genus *Emmonsia* have been described, owing to the absence of adiaspores production. The isolates that produce merely yeast-like forms or hyphal forms in tissue are not included under this genus.

Besides the vicissitudes in nomenclature of *Emmonsia*, a simultaneous rise in number of cases due to *Emmonsia*-like fungi was noted in immunocompromised individuals, which mandated the prerequisite to untangle the phylogenetic relationship between *Emmonsia* and *Emmonsia*-like fungi. The same was achieved in 2017 using concatenated sequencing data of internal transcribed spacer (ITS), large subunit rDNA (LSU), elongation factor-3 (*TEF3*), β-tubulin (*TUB2*) and RNA polymerase II (*rPB2*). The sequencing led to drastic modifications in the genus *Emmonsia* and to the origin of a new genus "*Emergomyces*'. Five species were placed under this new genus, including *Emergomyces pasteurianus*, *Emergomyces orientalis*, *Emergomyces europaeus*, *Emergomyces africanus* and *Emergomyces canadensis*. Moreover, *Ea. parva* was also transferred from genus *Emmonsia* to *Blastomyces*, while *Ea. crescens* and *Ea. sola* were retained within the genus *Emmonsia*. The major discriminating physiognomies between *Emmonsia* and *Emergomyces* were noted to be thermotolerance and the presence of small budding yeast cells. However, this distinction again blurred over time, and newer modifications were made in the nomenclature of *Emmonsia*, since species of *Emergomyces* may produce cells varying from small budding yeast to adiaspores. Subsequently, after extensive studies, these members of *Emmonsia* were reclassified under *Emergomyces* in 2020. Of these, *Es. pasteurianus* is considered a type species, and *Es. sola* is non-pathogenic to humans.

Genomics and Proteomics

The genomic and proteomic characterisation of *Emmonsia* infections is pivotal for elucidating molecular mechanisms underlying its pathogenicity and host interactions. Genomic studies have played a crucial role in deciphering the complete DNA sequences of various *Emmonsia* species, like *Es. crescens*, providing a foundation for understanding their genetic makeup. The identification of genes associated with virulence, pathogenicity and adaptation to host environments has been facilitated through comparative genomics.

In tandem with genomics, proteomic analyses have offered insights into the dynamic expression of proteins during *Emmonsia* infections. High-throughput techniques, including mass spectrometry and 2D gel electrophoresis, have been employed to identify and quantify the proteome of *Emmonsia* species. Such studies have revealed key virulence factors, cell wall components and secreted proteins that contribute to the pathogenicity of *Emmonsia* species. Understanding the interplay between genomics and proteomics is essential for developing targeted diagnostic tools and therapeutic interventions. The integration of these omics approaches allows for a comprehensive view of the molecular landscape of the pathogen, aiding in the identification of potential drug targets and vaccine candidates.

The Fungus Morphology

B. parva and *Es. crescens* are dimorphic fungi with mycelial growth at 25 °C and the formation of adiaspores at 40 °C. Sporulation with the formation of adiaspores is enhanced by culture on potato dextrose agar or phytone yeast-extract agar. The growth appears as white-to-buff-coloured colonies with a glabrous-to-velvety texture. There is no significant morphological difference between the mycelial phases of *B. parva* and *Es. crescens*. The growth appears as moderately spreading colonies, ranging in colour from white to buff, with a texture that can vary from glabrous to velvety. Spherical, slightly flattened aleurioconidia are borne on simple or complex conidiophores. One to three smooth, thin-walled conidia, measuring 2.5–4 μm in length and 3–5 μm in width, arise from the tips of slightly swollen stalks or the sides of the hyphae (sessile). After being stored for several weeks, the colonies undergo changes and acquire a waxy and cerebriform texture, and the conidia transform into thick-walled diaspores.

Differentiation of *B. parva* and *Es. crescens*: *B. parva* and *Es. crescens* adiaspores differ from each other in many ways, including their temperature requirements. Adiaspores of *B. parva* are mononucleate and typically range in size from 20 to 40 μm. They are produced at temperatures around 40 °C. On the other hand, adiaspores of *Es. crescens* are multinucleate and can reach sizes of up to 500 μm. These are produced at temperatures between 30 °C and 37 °C. *Es. crescens* is inhibited in the presence of cycloheximide, whereas the growth of *B. parva* is not inhibited with cycloheximide. Similarly, the wall thickness of the spherule is higher for *Es. crescens* (20–70 μm) than for *B. parva* (2 μm) (Fig. 22.2).

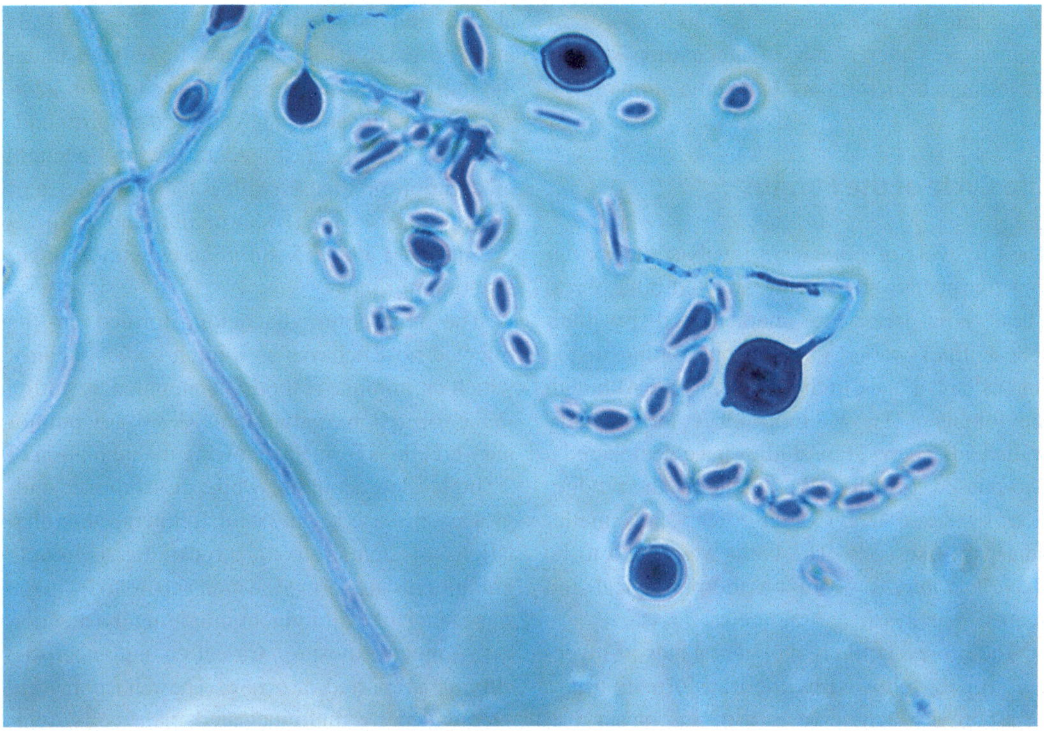

Fig. 22.2 Photomicrograph depicts both fungal spores and adiaspores, of what may have been an *Emmonsia* sp., or *Chrysosporium* sp. organism. (Courtesy CDC)

Furthermore, the adiaconidia of *Es. crescens* become multinucleate with increased size, while adiaconidia of *B. parva* remains mononucleate, even after attaining full size.

Cultivation of Fungi

Culture is done on Sabouraud's dextrose agar or brain heart infusion agar at 37 °C and 25 °C without cycloheximide. *B. parva* and *Es. crescens* grow rapidly as white-to-buff-coloured colonies and glabrous-to-velvety texture, with the production of aerial mycelia on the obverse and cream-to-pale brown on the reverse. Glucose-peptone agar has also been used as a culture medium.

Sporulation can be enhanced on potato dextrose agar or phytone yeast-extract agar. Transfer of the culture to 37 °C leads to the formation of multinucleate adiapores. Solid culture media often become dry during long-term storage. In a study on adiaspore development in a mouse model, foetal bovine serum (FBS) was employed as a culture medium. The study concluded that liquid media, such as FBS, have advantages over solid media in terms of their suitability for long-term culturing and ease of adiaspore collection.

Animal Models

Numerous studies have been conducted employing animal models to investigate the progression of adiaspiromycosis, analyse the immunological response, assess pathogenicity, examine the morphology of adiaspores and evaluate the efficacy of antifungal agents. The unavailability of clinical trial data due to the rarity of this disease makes animal models the optimal choice for studying various aspects of adiaspiromycosis. Mouse is the most used animal model, followed by rabbits. Most studies utilised the intraperitoneal route for inoculation of conidiospores of *Emmonsia*. Other routes of inoculation studied are intratracheal, intranasal and inhalation routes.

Life Cycle of the Fungus

The life cycle of *Emmonsia* involves the formation of specialised structures called adiaspores. These adiaspores are characterised by their robust, thick-walled nature, contributing to the fungus's resilience in the environment. Although the fungus is known to produce large, thick-walled adiaspores, few studies have deciphered a two-phased growth cycle, leading to the production of both larger adiaspores and smaller adiaspore-like cells. The latter was found to be like *Blastomyces* based on morphological and sequence analysis.

The life cycle likely includes stages of mycelial growth, followed by the transformation into adiaspores, which serve as a means of survival and dissemination. This unique life cycle underscores the adaptive strategies employed by *Emmonsia* species, highlighting their ability to persist in diverse environmental conditions and potentially pose challenges in the realm of public health. Further research continues to unveil the intricacies of *Emmonsia*'s life cycle, shedding light on the complex interactions between the fungus and its hosts.

Pathogenesis and Pathology

Emmonsia species produce many virulence factors, including haemolysis, phospholipases, proteinases, esterase and urease. No variation has been noted between both the species in terms of production of these virulence factors, except for urease production, which was produced more by *Es. crescens* at higher temperatures.

Emmonsia/Emergomyces is known to produce aleuroconidia (2–4 μm). These aleuroconidia are present in significant numbers in the air, usually at construction sites where there is more dust. These aleuroconidia, when inhaled, reach the alveoli, and enlarge in size up to 500 μm. They undergo a large-fold increase in volume under the influence of raised temperature to form adiaspores that appear as pale yellow to tan firm nodules. The size of adiaspore varies with the length of duration of illness, with a mean diameter of 101 μm in cases with a shorter duration of illness (<2 months) and 155 μm in cases with a longer duration of illness.

The adiaspores/adiaconidia formed in the infected host behaves as foreign bodies, against which an immune response is mounted, resulting in the formation of granuloma. A solitary pulmonary nodule is formed following the inhalation of a single aleuroconidia, while disseminated pulmonary disease might occur following the inhalation of numerous aleuroconidia. Multiple granulomas are formed, which appear to be in the same stage of evolution, highlighting the reaction to a single exposure. Caseation and necrosis, although reported, are rare. The adiaspores are surrounded by multinucleated giant cells, lymphocytes, macrophages, and connective tissue and might get calcified after the long incubation period. Pulmonary function gets compromised when the granuloma compresses the small airways.

Dissemination beyond the lungs has not been reported, confirming the unique characteristic of this fungus, namely its inability to replicate and spread to other organs. Nevertheless, extension to the nearby pulmonary lymph nodes has been documented in both humans and animals. It has been hypothesised by a few authors that *Emmonsia* might exhibit the Splendore-Hoeppli phenomenon with proteinaceous material radiating peripherally to the outer zones of adiaconidia.

Immunology

The host's immune response to adiaspiromycosis is marked by antibody responses and the development of delayed hypersensitivity reactions. Antibodies produced by the immune system play a vital role in recognising and neutralising the fungal pathogens responsible for adiaspiromycosis. Concurrently, the presence of delayed hypersensitivity reactions indicates the involvement of T cells and underscores the importance of cell-mediated immunity in the host's defence against the infection. Despite the significance of cell-mediated immunity, studies in this particular aspect of adiaspiromycosis remain limited.

Infection in Humans

Humans acquire the infection accidentally, by inhalation of conidia, after exposure to dust since *Emmonsia* is a saprobic soil fungus. The fungus most commonly causes respiratory infections in humans.

Respiratory Infection

Pulmonary infection is the most common. Asymptomatic cases are diagnosed with the detection of pulmonary nodules as an incidental finding on imaging or histopathological investigation carried out for other conditions. Symptomatic cases can present an indolent course, with fever, cough, haemoptysis, dyspnoea and weight loss. Adiaspiromycosis cases mimic tuberculosis, with both presenting as bilateral reticuloendothelial infiltrates in the lungs on imaging. Respiratory distress, because of granulomas compressing the smaller airways, may lead to displacement of pulmonary tissue. The condition may prove fatal, especially in cases with dissemination in immunocompromised patients.

Rare non-pulmonary manifestations include adiaspiromycosis of bone marrow, appendix and eye. Bone marrow involvement was reported in one AIDS patient. Although the dissemination is a rare outcome, the infection might extend to local lymph nodes. Few reports of dissemination with *B. parva* have been reported in immunocompromised patients. In one case, adiaspores were noted in pus debrided from bony lesions and subsequent dissemination to multiple bones, lungs and bone marrow.

Infection in Animals

Adiaspiromycosis in animals is predominantly asymptomatic, diagnosed mostly during post-mortem examinations. Nevertheless, a few cases of symptomatic and fatal infection have also been reported, typically associated with a substantial burden of conidia.

Epidemiology and Public Health

Adiaspiromycosis is primarily an infection of mammals and has been isolated from varied species of rodents worldwide. It is considered a non-arthropod-transmitted fungal infection in a lower vertebrate class of mammals. The organism has been isolated from more than 100 mammalian species, which include carnivores, insectivores, marsupials, lagomorphs and rodents. Mammals acquire the infection after inhalation of asexual conidia, from the soil while burrowing. These conidia eventually develop into large adiaconidia in the lungs of the host that might persist even for a lifetime. These are referred to as aleuroconidia, since conidia are released singly after the lytic disintegration of hyphal cells. These conidia have a unique property of not undergoing replication within the host tissue. Eventually, conidia develop into large adiaspores in the lungs which are later encapsulated by granulomatous inflammation.

The infection was first described in rodents in the United States in the 1940s, two decades before reports of human infections. *Es. crescens* has a broader host range and a wider geographic distribution compared to *Es. parvum.* Humans acquire the infection accidentally, via inhalation of conidia, after exposure to dust since *Emmonsia/Emergomyces* is a saprobic soil fungus. Although the disease has zoonotic potential, no case of transmission from rodents or other mammals has been reported to date. Occupational exposure to dust in immunocompetent as well as immunocompromised hosts is the predisposing factor.

Human cases of adiaspiromycosis have been reported from Brazil, Argentina, Venezuela

(South America), Guatemala, Honduras (North America), South Africa, Arizona, Texas, New York, Georgia, Oklahoma, North Carolina (United States), Japan, France, former Czechoslovakia, Germany, Spain, the former Soviet Union (Europe), United Kingdom, Finland and Asia. Most of these cases have been reported from Brazil, although the geographical association has not been established yet. Moreover, *Emmonsia* has been isolated from mammals across the globe, except in Australia and Africa, underlining the risk of transmission to humans worldwide. *Es. crescens* is more commonly associated with human infections than *B. parva*, which is more commonly implicated in disease in animals. *Es. crescens* is prevalent in temperate zones globally, while *B. parva* is geographically restricted to a few xerothermic regions like Eastern Europe, Australia, Kenya and the southwestern US. Recently, two subgroups have been identified within *Es. crescens* viz. North American population group and pan boreal Eurasian population group. However, the differences are not colossal enough to designate them as new species.

Non-inhalational transmission: Adiaspiromycosis in humans has been reported to be transmitted by a non-inhalational route. Two cases of adiaspiromycosis of the appendix after oral ingestion of fungus have been documented. An outbreak of ocular adiaspiromycosis was reported in Brazil in 2006, in which conjunctival nodules were noted in 18 children. A total of 64 out of 5084 screened children were reported to have active ocular infection. *Es. crescens* was reported to be the causative agent and was isolated from 2 of 14 biopsies from ocular nodules. It was suggested that conjunctival irritation with freshwater sponge spicules might have facilitated the entry of airborne aleuroconidia.

Adiaspiromycosis is a neglected zoonotic disease, with rodents serving as the reservoir of infection. However, animal-to-human transmission has not been reported yet. Despite the infection affecting the diversity of mammals and its propensity to infect humans as well, the reason for the smaller number of cases amongst humans was studied. It was noted that several cases of adiaspiromycosis are inversely related to the body temperature of the host. The fungi had greater predilection to infect the animal hosts with relatively lower body temperature, as was noted in more commonly infected animals such as rodents and nine-banded armadillos (body temperatures of 34–37 °C) compared to higher mammals such as Leporidae (body temperatures of ~40 °C). The same concept might support the suggested less frequent infection in humans with a body temperature of 37 °C.

Diagnosis

Diagnosis of adiaspiromycosis is primarily based on radiology, histopathology and molecular methods. *Emmonsia/Emergomyces* has been rarely cultured from human tissue samples; however, culture from environmental and animal sources is more common.

Histopathology

Biopsy or fine-needle aspiration cytology (FNAC) of tissues is frequently used. The sputum, bronchoalveolar lavage (BAL) and bronchial washings are not tested for the fungi, as they test negative for the same.

Haematoxylin-eosin staining is used to demonstrate adiaspores, which are symmetrical, round-to-oval in shape and thick-walled double-layered structures; the two layers consist of an inner chitinous hyaline layer covered by a narrow eosinophilic outer layer. The thick walls of the adiaspores are positive with periodic acid-Schiff (PAS) and Gomori's methenamine silver (GMS) stains. They appear trilaminar when stained with these dyes. GMS demonstrates uniformly stained, thick-walled structures. Adiaspores are usually empty and occasionally may contain eosinophilic globules along the inner surface of the wall, presenting as honeycomb structures. Periodic acid-Schiff (PAS), Papanicolaou and Gridley stains are also used, but not Mayer's mucicarmine stain, as

Table 22.1 Differentiating features of *Emergomyces crescens, Blastomyces parva, Coccidioides immitis* and *Rhinosporidium seeberi*

Characteristics	*Emergomyces crescens*	*Blastomyces parva*	*Coccidioides immitis*	*Rhinosporidium seeberi*
Spherules				
Size (μm)	50–500	10–25	60–100	10–500
Wall thickness (μm)	10–70	2	1–2	3–5
Endospores				
Number	NA	NA	200–300	10,000–16,000
Size (μm)	NA	NA	2–5	5–10
Staining and culture				
Mucicarmine stain	Negative	Negative	Negative	Positive
Culture on fungal media	Adiaconidia at 37 °C and inhibited at 40 °C	Mycelia at 37 °C and adiaconidia at 40 °C	Mycelia with arthroconidia	Cannot be cultivated

it fails to stain the conidial walls. Histopathology also helps to stain the granuloma consisting of giant cells and epitheliomas, caseation and scanty or absence of necrosis.

The histopathological characteristics of adiaspores may be confused with those of *Coccidioides* and *Rhinosporidium seeberi*. *Coccidioides* are differentiated by their smaller size, thinner walls and the presence of endospores within the spherules. On the contrary, *Rhinosporidium* has a larger sporangium (100–200 μm) with thinner walls. In addition, trophocyte with a single nucleus and flocculent cytoplasm might also be present (Table 22.1). Adiaspores may also be mistaken for *Sporothrix lurei*, which is identified by the presence of spectacle-shaped, thick-walled cells. Rarely, *Strongyloides* and *Dirofilaria* can mimic adiaspores.

Culture

Culture is done on Sabouraud's dextrose agar and brain heart infusion agar at 37 °C and 25 °C. *B. parva* and *Es. crescens* grow rapidly as white-to-buff-coloured colonies and glabrous-to-velvety texture, with the production of aerial mycelia on the obverse and cream-to-pale brown on the reverse. Sporulation can be enhanced on potato dextrose agar or phytone yeast-extract agar.

Serology

The serodiagnosis of adiaspiromycosis is challenging due to the lack of sensitive and specific serological tests or biomarkers. Immunological cross-reactivity has been observed with *Blastomyces dermatitidis* and *Histoplasma capsulatum*. In the past, delayed-type hypersensitivity reactions to antigens, immunodiffusion tests and passive cutaneous anaphylaxis tests have been employed; however, they are used rarely at present.

Molecular Diagnosis

Since there are no cardinal signs to suspect the infection clinically, molecular modalities are performed in patients with pulmonary involvement, in culture-negative cases or in cases of isolation of unknown organisms that cannot be identified phenotypically. Molecular diagnosis is also undertaken for patients with pulmonary/disseminated disease who are non-responsive to antifungals and empirical therapy. Molecular detection is based on polymerase chain reaction (PCR) and sequencing of rDNA, ITS and LSU. However, there are no commercial molecular diagnostic tests available for diagnosing this condition.

Radiological Diagnosis

Chest X-ray shows pronounced bilateral interstitial infiltrates, while thorax CT reveals bilateral diffuse peri-bronchial and interstitial reticulonodular infiltrates. The radiological features of the lungs are like those observed in pulmonary tuberculosis, sarcoidosis and histoplasmosis. The extent of radiological involvement in the lungs depends on the volume of dust-borne conidia inhaled.

Treatment

The infection is self-limiting in some cases. Currently, antifungal therapy along with corticosteroids, with or without surgical resection of the lesion, is the commonly used modality in the treatment of symptomatic adiaspiromycosis in immunocompetent hosts. Most cases are treated by ketoconazole, thiabendazole, itraconazole, levamisole, pimafucin and amphotericin B. Even though more than half of the patients reported to date have been treated by antifungal agents, their precise role in the outcome of the patient is still doubtful, since recovery might merely be a phase in the natural course of illness. The use of steroids in management is still controversial. Since granuloma formation and the host immune response can be relieved by the use of steroids, the risk of flaring up of infection may be of concern.

Prevention and Control

Prevention and control strategies are paramount in mitigating the impact of these fungal pathogens on public health. Given the opportunistic nature of *Emmonsia* species, especially in immunocompromised individuals, primary prevention efforts should focus on minimising exposure to the fungi. Occupational settings where individuals may come into contact with contaminated soil or organic matter should implement stringent hygiene practices, including the use of personal protective equipment.

Public health education programs should raise awareness about the risks associated with certain activities and environments. Additionally, surveillance and early detection mechanisms can play a crucial role in preventing severe infections. For those at higher risk, such as immunocompromised patients, prophylactic antifungal therapy may be considered under the guidance of healthcare professionals. Overall, a multifaceted approach integrating education, environmental management and targeted medical interventions is essential for effective prevention and control.

Case Study

A 65-year-old male farmer by occupation presented with complaints of shortness of breath and increased fatigue for the past 20 days. He is a smoker (20 pack years) and has no other relevant history. A physical examination of the patient revealed no abnormality with a temperature of 37 °C, pulse of 84 beats/min and respiratory rate of 16/min. Respiratory examination showed bilateral inspiratory crackles, while the cardiovascular system showed no abnormality. Laboratory investigations revealed a haemoglobin level of 11.1 g/dL, an erythrocyte sedimentation rate of 70 mm/h and a total leukocyte count of 11,200 cells/mm^3. Pulmonary function tests showed FVC of 2440 mL (77% of predicted), FEV1 of 2130 mL (84% of predicted) and FEV1/FVC of 87%. A chest X-ray was performed which revealed bilateral interstitial micronodular infiltrates. Computerised tomography showed diffuse, bilateral peribranchial and interstitial reticulonodular infiltrates. A bronchoscopy was performed, and bronchoalveolar lavage (BAL) was sent for cytological examination. The cytological examination did not reveal any malignancy, and the examination of transbronchial lung biopsy was also non-diagnostic. Subsequently, a biopsy with video-assisted thoracoscopic surgery (VATS) was performed. Histopathologic examination of the biopsy showed granulomas with parasite-like structures. Oral albendazole 800 mg/day was administered to the patient for 3 weeks; however, the patient did not recover fully. The biopsy was then sent for histopathology study. Grocott's methenamine silver staining of the same revealed the presence of characteristic 'adiaconidia'. The patient was managed with amphotericin B and corticosteroids and recovered fully with no recurrences.

1. How will you confirm the diagnosis of adiaspiromycosis?
2. How can you differentiate different species implicated in adiaspiromycosis?
3. What are the other diagnostic modalities available?
4. How will you manage cases of pulmonary and disseminated adiaspiromycosis?

Research Questions

1. How do you diagnose adiaspiromycosis in a clinical setting?
2. Which test is the gold standard in the diagnosis of adiaspiromycosis?
3. What is the role of antifungals and steroids in the management of cases?
4. What are the clinical breakpoints and susceptibility profiles for *Es. crescens*?

Further Reading

Anstead GM, Sutton DA, Graybill JR (2012) Adiaspiromycosis causing respiratory failure and a review of human infections due to *Emmonsia* and *Chrysosporium* spp. J Clin Microbiol 50(4):1346–1354

Carmichael JW (1951) The pulmonary fungus *Haplosporangium Parvum*. II. Strain and generic relationships. Mycologia 43(6):605–624

Carmichael JW (1962) *Chrysosporium* and some other Aleuriosporic *Hyphomycetes*. Botany 40(8):1137–1173

Doby-Dubois M, Cheberel ML, Doby JM, Louvet M (1964) Premier Cas Humain D'adiaspiromycose, par *Emmonsia crescens*, Emmons et Jellison, 1960 [1st human case of adiaspiromycosis, caused by *Emmonsia crescens*, Emmons and Jellison, 1960]. Bulletin de la Societe de pathologie exotique et de ses filiales 57:240–244

Dowding ES (1947) The pulmonary fungus, *Haplosporangium parvum*, and its relationship with some human pathogens. Can J Res 25(5):195–206

Drápela J, Viklický J, Novák J, Tousek J, Vána M (1980) Peritoneale Form der Adiasporose [peritoneal form of adiaspiromycosis (author's transl)]. Zeitschrift fur Erkrankungen der Atmungsorgane 155(3):393–398

Emmons CW, Ashburn LL (1942) The isolation of *Haplosporangium parvum* n. sp. and *Coccidioides immitis* from wild rodents. Their relationship to Coccidioidomycosis. Public Health Rep 57(46):1715

Emmons CW, Jellison WL (1960) *Emmonsia crescens* sp. n. and adiaspiromycosis (haplomycosis) in mammals. Ann N Y Acad Sci 89:91–101

Gori S, Drohuet E, Guebo E, Huerre M, Lofaro A, Parenti M, Dupont B (1998) Cutaneous disseminated mycosis in a patient with AIDS due to a new dimorphic fungus. J Med Mycol 8(2):57–63

Jagdish CJ (2002) Medical Mycology, vol 7, 2nd edn. Mehta Publishers, New Delhi, pp 707–711

Jiang Y, Tsui CKM, Ahmed SA, Hagen F, Shang Z, Gerrits van den Ende AHG, Verweij PE, Lu H, de Hoog GS (2020) Intraspecific diversity and taxonomy of *Emmonsia crescens*. Mycopathologia 185(4):613–627

Kodousek R (1972) Finding of isolated spherules probably of fungus *Emmonsia crescens* in a surgical specimen of the appendix in a 7-year-old boy. Ceskoslovenska Patologie 8(3):160–162

Mendes MO, Moraes MA, Renoiner EI, Dantas MH, Lanzieri TM, Fonseca CF, Luna EJ, Hatch DL (2009) Acute conjunctivitis with episcleritis and anterior uveitis linked to adiaspiromycosis and freshwater sponges, Amazon region, Brazil, 2005. Emerg Infect Dis 15(4):633–639

Peterson SW, Sigler L (1998) Molecular genetic variation in *Emmonsia crescens* and *Emmonsia parva*, etiologic agents of adiaspiromycosis, and their phylogenetic relationship to *Blastomyces dermatitidis* (*Ajellomyces dermatitidis*) and other systemic fungal pathogens. J Clin Microbiol 36(10):2918–2925

Pelegrín I, Ayats J, Xiol X, Cuenca-Estrella M, Jucglà A, Boluda S, Fernàndez-Sabé N, Rafecas A, Gudiol F, Cabellos C (2011) Disseminated adiaspiromycosis: case report of a liver transplant patient with human immunodeficiency infection, and literature review. Transpl Infect Dis 13(5):507–514

Samaddar A, Sharma A (2021) Emergomycosis, an emerging systemic mycosis in immunocompromised patients: current trends and future prospects. Front Med 8:670731

Schwartz IS, Kenyon C, Feng P, Govender NP, Dukik K, Sigler L, Jiang Y, Stielow JB, Muñoz JF, Cuomo CA, Botha A, Stchigel AM, de Hoog GS (2015) 50 years of emmonsia disease in humans: the dramatic emergence of a cluster of novel fungal pathogens. PLoS Pathog 11(11):e1005198

Seyedmousavi S, Guillot J, Tolooe A, Verweij PE, de Hoog GS (2015) Neglected fungal zoonoses: hidden threats to man and animals. Clin Microbiol Infect 21(5):416–425

Sigler L (2005) Adiaspiromycosis and other infections caused by *Emmonsia* species. In: Merz WG, Hay RJ (eds) Topley and Wilson's microbiology and microbial infections, 10th edn. ASM Press, Washington, DC, pp 810–824

Untereiner WA, Scott JA, Naveau FA, Sigler L, Bachewich J, Angus A (2004) The Ajellomycetaceae, a new family of vertebrate-associated Onygenales. Mycologia 96(4):812–821

23 Talaromycosis

Ariya Chindamporn, Kornvalee Meesilpavikkai, and Navaporn Worasilchai

Learning Objectives
By the end of the chapter, the readers should be able to:

1. Determine the etiologic agent of talaromycosis and its epidemiology.
2. Understand the pathogenesis and clinical presentation of talaromycosis.
3. Differentiate the diagnosis of talaromycosis.
4. Learn talaromycosis management.

Introduction

Talaromycosis (penicilliosis marneffei) is caused by *Talaromyces* (*Penicillium*) *marneffei*, a thermally dimorphic fungus, by inhalation. Bamboo rat is the natural reservoir for the fungus. Southeast Asia including Thailand, Vietnam, Myanmar, and Southern China is endemic for the infection. The first natural human infection was reported in 1973. During the AIDS era, between 1988 and mid-1990s, the incidence of talaromycosis was signaficantly high as one of the opportunistic infection in the HIV-infected patients, especially those with CD4+ cells ≤100 cells/mm^3.

After the introduction of highly active antiretroviral therapy (HAART) to control HIV infection, along with the advent of newer and global travel, this infection has also been diagnosed in non-HIV patients with primary and secondary immunodeficiency syndromes, patients with solid organ transplant, hematological malignancies, and those receiving novel anticancer targeted therapies. Reported cases have been found not only in the endemic zone but also in the United States and Europe, including France, Germany, Belgium, The Netherlands, Japan, etc. Accurate and rapid assay to differentiate talaromycosis is most important for the treatment and management of cases to reduce morbidity and mortality due to the condition.

History

P. marneffei, currently named *T. marneffei*, was first discovered by Capponi et al. in 1956. It was isolated from the liver abscesses of a Chinese

A. Chindamporn (✉) · K. Meesilpavikkai
Mycology Unit, Department of Microbiology, Faculty of Medicine, Chulalongkorn University, Bangkok, Thailand
e-mail: ariya.c@chula.ac.th

N. Worasilchai
Department of Transfusion Medicine and Clinical Microbiology, Faculty of Allied Health Sciences, Chulalongkorn University, Bangkok, Thailand

S. C. Parija, S. M. Rudramurthy (eds.), *Textbook of Fungal Zoonoses and Sapronoses*, Microbial Zoonoses, https://doi.org/10.1007/978-981-97-3214-2_23

bamboo rat (*Rhizomys sinensis*) that lived in the animal house at the Pasteur Institute, Dalat, South Vietnam and died spontaneously after getting the infection. Later, this fungus was also isolated from other species of bamboo rats belonging to the subfamily of Rhizomyinae, including *Rhizomys pruinosus* (hoary bamboo rats), *Rhizomys sumatrensis* (large bamboo rats), and *Cannomys badius* (lesser bay bamboo rats). Three years after its first discovery, the characteristics of *P. marneffei*, together with its human pathogenicity, were formally described by Professor Gabriel Segretain after he got a nodule at the infection site. This agent was named *Penicillium marneffei*, in recognition of Dr. Hubert Marneffe of the Pasteur Institute in Indochina. The first human case was reported in 1973 in a Hodgkin's lymphoma patient with a travel history to Southeast Asia 3 years before penicilliosis (presently named talaromycosis) was diagnosed. In the early 1990s, prior to the acquired immune deficiency syndrome (AIDS) era in Asia, talaromycosis was recognized as a rare fungal infection in Southeast Asia, with an estimated fewer than 40 cases reported.

Taxonomy

P. marneffei was earlier classified as a member of Deuteromycotina, Hyphomycetes, subgenus Penicillium, and subgenus Biverticillium of Halomycetes based on its macro- and microscopic characteristics. However, based on the phylogenetic analysis using its mitochondrial genomes, Internal Transcribed Spacer (ITS) region, and the RNA polymerase II largest subunit (rpb2), in 2011, *P. marneffei* was reclassified into the kingdom Fungi, subkingdom Dikarya, phylum Ascomycota, class Eurotiomycetes, order Eurotiales, family Trichocomaceae, and renamed as *Talaromyces marneffei*. Currently, the name of the disease, penicilliosis marneffei has been changed to talaromycosis, following the name of the causative agent. *T. marneffei*, the only unique species among over 200 species, is the thermal dimorphic fungus, a dual form of growth based on different temperatures. At 25–30 °C and 35–37 °C, it grows as a mold and yeast colony, respectively.

Genomics and Proteomics

The study of genome sequence in *T. marneffei* is quite limited when compared to other common medically important fungi. The draft genome sequence of *T. marneffei*, isolated from a non-HIV patient in Hong Kong, was obtained by the whole-genome shotgun method in 2011. This draft genome consists of 2780 sequence contigs with a total length of 28,887 bp. The obtained contigs were assembled into 273 supercontigs with a total length of 28.42 Mb containing the complete mitochondrial genome, genes of polyketide synthase, sexual cycle, and Mp1p homologs. *T. marneffei* is closely related to molds than yeasts based on the mitochondrial genome sequence. There are all meiotic genes (except HOP1) and genes encoding putative pheromone processing enzymes, pheromone receptors, and pheromone response pathway proteins, which were reported in *Aspergillus fumigatus* and *A. nidulans*, and a putative MAT-1-alpha box mating-type gene. This finding indicates that *T. marneffei* has the properties of heterothallic fungi without mating-type switching. Regarding genes involved in the mode of growth, high expression of dicer protein (dcl-2) associated with the biogenesis of microRNA-like RNAs (milRNAs), PM-milR-M1, and PM-milR-M2 is found in the hyphal phase along with glyceraldehyde-3-phosphate dehydrogenase, which plays a role as a factor for conidial attachment. On the other hand, overexpression of genes involved in host immunity evasion is indicated in the yeast form. The homologs of Mp1p were another encoding gene investigated and used as a template for multilocus sequence typing construction. Mp1p is claimed as one of the virulence factors of these organisms, binding to various organic substances such as proteins, fatty acids, and lipids of the host. Even though the homologs of Mp1p are also found in *Aspergillus* spp., there are 13 additional special Mp1p homologs that consist of only one lipid-binding domain found only in *T. marneffei*.

The polyketide synthase gene is another virulence factor related to melanin pigment formation, mitorubrinic acid/mitorubrinol biosynthesis, and biogenesis of the red soluble pigment. Twenty-three putative polyketide synthase (PKS) genes and two putative PKS-non-ribosomal peptide synthase hybrid genes have been identified. More diversity of the PKS gene in *T. marneffei* than in other medically thermal dimorphic fungi such as *Histoplasma capsulatum* (one PKS gene) and *Coccidioides immitis* (ten PKS genes) is noted.

Complete data on the genomic and proteomic of *T. marneffei* might help in the better understanding of the pathogenesis of talaromycosis culminating in the better management of the patient with survival rate.

Fungal Morphology

T. marneffei can present its unique form at both 25 °C and 35 °C. In general, at 25–30 °C, an off-white minute filamentous colony on the agar plate can be observed approximately 2–3 days after inoculation and 3–5 days later, a yellow-green or gray colony with soluble red pigment (Fig. 23.1a) is expected whereas the colony at 35–37 °C is a yeast-like colony. A typical mold colony, together with the fission yeast structure in the human samples (Fig. 23.1b) supports the identification as *T. marneffei*. Not only does temperature influence the growth morphology, but the acidity of the medium is also affected. This mold is insensitive to cycloheximide, and metabolomics studies reveal mechanisms for environmental adaptations and virulence.

Mold characteristics are an identification key since the conversion to yeast form in vitro is not commonly performed. In addition, recently, new species of *Talaromyces* have been isolated from clinical specimens: *Talaromyces alveolaris*, *Talaromyces albobiverticillius*, *Talaromyces kabodanensis*, *Talaromyces rapidus*, *Talaromyces Subaurantiacus*, and *Talaromyces minnesotensis*. Therefore, rapid and accurate identification is an important factor for patient management. Moreover, sequencing of the PCR products and blast analysis from databanks, such as NCBI and ISHAM databanks, are required to reveal the genus and species. Another approach, matrix-assisted laser desorption/ionization-time of flight (MALDI-TOF) mass spectrometry, is a less time-consuming tool if the database is available.

One of the typical phenotypic characteristics of *T. marneffei*, which is an applicable key for the preliminary detection or identification, is the presence of diffuse or soluble red pigment in the mold colony at temperatures <30 °C. In 2007, this purified pigment was investigated for its structure and composition by reverse-phase liquid chromatography. The herquinone-like structure that contains a phenalene carbon framework and the dauphine bond dimer form was found. In 2014, the *pks3*-KD mutant (the mutant with a complete loss of the red pigment secretion) was retrieved from the systematically knocked down 25*pks* genes of *T. marneffei*. After metabolomic analysis, the derivatives of monascorubrin and rubropunctatin are revealed. Both compounds, the well-known red pigment produced from *Monascus purpureus*, are used in traditional oriental foods, such as bean curd, and have recently been used in coloring in Western foods, such as sausages. Moreover, the toxin, citrinin, and the yellow pigment, ankaflavin, also produced by the PKS3 gene (polyketide synthase3) are found. The researchers claimed that no compound with herquinone-like structure was found. In addition to the red pigment biosynthesis gene clusters, PKS3, the other four genes are a transcription activator (RP1), a β-subunit fatty acid synthase (RP2), an α-subunit fatty acid synthase (RP3), and an oxidoreductase (RP4). A variety of pigment production has been found in the different morphologies and stages of *T. marneffei*, such as black in young conidia and yellow in maturation conidia. In terms of this variety, the appearance of yellow to cream color in the yeast phase and red color in the mold phase are also observed. In 2010, it was found that black pigment is produced from the dihydroxynaphthalene (DHN)-melanin gene cluster, and its biosynthesis is related to the PKS pathway. These DHN-melanin genes in *T. marneffei* are like those of *A. fumigatus* and some dematiaceous pathogens, such as

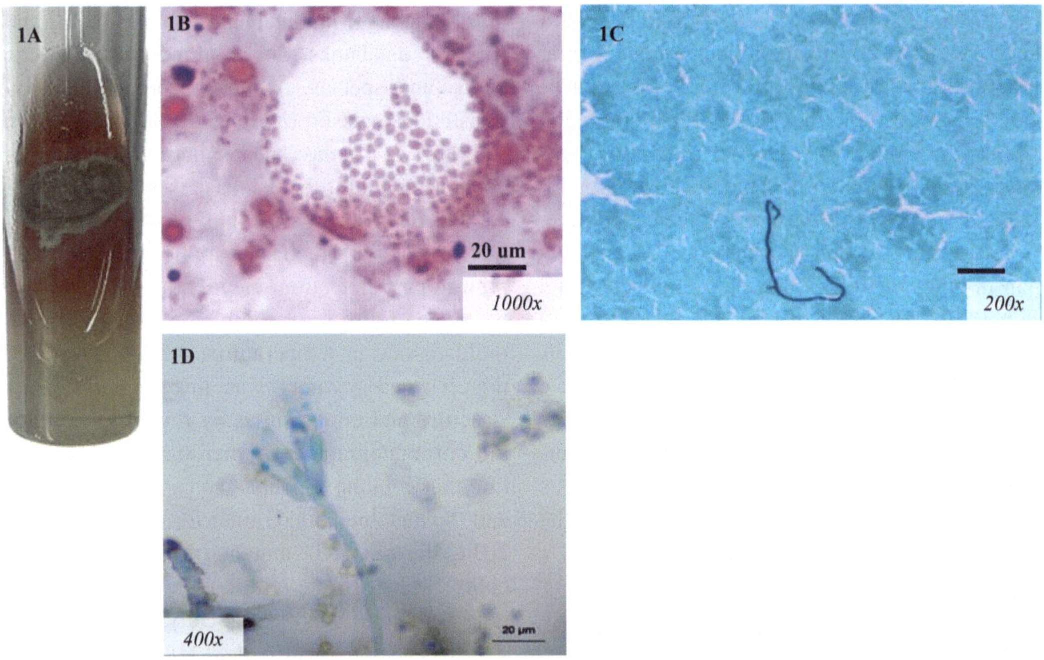

Fig. 23.1 Morphology of *Talaromyces marneffei* at 35 °C and 25 °C: (**a**) red diffuse mold colony on Sabouraud dextrose agar incubated 10 days at 25 °C, (**b**) Gram stain from skin imprint showing pleomorphic cells and binary fission cells, (**c**) Gomori Methenamine Silver stain showing arthrospore-like and septate hyphae in 38 h positive hemoculture bottle, and (**d**) Lactophenol Cotton Blue wet mount showing penicillium-like conidiogenesis of the mold form

Cladosporium carrionii, *Exophiala jeanselmei*, *Sporothrix schenckii*, and *Cladophialophora bantiana*. Polyketides are the secondary metabolites involved in pigment production, toxin synthesis, and other compounds from fungi, bacteria, and plants, and are commonly biosynthesized by polyketide synthases. However, the association between *T. marneffei*-produced polyketides and its virulence, antimicrobial properties, and other metabolites needs to be explored in the future.

Red pigment production in *Talaromyces* is not restricted to *T. marneffei* only. It is also produced by some other species of *Talaromyces*, such as *T. albobiverticillius* (some strains), *Talaromyces aculeatus*, *Talaromyces pinophilus*, *Talaromyces purpurogenus (Penicillium rubrum)*, *Talaromyces funiculosus*, *Talaromyces atroroseus*, and *Talaromyces minioluteus* (some strains). Some of these *Talaromyces* are isolated from soil, indoor environments, various foods, and are involved in industrial pigment production and some are isolated from patients. Recently, with global warming, these environmental/plant species have the potential to cause infection in humans. Isolated cultures from clinical samples with red pigment-diffusing phenotypic colony may lead to the misidentification as *T. marneffei* if no further confirmation of the colony or evidence of intracellular binary fission yeast from the lesion is not available.

Animal Model

Several animal models have been used to describe the pathogenicity of *T. marneffei* in the early years after the discovery of talaromycosis. These animal models include mice, rats, hamsters, guinea pigs, and zebra fish (*Danio rerio*). The pathophysiology and virulence of *T. marneffei* have been mainly studied in murine models (*Mus musculus*) including BALB/c mice. Different in vivo experimental models with invertebrate animals such as honeycomb moth (*Galleria mellonella*) and roundworm (*Caenorhabditis *elegans*) have also been proposed to investigate virulence of the pathogen, innate immune response, and antifungal agents.

Pathogenesis and Pathology

It is believed that a large amount of hydrophobic and readily aerosolized conidia supports the process of both lung invasion of the host through the respiratory tract and dissemination in the blood circulation system. When cultured at 35–37 °C or living in the human body, the conidia of *T. marneffei* convert into a fission yeast form, called parasitic form.

So far, the information on talaromycosis pathogenesis is very limited, especially regarding its mechanism. Most of the publications focus on clinically based information. The localized type of infection is very rare and the disseminated form is more frequently reported in both immunocompromised and immunocompetent hosts. Inhalation of its conidia is the main route of infection, followed by invasion through the skin. Once the invasion succeeds, it is recognized by host immune cells as a pathogen leading to phagocytosis, primarily by macrophages. Instead of being killed by macrophage, *T. marneffei* uses it as a host for cell multiplication and transformation into the yeast form. The infection can remain dormant in the host and reactivate once the host's immunity drops. Several systems, such as the lymphatics, hematogenous, and reticuloendothelial systems, are the main routes for dissemination. In a mouse model, T helper 1 (Th1) response is suggested as a major immune response, while no role has been revealed for antibody production. In the case of talaromycosis-associated HIV infection patients, the severity of the disease likely depends on CD4 numbers. Patients with a low number of CD4 cells tend to have more invasive, anergic, and necrotizing tissue reactions.

The pathology trends in talaromycosis relate to the immunity of the host. Granulomatous formation trends support the localized form and prevent further dissemination. The most frequent organs involved in this disease are the liver and lungs. However, infection in these two organs also affects lymph nodes, bone marrow, skin, and intestines. In the liver, the pathology usually presents as histiocytic infiltration of the sinusoids, parenchyma, and epithelioid granulomas. It is noted that there is no correlation between these histological characteristics and liver function test. In lymph nodes, lymphoid depletion with histiocytic proliferation and focal necrosis can be seen. Like bone marrow, histiocytic proliferation with or without granuloma formation can be observed.

Immunology

Without normal immunological function, clearance of the conidia is defective and leads to infection or dissemination of the organisms. *T. marneffei* survives and replicates inside the phagosome of macrophages by exploiting several survival strategies. These strategies include the production of heat shock proteins, MAPK signaling cascades, and antioxidant molecules such as melanin, catalase-peroxidase, superoxide dismutase, yeast activating protein-1 transcription factor, and basic region leucine zipper-type transcription factor. Progression and expansion of *T. marneffei* infection result in the recognition of pattern recognition receptors, production of several cytokines, and activation of an adaptive immune response. The study found that the cytokines secreting from innate immune cells of the HIV-associated talaromycosis include IL-6, TNF-α, IL-8, IL-18, IL-12, IL-10, TNF-α, and IFN-γ.

For adaptive immune response, CD4+ T lymphocytes, particularly Th1 lymphocytes, play important roles in controlling intracellular pathogens. Interferon-gamma (IFN-γ), Th1 signature cytokine, promotes macrophage activation and reactive oxygen species production, which are crucial for intracellular *T. marneffei* elimination. The contribution of Th1 lymphocyte demonstrated in BALB/c mice also reveals elevated levels of IL-12, IFN-γ, and TNF-α upon infection of *T. marneffei*.

The importance of IL-12/IFN-γ cascade in the eradication of intramacrophagic *T. marneffei* is illustrated by reports of talaromycosis in patients with inborn errors of IFN-γ-dependent immunity that include patients with CD40 ligand deficiency, autoantibodies against IFN-γ, and IFNGR1 mutation. Notably, talaromycosis with the underlying disease of chronic granulomatous disease (CGD) has never been reported. This fact might suggest that solitary impairment in an oxidative burst is not sufficient for susceptibility to *T. marneffei* infection. Although Th17 has been described to contribute to immune defense against several fungal pathogens, including *Candida* spp., Th17 antifungal immunity against *T. marneffei* infection is still not clearly established. However, susceptibility to *T. marneffei* in patients with Th17-mediated immunity errors inborn, such as hyper-IgE syndrome (HIES) and autosomal dominant gain-of-function signal transducer and activator of transcription 1 (STAT1) mutation, proposes as an essential role of Th17 for *T. marneffei* infection.

Infection in Humans

The incubation period of acute infection ranges from 1 to 3 weeks; however, reactivation of latent infection could occur several years after exposure.

Talaromycosis typically causes an opportunistic infection in immunocompromised hosts, particularly in HIV patients with a low CD4+ T-cell count defined as below 100 cells/mm^3. Talaromycosis is listed as the third-ranking AIDS-defining illness in Southeast Asia after tuberculosis and cryptococcosis. In addition, the disease is also reported in non-HIV cases with immunodeficiencies, diabetes mellitus, lupus, IFN-gamma autoantibody, cancer, solid organ or hematopoietic stem cell transplantation, and treatment with immunosuppressive therapy. Due to advances in antiretroviral therapy, the incidence of talaromycosis in HIV cases markedly declined. However, the reports of talaromycosis in non-HIV immunocompromised patients have increased in the past decade. The disease has also been described in immunocompetent patients without underlying disease, yet it is accounted rare.

The severity level of talaromycosis depends on the underlying patient condition and the timing of the diagnosis. In immunocompetent individuals, *T. marneffei* infection is usually asymptomatic. Talaromycosis can present in two major clinical forms: local infection and disseminated infection, the latter being the most common form. *T. marneffei* usually spreads through a hematogenous or lymphatic route to multiple organs. The patient generally presents with constitutional symptoms such as fever, fatigue, and weight loss, which are nonspecific for talaromycosis. Anemic symptoms, lymphadenopathy, hepatomegaly, splenomegaly, cutaneous lesions, respiratory, and gastrointestinal symptoms also develop in talaromycosis. Mucocutaneous lesions are also frequently present in the disseminated form and are usually found as multiple facial papules with necrotic lesions on the arms, trunk, and legs. However, other morphologies of skin lesions have been reported, including chronic cutaneous ulcers, pustular psoriatic-like lesions, and subcutaneous nodules. Oral lesions include papules, erosions, or ulcers that can extend into the oropharynx. Abnormalities in routine biochemical and hematological assessments include hypoalbuminemia, anemia, thrombocytopenia, and abnormal liver function tests. The disease could be misdiagnosed as histoplasmosis, cryptococcosis, or tuberculosis, as some of the clinical manifestations of these diseases overlap, and they are found in the same endemic area.

Clinical characteristics of talaromycosis are different in AIDS-related and non-AIDS-related patients. While HIV-infected patients generally develop a disseminated infection, non-HIV-infected patients may develop either local infection or disseminated infection. HIV-infected patients also tend to develop fever, skin lesions, and splenomegaly more frequently than non-HIV-infected patients. Conversely, talaromycosis in non-HIV-infected patients is more prone to involve bones and joints and is more likely to develop Sweet's syndrome (acute febrile neutrophilic dermatosis) than HIV-infected patients. Interestingly, the blood culture-positive rate in AIDS patients is higher than that in non-AIDS patients. The concrete

Table 23.1 Clinical features of *Talaromyces marneffei* infections in HIV-infected and non-HIV-infected patients

Clinical features	Case[a]	HIV[a]	Non-HIV[a]	*P*-value	Non-HIV[b]
		Number of patients (*N*) (%)			*N* (%)
	92	73	19		119
Fever	75 (81.5)	63 (86.3)	12 (63.2)	0.047	89 (74.8)
Cough	58 (63.0)	42 (57.5)	16 (84.2)	0.032	50 (42.0)
Expectoration	43 (46.7)	28 (38.4)	15 (78.9)	0.002	
Dyspnea	18 (19.6)	13 (17.8)	5 (26.3)	0.611	33 (27.7)
Abdominal pain	14 (15.2)	13 (17.8)	1 (5.3)	0.318	15 (12.6)
Diarrhea	17 (18.5)	14 (19.2)	3 (15.8)	0.994	
Weight loss	41 (44.6)	36 (49.3)	5 (26.3)	0.072	34 (28.6)
Skin lesion	45 (48.9)	39 (53.4)	6 (31.6)	0.09	53 (44.5)
Bone and joint pain	3 (3.3)	2 (2.7)	1 (5.3)	1	25 (21.0)
Lymphadenopathy	74 (80.4)	58 (79.5)	16 (84.2)	0.888	ND
Hepatomegaly	7 (7.6)	7 (9.6)	0	0.358	23 (19.3)
Splenomegaly	45 (48.9)	43 (58.9)	2 (10.5)	<0.001	19 (16)
Abnormal pulmonary breathing sounds	48 (52.2)	34 (46.6)	14 (73.7)	0.035	ND
Thrombocytopenia	ND	ND	ND		9 (7.6)
Leukocytosis	ND	ND	ND		66
Immunosuppressive agents' history or hormone use	14 (15.2)	5 (6.8)	9 (47.4)	<0.001	

ND No data

[a]Retrospective study in a non-endemic area, Zhejiang province, China (Ref: Shen Q., et al. 2022:50:169–178

[b]Review English language literature using the keyword "talaromycosis, Penicillosis, *Penicillium, marneffei*" in a PubMed search on October 1, 2015 (Ref: Chan JF., et al., 2016:5:e19–27

evidence of clinical feature data of AIDS-related and non-AIDS-related patients has been hardly comparable. However, some data of *T. marneffei* from non-endemic zones and from PubMed review literature are listed in Table 23.1.

Infection in Animals

Epidemiology and the natural reservoir of *T. marneffei* are still unclear. So far, bamboo rats living across Thailand, Vietnam, India, and China are known as natural reservoirs of *T. marneffei.* The infected bamboo rats are asymptomatic without any pathological change in internal organs. Since the organisms were mostly isolated from bamboo rat lung tissues but not embryonic tissue, intestinal tissues, nor intestinal contents of bamboo rats, inhalation of the conidia derived from the surrounding environment is suggested as a cause of infection among these animals. *T. marneffei* is also identified from bamboo rat stool and the natural environments surrounding the bamboo rat burrows, including soil and plant roots. Therefore, exposure to soil and the environment containing pathogens is considered one of the risk factors for infection. There is no report about human infection from the consumption of bamboo rats.

Epidemiology and Public Health

It is an AIDS-defining illness, which is mainly found in patients with HIV infection, due to its host. So far, most cases of *T. marneffei* infection in HIV-infected patients have been reported from the endemic countries or among people with a history of travel to those endemic areas. An exception was one case that was reported in Ghana involving an African man with no history of traveling to South or Southeast Asia. Among endemic areas, the highest incidence of talaromycosis cases was reported from the Northern part of Thailand, where it presents as the third most common disease in HIV-positive patients. The isolation of *T. marneffei* from the soil is also possible. The notion that exposure to the infected soils might be the cause of infection was supported by the case-control study in Thailand. The study demonstrated no risk of bamboo rat exposure or consumption leading to infection of *T. marneffei*. On the other hand, exposure to infected soil, particularly in the rainy season, was found to be one of the risk factors. However, it is important to note that soil samples from the residences of both infected bamboo rats and patients with penicilliosis rarely test culture-positive for *T. marneffei*. Air-borne transmission and inhalation of conidia from an environmental source are routes that need to be considered, especially in immunocompromised hosts. This route of transmission was emphasized by the case of the undiagnosed HIV-positive patient who got an infection after exposure to the conidia of *T. marneffei* while handling them on an open bench in the laboratory. It may be noted that age and host immunity are also factors associated with those periods. The season is associated with the rate of infection, as supported by the case number reported in the Northern part of Thailand during the rainy season.

During 1974–1982, five cases of non-HIV-infected patients were reported from a tertiary care hospital in Bangkok, Thailand. Three of them presented with the underlying diseases as systemic lupus erythematosus (SLE), a lymphoproliferative disorder, and tuberculosis, while the other two were immunocompetent hosts. During 1984–1985, there were several reported cases in Guangxi, China (n = 8), Thailand (n = 5), Hong Kong (n = 1), and an imported case in the USA. In Southeast Asia, the first case of HIV-infected patient was reported in Bangkok in 1989. In the AIDS era, the talaromycosis associated with HIV-infection incidence increased dramatically. The cases were reported from several countries in Southeast Asia, especially in Southern China (Guangdong; n = 1079, during 1984–2009) and in Vietnam (Ho Chi Minh City; n = 719, during 2004–2010). The cases expanded to Northeast India, Myanmar, Indonesia, and the other non-endemic zones. This infection became one of the most common AIDS-defining illnesses. This drives the concern in world health care management, and finally, it was recognized as a major fungal disease in the endemic area. In 2008, after the launch of the policy of antiretroviral treatment by WHO, the targeted group for talaromycosis was shifted from HIV-associated talaromycosis to the non-HIV-associated group.

The epidemiology of talaromycosis has been changing. It is not limited to the endemic zone only but also spreading beyond endemic areas. Also, an increased number of cases of talaromycosis are reported, possibly due to the increasing number of immunocompromised and immunosuppressive hosts. The enlarging number of patients with organ transplantation and those who are on targeted therapies are also at risk of infection. The world distribution of *T. marneffei* infection in both HIV and non-HIV-infected patients and in the endemic and non-endemic zones is shown in Fig. 23.2. Controlling exposure of people at risk to the source of the pathogen is difficult due to paucity of information about how *T. marneffei* is transmitted to humans.

Diagnosis

Mostly, clinical samples are collected from patients after *T. marneffei* has spread. Like other fungal infections, direct examination and cultivation will be very helpful rapid diagnosis but requires skill. Samples are suitable for cultivation at both 25 °C and 35 °C for 5–14 days, at least. *T. marneffei,* specific monoclonal antibodies against this thermal dimorphic fungus were prepared and used as a serological diagnostic tool. However, this tool has been used at the research level.

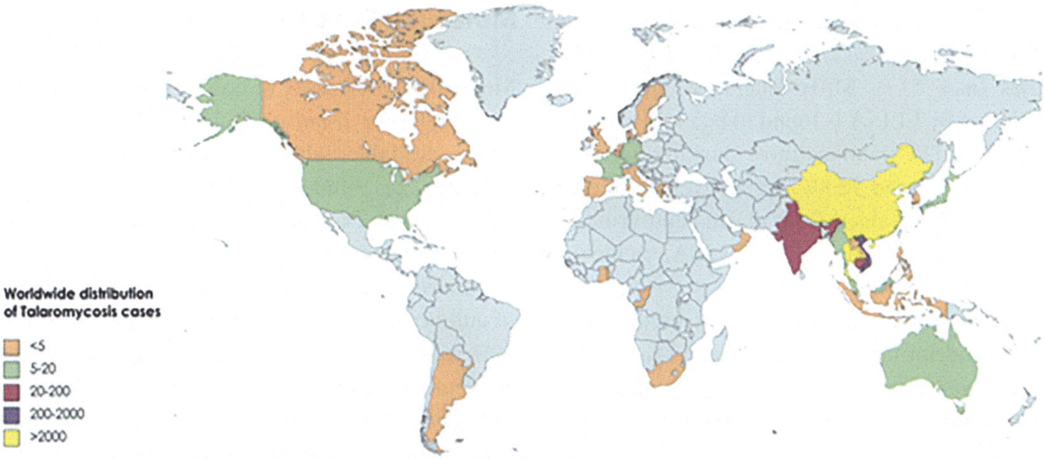

Fig. 23.2 The worldwide distribution of *Talaromyces marneffei* reported cases from 34 countries by 2020s

Microscopy

Gram stain is not a specific stain for the detection of *T. marneffei*. However, it is the fundamental stain in microbiology laboratories for most of the clinical specimens (Fig. 23.1b). Wright stain, Gomori methenamine Silver (GMS) stain, and hematoxylin and eosin (H&E) stain are used to reveal the unique morphologies in polymorphonuclear cells. Imprints of fingertip blood, skin lesions, especially umbilicated lesions, and pus are adequate for direct examination by microscopy.

Microscopy using Wright stain and other histological staining of the specimen is useful. The detection of yeast by its sausage-shaped and pleomorphic shape with cross-wall, approximately 2.5–4.5 μm in diameter, with elongated forms 1–2 × 3–6 μm, and presence in both intra- and extra-white blood cells, is diagnostic of *T. marneffei* (Fig. 23.1b). The parasitic stage of *Histoplasma* is characterized by the uniformity of oval yeast cells, whereas the presence of pleomorphic shape of fission yeast cells of *Talaromyces* is characteristic.

Culture

Isolation of the fungi by in vitro culture is the gold standard. Among clinical specimens, bone marrow specimen provide the highest culture-positive rate, estimated to be 100%, followed by skin biopsy and hemoculture, estimated to be 62%–90% and 55%–76%, respectively. The sensitivity of the fungal isolation varies and depends on various factors related to the clinical course of the diseases. For example, a higher incidence rate of fungemia, indicating disseminated form, is found in HIV-positive patients, estimated as 76.7% compared to 47.1% in HIV-negative cases.

An automated blood culture system is used to detect *T. marneffei* in bloodstream, in the phase of fungemia. After 2–7 days of incubation at 37 °C, *T. marneffei* shows a septate hyphae-like structure, and arthroconidia-like yeast cells (Fig. 23.1c) are detected in a positive hemoculture bottle; even though yeast form is usually found in the human body at the same temperature. To isolate *T. marneffei* from monomorphic fungi, dual incubation in basic fungal media (Sabouraud dextrose agar) at 25–30 °C (Fig. 23.1a, d) and at 35–37 °C for 5–14 days is carried out to observe its hyphal/yeast phases.

Serodiagnosis

Serology, both antibody and antigen tests, has been developed and evaluated for the diagnosis of *Talaromyces* infections. Based on the retrospective study in plasma obtained from culture-proven talaromycosis (n = 372) and non-talaromycosis cases (n = 517), Mp1p mannoprotein antigen detection by ELISA is found to be more sensitive than BACTEC hemoculture, showing sensitivities of 86.3% and 72.8%, respectively. The combined result obtained from plasma and urine led to a higher sensitivity of ELISA, achieving 88.8%.

Urine, a noninvasive sample, is another specimen used for talaromycosis diagnosis. Bedside strip tests using 4D1 monoclonal antibodies have been developed as the diagnostic test of choice, with 87.87%, 100%, and 95.5% for sensitivity, specificity, and accuracy, respectively. However, it may be noted that various levels of antibodies and antigens are found in each group of patients. A lower level of antibody and a higher level of antigen have been demonstrated in the HIV-positive patient compared to the negative group. Galactomannan assay for *Aspergillus* is another test used for the diagnosis of talaromycosis with a sensitivity range from 73.3% to 80%. The serological assays for the detection of *T. marneffei* antigens in yeast formare listed in Table 23.2.

Molecular Diagnosis

Molecular-based assay or polymerase chain reaction (PCR) for *T. marneffei* identification is currently available; although, most of them are in-house or limited to research use only. Conventional PCR, along with sequencing, is being used frequently. The TaqMan real-time PCR used to detect 5.8S rRNA from DNA extract of cultures and peripheral blood samples showed 60% sensitivity, which surpasses that achieved by conventional blood culture, calculated to be 55% sensitivity. Real-time PCR compared to the hemoculture automated system helps in the earlier detection of *T. marneffei*. In 2019, the surface mannoprotein named Mp1p was applied in a real-time PCR assay for the identification and diagnosis of *T. marneffei* using human plasma. It was shown that the sensitivity of the samples prior to antifungal treated and 48 h after treatment was 70.4% and 52.2%, respectively. The limitation of detection of this assay was ten times lower than that of 5.8s RNA TaqMan real-time PCR.

Even though several effective options are currently available, rapid and accurate diagnostic tools with high sensitivity and specificity are

Table 23.2 Summarized serological assays for yeast form for antigen detection of *Talaromyces marneffei* during 2000–2018

Method	Target	Remark
Dot Blot ELISA	Heat-killed fission cells	94.6% sensitivity and 97.3% specificity
Sandwich ELISA	Heat-killed fission cells	97.3% sensitivity and 98% specificity
Reverse passive Latex agglutination	Heat-killed fission cells	100% sensitivity and 99.3% specificity
Antigen capture ELISA	Heat-killed fission cells	Using MAb 8C3 (IgM) against mycelial culture filtrate; 72% sensitivity and 100% specificity
ELISA, Platelia *Aspergillus*	Galactomannan (cutoff index:0.5)	Using rat MAb EB-A1; 73.3% sensitivity and 13.6% specificity
ELISA, Platelia *Aspergillus*	Galactomannan (cutoff index:1.0)	Using rat MAb EB-A1; 95.8% sensitivity and 90.9% specificity
Indirect inhibition ELISA	Cytoplasmic yeast antigen	Using MAb 4D1, 100% sensitivity and specificity
Lateral flow immune-chromatography	Cytoplasmic yeast antigen	Using MAb 4D1, 87.9% sensitivity and 100% specificity

(Ref.: Modified Table 2, Pruksaphon K et al., 2020;96(3):114959)

most important for patients' survival. *T. marneffei*-specific antibodies might aid in the detection of *T. marneffei* in tissue samples by microscopy along with the increased use of bedside diagnostic kits. Serological assays for the yeast form antigen detection of *T. marneffei* are listed in Table 23.2.

Treatment

Currently, the treatment protocol is recommended mainly based on studies conducted in areas endemic for talaromycosis. Without treatment, disseminated talaromycosis is always fatal.

The therapeutic strategy for talaromycosis comprises three phases: induction, consolidation, and maintenance. Induction phases with 2 weeks of amphotericin B (liposomal amphotericin B 3–5 mg/kg body weight/day or amphotericin B deoxycholate 0.7 mg/kg body weight/day, intravenous) followed by a consolidation phase with 10 weeks of itraconazole (200 mg every 12 h) is currently considered the treatment of choice. Induction with amphotericin B is more effective than itraconazole in blood fungal clearance with a lower rate of disease complication. Voriconazole can be used for induction treatment in patients contraindicated for amphotericin B or patients who cannot tolerate the drug's side effects. After the consolidation phase, continuation with oral itraconazole 200 mg daily is recommended for the maintenance phase (secondary prophylaxis) until the CD4 count reaches the target of above 100 cells/mm^3 for at least 6 months to prevent the recurrence of talaromycosis. The recurrence rate is lower than 10% if the patients have good medication compliance with both maintenance treatment and effective antiretroviral treatment. A sustained level of CD4 counts of more than 100 cells/mm^3 and virological reduction for at least 6 months are indications to discontinue the secondary prophylaxis. Itraconazole is a potent inhibitor as well as a substrate of cytochrome P450 3A4 (CYP3A4). Hence, depletion of itraconazole exposure is of concern when coadministered with antiretroviral medications. The latter is classified in the group of non-nucleoside reverse transcriptase inhibitors (NNRTIs), which are also metabolized in the liver by CYP3A isoenzymes. Therefore, to prevent underdosing of azole antifungal drugs in HIV-associated endemic mycoses, concomitant administration of itraconazole with NNRTIs antiretroviral medications needs to be avoided without itraconazole drug level monitoring.

Initiation of antiretroviral treatment in HIV-infected patients is associated with the development of an immune reconstitution inflammatory syndrome (IRIS), resulting in imbalanced T lymphocyte responses. The aberrancy of inflammatory responses in IRIS usually leads to the clinical deterioration of a treated infection (paradoxical IRIS) or an undiagnosed infection (unmasking IRIS). IRIS associated with *T. marneffei* infection could develop in both forms of IRIS.

Prevention and Control

A study in Chiang Mai province, Thailand, documented a significant reduction in the incidence of talaromycosis in patients with a CD4 count <200 cells/mm^3, who were treated with itraconazole 200 mg daily as primary prophylaxis. However, this prevention strategy has not been widely used due to the high cost of the therapy, drug toxicity effects, and development of resistance.

Currently, the indication for primary prophylaxis is only recommended for patients with HIV who inhabit the endemic regions such as Northern Thailand, Southern China, and Northern and Southern Vietnam. These patients are unable to access antiretroviral treatment or have treatment failure with CD4 counts <100 cells/mm^3. Oral itraconazole 200 mg daily or oral fluconazole 400 mg once weekly is given as an alternative prophylaxis medication. Primary prophylaxis is not recommended for patients who are on antiretroviral treatment or impending to start effective antiretroviral treatment. For the patients living in non-endemic areas, prophylaxis treatment only applies to those who are planning to travel to highly endemic areas, who are not on effective antiretroviral treatment, and who have a CD4 count <100 cells/mm^3. The prophylaxis with either itraconazole or fluconazole, hence, needs to be initiated 3 days before the trip and maybe continued for 1 week after the trip.

Case Study

A male PLHA in his early 60 s presents with chief complaints of cough associated with the production of sputum for a period of 8 months, raised body temperature, and difficulty in breathing for 1 month. He also complained of decreased urine output, difficulty in taking oral feeds, and weight loss of more than 10 kg. The patient appeared emaciated and had molluscum contagiosum-like lesions on the face and trunk.

Investigations revealed a white blood cell count of 4200/mm^3, a platelet count of 58,500/mm^3, and a hemoglobin of 8 g%. The liver function test showed cholestasis with a total bilirubin of 8.8 mg/dL, direct bilirubin of 8.2 mg/dL, AST of 124 U/L, ALT of 84 U/L, ALP level of 459 U/L, and GGT level of 82 U/L. The renal function tests indicate renal impairment with a blood urea level of 128 mg/dL and a serum creatinine level of 2.69 mg/dL. Chest X-ray revealed bilateral reticular nodular patches, whereas computed tomography (CT) of the chest revealed bilateral opacification. Abdominal ultrasound showed mild-to-moderate ascites with an enlarged spleen. CD4 count at presentation was 13 cells/mm^3, indicating severe immunosuppression. A peripheral blood smear revealed neutrophils containing yeast cells with septa instead of budding. Blood culture and skin biopsy from molluscum-like lesions grew mold which was identified as *T. marneffei* based on sequencing of the rDNA.

He was treated with amphotericin B deoxycholate (50 mg) for 2 weeks and switched to itraconazole for the next 10 weeks to which he responded well clinically and radiologically. He was further advised to take itraconazole 200 mg once daily and report after 6 months.

1. What significant points can be learned from the above clinical vignette?
2. How to identify the pathogen by conventional and molecular methods?
3. What are the risk factors for *T. marneffei* infection?
4. How do you treat this infection?

Research Questions

1. Which group of fungi does *T. marneffei* belongs to, and why was the name changed from *Penicillium marneffei* to the current one?
2. Given that *T. marneffei* is a dimorphic fungus, what is the current state of antifungal susceptibility for this group of fungi?
3. What pigments are produced by *T. marneffei*, and their role in pathogenicity and identification?

Further Reading

Castro-Lainez MT, Sierra-Hoffman M, LLompart-Zeno J, Adams R, Howell A, Hoffman-Roberts H et al (2018) *Talaromyces marneffei* infection in a non-HIV non-endemic population. IDCases 12:21–24

Ching-López R, Rodríguez PS (2020) Talaromycosis in a lung cancer patient: a rare case. Cureus 12(9):e10615

Cuomo CA, Shea T, Nguyen T, Ashton P, Perfect J, Le T (2020) Complete genome sequences for two *Talaromyces marneffei* clinical isolates from northern and southern Vietnam. Microbiol Resour Announc 9(2):e01367

Hilmarsdottir I, Coutellier A, Elbaz J, Klein JM, Datry A, Gueho E et al (1994) A French case of laboratory-acquired disseminated *Penicillium marneffei* infection in a patient with AIDS. Clin Infect Dis 19(2):357–358

Jones PD, See J (1992) *Penicillium marneffei* infection in patients infected with human immunodeficiency virus: late presentation in an area of nonendemicity. Clin Infect Dis 15(4):744

Lau SKP, Tsang CC, Woo PCY (2017) *Talaromyces marneffei* genomic, transcriptomic, proteomic and metabolomic studies reveal mechanisms for environmental adaptations and virulence. Toxins (Basel) 9(6):192

Lau SKP, Lo GCS, Chow FWN, Fan RYY, Cai JJ, Yuen K-Y et al (2018) Novel partitivirus enhances virulence of and causes aberrant gene expression in *Talaromyces marneffei*. MBio 9(3):e00947–e00918

Le T, Wolbers M, Chi NH, Quang VM, Chinh NT, Lan NP et al (2011) Epidemiology, seasonality, and predictors of outcome of AIDS-associated *Penicillium marneffei* infection in Ho Chi Minh City, Viet Nam. Clin Infect Dis 52(7):945–952

Peng J, Chen Z, Cai R, Huang X, Lin L, Liang W et al (2017) Recovery from *Talaromyces marneffei* involving the kidney in a renal transplant recipient: a case report and literature review. Transpl Infect Dis 19(4). https://doi.org/10.1111/tid.12710

Pruksaphon K, Intaramat A, Ratanabanakoon K, Nosanchuk J, Vanittanakom N, Youngchim S (2020) Diagnostic laboratory immunology for talaromycosis (penicillosis): review from the bench-top techniques to the point-of-care testing. Diagn Microbiol Infect Dis 96(3):114959

Qin Y, Huang X, Chen H, Liu X, Li Y, Hou J et al (2020) Burden of *Talaromyces marneffei* infection in people living with HIV/AIDS in Asia during ART era: a systematic review and meta-analysis. BMC Infect Dis 20(1):551

Samson RA, Yilmaz N, Houbraken J, Spierenburg H, Seifert KA, Peterson SW et al (2011) Phylogeny and nomenclature of the genus *Talaromyces* and taxa accommodated in *Penicillium* subgenus *Biverticillium*. Stud Mycol 70(1):159–183

Shi N, Kong J, Wang K, Cao C (2019) Coinfection with *Talaromyces marneffei* and other pathogens associated with acquired immunodeficiency. JAMA Dermatol 155(10):1195–1197

Sirisanthana T, Supparatpinyo K (1998) Epidemiology and management of penicilliosis in human immunodeficiency virus-infected patients. Int J Infect Dis 3(1):48–53

Supparatpinyo K, Sirisanthana T (2008) CHAPTER 35—infection due to *Penicillium marneffei*. In: Volberding PA, Sande MA, Greene WC, Lange JMA, Gallant JE, Walsh CC (eds) Global HIV/AIDS medicine. W.B. Saunders, Edinburgh, pp 393–398

Tang BS, Chan JF, Chen M, Tsang OT, Mok MY, Lai RW et al (2010) Disseminated penicilliosis, recurrent bacteremic nontyphoidal salmonellosis, and burkholderiosis associated with acquired immunodeficiency due to autoantibody against gamma interferon. Clin Vaccine Immunol 17(7):1132–1138

Tang Y, Zhang H, Xu H, Zeng W, Qiu Y, Tan C et al (2020) Dendritic cells promote Treg expansion but not Th17 generation in response to *Talaromyces marneffei* yeast cells. Infect Drug Resist 13:805–813

Weerasinghe H, Payne M, Beard S, Andrianopoulos A (2016) Organism-wide studies into pathogenicity and morphogenesis in *Talaromyces marneffei*. Future Microbiol 11(4):511–526

Woo PC, Lau SK, Liu B, Cai JJ, Chong KT, Tse H et al (2011) Draft genome sequence of *Penicillium marneffei* strain PM1. Eukaryot Cell 10(12):1740–1741

Woo PC, Tam EW, Chong KT, Cai JJ, Tung ET, Ngan AH et al (2010) High diversity of polyketide synthase genes and the melanin biosynthesis gene cluster in *Penicillium marneffei*. FEBS J 277(18):3750–3758

Yang D, Shen LX, Chen RF, Fu Y, Xu HY, Zhang LN et al (2021) The effect of *Talaromyces marneffei* infection on CD86 expression in THP-1 cells. Infect Drug Resist 14:651–660

Zheng YQ, Pan KS, Latgé JP, Andrianopoulos A, Luo H, Yan RF et al (2019) Calcineurin a is essential in the regulation of asexual development, stress responses and pathogenesis in *Talaromyces marneffei*. Front Microbiol 10:3094

Pneumocystosis

24

Vinaykumar Hallur and Ankita Das

Learning Objectives

By the end of the chapter, the readers should be able to:

1. Learn about the etiological agents of pneumocystosis.
2. Understand the pathogenesis and clinical presentation of pneumocystosis.
3. Evaluate strategies for sample collection and diagnosis of pneumocystosis.
4. Discuss the management strategies for pneumocystosis.

Introduction

Pneumocystosis is a pulmonary infection in mammals caused by fungi belonging to the genus *Pneumocystis*. Members of the *Pneumocystis* are so named because of cysts (asci) in the lungs of infected animals. They are characterized by their affinity to lungs and inability to be reliably cultured in culture media. While considered as zoonoses in the past, cross infection and molecular studies have demonstrated that these organisms have an extremely narrow host range to the extent that *Pneumocystis* species could be delineated based on the host being infected. The disease itself is self-limiting or asymptomatic in healthy individuals. However, in immunocompromised hosts, it causes interstitial pneumonia with alveolar exudate described as honeycomb on histological examination. Animal studies are limited to case reports from dogs, rabbits, horses, monkeys, pigs, and cats. In humans, dysfunctional or reduced cell-mediated immunity due to HIV is the major risk factor for infection. Other risk factors include steroid therapy, immunosuppression following an organ transplant, hematological malignancies, solid tumors, and autoimmune disorders. Trimethoprim-sulfamethoxazole is used as a first-line treatment and for prevention in at-risk individuals. Despite the best care, mortality rates are high, with a case fatality rate of 50% in non-HIV patients and 15% in HIV-infected patients. Although its prevalence is declining in developed countries following the introduction of antiretroviral therapy, it is expected to be high in developing countries without access to diagnostics and antiretroviral treatments.

History

Discovered by Chagas in Brazil in 1909, *Pneumocystis* was initially thought to be a stage in the life cycle of trypanosomes in guinea pigs. Carinii, who found similar structures in rat lung

V. Hallur (✉) · A. Das
Department of Microbiology, ICMR Advanced Molecular and Diagnostics Research Centre for Fungi, All India Institute of Medical Sciences, Bhubaneswar, Odisha, India
e-mail: vinay118@gmail.com

S. C. Parija, S. M. Rudramurthy (eds.), *Textbook of Fungal Zoonoses and Sapronoses*, Microbial Zoonoses, https://doi.org/10.1007/978-981-97-3214-2_24

tissue, sent them to the Delanoe couples working in Instituit Pasteur. The Delanoes confirmed that this was a new organism after finding a consistent lack of other trypanosome stages in tissue sent to them from Brazil and the presence of similar structures in the lungs of Parisian rats never exposed to *Trypanosomes*. Because the organism preferred lung tissue and had cystic structures, and to honor Carinii, the organism was formally named *Pneumocystis carinii* (PC). Subsequently, during World War II, *Pneumocystis* first gained attention as a cause of interstitial pneumonia among undernourished and premature infants in orphanages. Later in the 1980s, with the advent of the AIDS epidemic, *Pneumocystis jirovecii* was recognized as a common cause of pneumonia with HIV and was identified as an AIDS-defining illness. The incidence of *P. jirovecii* infections has declined after highly active antiretroviral therapy (HAART) and anti-*Pneumocystis* prophylaxis.

Taxonomy

P. carinii was initially thought to be a parasite due to its morphology, presence of a single 18S rRNA gene, lack of ergosterol, no response to amphotericin B, and response to antiparasitic drugs like trimethoprim-sulfamethoxazole. Nevertheless, it was conclusively proven in 1988 to be a fungus by using advanced molecular biology techniques. Other reports, including multigene and whole genome-based phylogeny, have also corroborated it is a fungus. Currently, it is placed in the phylum Ascomycota under the monophyletic subphylum Taphrinomycotina. While several species of *Pneumocystis* have been discovered in the different mammals studied, only five members, namely *P. jirovecii* (humans), *Pneumocystis murina* (rats), *Pneumocystis wakefieldiae* (mice), and *Pneumocystis oryctolagi* (rabbits), have been described as per the International Council of Botanical Nomenclature rules and the rest are described using trinomial or other invalid nomenclature systems. Previously, the members of the genus were considered monoxenous, and it was assumed that species could be delineated based on the infected host. The organism infecting humans is currently known as *P. jirovecii* (PJ). However, this view is increasingly challenged because only a few mammals have been sampled for *Pneumocystis*. Future studies may confirm or refute if the organism is human specific or is a zoonosis.

Genomics and Proteomics

The absence of reliable in vitro culture methods makes the genomic studies of *Pneumocystis* challenging. Notwithstanding the challenges, whole genomes of *P. carinii*, *P. jirovecii*, *P. canis*, *P. murina*, *P. oryctolagi*, *P.* sp. "*macacae*," and *P. wakefieldiae* have been sequenced and are available in public databases. To offset the problem due to the non-cultivable nature of *Pneumocystis*, cellular immunoprecipitation was used to enrich *P. jirovecii* DNA from bronchoalveolar lavage (BAL) samples of patients with PCP. The whole-genome data of *P. jirovecii* was obtained by de novo sequencing using pyrosequencing and published in 2012. Studies from the past have found that *P. jirovecii* has 20 chromosomes with around 3761 genes, a genome length of 8.4 Mb, and a low GC content of 28%. The genomes of other *Pneumocystis* range between 7.3 and 8.2 Mb. All the *Pneumocystis* genomes have shown evidence of massive genome reduction of up to 40% compared to other members of subphylum Taphrinomycotina. Several genes linked to the amino acid, lipid, glycerol, and cofactor metabolism, along with virulence and transporter proteins, are either reduced or completely absent. Simultaneously, the major surface glycoprotein (MSG) superfamily genes are enriched. MSG proteins undergo genetic variation and help the pathogen evade the host's immune response. The reductive evolution of this pathogen and selective gene enrichment could explain the host preference, immune evasion, and predominant pulmonary niche of this unique extracellular pathogen.

A search on the UniProt database gives two hits of available PCP proteome with a count of approximately 3746 and 3419 proteins. A total of 2103 proteins out of 3746 and 1880 out of 3419 were domain-containing proteins, which aligns with the finding that the major surface glycoproteins (MSG), a domain-containing protein, are enriched in PC.

Table 24.1 Characteristic features of three morphological forms of *Pneumocystis jirovecii*

Character	Trophic form (small and large)	Sporocyte	Ascus
Size	Large-1.5–5 μm Small-1–2 μm	~4–8 μm	5–8 μm
Shape	Ellipsoidal in fresh preparation	Oval-spherical shape	Spherical shape
Rapid Wright-Giemsa	Purple-red nucleus and light to blue cytoplasm	Purple-red nucleus and light to blue cytoplasm	Spores/intracystic bodies seen around a clear halo
Methenamine silver	Not stained	Stained	Stained ping pong ball or comma-shaped or cup-shaped diagnostic form
Electron microscopy	Ameboid in shape. Single nucleus and mitochondria found. Pseudohyphal structures found	2–8 nuclei along with groups of mitochondria are found. Pseudohyphae are mostly absent	Thick-walled containing eight intracystic bodies found have a thick cell wall

The Fungus Morphology

As the fungus cannot be reliably cultured outside the mammalian lungs, morphological observations are made on tissue forms of the fungus. It exhibits three morphological structures in the tissue: trophic, sporocyte, and ascus (cyst) forms. It is challenging to identify structures as trophic form, precyst, and asci distinctively in clinical samples as these are found in aggregates or tight clusters. Table 24.1 compares the three morphotypes.

Laboratory Animals

A stable laboratory animal model for PCP is difficult to establish since the pathogen is highly host specific. The model should mimic similar transmission and pathophysiology of the pathogen in a human host. PC is inoculated into the model organisms either by passive cohousing or active instillation with filtered lung shreds of infected animals through the oropharyngeal, intranasal, transtracheal, and intratracheal routes. Ten or more animal models have been investigated for in vivo studies of PCP so far. Mouse models are generally preferred in 75% of studies reported due to their genetic homology and well-characterized physiology. They are said to be one of the best models with various genetic modifications (e.g., Severe Combined Immunodeficiency Disease or SCID mice). Other major laboratory animals for PCP include rats, rabbits, and nonhuman primates. Mice are preferred for studying PCP immunology, rabbits for transmission studies, and rats for preclinical therapy studies. The use of nonhuman primates and rabbits is limited for ethical reasons.

Cultivation of Fungi

In vivo, PC grows in the "hypophase" of the lung alveolar surface, a thin continuous layer of about 200 nm. Pulmonary surfactant is the major component of the hypophase which is regulated by a myriad of factors like calcium content, pH, and the secretions from the alveolar epithelial cells. Attempts to mimic in vivo conditions in culture systems have been futile. Ex vivo culturing of *Pneumocystis* is challenging due to the lack of a continuous passage in the in vitro culture system and limited growth in primary culture. Previously reported methods included inoculating rodent-derived and human-derived *Pneumocystis* on cell monolayers flooded by liquid or cell-free media. However, these have not been reproducible. Commercially available alveolar lung chips provide platforms for epithelial and vascular channels, cell–cell interactions, immune cells, extracellular matrix, and alveolar epithelial cell culture. These may be used for drug response and inflammatory response validation tests.

However, alveolar chips have a lifespan of 2 weeks, which limits their potential application on a larger scale. Techniques like metabolomics and metabolic flux analysis are being investigated to understand the in vivo milieu of the *Pneumocystis* lung culture system.

Life Cycle of Fungi

The life cycle of *Pneumocystis* species occurs in the alveolus within the lungs. It is hypothesized that the fungus undergoes an asexual life cycle wherein the trophozoites divide by binary fission and a sexual cycle. The sexual cycle begins when the opposite mating types encounter each other in the alveolus and combine their genetic material through karyogamy resulting in the zygote. The zygote then undergoes meiosis to form four nuclei. Additional postmeiotic mitosis leads to the formation of eight nuclei. Folding of the inner plasma membrane leads to compartmentalization of newly formed nuclei along with mitochondria resulting in eight spores within the asci. The spores may then enter the asexual life cycle.

Pathogenesis and Pathology

The exact mechanisms of pathogenesis of pneumocystis pneumonia (PCP) pneumonia are not well understood. The infection is self-limited in immunocompetent individuals, and in patients with T cell defects, it causes pneumonia, which is fatal if left untreated. In such patients, the infection is initiated by inhaling the infectious asci, which may release trophic forms in the alveoli. The trophic form attaches to the type 1 alveolar epithelial cells using pseudopodal structures. This process is facilitated by extracellular proteins, viz. collagen 1, fibronectin, and vitronectin, leading to their enhanced proliferation and subsequent contact with type 2 alveolar epithelial cells. Continued proliferation leads to direct lung injury and activation of innate and adaptive immune responses, resulting in immune-mediated inflammatory lung damage. The type of immune deficiency determines the course of the disease, with relatively more competent individuals clearing infection better but suffering higher lung injury. This explains why the condition has a subacute course and less mortality in patients with HIV compared to a rapid downhill course and increased mortality in those without HIV.

Immunology

The immune events leading to PCP or its clearance are unknown. However, it is well known that (a) defects in cellular immunity predispose to disease, (b) the role of B cells is being investigated, and (c) more than one mechanism leads to its clearance. After *Pneumocystis* reaches the alveoli, β D glucan stimulates dendritic cells and other antigen-presenting cells (APC); MSG stimulates APC other than dendritic cells by using toll-like receptors (TLR2& 4) and/or C-type lectin receptors (dectin-1 and mincle) leading to activation of the NFκß pathway and acute inflammatory response. Additionally, damage-associated molecular patterns released from tissue also potentiate the influx of various immune cells.

It is hypothesized that in an immunocompetent host, the APCs activate CD4 T helper cells, which in turn co-activate B lymphocytes (BL) within inducible bronchus-associated lymphoid tissue (iBALT). The CD4 T cells then differentiate to a follicular helper T cell profile, which, along with BLs, would activate the macrophages and favor their polarization to the M2 lineage. Thus, activated M2 macrophages then phagocytose and destroy *Pneumocystis* with or without opsonization.

In immunocompromised hosts, proinflammatory cells of innate immunity (NK) and adaptive immunity (CD4 Th1, Tc1 CD8) can help clear *Pneumocystis* by releasing cytokines like IFNγ. IFNγ. This leads to the polarization of macrophages to the M1 lineage, which phagocytoses and eliminates *Pneumocystis*.

Infection in Humans

Patients with PCP usually present with fever, progressive exertional dyspnea, and nonproductive cough. Occasionally, patients may produce sputum, and rarely hemoptysis may be present. However, the progression of the disease is markedly different in patients with and without HIV. Patients with HIV follow an insidious course spread over 3–8 weeks with mild clinical signs and symptoms, with higher pathogen loads. The disease follows a much more rapid (<1 week) and aggressive course in patients without HIV. These patients needing intensive care and a high case fatality rate (50%) despite having a lower fungal burden. The discordance in fungal burden and outcome may be related to lower lung injury and lesser impairment of gas exchange in patients with HIV than in those without it. On general physical examination, there may be varying degrees of respiratory distress like tachypnea, tachycardia, central cyanosis, etc. Auscultation reveals no abnormality in two thirds of patients, while crepitations may be found in the rest. Table 24.2 shows the major differences between HIV and non-HIV patients with PJP.

Infection in Animals

Different *Pneumocystis* species have been described in the lungs of murine, primate, canine, feline, equine, porcine, caprine, and small laboratory animals (rats, mice, and rabbits). It is hypothesized that these animals acquire the infections via the airborne route or vertical transmission from the infected mother. Among the dogs, it has been mainly reported in dachshunds, Cavalier King Charles Spaniels, and Pomeranians. Arabian foals with autosomal recessive severe combined immunodeficiency and cats with feline immunodeficiency virus infections develop this disease. The disease is unique in rabbits as a self-limiting illness with an abrupt onset and spontaneous healing seen exclusively during the weaning period. The animals usually present with respiratory symptoms like dyspnea, tachypnea, exercise intolerance, with or without coughing, and weight loss. Physical examination confirms tachypnea, respiratory distress, or increased respiratory effort, while cyanosis can often be seen in severe cases.

Epidemiology and Public Health

P. jirovecii has a global presence, and serological evidence of PJ is seen among individuals by 2 years of age. PJ seems to infect early in life among HIV-infected infants. Thus, improved maternal screening of HIV-infected mothers would be more important for disease prevention. The AIDS epidemic of the 1980s in the United States saw a rise in the cases of HIV with PCP. This was followed by a decline from 53% in 1989 to 3.4% after 1992 with the introduction of antiretroviral therapies (ART) and anti-*Pneumocystis* prophylaxis. A similar reduction in cases was reported in Europe and other industrialized countries, but PCP elimination has not been possible yet. The estimated burden of pneumocystosis in the five countries with the highest burden and India is given in Table 24.3. The world map showing the burden of pneumocystosis is given in Fig. 24.1.

Table 24.2 Major differences between PCP in HIV and non-HIV-infected patients

Features	HIV with PCP	Non-HIV with PCP
Risk factors	Low CD4 count (absolute/percentage), oral candidiasis, noncompliance to antiretroviral therapy	CD4 count <200/μL, solid organ transplantation, hematologic malignancies, prolonged corticosteroid therapy, and other immunosuppressive conditions
Onset	Slower	Rapid
Severity	Milder	Severe disease
Mortality	10%–12%	30%–50%
Prognosis	Good with early diagnosis and treatment	Worse than HIV-infected patients

P. jirovecii pneumonia cases occur across the world except in Antarctica. Globally, it is estimated that ~500,000 cases of *P. jirovecii* pneumonia occur annually, with high mortality rates ranging from 15% to 50%. The colonization rates vary across different study groups and geographically, in the range of 4.4%–44% in patients receiving immunosuppressive drugs, 15.5%–48% in pregnant women, up to 85% in children, 20%–69% in HIV-infected patients with *P. jirovecii* pneumonia, and up to 100% of patients with chronic lung diseases. Risk factors for infection in HIV include CD4 counts <200/μL due to undiagnosed HIV, nonadherence to antiretroviral therapy, or suboptimal immune recovery. In the HIV-uninfected population, chemotherapy, bone marrow transplantation, and treatment using corticosteroids, monoclonal antibodies for autoimmune disease, hematological malignancies, solid organ tumors, or transplantation predispose the patient to PCP.

Table 24.3 India and five countries with the highest burden of pneumocystosis and India (Adapted from Bongomin et al. J Fungi (Basel). 2017 Oct 18;3(4):57 and Ray et al. Open Forum Infect Dis. 2022 Dec 26;9(12): ofac603)

Country	Burden	Rate/100,000
Nigeria	74,595	48.2
Kenya	17,000	43
Trinidad and Tobago	400	30
Tanzania	9600	22
Ukraine	6152	13.5
India	58,378	4.19

The epidemiology of PCP in the community is poorly understood and remains limited to drug resistance and outbreak investigation. Though the environmental reservoir and transmission route remain unclear, it is known that the primary transmission route is airborne, and the asci(cyst) is the infectious form. *Pneumocystis* genetic material has been reported in a few studies in environmental samples, e.g., in air samples from hospital sites and beds that harbor patients with PCP. In contrast, other studies report no evidence of it in the samples collected outdoors. Further studies are

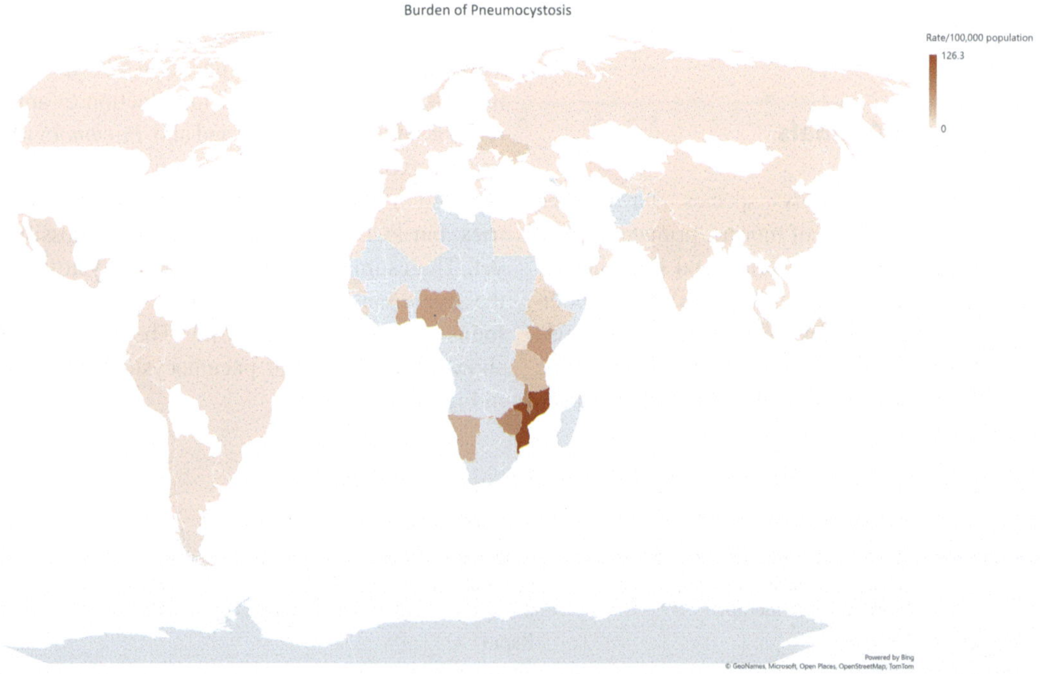

Fig. 24.1 World map showing the burden of pneumocystosis extracted from various studies on the burden of fungal infections by Leading International Fungal Education organization. Note that the infection occurs globally. (Is it copyright (permission to be obtained) or Copy left)

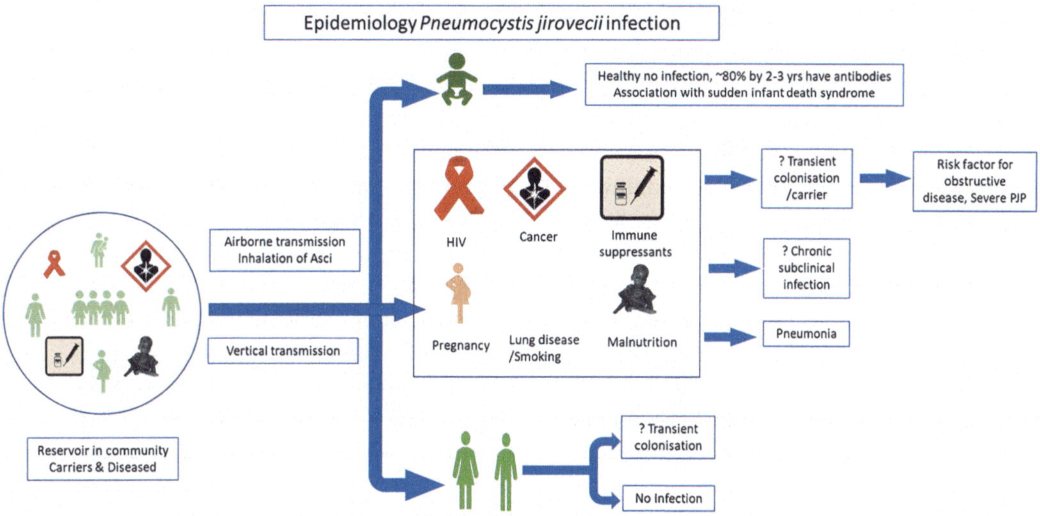

Fig. 24.2 Current understanding of the epidemiology transmission and natural history of *Pneumocystis jirovecii* pneumonia

therefore required to understand the transmission and spread of PCP using more appropriate molecular methodological designs. Several outbreaks of pneumocystosis have been described mainly in solid organ transplant recipients with the possibility of a person-to-person transmission in hospital setup. A summary showing the current understanding of transmission and epidemiology of PCP is given in Fig. 24.2.

Different *Pneumocystis* species have been described in the lungs of murine, primate, canine, feline, equine, porcine, caprine, and small laboratory animals (rats, mice, and rabbits). It is hypothesized that these animals acquire the infections via the airborne route or vertical transmission from the infected mother. As in humans, malnourishment or immunodeficiency (primary or acquired) predisposes them to disease. While animal models are used to study the pathogenesis of the disease in the laboratory, natural infections are limited to small studies or case reports from dogs, rabbits, horses, pigs, and monkeys. Even in these animals, the lungs are the primary organ involved in the disease process.

Diagnosis

Presumptive diagnosis of PCP in an immunocompromised (HIV or non-HIV) patient with clinical suspicion of pneumocystosis is achieved by tests like oxygenation at rest and after exertion, pulmonary function tests, ABG, and radiology. Radiologic signs on chest radiography include bilateral interstitial and alveolar infiltrates in HIV patients and diffuse infiltrates with consolidation and occasional pulmonary cysts or pneumatoceles. The earlier mentioned tests and radiology findings are nonspecific and traditional culture is not helpful as *Pneumocystis* cannot be reliably cultured from human samples. Hence, laboratory methods including microscopy, serology, and molecular tests are carried out to confirm the diagnosis of pneumocystosis.

Specimens

Various specimens for microscopy and molecular detection for laboratory diagnosis of PCP include open lung biopsy, transbronchial biopsy, bronchoalveolar lavage (BAL) fluid, bronchial secretions, induced sputum, spontaneous sputum, nasopharyngeal aspirate, and oral washes. Induced sputum is the most common specimen used in hospitals catering to HIV patients due to its relatively noninvasive nature and the ability to detect up to 80% of cases. BAL is the recommended sample for non-HIV and HIV patients in whom a diagnosis could not be achieved using induced sputum because BAL has a higher diagnostic yield (>80% in non-HIV and >95% in HIV patients). Invasive specimens such as transbronchial lung biopsy or, on rare occasions, open lung biopsy are resorted to in patients without HIV and in HIV coinfection when BAL fluid is negative for *Pneumocystis*. These can also be used to detect other lung pathologies that may mimic pneumocystosis. *Pneumocystis* is rarely detected in spontaneous sputum. Non invasive specimens for the detection of PJ include oral wash specimens. Nasopharyngeal aspirates are suitable for use in pediatric patients who cannot produce sputum. When used with highly sensitive PCR detection systems, nasopharyngeal aspirates and oral washes demonstrated a 75%–91% sensitivity and a 68%–100% specificity compared to sputum and BAL. Tissue biopsies are also used in the diagnosis of extrapulmonary pneumocystosis. Blood or serum is not currently recommended for the detection of *Pneumocystis*.

Microscopy

Microscopic detection and identification of *P. jirovecii* in respiratory samples using conventional or immunofluorescence techniques using invasive respiratory samples is the gold standard. Microscopic identification involves the detection of mainly the ascus (cysts containing ascospores) in respiratory specimens or tissues by using conventional stains and immunofluorescent staining. Different traditional staining methods like Gomori's methenamine silver (GMS) stain (Fig. 24.3), Toluidine blue O (TBO), and Calcofluor white (CW) stain the cell wall of the ascus. These stains are preferred due to the ability for rapid analysis and minimal expertise needed for interpretation. Some other staining methods, viz., Giemsa, Diff-Quik, Wright, Modified Papanicolaou, and Gram Weigert, also stain the trophic forms and asci. These methods are easy to perform, but results depend on the sample's quality and the microscopist's expertise. Because of the asci being always present with the trophic and precyst forms in clusters and the variability in ascus-to-trophic form ratio during different stages of the disease, staining and microscopy cannot be used for quantitation. All the traditional staining methods have high specificity but, if negative, cannot be used to rule out the disease.

Further, all staining methods have low sensitivity when the organism load is low. Direct immunofluorescence (IF) microscopy (Fig. 24.4)

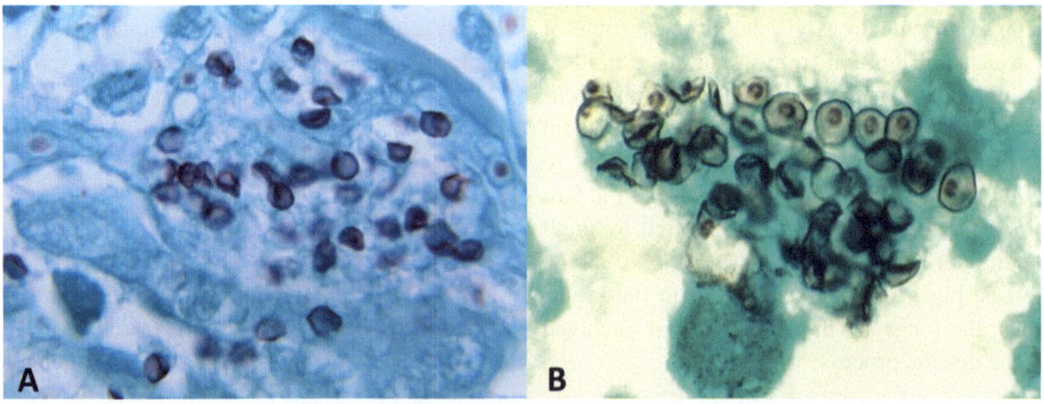

Fig. 24.3 Gomori's methanamine silver stained preparations of (**a**) lung tissue and (**b**) bronchoalveolar lavage showing the presence of cysts of *Pneumocystis*. Source: CDC Public Health Image Library

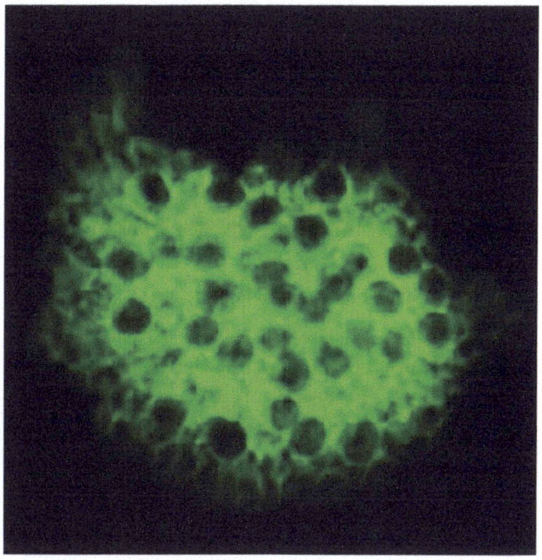

Fig. 24.4 Direct immunofluorescence of bronchoalveolar lavage fluid for *Pneumocystis* showing apple-green fluorescence of clumps containing cysts and trophozoites of *P. jirovecii*. Source: CDC (https://www.cdc.gov/dpdx/pneumocystis/images/3/Pneumocystis_IFA_B.jpg)

using anti-*P. jirovecii* monoclonal antibodies has superseded conventional staining. Commercial kits that detect ascus and trophic forms have high sensitivity but lack specificity. Hence, respiratory samples should be screened with a highly sensitive IF method, followed by confirmation using a specific test like conventional staining.

Serodiagnosis

The detection of several markers like (1,3)-Beta D-Glucan (BDG), lactate dehydrogenase (LDH), Krebs von den Lungen-6 (KL-6), and S-adenosylmethionine (SAM) in serum has been evaluated for diagnosis of pneumocystosis. Of these, only BDG and LDH have been assessed for laboratory diagnosis in multiple studies, even though both of them are not specific for diagnosing pneumocystosis. In one study, the best serological tests for diagnosis of PCP were serum BDG, followed by KL-6, LDH, and SAM in descending order. Given their low specificity, serological tests should always be followed with microscopic examination/PCR of respiratory samples for a definite diagnosis of pneumocystosis.

Molecular Diagnosis

In the early years following the AIDS pandemic, nested PCR-based protocols were explored to detect PCP, but these methods were labor-intensive, expensive, not quantitative, less specific, and time-consuming, with an inherent propensity for amplicon contamination.

Currently, nested PCR testing is increasingly being replaced by qPCR for confirming the diagnosis of PCP. qPCR allows accurate and specific quantification of DNA and has the potential ability to discriminate between colonization and clinical disease based on pathogen load without the risk of amplicon contamination. While various targets like mitochondrial large-subunit ribosomal RNA (*mtLSU* rRNA), mitochondrial small-subunit ribosomal RNA (*mtSSU* rRNA), Kex-1, *dihydropteroate synthase* (DHPS), 5S rRNA, *major surface glycoprotein* (MSG), and internal transcribed spacer (ITS) sequence have been investigated. Studies have found that multi-copy genes like *mtLSU* have higher sensitivity compared to single-copy genes like the ITS. PCR based on *mtLSU* rRNA is the standard method for diagnosing pneumocystosis. PCR using other targets can be used to study transmission and strain typing. DHPS and dihydrofolate reductase (DHFR) as gene targets can also detect resistance to sulfa-based drugs, such as sulfamethoxazole and dapsone, used to manage pneumocystosis.

Meta-analyses on the utility of PCR on BAL for pneumocystosis diagnosis involving both HIV-positive and HIV-negative populations have reported a pooled sensitivity of 98%, 99%, and 97% and a pooled specificity of 91%, 90%, and 94%. However, these studies have noted a heterogeneity among the studies analyzed because of the nonuniformity in sample type, testing process, gene targets, etc. The availability of commercial assays may help overcome these problems. A study comparing three commercial tests, *P. jirovecii* (*carinii*)—FRT PCR Kit (AmpliSens), MycAssay *Pneumocystis* (Myconostica), and real-time PCR *P. jirovecii* (Bio-Evolution) with in-house PCR, found a sensitivity and specificity of 100%, 100%, 95% and 83%, 93%, and 100%, respectively, when testing proven/probable PCP.

A commercial kit (PathoNostics PneumoGenius® real-time PCR) is available, which can detect *Pneumocystis* and a mutation in the DHFS gene leading to resistance to sulfa drugs. However, more studies are required before implementing such PCRs to detect mutations. Only a few studies are available on loop-mediated isothermal amplification (LAMP)-based assays for *Pneumocystis,* and these have found sensitivity and specificity comparable to PCR. Given their operational simplicity, less reliance on electricity, single-step amplification, immediate visual detection, and low turnaround time, they may be well-suited alternatives to qPCR in resource-limited settings.

Table 24.4 summarizes the findings of different tests used to manage PCP.

Table 24.4 Summary of different tests for diagnosis of pneumocystosis

Test	Observations	Comments
Microscopy		
Calcofluor white stain	Cysts fluoresce blue-green and trophozoites are unstained	Simple, fast, and inexpensive but require a fluorescent microscope. Host tissue reaction cannot be deciphered.
Silver stains (Gomori's or Grocott's methenamine silver)	Cysts are stained black to brown with cup and saucer or ping pong ball appearance	Time consuming. Host tissue reaction cannot be deciphered.
Toluidine blue	Cysts are stained violet-purple	Time consuming. Host tissue reaction cannot be deciphered.
Romanowsky stains such as Giemsa or its rapid adaptations	Trophozoite nucleus satin purple-blue if organism aggregates are loose, intracystic bodies-stained cyst wall remains unstained	All life stages stained, require a trained microscopist. Diff quick is not recommended.
Immunofluorescence	Fluorescence depends on the fluorophore used	All life stages are stained, best for inexperienced microscopists, expensive. High sensitivity and specificity (>95%).
Molecular tests		
Diagnostic PCR—conventional PCR	Different targets used. Commonest is nested *mtLSUrRNA*—positive or negative	Multicopy mitochondrial genes have higher sensitivity than different single-copy genes. Nested PCR is superior to single PCR, however, suffers from contamination issues and low specificity.
Diagnostic PCR—Real-Time	Different genes *MSG KEX, DHPS,* etc. used—positive or negative	High sensitivity and specificity. Avoids amplicon contamination. Cut-offs have been proposed to differentiate colonization from infections.

(continued)

Table 24.4 (continued)

Test	Observations	Comments
PCR for resistance detection	Mutations in *DHPS or DHFR* detected by melt curve analysis or after amplicon sequencing	Used in a research setting.
Loop-mediated isothermal amplification	*Pneumocystis jirovecii* present or absent	Direct sample lysis followed by LAMP in a commercial kit. Cannot be used in quantitation.
Nonspecific tests		
Oxygenation at rest and after exercise	Increased or decreased oxygen levels after exercise	*Pneumocystis* is unlikely if there is an increase in oxygen levels after exercise.
Arterial blood gas analysis	PaO_2 ≤9.3 kPa	10%–20% of patients have normal levels of PaO_2.
Radiology (chest X-ray or computed tomography scan)	B/L interstitial pneumonia, GGOs, and consolidations	Nonspecific, 6%–20% have normal radiology.
Serum lactate dehydrogenase	Elevated in PCP	Nonspecific can be used to monitor therapy.
CD4 count	≤200/μL blood	Risk factor for PCP in both HIV and non-HIV patients used to start and stop prophylaxis.
β D glucan	Elevated	Nonspecific can be raised in any invasive fungal infection. Sensitivity of 94% and 86% in HIV and non-HIV patients. Specificity of 83% on both HIV and non-HIV patients.

Notes: PCR Polymerase chain reaction, *mtLSUrRNA* Mithochondrial large subunit RNA, *PJP Pneumocystis jirovecii, DHPS* Dihydropteroate synthase, *DHFR* Dihydrofolate reductase, *MSG* Major surface glycoprotein, *KEX* Kexin

Treatment

Guidelines for the treatment and prevention of PCP are available mainly for HIV-infected and hematology patients. As per these guidelines, trimethoprim (TMP) plus sulfamethoxazole (SMX) is the first-line drug for mild-to-moderate and moderate-to-severe disease. TMP/SMX can still be given to patients with a TMP/SMX prophylaxis history. Alternatives to TMP plus SMX include clindamycin plus primaquine or IV pentamidine for treatment of moderate-to-severe illness and dapsone plus TMP or clindamycin plus primaquine and atovaquone suspension for treatment of mild-to-moderate disease in HIV and non-HIV patients. HIV patients with moderate-to-severe illness need to receive adjunctive corticosteroids within 72 h of starting specific PCP therapy. Details of the medical management of pneumocystis are given in Table 24.5.

Prevention and Control

Primary prophylaxis is provided to HIV patients with CD4 cell counts of <200/μL or a percentage of <14%. Additionally, those with other AIDS-defining illnesses should receive prophylaxis. Primary prophylaxis is stopped when the CD4 count is consistently more than 200/μL for >3 months or on two consecutive occasions for 6 months. Non-HIV patients with acute leukemoid leukemia, allogeneic hematopoietic stem cell transplant, those who have received prednisone (>20 mg/day) for a month and are receiving alemtuzumab, fludarabine/cyclophosphamide/rituximab need primary prophylaxis until the end of immunosuppression as defined in guidelines. Trimethoprim plus sulfamethoxazole is the drug of choice for primary and secondary prophylaxis in HIV and non-HIV hematology patients. Alternative drugs for primary prophylaxis in HIV and non-HIV are dapsone, atovaquone, and aero-

Table 24.5 Treatment strategies for pneumocystis pneumonia

Drug	Patient group	Purpose	Dosage	Strength of recommendations	Comments
Preferred/First-line treatment					
Trimethoprim plus Sulfamethoxazole	HIV	Treatment in moderate-to-severe disease	TMP 15–20 mg and SMX 75–100 mg/kg/day IV given qid or tid for 21 days	A1	Adjunctive prednisone Days 1–5: 40 mg PO b.d. Days 6–10: 40 mg PO o.d., days 11–21: 20 mg PO o.d.
		Treatment in mild-to-moderate disease	TMP 15–20 mg and SMX 75–100 mg/kg/day PO tid for 21 days or tablet TMP-SMX double strength PO t.i.d		–
	Non-HIV	Therapy in moderate-to-severe disease	TMP 15–20 mg and SMX 75–100 mg/kg/day IV for 21 days	A-IIu	Routine use of steroids not recommended. Decision to use on case-to-case basis
		Therapy in mild-to-moderate disease	TMP 15–20 mg and SMX 75–100 mg/kg/day PO for 21 days	B-IIt	Compliance and enteral absorption need to be confirmed. In mild cases, duration should be at least 14 days
Alternative first-line treatment					
Pentamidine	HIV	Treatment in moderate-to-severe disease	4 mg/kg IV given o.d. over 1 h for 21 days	A1	Adjunctive steroids as above
			3 mg/kg IV given o.d. over 1 h for 21 days	B1	Dose reduced when toxicity develops
	Non-HIV	Therapy	4 mg/kg IV/day for 21 days	C-IIt	
Primaquine and Clindamycin	HIV	Treatment in moderate-to-severe disease	Primaquine 30 mg PO o.d. plus clindamycin IV 600 mg qid or 900 mg t.i.d for 21 days	A1	Check for G6PD deficiency before administration of primaquine
			Primaquine 30 mg PO o.d. plus clindamycin PO 450 mg q.i.d or 600 mg t.i.d for 21 days		
		Treatment in mild-to-moderate disease	Primaquine 30 mg PO o.d. plus clindamycin PO 450 mg q.i.d or 600 mg PO t.i.d	B1	
	Non-HIV	Treatment	Primaquine 30 mg PO o.d. plus or 600 mg PO t.i.d. IV or Oral for 21 days	C-IIt	

(continued)

Table 24.5 (continued)

Drug	Patient group	Purpose	Dosage	Strength of recommendations	Comments
Dapsone with trimethoprim	HIV	Treatment in mild-to-moderate disease	Dapsone 100 mg PO/with TMP 15 mg/kg/day PO in three divided doses	BI	Check for G6PD deficiency before administration of dapsone
Atovaquone	HIV	Treatment in mild-to-moderate disease	750 mg PO b.d. for 21 days	B1	
	Non-HIV	Treatment	750 mg PO b.d. or t.i.d for 21 days	C-IIt	
Second-line/Salvage therapy					
Pentamidine	HIV	Treatment in moderate-to-severe disease	4 mg/kg IV given o.d. over 1 h	BII	–
	Non-HIV	Alternative to primaquine and clindamycin	4 mg/kg IV/day	BIII	–
Primaquine and Clindamycin	HIV	Treatment in moderate-to-severe disease	Primaquine 30 mg PO o.d. plus Clindamycin IV 600 mg q.i.d or 900 mg t.i.d	BII	Check for G6PD deficiency before administration
	Non-HIV	Treatment	Primaquine 30 mg PO o.d. plus or 600 mg PO t.i.d. IV	BIIt	First choice
Atovaquone	HIV	Treatment in mild disease	750 mg PO b.d.	BII	–
Trimethoprim plus sulfamethoxazole with caspofungin	Non-HIV	Alternative to IV pentamidine	TMP 15–20 mg and SMX 75–100 mg/kg/day IV given q.i.d or t.i.d and Caspofungin 70–50 mg/day	C-IIu	–

Notes: HIV Human Immunodeficiency Virus, *TMP* Trimethoprim, *SMX* Sulfamethoxazole, *o.d.* One time a day, *b.d.* Two times a day, *t.i.d.* Three times a day, *PO* By mouth IV-Intravenous, *q.i.d.* Four times a day

solized pentamidine. A combination of dapsone with pyrimethamine and leucovorin is used as an alternative only in HIV patients. While recurrent PCR is uncommon in non-HIV patients, secondary prophylaxis using the same drugs as primary prophylaxis is given to patients till CD4 counts are >200/μL for >3 months or in patients with CD4 counts between 100 and 200/μL and the viral load below the detection limits in the plasma.

In conclusion, *P. jirovecii* is a cosmopolitan fungus that causes opportunistic infections in immunocompromised hosts. The infection is acquired by inhalation and may cause outbreaks in centers catering to patients with deficient immunity. Diagnosis is achieved by using microscopic methods like immunofluorescence and silver staining. PCR is also widely recognized as a valuable tool in diagnosis and can be applied for typing and antimicrobial resistance detection and typing. Trimethoprim/sulfamethoxazole is the drug of choice for prevention and curative therapy. The inability to culture the organism in a reliable manner has hindered our ability to understand the pathogenesis and develop diagnostics and effective vaccines. More research is needed to reveal environmental sources (if any), confirm if the organism is stenoxenous or if animal reservoirs are present, and develop better diagnostic and therapeutic tools for pneumocystosis.

Case Study

A 35-year-old truck driver presents with fever, nonproductive cough, and progressive dyspnea for 2 weeks. On inquiry, he also gave a history of unintentional significant weight loss. He has tachypnea (20 breaths/min), tachycardia (90 beats/min), and generalized lymphadenopathy on general physical examination. Systemic examination was normal except for end-inspiratory crackles. Chest X-ray shows bilateral diffuse interstitial infiltrate extending from the peripheral region. Laboratory investigations revealed that he had a hemoglobin of 8.8 g/L, a total WBC count of 7.8×10^9/L, and was HIV reactive, with a CD4 count of 16/μL and PaO_2 of 7.8 kPa. A BAL obtained subjected to direct immunofluorescence for *Pneumocystis* showed apple-green fluorescence of clumps containing cysts and trophozoites of *P. jirovecii*. The patient underwent successful treatment and prophylaxis with TMP/SMX and was started concomitantly with HAART and is currently doing well.

1. What are the telltale pointers in this case?
2. How will you confirm the diagnosis of pneumocystosis?
3. What are the other diagnostic modalities available?
4. How will you manage the case of pneumocystosis?

Research Questions

Further research is needed to establish a culture system that can help better understand the pathogenesis aid in vaccine and new drug development.

1. Is there an environmental reservoir of *Pneumocystis*?
2. Can whole genome analysis provide clues for
   (a) Development of a reliable culture system for *Pneumocystis*?
   (b) Understanding pathogenesis?
   (c) Antifungal resistance testing?
   (d) Development of new treatments and vaccines?
3. Can other mammals be sampled to confirm or refute the monoxenous nature of the pathogen?
4. Can newer diagnostic tools be developed or tested on noninvasive samples like saliva, urine, etc., for diagnosis of pneumocystosis?

Further Reading

2019 Panel on Guidelines for the Prevention and Treatment of Opportunistic Infections in Adults and Adolescents with HIV (n.d.) Guidelines for the prevention and treatment of opportunistic infections in HIV-infected adults and adolescents: recommendations from the Centers for Disease Control and Prevention, the National Institutes of Health, and the HIV Medicine Association of the Infectious Diseases Society of America. [Online]. https://clinicalinfo.hiv.gov/en/guidelines/adult-and-adolescent-opportunistic-infection. Accessed 20 Feb 2022

Babb-Biernacki SJ, Esselstyn JA, Doyle VP (2020) Rethinking host range in *Pneumocystis*. PLoS Pathog 16:e1008824

Bateman M, Oladele R, Kolls JK (2020) Diagnosing *Pneumocystis jirovecii* pneumonia: A review of current methods and novel approaches. Med Mycol 58:1015–1028

Bongomin F, Gago S, Oladele RO, Denning DW (2017) Global and multi-national prevalence of fungal diseases-estimate precision. J Fungi (Basel) 3:57

Chabe M, Aliouat-Denis CM, Delhaes L, Aliouat El M, Viscogliosi E, Dei-Cas E (2011) *Pneumocystis*: from a doubtful unique entity to a group of highly diversified fungal species. FEMS Yeast Res 11:2–17

Charpentier E, Menard S, Marques C, Berry A, Iriart X (2021) Immune response in *Pneumocystis* infections according to the host immune system status. J Fungi (Basel) 7:625

Cilloniz C, Dominedo C, Alvarez-Martinez MJ, Moreno A, Garcia F, Torres A, Miro JM (2019) *Pneumocystis* pneumonia in the twenty-first century: HIV-infected versus HIV-uninfected patients. Expert Rev. Anti-Infect Ther 17:787–801

Cisse OH, Pagni M, Hauser PM (2012) De novo assembly of the *Pneumocystis jirovecii* genome from a single bronchoalveolar lavage fluid specimen from a patient. MBio 4:e00428–e00412

Cushion MT, Tisdale-Macioce N, Sayson SG, Porollo A (2021) The persistent challenge of *Pneumocystis*

growth outside the mammalian lung: past and future approaches. Front Microbiol 12:681474
Del Corpo O, Butler-Laporte G, Sheppard DC, Cheng MP, Mcdonald EG, Lee TC (2020) Diagnostic accuracy of serum (1-3)-beta-D-glucan for *Pneumocystis jirovecii* pneumonia: a systematic review and meta-analysis. Clin Microbiol Infect 26:1137–1143
Demanche C, Guillot J, Chabé M (2019) *Pneumocystis* species co-evolution: state-of-the-art review. OBM Genet 3:1–1
Esteves F, Calé SS, Badura R, De Boer MG, Maltez F, Calderón EJ, Van Der Reijden TJ, Márquez-Martín E, Antunes F, Matos O (2015) Diagnosis of *Pneumocystis* pneumonia: evaluation of four serologic biomarkers. Clin Microbiol Infect 21:379.e1–379.e10
Hughes W (2005) Historical overview. In: Walzer PD, Cushion MT (eds) *Pneumocystis* pneumonia, 3rd edn. Marcel Dekker, New York
Kim KS, Jung JY, Kim JH, Kang SC, Hwang EK, Park BK, Kim DY, Kim JH (2011) Epidemiological characteristics of pulmonary pneumocystosis and concurrent infections in pigs in Jeju Island, Korea. J Vet Sci 12:15–19
Maertens J, Cesaro S, Maschmeyer G, Einsele H, Donnelly JP, Alanio A, Hauser PM, Lagrou K, Melchers WJ, Helweg-Larsen J, Matos O, Bretagne S, Cordonnier C, The European Conference on Infections in Leukaemia, a joint venture of the European Group for Blood Marrow Transplantation, the European Organisation for Research Treatment of Cancer, the Immunocompromised Host Society the European, LeukemiaNet (2016) ECIL guidelines for preventing *Pneumocystis jirovecii* pneumonia in patients with haematological malignancies and stem cell transplant recipients. J Antimicrob Chemother 71:2397–2404
Maschmeyer G, Helweg-Larsen J, Pagano L, Robin C, Cordonnier C, Schellongowski P, on behalf of the 6th European Conference on Infections in Leukemia, a joint venture of The European Group for Blood Marrow Transplantation, The European Organization for Research Treatment of Cancer, the International Immunocompromised Host Society The European LeukemiaNet (2016) ECIL guidelines for treatment of *Pneumocystis jirovecii* pneumonia in non-HIV-infected haematology patients. J Antimicrob Chemother 71:2405–2413
Matos O, Esteves F (2016) Chapter 13—Laboratory diagnosis of *Pneumocystis jirovecii* pneumonia. In: Kon K, Rai M (eds) The microbiology of respiratory system infections. Academic Press
Matos O, Tomás AL, Antunes F (2017b) *Pneumocystis jirovecii* and Pneumocystosis. In: Mora-Montes HM, Lopes-Bezerra LM (eds) Current progress in medical mycology. Springer International Publishing, Cham
Ray A, Aayilliath K, Banerjee S, Chakrabarti A, Denning DW (2022) Burden of serious fungal infections in India. Open Forum Infect Dis 9:ofac603
Sasso M, Chastang-Dumas E, Bastide S, Alonso S, Lechiche C, Bourgeois N, Lachaud L (2016) Performances of four real-time PCR assays for diagnosis of *Pneumocystis jirovecii* Pneumonia. J Clin Microbiol 54:625–630
Vera C, Rueda ZV (2021) Transmission and colonization of *Pneumocystis jirovecii*. J Fungi 7:979

Index

S. C. Parija, S. M. Rudramurthy (eds.), *Textbook of Fungal Zoonoses and Sapronoses*, Microbial Zoonoses, https://doi.org/10.1007/978-981-97-3214-2

R

SPRINGER NATURE

GPSR Compliance

The European Union's (EU) General Product Safety Regulation (GPSR) is a set of rules that requires consumer products to be safe and our obligations to ensure this.

If you have any concerns about our products, you can contact us on ProductSafety@springernature.com

In case Publisher is established outside the EU, the EU authorized representative is:

Springer Nature Customer Service Center GmbH
Europaplatz 3
69115 Heidelberg, Germany

The manufacturer's authorised representative in the EU is Springer Nature Customer Service Centre GmbH, Europaplatz 3, 69115 Heidelberg, Germany. If you have any concerns regarding our products, please contact ProductSafety@springernature.com

Printed and bound by CPI Group (UK) Ltd, Croydon, CR0 4YY
01/10/2026
02225857-0020